INTERIOR POINT METHODS FOR LINEAR OPTIMIZATION

Revised Edition

INTERIOR POINT METHODS FOR LINEAR OPTIMIZATION

Revised Edition

By

CORNELIS ROOS
Delft University of Technology, The Netherlands

TAMÁS TERLAKY
McMaster University, Ontario, Canada

JEAN-PHILIPPE VIAL
University of Geneva, Switzerland

 Springer

Library of Congress Cataloging-in-Publication Data

Roos, Cornelis, 1941–
    Interior point methods for linear optimization / by C. Roos, T. Terlaky, J.-Ph. Vial.
        p. cm.
    Rev. ed. of: Theory and algorithms for linear optimization. c1997.
    Includes bibliographical references and index.

    ISBN 978-1-4419-3887-9   e-ISBN 978-0-387-26379-3

    1. Linear programming. 2. Interior-point methods. 3. Mathematical optimization. 4.
Algorithms.
    I. Terlaky, Tamás. II. Vial, J.P. III. Roos, Cornelis, 1941– Theory and algorithms for linear
optimization. IV. Title.

T57.74.R664 2005
519.7'2—dc22

                                                                          2005049785

AMS Subject Classifications: 90C05, 65K05, 90C06, 65Y20, 90C31

Printed in the United States of America.

9 8 7 6 5 4 3 2 1

springeronline.com

Dedicated to our wives

Gerda, Gabriella and Marie

and our children

Jacoline, Geranda, Marijn

Viktor

Benjamin and Emmanuelle

Contents

Contents

List of Figures

List of Tables

Preface

Linear Optimization[1] (LO) is one of the most widely taught and applied mathematical techniques. Due to revolutionary developments both in computer technology and algorithms for linear optimization, 'the last ten years have seen an estimated six orders of magnitude speed improvement'.[2] This means that problems that could not be solved 10 years ago, due to a required computational time of one year, say, can now be solved within some minutes. For example, linear models of airline crew scheduling problems with as many as 13 million variables have recently been solved within three minutes on a four-processor Silicon Graphics Power Challenge workstation. The achieved acceleration is due partly to advances in computer technology and for a significant part also to the developments in the field of so-called *interior-point methods* for linear optimization.

Until very recently, the method of choice for solving linear optimization problems was the Simplex Method of Dantzig [59]. Since the initial formulation in 1947, this method has been constantly improved. It generally recognized to be very robust and efficient and it is routinely used to solve problems in Operations Research, Business, Economics and Engineering. In an effort to explain the remarkable efficiency of the Simplex Method, people strived to prove, using the theory of complexity, that the computational effort to solve a linear optimization problem via the Simplex Method is polynomially bounded with the size of the problem instance. This question is still unsettled today, but it stimulated two important proposals of new algorithms for LO. The first one is due to Khachiyan in 1979 [167]: it is based on the ellipsoid technique for nonlinear optimization of Shor [255]. With this technique, Khachiyan proved that LO belongs to the class of polynomially solvable problems. Although this result has had a great theoretical impact, the new algorithm failed to deliver its promises in actual computational efficiency. The second proposal was made in 1984 by Karmarkar [165]. Karmarkar's algorithm is also polynomial, with a better complexity bound

---

[1] The field of Linear Optimization has been given the name *Linear Programming* in the past. The origin of this name goes back to the Dutch Nobel prize winner Koopmans. See Dantzig [60]. Nowadays the word 'programming' usually refers to the activity of writing computer programs, and as a consequence its use instead of the more natural word 'optimization' gives rise to confusion. Following others, like Padberg [230], we prefer to use the name *Linear Optimization* in the book. It may be noted that in the nonlinear branches of the field of Mathematical Programming (like *Combinatorial Optimization, Discrete Optimization, Semidefinite Optimization*, etc.) this terminology has already become generally accepted.

[2] This claim is due to R.E. Bixby, professor of Computational and Applied Mathematics at Rice University, and director of CPLEX Optimization, Inc., a company that markets algorithms for linear and mixed-integer optimization. See the news bulletin of the Center For Research on Parallel Computation, Volume 4, Issue 1, Winter 1996. Bixby adds that parallelization may lead to 'at least eight orders of magnitude improvement—the difference between a year and a fraction of a second!'

than Khachiyan, but it has the further advantage of being highly efficient in practice. After an initial controversy it has been established that for very large, sparse problems, subsequent variants of Karmarkar's method often outperform the Simplex Method.

Though the field of LO was considered more or less mature some ten years ago, after Karmarkar's paper it suddenly surfaced as one of the most active areas of research in optimization. In the period 1984–1989 more than 1300 papers were published on the subject, which became known as Interior Point Methods (IPMs) for LO.[3] Originally the aim of the research was to get a better understanding of the so-called Projective Method of Karmarkar. Soon it became apparent that this method was related to classical methods like the Affine Scaling Method of Dikin [63, 64, 65], the Logarithmic Barrier Method of Frisch [86, 87, 88] and the Center Method of Huard [148, 149], and that the last two methods could also be proved to be polynomial. Moreover, it turned out that the IPM approach to LO has a natural generalization to the related field of convex nonlinear optimization, which resulted in a new stream of research and an excellent monograph of Nesterov and Nemirovski [226]. Promising numerical performances of IPMs for convex optimization were recently reported by Breitfeld and Shanno [50] and Jarre, Kocvara and Zowe [162]. The monograph of Nesterov and Nemirovski opened the way into another new subfield of optimization, called Semidefinite Optimization, with important applications in System Theory, Discrete Optimization, and many other areas. For a survey of these developments the reader may consult Vandenberghe and Boyd [48].

As a consequence of the above developments, there are now profound reasons why people may want to learn about IPMs. We hope that this book answers the need of professors who want to teach their students the principles of IPMs, of colleagues who need a unified presentation of a desperately burgeoning field, of users of LO who want to understand what is behind the new IPM solvers in commercial codes (CPLEX, OSL, ...) and how to interpret results from those codes, and of other users who want to exploit the new algorithms as part of a more general software toolbox in optimization.

Let us briefly indicate here what the book offers, and what does it not. Part I contains a small but complete and self-contained introduction to LO. We deal with the duality theory for LO and we present a first polynomial method for solving an LO problem. We also present an elegant method for the initialization of the method, using the so-called self-dual embedding technique. Then in Part II we present a comprehensive treatment of Logarithmic Barrier Methods. These methods are applied to the LO problem in standard format, the format that has become most popular in the field because the Simplex Method was originally devised for that format. This part contains the basic elements for the design of efficient algorithms for LO. Several types of algorithm are considered and analyzed. Very often the analysis improves the existing analysis and leads to sharper complexity bounds than known in the literature. In Part III we deal with the so-called Target-following Approach to IPMs. This is a unifying framework that enables us to treat many other IPMs, like the Center Method, in an easy way. Part IV covers some additional topics. It starts with the description and analysis of the Projective Method of Karmarkar. Then we discuss some more

---

[3] We refer the reader to the extensive bibliography of Kranich [179, 180] for a survey of the literature on the subject until 1989. A more recent (annotated) bibliography was given by Roos and Terlaky [242]. A valuable source of information is the World Wide Web interior point archive: http://www.mcs.anl.gov/home/otc/InteriorPoint.archive.html.

interesting theoretical properties of the central path. We also discuss two interesting methods to enhance the efficiency of IPMs, namely Partial Updating, and so-called Higher-Order Methods. This part also contains chapters on parametric and sensitivity analysis and on computational aspects of IPMs.

It may be clear from this description that we restrict ourselves to Linear Optimization in this book. We do not dwell on such interesting subjects as Convex Optimization and Semidefinite Optimization, but we consider the book as a preparation for the study of IPMs for these types of optimization problem, and refer the reader to the existing literature.[4]

Some popular topics in IPMs for LO are not covered by the book. For example, we do not treat the (Primal) Affine Scaling Method of Dikin.[5] The reason for this is that we restrict ourselves in this book to polynomial methods and until now the polynomiality question for the (Primal) Affine Scaling Method is unsettled. Instead we describe in Appendix E a primal-dual version of Dikin's affine-scaling method that is polynomial. Chapter 18 describes a higher-order version of this primal-dual affine-scaling method that has the best possible complexity bound known until now for interior-point methods.

Another topic not touched in the book is (Primal-Dual) Infeasible Start Methods. These methods, which have drawn a lot of attention in the last years, deal with the situation when no feasible starting point is available.[6] In fact, Part I of the book provides a much more elegant solution to this problem; there we show that any given LO problem can be embedded in a self-dual problem for which a feasible interior starting point is known. Further, the approach in Part I is theoretically more efficient than using an Infeasible Start Method, and from a computational point of view is not more involved, as we show in Chapter 20.

We hope that the book will be useful to students, users and researchers, inside and outside the field, in offering them, under a single cover, a presentation of the most successful ideas in interior-point methods.

Kees Roos
Tamás Terlaky
Jean-Philippe Vial

Preface to the 2005 edition

Twenty years after Karmarkar's [165] epoch making paper *interior point methods (IPMs)* made their way to all areas of optimization theory and practice. The theory of IPMs matured, their professional software implementations significantly pushed the boundary of efficiently solvable problems. Eight years passed since the first edition of this book was published. In these years the theory of IPMs further crystallized. One of the notable developments is that the significance of the self-dual embedding

---

4 For Convex Optimization the reader may consult den Hertog [140], Nesterov and Nemirovski [226] and Jarre [161]. For Semidefinite Optimization we refer to Nesterov and Nemirovski [226], Vandenberghe and Boyd [48] and Ramana and Pardalos [236]. We also mention Shanno and Breitfeld and Simantiraki [252] for the related topic of barrier methods for nonlinear programming.

5 A recent survey on affine scaling methods was given by Tsuchiya [272].

6 We refer the reader to, e.g., Potra [235], Bonnans and Potra [45], Wright [295, 297], Wright and Ralph [296] and the recent book of Wright [298].

model –that is a distinctive feature of this book– got fully recognized. Leading linear and conic-linear optimization software packages, such as MOSEK[7] and SeDuMi[8] are developed on the bedrock of the self-dual model, and the leading commercial linear optimization package CPLEX[9] includes the embedding model as a proposed option to solve difficult practical problems.

This new edition of this book features a completely rewritten first part. While keeping the simplicity of the presentation and accessibility of complexity analysis, the featured IPM in Part I is now a standard, primal-dual path-following Newton algorithm. This choice allows us to reach the so-far best known complexity result in an elementary way, immediately in the first part of the book.

As always, the authors had to make choices when and how to cut the expansion of the material of the book, and which new results to include in this edition. We cannot resist mentioning two developments after the publication of the first edition.

The first development can be considered as a direct consequence of the approach taken in the book. In our approach properties of the univariate function $\psi(t)$, as defined in Section 5.5 (page 92), play a key role. The book makes clear that the primal-, dual- and primal-dual logarithmic barrier function can be defined in terms of $\psi(t)$, and as such $\psi(t)$ is at the heart of all logarithmic barrier functions; we call it now the kernel function of the logarithmic barrier function. After the completion of the book it became clear that more efficient large-update IPMs than those considered in this book, which are all based on the logarithmic barrier function, can be obtained simply by replacing $\psi(t)$ by other kernel functions. A large class of such kernel functions, that allowed to improve the worst case complexity of large-update IPMs, is the family of self-regular functions, which is the subject of the monograph [233]; more kernel functions were considered in [32].

A second, more recent development, deals with the complexity of IPMs. Until now, the best iteration bound for IPMs is $O(\sqrt{n}L)$, where n denotes the dimension of the problem (in standard from), and L the binary input size of the problem. In 1996, Todd and Ye showed that $O(\sqrt[3]{n}L)$ is a lower bound for the iteration complexity of IPMs [267]. It is well known that the iteration complexity highly depends on the curliness of the central path, and that the presence of redundancy may severely affect this curliness. Deza et al. [61] showed that by adding enough redundant constraints to the Klee-Minty example of dimension n, the central path may be forced to visit all 2^n vertices of the Klee-Minty cube. An enhanced version of the same example, where the number of inequalities is $N = O(2^{2n}n^3)$, yields an $O(\sqrt{N}/\log N)$ lower bound for the iteration complexity, thus almost closing (up to a factor of $\log N$) the gap with the best worst case iteration bound for IPMs [62].

Instructors adapting the book as textbook in a course may contact the authors at <terlaky@mcmaster.ca> for obtaining the "Solution Manual" for the exercises and getting access to a user forum.

March 2005

Kees Roos
Tamás Terlaky
Jean-Philippe Vial

---

[7] MOSEK: http://www.mosek.com

[8] SeDuMi: http://sedumi.mcmaster.ca

[9] CPLEX: http://cplex.com

Acknowledgements

The subject of this book came into existence during the twelve years following 1984 when Karmarkar initiated the field of interior-point methods for linear optimization. Each of the authors has been involved in the exciting research that gave rise to the subject and in many cases they published their results jointly. Of course the book is primarily organized around these results, but it goes without saying that many other results from colleagues in the 'interior-point community' are also included. We are pleased to acknowledge their contribution and at the appropriate places we have strived to give them credit. If some authors do not find due mention of their work we apologize for this and invoke as an excuse the exploding literature that makes it difficult to keep track of all the contributions.

To reach a unified presentation of many diverse results, it did not suffice to make a bundle of existing papers. It was necessary to recast completely the form in which these results found their way into the journals. This was a very time-consuming task: we want to thank our universities for giving us the opportunity to do this job.

We gratefully acknowledge the developers of LATEX for designing this powerful text processor and our colleagues Leo Rog and Peter van der Wijden for their assistance whenever there was a technical problem. For the construction of many tables and figures we used MATLAB; nowadays we could say that a mathematician without MATLAB is like a physicist without a microscope. It is really exciting to study the behavior of a designed algorithm with the graphical features of this 'mathematical microscope'.

We greatly enjoyed stimulating discussions with many colleagues from all over the world in the past years. Often this resulted in cooperation and joint publications. We kindly acknowledge that without the input from their side this book could not have been written. Special thanks are due to those colleagues who helped us during the writing process. We mention János Mayer (University of Zürich, Switzerland) for his numerous remarks after a critical reading of large parts of the first draft and Michael Saunders (Stanford University, USA) for an extremely careful and useful preview of a later version of the book. Many other colleagues helped us to improve intermediate drafts. We mention Jan Brinkhuis (Erasmus University, Rotterdam) who provided us with some valuable references, Erling Andersen (Odense University, Denmark), Harvey Greenberg and Allen Holder (both from the University of Colorado at Denver, USA), Tibor Illés (Eötvös University, Budapest), Florian Jarre (University of Würzburg, Germany), Etienne de Klerk (Delft University of Technology), Panos Pardalos (University of Florida, USA), Jos Sturm (Erasmus University, Rotterdam), and Joost Warners (Delft University of Technology).

Finally, the authors would like to acknowledge the generous contributions of numerous colleagues and students. Their critical reading of earlier drafts of the manuscript helped us to clean up the new edition by eliminating typos and using their constructive remarks to improve the readability of several parts of the books. We mention Jiming Peng (McMaster University), Gema Martinez Plaza (The University of Alicante) and Manuel Vieira (University of Lisbon/University of Technology Delft).

Last but not least, we want to express warm thanks to our wives and children. They also contributed substantially to the book by their mental support, and by forgiving our shortcomings as fathers for too long.

1

Introduction

1.1 Subject of the book

This book deals with linear optimization (LO). The object of LO is to find the optimal (minimal or maximal) value of a *linear function* subject to *linear constraints* on the variables. The constraints may be either equality or inequality constraints.[1] From the point of view of applications, LO possesses many nice features. Linear models are relatively simple to create. They can be realistic enough to give a proper account of the problems at hand. As a consequence, LO models have found applications in different areas such as engineering, management, logistics, statistics, pattern recognition, etc. LO is also very relevant to economic theory. It underlies the analysis of linear activity models and provides, through duality theory, a nice insight into the price mechanism.

However, we will not deal with applications and modeling. Many existing textbooks teach more about this.[2]

Our interest will be mainly in methods for solving LO problems, especially Interior Point Methods (IPM's). Renewed interest in these methods for solving LO problems arose after the seminal paper of Karmarkar [165] in 1984. The overwhelming amount of research of the last ten years has been tremendously prolific. Many new algorithms were proposed and almost all of these algorithms have been shown to be efficient, at least from a theoretical point of view. Our first aim is to present a comprehensive and unified treatment of many of these new methods.

It may not be surprising that exploring a new method for LO should lead to a new view of the theory of LO. In fact, a similar interaction between method and theory is well known for the Simplex Method; in the past the theory of LO and the Simplex Method were intimately related. The fundamental results of the theory of LO concern strong duality and the existence of a strictly complementary solution. Our second aim will be to derive these results from limiting properties of the so-called *central path* of an LO problem.

Thus the very theory of LO is revisited. The central path appears to play a key role both in the development of the theory and in the design of algorithms.

---

[1] The more general optimization problem arising when the objective function and/or the constraints are nonlinear is not considered. It may be pointed out that LO is the first building block in the development of the theory of nonlinear optimization. Algorithmically, LO is also widely used in nonlinear and integer optimization, either as a subroutine in a more complicated algorithm or as a starting point of a specialized algorithm.

[2] The book of Williams [293] is completely devoted to the design of mathematical models, including linear models.

As a consequence, the book can be considered a self-contained treatment of LO. The reader familiar with the subject of LO will easily recognize the difference from the classical approach to the theory. The Simplex Method in essence explores the polyhedral structure of the domain (or feasible region) of an LO problem. Accordingly, the classical approach to the theory of LO concentrates on the polyhedral structure of the domain. On the other hand, the IPM approach uses the central path as a guide to the set of optimal solutions, and the theory follows by studying the limiting properties of this path.[3] As we will see, the limit of the central path is a strictly complementary solution. Strictly complementary solutions play a crucial role in the theory as presented in Part I of the book. Also, in general, the output of a well-designed IPM for LO is a strictly complementary solution. Recall that the Simplex Method generates a so-called *basic solution* and that such solutions are fundamental in the classical theory of LO.

From the practical point of view it is most important to study the sensitivity of an optimal solution under perturbations in the data of an LO problem. This is the subject of Sensitivity (or Parametric or Postoptimal) Analysis. Our third aim will be to present some new results in this respect, which will make clear the well-known fact that the classical approach has some inherent weaknesses. These weaknesses can be overcome by exploring the concept of the *optimal partition* of an LO problem which is closely related to a strictly complementary solution.

1.2 More detailed description of the contents

As stated in the previous section, we intend to present an interior point approach to both the theory of LO and algorithms for LO (design, convergence, complexity and asymptotic behavior). The common thread through the various parts of the book will be the prominent role of strictly complementary solutions; this notion plays a crucial role in the IPM approach and distinguishes the new approach from the classical Simplex based approach.

Part I of the book consists of Chapters 2, 3 and 4. This part is a self-contained treatment of LO. It provides the main theoretical results for LO, as well as a polynomial method for solving the LO problem. The theory of LO is developed in Chapter 2. This is done in a way that is probably new for most readers, even for those who are familiar with LO. As indicated before, in IPM's a fundamental element is the central path of a problem. This path is introduced in Chapter 2 and the duality theory for LO is derived from its properties. The general theory turns out to follow easily when considering first the relatively small class of so-called self-dual problems. The results for self-dual problems are extended to general problems by embedding any given LO problem in an appropriate self-dual problem. Chapter 3 presents an algorithm that solves self-dual problems in polynomial time. It may be emphasized that this algorithm yields a so-called *strictly complementary solution* of the given problem. Such a solution, in general, provides much more information on the set of

---

[3] Most of the fundamental duality results for LO will be well known to many of the readers; they can be found in any textbook on LO. Probably the existence of a strictly complementary solution is less well known. This result has been shown first by Goldman and Tucker [111] and will be referred to as the Goldman–Tucker theorem. It plays a crucial role in this book. We get it as a byproduct of the limiting behavior of the central path.

optimal solutions than an optimal basic solution as provided by the Simplex Method. The strictly complementary solution is obtained by applying a *rounding procedure* to a sufficiently accurate approximate solution. Chapter 4 is devoted to LO problems in canonical format, with (only) nonnegative variables and (only) inequality constraints. A thorough discussion of the special structure of the canonical format provides some specialized embeddings in self-dual problems. As a byproduct we find the central path for canonical LO problems. We also discuss how an approximate solution for the canonical problem can be obtained from an approximate solution of the embedding problem.

The two main components in an iterative step of an IPM are the search direction and the step-length along that direction. The algorithm in Part I is a rather simple primal-dual algorithm based on the primal-dual Newton direction and uses a very simple step-length rule: the step length is always 1. The resulting Full-Newton Step Algorithm is polynomial and straightforward to implement. However, the theoretical iteration bound derived for this algorithm, although polynomial, is relatively poor when compared with algorithms based on other search strategies. Therefore, more efficient methods are considered in Part II of the book; they are so-called Logarithmic Barrier Methods. For reasons of compatibility with the existing literature, on both the Simplex Method and IPM's, we abandon the canonical format (with nonnegative variables and inequality constraints) in Part II and use the so-called standard format (with nonnegative variables and equality constraints).

In order to make Part II independent of Part I, in Chapter 5 we revisit duality theory and discuss the relevant results for the standard format from an interior point of view. This includes, of course, the definition and existence of the central paths for the (primal) problem in standard form and its dual problem (which has free variables and inequality constraints). Using a symmetric formulation of both problems we see that any method for the primal problem induces in a natural way a method for the dual problem and vice versa. Then, in Chapter 6, we focus on the Dual Logarithmic Barrier Method; according to the previous remark the analysis can be naturally, and easily, transformed to the primal case. The search direction here is the Newton direction for minimizing the (classical) dual logarithmic barrier function with barrier parameter μ. Three types of method are considered. First we analyze a method that uses full Newton steps and small updates of the barrier parameter μ. This gives another central-path-following method that admits the best possible iteration bound. Secondly, we discuss the use of adaptive updates of μ; this leaves the iteration bound unchanged, but enhances the practical behavior. Finally, we consider methods that use large updates of μ and a bounded number of damped Newton steps between each pair of successive barrier updates. The (theoretical worst-case) iteration bound is worse than for the full Newton step method, but this seems to be due to the poor analysis of this type of method. In practice large-update methods are much more efficient than the full Newton step method. This is demonstrated by some (small) examples. Chapter 7, deals with the Primal-Dual Logarithmic Barrier Method. It has basically the same structure as Chapter 6. Having defined the primal-dual Newton direction, we deal first with a full primal-dual Newton step method that allows small updates in the barrier parameter μ. Then we consider a method with adaptive updates of μ, and finally methods that use large updates of μ and a bounded number of damped primal-dual Newton steps between each pair of successive barrier updates. In-between we also deal with the Predictor-Corrector Method. The nice feature of this method is its asymptotic quadratic convergence rate. Some small computational examples are included that highlight the better performance of the primal-dual Newton method compared with the dual (or primal) Newton method. The methods used in Part II need to be initialized with a *strictly feasible* solution.[4] Therefore, in Chapter 8 we discuss how to meet this condition. This concludes the description of Part II.

At this stage of the book, the reader will have encountered the main theoretical ideas underlying efficient implementations of IPM's for LO. He will have been exposed to many variants of IPM's, dual and primal-dual methods with either full or damped Newton steps.[5] The search directions in these methods are Newton directions. All these methods, in one way or another, use the central path as a guideline to optimality. Part III is devoted to a broader class of IPM's, some of which also follow the central path but others do not. In Chapter 9 we introduce the unifying concepts of *target sequence* and *Target-following Methods*. In the Logarithmic Barrier Methods of Part II the target sequence always consists of points on the central path. Other IPM's can be simply characterized by their target sequence. We present some examples in Chapter 11, where we deal with *weighted-path-following methods*, a *Dikin-path-following method*, and also with a *centering method* that can be used to compute the so-called *weighted-analytic center of a polytope*. Chapters 10, 12 and 13 present respectively primal-dual, dual and primal versions of Newton's method for following a given target sequence. Finally, concluding Part III, in Chapter 14 we describe a famous interior-point method, due to Renegar and based on the center method of Huard; we show that it nicely fits in the framework of target-following methods, with the targets on the central path.

Part IV is entitled *Miscellaneous Topics*: it contains material that deserves a place in the book but did not fit well in any of the previous three parts. The reader will have noticed that until now we have not discussed the very first polynomial IPM, the Projective Method of Karmarkar. This is because the mainstream of research into IPM's diverged from this method soon after 1984.[6] Because of the big influence this algorithm had on the field of LO, and also because there is still a small ongoing stream of research in this direction, it deserves a place in this book. We describe and analyze Karmarkar's method in Chapter 15. Surprisingly enough, and in contrast with all other methods discussed in this book, both in the description and the analysis of Karmarkar's method we do not refer to the central path; also, the search direction differs from the Newton directions used in the other methods. In Chapter 16 we return to the central path. We show that the central path is differentiable and study the asymptotic

---

[4] A feasible solution is called strictly feasible if no variable or inequality constraint is at (one of) its bound(s).

[5] In the literature, full-step methods are often called *short-step methods* and damped Newton step methods *long-step methods* or *large-step methods*. In damped-step methods a line search is made in each iteration that aims to (approximately) minimize a barrier (or potential) function. Therefore, these methods are also known as *potential reduction methods*.

[6] There are still many textbooks on LO that do not deal with IPM's. Moreover, in some other textbooks that pay attention to IPM's, the authors only discuss the Projective Method of Karmarkar, thereby neglecting the important developments after 1984 that gave rise to the efficient methods used in the well-known commercial codes, such as CPLEX and OSL. Exceptions, in this respect, are Bazaraa, Sherali and Shetty [37], Padberg [230] and Fang and Puthenpura [74], who discuss the existence of other IPM's in a separate section or chapter. We also mention Saigal [249], who gives a large chapter (of 150 pages) on a topic not covered in this book, namely (primal) affine-scaling methods. A recent survey on these methods is given by Tsuchiya [272].

behavior of the derivatives when the optimal set is approached. We also show that we can associate with each point on the central path two homothetic ellipsoids centered at this point so that one ellipsoid is contained in the feasible region and the other ellipsoid contains the optimal set. The next two chapters deal with methods for accelerating IPM's. Chapter 17 deals with a technique called *partial updating*, already proposed in Karmarkar's original paper. In Chapter 18 we consider so-called *higher-order methods*. The Newton methods used before are considered to be first-order methods. It is shown that more advanced search directions improve the iteration bound for several first order methods. The complexity bound achieves the best value known for IPM's nowadays. We also apply the higher-order-technique to the Logarithmic Barrier Method.

Chapter 19 deals with Parametric and Sensitivity Analysis. This classical subject in LO is of great importance in the analysis of practical linear models. Almost any textbook includes a section about it and many commercial optimization packages offer an option to perform post-optimal analysis. Unfortunately, the classical approach, based on the use of an optimal basic solution, has some inherent weaknesses. These weaknesses are discussed and demonstrated. We follow a new approach in this chapter, leading to a better understanding of the subject and avoiding the shortcomings of the classical approach. The notions of optimal partition and strictly complementary solution play an important role, but to avoid any misunderstanding, it should be emphasized that the new approach can also be performed when only an optimal basic solution is available.

After all the efforts spent in the book to develop beautiful theorems and convergence results the reader may want to get some more evidence that IPM's work well in practice. Therefore the final chapter is devoted to the implementation of IPM's. Though most implementations more or less follow the scheme prescribed by the theory, there is still a large stretch between the theory and an efficient implementation. Chapter 20 discusses some of the important implementation issues.

1.3 What is new in this book?

The book offers an approach to LO and to IPM's that is new in many aspects.[7] First, the derivation of the main theoretical results for LO, like the duality theory and the existence of a strictly complementary solution from properties of the central path, is new. The primal-dual algorithm for solving self-dual problems is also new; equipped with the rounding procedure it yields an exact strictly complementary solution. The derivation of the polynomial complexity of the whole procedure is surprisingly simple.[8] The algorithms in Part II, based on the logarithmic barrier method, are known from the literature, but their analysis contains many new elements, often resulting in much sharper bounds than those in the literature. In this respect an important (and new) tool is the function ψ, first introduced in Section 5.5 and used through the rest of the book. We present a comprehensive discussion of all possible variants of these algorithms (like dual, primal and primal-dual full-step, adaptive-update and

---

[7] Of course, the book is inspired by many papers and results of many colleagues. Thinking over these results often led to new insights, new algorithms and new ways to analyze these algorithms.

[8] The approach in Part I, based on the embedding of a given LO problem in a self-dual problem, suggests some new and promising implementation strategies.

large-update methods). We also deal with the — from the practical point of view very important — predictor-corrector method, and show that this method has an asymptotically quadratic convergent rate. We also discuss the techniques of partial updating and the use of higher-order methods. Finally, we present a new approach to sensitivity analysis and discuss many computationally aspects which are crucial for efficient implementation of IPM's.

1.4 Required knowledge and skills

We wanted to write a book that presents the most prominent results on IPM's in a unified and comprehensive way, with a full development of the most important items. Especially Part I can be considered as an elementary introduction to LO, containing both a complete derivation of the duality theory as well as an easy-to-analyze polynomial algorithm.

The mathematical tools that are used do not go beyond standard calculus and linear algebra. Nevertheless, people educated in the Simplex based approach to LO will need some effort to get acquainted with the formalism and the mathematical manipulations. They have struggled with the algebra of pivoting, the new methods do not refer to pivoting.[9] However, the tools used are not much more advanced than those that were required to master the Simplex Method. We therefore expect that people will quickly get acquainted with the new tools, just as many generations of students have become familiar with pivoting.

In general, the level of the book will be accessible to any student in Operations Research and Mathematics, with 2 to 3 years of basic training in calculus and linear algebra.

1.5 How to use the book for courses

Owing to the importance of LO in theory and in practice, it must be expected that IPM's will soon become a popular topic in Operations Research and other fields where LO is used, such as Business, Economics and Engineering. More and more institutions will open courses dedicated to IPM's for LO. It has been one of our purposes to collect in this book all relevant material from research papers, survey papers, etc. and to strive for a cohesive and easily accessible source for such courses.

The dependence between the chapters is demonstrated in Figure 1.1. This figure indicates some possible reading paths through the book. For newcomers in the field we recommend starting with Part I, consisting of Chapters 2, 3 and 4. This part of the book can be used for a basic course in LO, covering duality theory and offering a first and easy-to-analyze polynomial algorithm: the Full-Newton Step Algorithm. Part II deals with LO problems in standard format. Chapter 5 covers the duality theory and Chapters 6 and 7 deal with several interesting variants of the Logarithmic

---

[9] However, numerical analysts who want to perform the actual implementation really need to master advanced sparse linear algebra, including pivoting strategies in matrix factorization. See Chapter 20.

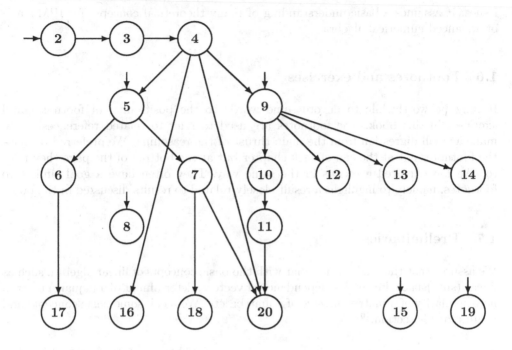

Figure 1.1 Dependence between the chapters.

Barrier Method that underly the efficient solvers in existing commercial optimization packages. For readers who know the Simplex Method and who are familiar with the LO problem in standard format, we made Part II independent of Part I; they might wish to start their reading with Part II and then proceed with Part I.

Part III, on the target-following approach, offers much new understanding of the principles of IPM's, as well as a unifying and easily accessible treatment of other IPM's, such as the method of Renegar (Chapter 14). This part could be part of a more advanced course on IPM's.

Chapter 15 contains a relatively simple description and analysis of Karmarkar's Projective Method. This chapter is almost independent of the previous chapters and hence can be read at any stage.

Chapters 16, 17 and 18 could find a place in an advanced course. The value of Chapter 16 is purely theoretical and is recommended to readers who want to delve more deeply into properties of the central path. The other two chapters, on the other hand, have more practical value. They describe and apply two techniques (partial updating and higher-order methods) that can be used to enhance the efficiency of some methods.

We consider Chapter 19 to be extremely important for users of LO who are interested in the sensitivity of their models to perturbations in the input data. This chapter is independent of almost all the previous chapters.

Finally, Chapter 20 is relevant for readers who are interested in implementation issues. It assumes a basic understanding of many theoretical concepts for IPM's and of advanced numerical algebra.

1.6 Footnotes and exercises

It may be worthwhile to devote some words to the positioning of footnotes and exercises in this book. The footnotes are used to refer to related references, or to make a small digression from the main thrust of the reasoning. We preferred to place the footnotes not at the end of each chapter but at the bottom of the page they refer to. We have treated exercises in the same way. They often have a goal similar to footnotes, namely to highlight a result closely related to results discussed in the book.

1.7 Preliminaries

We assume that the reader is familiar with the basic concepts of linear algebra, such as linear (sub-)space, linear (in-)dependence of vectors, determinant of a (square) matrix, nonsingularity of a matrix, inverse of a matrix, etc. We recall some basic concepts and results in this section.[10]

1.7.1 Positive definite matrices

The space of all square $n \times n$ matrices is denoted by $\mathbb{R}^{n \times n}$. A matrix $A \in \mathbb{R}^{n \times n}$ is called a *positive definite matrix* if A is symmetric and each of its eigenvalues is positive.[11] The following statements are equivalent for any symmetric matrix A:

(i) A is positive definite;
(ii) $A = C^T C$ for some nonsingular matrix C;
(iii) $x^T A x > 0$ for each nonzero vector x.

A matrix $A \in \mathbb{R}^{n \times n}$ is called a *positive semi-definite matrix* if A is symmetric and its eigenvalues are nonnegative. The following statements are equivalent for any symmetric matrix A:

(i) A is positive semi-definite;
(ii) $A = C^T C$ for some matrix C;
(iii) $x^T A x \geq 0$ for each vector x.

1.7.2 Norms of vectors and matrices

In this book a vector x is always an n-tuple $(x_1, x_2, \ldots, x_n)$ in $\mathbb{R}^n$. The numbers x_i $(1 \leq i \leq n)$ are called the *coordinates* or *entries* of x. Usually we think of x as a

---

[10] For a more detailed treatment we refer the reader to books like Bellman [38], Birkhoff and MacLane [41], Golub and Van Loan [112], Horn and Johnson [147], Lancester and Tismenetsky [181], Ben-Israel and Greville [39], Strang [259] and Watkins [289].

[11] Some authors do not include symmetry as part of the definition. For example, Golub and Van Loan [112] call A positive definite if (iii) holds without requiring symmetry of A.

column vector and of its transpose, denoted by x^T, as a row vector. If all entries of x are zero we simply write $x = 0$. A special vector is the all-one vector, denoted by e, whose coordinates are all equal to 1. The scalar product of x and $s \in \mathbb{R}^n$ is given by

$$x^T s = \sum_{i=1}^{n} x_i s_i.$$

We recall the following properties of norms for vectors and matrices. A norm (or vector norm) on $\mathbb{R}^n$ is a function that assigns to each $x \in \mathbb{R}^n$ a nonnegative number $\|x\|$ such that for all $x, s \in \mathbb{R}^n$ and $\alpha \in \mathbb{R}$:

$$\|x\| > 0, \quad \text{if } x \neq 0$$
$$\|\alpha x\| = |\alpha| \, \|x\|$$
$$\|x + s\| \leq \|x\| + \|s\| \, .$$

The Euclidean norm is defined by

$$\|x\|_2 = \sqrt{\sum_{i=1}^{n} x_i^2}.$$

When the norm is not further specified, $\|x\|$ will always refer to the Euclidean norm. The Cauchy–Schwarz inequality states that for $x, s \in \mathbb{R}^n$:

$$x^T s \leq \|x\| \, \|s\| \, .$$

The inequality holds with equality if and only if x and s are linearly dependent.

For any positive number p we also have the p-norm, defined by

$$\|x\|_p = \left(\sum_{i=1}^{n} |x_i|^p \right)^{\frac{1}{p}}.$$

The Euclidean norm is the special case where $p = 2$ and is therefore also called the 2-norm. Another important special case is the 1-norm:

$$\|x\|_1 = \sum_{i=1}^{n} |x_i|.$$

Letting p go to infinity we get the so-called infinity norm:

$$\|x\|_\infty = \lim_{p \to \infty} \|x\|_p \, .$$

We have

$$\|x\|_\infty = \max_{1 \leq i \leq n} |x_i| \, .$$

For any positive definite $n \times n$ matrix A we have a vector norm $\|.\|_A$ according to

$$\|x\|_A = \sqrt{x^T A x}.$$

For any norm the *unit ball in* $\mathbb{R}^n$ is the set

$$\{x \in \mathbb{R}^n \ : \ \|x\| = 1\}.$$

By concatenating the columns of an $n \times n$ matrix A (in the natural order), A can be considered a vector in $\mathbb{R}^{n^2}$. A function assigning to each $A \in \mathbb{R}^{n \times n}$ a real number $\|A\|$ is called a *matrix norm* if it satisfies the conditions for a vector norm and moreover

$$\|AB\| \leq \|A\| \, \|B\|,$$

for all $A, B \in \mathbb{R}^{n \times n}$. A well-known matrix norm is the *Frobenius norm* $\|.\|_F$, which is simply the vector 2-norm applied to the matrix:

$$\|A\|_F = \sqrt{\sum_{i=1}^n \sum_{j=1}^n A_{ij}^2}.$$

Every vector norm induces a matrix norm according to

$$\|A\| = \max_{\|x\|=1} \|Ax\|.$$

This matrix norm satisfies

$$\|Ax\| \leq \|A\| \, \|x\|, \quad \forall x \in \mathbb{R}^n.$$

The vector 1-norm induces the matrix norm

$$\|A\|_1 = \max_{1 \leq j \leq n} \sum_{i=1}^n |A_{ij}|,$$

and the vector ∞-norm induces the matrix norm

$$\|A\|_\infty = \max_{1 \leq i \leq n} \sum_{j=1}^n |A_{ij}|.$$

$\|A\|_1$ is also called the *column sum norm* and $\|A\|_\infty$ the *row sum norm*. Note that

$$\|A\|_\infty = \|A^T\|_1.$$

Hence, if A is symmetric then $\|A\|_\infty = \|A\|_1$. The matrix norm induced by the vector 2-norm is, by definition,

$$\|A\|_2 = \max_{\|x\|_2=1} \|Ax\|_2.$$

This norm is also called the *spectral matrix norm*. Observe that it differs from the Frobenius norm (consider both norms for $A = I$, where $I = \text{diag}\,(e)$). In general,

$$\|A\|_2 \leq \|A\|_F.$$

1.7.3   Hadamard inequality for the determinant

For an $n \times n$ matrix A with columns $a_1, a_2, \ldots, a_n$ its determinant satisfies

$$\det(A) = \text{volume of the parallelepiped spanned by } a_1, a_2, \ldots, a_n.$$

This interpretation of the determinant implies the inequality

$$\det(A) \leq \|a_1\|_2 \|a_2\|_2 \cdots \|a_n\|_2,$$

which is known as the *Hadamard inequality*.[12]

1.7.4   Order estimates

Let f and g be functions from the positive reals to the positive reals. In many estimates the following definitions will be helpful.

- We write $f(x) = \mathcal{O}(g(x))$ if there exists a positive constant c such that $f(x) \leq cg(x)$, for all $x > 0$.
- We write $f(x) = \Omega(g(x))$ if there exists a positive constant c such that $f(x) \geq cg(x)$, for all $x > 0$.
- We write $f(x) = \Theta(g(x))$ if there exist positive constants c_1 and c_2 such that $c_1 g(x) \leq f(x) \leq c_2 g(x)$, for all $x > 0$.

1.7.5   Notational conventions

The identity matrix usually is denoted as I; if the size of I is not clear from the context we use a subscript like in I_n to specify that it is the $n \times n$ identity matrix. Similarly, zero matrices and zero vectors usually are denoted simply as 0; but if the size is ambiguous, we use subscripts like in $0_{m \times n}$ to specify the size. The all-one vector is always denoted as e, and if necessary the size is specified by a subscript.

For any $x \in \mathbb{R}^n$ we often denote the diagonal matrix diag (x) by the corresponding capital X. For example, $D = $ diag (d). The componentwise product of two vectors $x, s \in \mathbb{R}^n$, known as the Hadamard product of x and s is denoted compactly by xs.[13] The i-th entry of xs is $x_i s_i$. In other words, $xs = Xs = Sx$. As a consequence we have for the scalar product of x and s,

$$x^T s = e^T(xs),$$

which will be used repeatedly later on. Similarly we use x/s for the componentwise quotient of x and s. This kind of notation is also used for unitary operations. For example, the i-th entry of x^{-1} is x_i^{-1} and the i-th entry of $\sqrt{x}$ is $\sqrt{x_i}$. This notation is consistent as long as componentwise operations are given precedence over matrix operations. Thus, if A is a matrix then $Axs = A(xs)$.

---

[12] See, e.g., Horn and Johnson [147], page 477.

[13] In the literature this product is known as the *Hadamard product* of x and s. It is often denoted by $x \bullet s$. Throughout the book we will use the shorter notation xs. Note that if x and s are nonnegative then $xs = 0$ holds if and only if $x^T s = 0$.

Part I

Introduction: Theory and Complexity

2

Duality Theory for Linear Optimization

2.1 Introduction

This chapter introduces the reader to the main theoretical results in the field of linear optimization (LO). These results concern the notion of *duality in LO*. An LO problem consists of optimizing (i.e., minimizing or maximizing) a linear *objective function* subject to a finite set of *linear constraints*. The constraints may be *equality constraints* or *inequality constraints*. If the constraints are inconsistent, so that they do not allow any *feasible solution*, then the problem is called *infeasible*, otherwise *feasible*. In the latter case the *feasible set* (or *domain*) of the problem is not empty; then there are two possibilities: the objective function is either unbounded or bounded on the domain. In the first case, the problem is called *unbounded* and in the second case *bounded*. The set of optimal solutions of a problem is referred to as the *optimal set*; the optimal set is empty if and only if the problem is infeasible or unbounded.

For any LO problem we may construct a second LO problem, called its *dual problem*, or shortly its *dual*. A problem and its dual are closely related. The relation can be expressed nicely in terms of the optimal sets of both problems. If the optimal set of one of the two problems is nonempty, then neither is the optimal set of the other problem; moreover, the optimal values of the objective functions for both problems are equal. These nontrivial results are the basic ingredients of the so-called *duality theory* for LO.

The duality theory for LO can be derived in many ways.[1] A popular approach in textbooks to this theory is constructive. It is based on the Simplex Method. While solving a problem by this method, at each iterative step the method generates so-

---

[1] The first duality results in LO were obtained in a nonconstructive way. They can be derived from some variants of Farkas' lemma [75], or from more general separation theorems for convex sets. See, e.g., Osborne [229] and Saigal [249]. An alternative approach is based on direct inductive proofs of theorems of Farkas, Weyl and Minkowski and derives the duality results for LO as a corollary of these theorems. See, e.g., Gale [91]. Constructive proofs are based on finite termination of a suitable algorithm for solving either linear inequality systems or LO problems. A classical method for solving linear inequality systems in a finite number of steps is Fourier-Motzkin elimination. By this method we can decide in finite time if the system admits a feasible solution or not. See, e.g., Dantzig [59]. This can be used to proof Farkas' lemma from which the duality results for LO easily follow. For the LO problem there exist several finite termination methods. One of them, the Simplex Method, is sketched in this paragraph. Many authors use such a method to derive the duality results for LO. See, e.g., Chvátal [55], Dantzig [59], Nemhauser and Wolsey [224], Papadimitriou and Steiglitz [231], Schrijver [250] and Walsh [287].

called *multipliers* associated with the constraints. The method terminates when the multipliers turn out to be feasible for the dual problem; then it yields an optimal solution both for the primal and the dual problem.[2]

Interior point methods are also intimately linked with duality theory. The key concept is the so-called *central path*, an analytic curve in the interior of the domain of the problem that starts somewhere in the 'middle' of the domain and ends somewhere in the 'middle' of the optimal set of the problem. The term 'middle' in this context will be made precise later. Interior point methods follow the central path (approximately) as a guideline to the optimal set.[3] One of the aims of this chapter is to show that the aforementioned duality results can be derived from properties of the central path.[4] Not every problem has a central path. Therefore, it is important in this framework to determine under which condition the central path exists. It happens that this condition implies the existence of the central path for the dual problem and the points on the dual central path are closely related to the points on the primal central path. As a consequence, following the primal central path (approximately) to the primal optimal set goes always together with following the dual central path (approximately) to the dual optimal set. Thus, when the primal and dual central paths exist, the interior-point approach yields in a natural way the duality theory for LO, just as in the case of the Simplex Method. When the central paths do not exist the duality results can be obtained by a little trick, namely by embedding the given problem in a larger problem which has a central path. Below this approach will be discussed in more detail.

We start the whole analysis, in the next section, by considering the LO problem in the so-called canonical form. So the objective is to minimize a linear function over a set of inequality constraints of greater-than-or-equal type with nonnegative variables.

Since every LO problem admits a canonical representation, the validity of the duality results in this chapter naturally extend to arbitrary LO problems. Usually the canonical form of an LO problem is obtained by introducing new variables and/or constraints. As a result, the number of variables and/or constraints may be doubled. In Appendix D.1 we present a specific scheme that transforms any LO problem that is not in the canonical form to a canonical problem in such a way that the total number of variables and constraints does not increase, and even decreases in many cases.

We show that solving the canonical LO problem can be reduced to finding a solution of an appropriate system of inequalities. In Section 2.4 we impose a condition on the system—the *interior-point condition*— and we show that this condition is not satisfied by our system of inequalities. By expanding the given system slightly however we get an equivalent system that satisfies the interior-point condition. Then we construct a *self-dual problem*[5] whose domain is defined by the last system. We further show that a solution of the system, and hence of the given LO problem, can easily be obtained

---

[2] The Simplex Method was proposed first by Dantzig [59]. In fact, this method has many variants due to various strategies for choosing the pivot element. When we refer to the Simplex Method we always assume that a pivot strategy is used that prevents cycling and thus guarantees finite termination of the method.

[3] This interpretation of recent interior-point methods for LO was proposed first by Megiddo [200]. The notion of central path originates from nonlinear (convex) optimization; see Fiacco and McCormick [77].

[4] This approach to the duality theory has been worked out by Güler et al. [133, 134].

[5] Problems of this special type were considered first by Tucker [274], in 1956.

from a so-called *strictly complementary solution* of the self-dual problem.

Thus the canonical problem can be embedded in a natural way into a self-dual problem and using the existence of a strictly complementary solution for the embedding self-dual problem we derive the classical duality results for the canonical problem. This is achieved in Section 2.9.

The self-dual problem in itself is a trivial LO problem. In this problem all variables are nonnegative. The problem is trivial in the sense that the zero vector is feasible and also optimal. In general the zero vector will not be the only optimal solution. If the optimal set contains nonzero vectors, then some of the variables must occur with positive value in an optimal solution. Thus we may divide the variables into two groups: one group contains the variables that are zero in each optimal solution, and the second group contains the other variables that may occur with positive sign in an optimal solution. Let us call for the moment the variables in the first group 'good' variables and those in the second group 'bad' variables.

We proceed by showing that the interior-point condition guarantees the existence of the central path. The proof of this fact in Section 2.7 is constructive. From the limiting behavior of the central path when it approaches the optimal set, we derive the existence of a strictly complementary solution of the self-dual problem. In such an optimal solution all 'good' variables are positive, whereas the 'bad' variables are zero, of course. Next we prove the same result for the case where the interior-point condition does not hold. From this we derive that every (canonical) LO problem that has an optimal solution, also has a strictly complementary optimal solution.

It may be clear that the nontrivial part of the above analysis concerns the existence of a strictly complementary solution for the self-dual problem. Such solutions play a crucial role in the approach of this book. Obviously a strictly complementary solution provides much more information on the optimal set of the problem than just one optimal solution, because variables that occur with zero value in a strictly complementary solution will be zero in any optimal solution.[6]

One of the surprises of this chapter is that the above results for the self-dual problem immediately imply all basic duality results for the general LO problem. This is shown first for the *canonical problem* in Section 2.9 and then for general LO problems in Section 2.10; in this section we present an easy-to-remember scheme for writing down the dual problem of any given LO problem. This involves first transforming the given problem to a canonical form, then taking the dual of this problem and reformulating the canonical dual so that its relation to the given problem becomes more apparent. The scheme is such that applying it twice returns the original problem. Finally, although the result is not used explicitly in this chapter, but because it is interesting in itself, we conclude this chapter with Section 2.11 where we show that the central path converges to an optimal solution.

---

[6] The existence of strictly complementary optimal solutions was shown first by Goldman and Tucker [111] in 1956. Balinski and Tucker [33], in 1969, gave a constructive proof.

2.2   The canonical LO-problem and its dual

We say that a linear optimization problem is in canonical form if it is written in the following way:

$$(P) \qquad \min \left\{ c^T x \ : \ Ax \geq b, \, x \geq 0 \right\}, \qquad\qquad (2.1)$$

where the matrix A is of size $m \times n$, the vectors c and x are in $\mathbb{R}^n$ and b in $\mathbb{R}^m$. Note that all the constraints in (P) are inequality constraints and the variables are nonnegative. Each LO-problem can be transformed to an equivalent canonical problem.[7] Given the above canonical problem (P), we consider a second problem, denoted by (D) and called the *dual problem* of (P), given by

$$(D) \qquad \max \left\{ b^T y \ : \ A^T y \leq c, \, y \geq 0 \right\}. \qquad\qquad (2.2)$$

The two problems (P) and (D) share the matrix A and the vectors b and c in their description. But the role of b and c has been interchanged: the objective vector c of (P) is the right-hand side vector of (D), and, similarly, the right-hand side vector b of (P) is the objective vector of (D). Moreover, the constraint matrix in (D) is the transposed matrix A^T, where A is the constraint matrix in (P). In both problems the variables are nonnegative. The problems differ in that (P) is a minimization problem whereas (D) is a maximization problem, and, moreover, the inequality symbols in the constraints have opposite direction.[8,9]

At this stage we make a crucial observation.

Lemma I.1 (Weak duality) *Let x be feasible for (P) and y for (D). Then*

$$b^T y \leq c^T x. \qquad\qquad (2.3)$$

Proof: If x is feasible for (P) and y for (D), then $x \geq 0, y \geq 0, Ax \geq b$ and $A^T y \leq c$. As a consequence we may write

$$b^T y \leq (Ax)^T y = x^T \left(A^T y \right) \leq c^T x.$$

This proves the lemma.   □

Hence, any y that is feasible for (D) provides a lower bound $b^T y$ for the value of $c^T x$, whenever x is feasible for (P). Conversely, any x that is feasible for (P) provides an upper bound $c^T x$ for the value of $b^T y$, whenever y is feasible for (D). This phenomenon is known as the *weak duality property*. We have as an immediate consequence the following.

Corollary I.2 *If x is feasible for (P) and y for (D), and $c^T x = b^T y$, then x is optimal for (P) and y is optimal for (D).*

---

[7] For this we refer to any text book on LO. In Appendix D it is shown that this can be achieved without increasing the numbers of constraints and variables.

[8] **Exercise 1** The dual problem (D) can be transformed into canonical form by replacing the constraint $A^T y \leq c$ by $-A^T y \geq -c$ and the objective $\max b^T y$ by $\min -b^T y$. Verify that the dual of the resulting problem is exactly (P).

[9] **Exercise 2** Let the matrix A be skew-symmetric, i.e., $A^T = -A$, and let $b = -c$. Verify that then (D) is essentially the same problem as (P).

The (nonnegative) difference

$$c^T x - b^T y \tag{2.4}$$

between the primal objective value at a primal feasible x and the dual objective value at a dual feasible y is called the *duality gap* for the pair (x, y). We just established that if the duality gap vanishes then x is optimal for (P) and y is optimal for (D). Quite surprisingly, the converse statement is also true: if x is an optimal solution of (P) and y is an optimal solution of (D) then the duality gap vanishes at the pair (x, y). This result is known as the *strong duality property* in LO. One of the aims of this chapter is to prove this most important result. So, in this chapter we will not use this property, but prove it!

Thus our starting point is the question under which conditions an optimal pair (x, y) exists with vanishing duality gap. In the next section we reduce this question to the question if some system of linear inequalities is solvable.

2.3   Reduction to inequality system

In this section we consider the question whether (P) and (D) have optimal solutions with vanishing duality gap. This will be true if and only if the inequality system

$$
\begin{aligned}
Ax &\geq b, & x \geq 0, \\
-A^T y &\geq -c, & y \geq 0, \\
b^T y - c^T x &\geq 0
\end{aligned}
\tag{2.5}
$$

has a solution. This follows by noting that x and y satisfy the inequalities in the first two lines if and only if they are feasible for (P) and (D) respectively. By Lemma I.1 this implies $c^T x - b^T y \geq 0$. Hence, if we also have $b^T y - c^T x \geq 0$ we get $b^T y = c^T x$, proving the claim.

If $\kappa = 1$, the following inequality system is equivalent to (2.5), as easily can be verified.

$$
\begin{bmatrix} 0_{m \times m} & A & -b \\ -A^T & 0_{n \times n} & c \\ b^T & -c^T & 0 \end{bmatrix}
\begin{bmatrix} y \\ x \\ \kappa \end{bmatrix}
\geq
\begin{bmatrix} 0_m \\ 0_n \\ 0 \end{bmatrix}, \quad x \geq 0,\ y \geq 0,\ \kappa \geq 0.
\tag{2.6}
$$

The new variable κ is called the *homogenizing variable*. Since the right-hand side in (2.6) is the zero vector, this system is *homogeneous*: whenever (y, x, κ) solves the system then $\lambda(y, x, \kappa)$ also solves the system, for any positive λ. Now, given any solution (x, y, κ) of (2.6) with $\kappa > 0$, $(x/\kappa, y/\kappa, 1)$ yields a solution of (2.5). This makes clear that, in fact, the two systems are completely equivalent unless every solution of (2.6) has $\kappa = 0$. But if $\kappa = 0$ for every solution of (2.6), then it follows that no solution exists with $\kappa = 1$, and therefore the system (2.5) cannot have a solution in that case. Evidently, we can work with the second system without loss of information about the solution set of the first system.

Hence, defining the matrix $\bar{M}$ and the vector $\bar{z}$ by

$$\bar{M} := \begin{bmatrix} 0 & A & -b \\ -A^T & 0 & c \\ b^T & -c^T & 0 \end{bmatrix}, \quad \bar{z} := \begin{bmatrix} y \\ x \\ \kappa \end{bmatrix}, \tag{2.7}$$

where we omitted the size indices of the zero blocks, we have reduced the problem of finding optimal solutions for (P) and (D) with vanishing duality gap to finding a solution of the inequality system

$$\bar{M}\bar{z} \geq 0, \quad \bar{z} \geq 0, \quad \kappa > 0. \tag{2.8}$$

If this system has a solution then it gives optimal solutions for (P) and (D) with vanishing duality gap; otherwise such optimal solutions do not exist. Thus we have proved the following result.

Theorem I.3 *The problems (P) and (D) have optimal solutions with vanishing duality gap if and only if system (2.8), with $\bar{M}$ and $\bar{z}$ as defined in (2.7), has a solution.*

Thus our task has been reduced to finding a solution of (2.8), or to prove that such a solution does not exists. In the sequel we will deal with this problem. In doing so, we will strongly use the fact that the matrix $\bar{M}$ is skew-symmetric, i.e., $\bar{M}^T = -\bar{M}$.[10] Note that the order of $\bar{M}$ equals $m + n + 1$.

2.4 Interior-point condition

The method we are going to use in the next chapter for solving (2.8) is an interior-point method (IPM), and for this we need the system to satisfy the *interior-point condition*.

Definition I.4 (IPC) *We say that any system of (linear) equalities and (linear) inequalities satisfies the interior-point condition (IPC) if there exists a feasible solution that strictly satisfies all inequality constraints in the system.*

Unfortunately the system (2.8) does not satisfy the IPC. Because if $z = (x, y, \kappa)$ is a solution then x/κ is feasible for (P) and y/κ is feasible for (D). But then $(c^T x - b^T y)/\kappa \geq 0$, by weak duality. Since $\kappa > 0$, this implies $b^T y - c^T x \leq 0$. On the other hand, after substitution of (2.7), the last constraint in (2.8) requires $b^T y - c^T x \geq 0$. It follows that $b^T y - c^T x = 0$, and hence no feasible solution of (2.8) satisfies the last inequality in (2.8) strictly.

To overcome this shortcoming of the system (2.8) we increase the dimension by adding one more nonnegative variable ϑ to the vector $\bar{z}$, and by extending $\bar{M}$ with one extra column and row, according to

$$M := \begin{bmatrix} \bar{M} & r \\ -r^T & 0 \end{bmatrix}, \quad z := \begin{bmatrix} \bar{z} \\ \vartheta \end{bmatrix}, \tag{2.9}$$

_____

[10] **Exercise 3** If S is an $n \times n$ skew-symmetric matrix and $z \in \mathbb{R}^n$, then $z^T S z = 0$. Prove this.

where

$$r = e_{m+n+1} - \bar{M}e_{m+n+1}, \qquad (2.10)$$

with e_{m+n+1} denoting an all-one vector of length $m + n + 1$. So we have

$$M = \begin{bmatrix} \begin{bmatrix} 0 & A & -b \\ -A^T & 0 & c \\ b^T & -c^T & 0 \end{bmatrix} & r \\ -r^T & & 0 \end{bmatrix}, \quad r = \begin{bmatrix} e_m - Ae_n + b \\ e_n + A^Te_m - c \\ 1 - b^Te_m + c^Te_n \end{bmatrix}, \quad z = \begin{bmatrix} y \\ x \\ \kappa \\ \vartheta \end{bmatrix} \quad (2.11)$$

The order of the matrix M is $m + n + 2$. To simplify the presentation, in the rest of this chapter we denote this number as $\bar{n}$:

$$\bar{n} = m + n + 2.$$

Letting q be the vector of length $\bar{n}$ given by

$$q := \begin{bmatrix} 0_{\bar{n}-1} \\ \bar{n} \end{bmatrix}, \qquad (2.12)$$

we consider the system

$$Mz \geq -q, \quad z \geq 0. \qquad (2.13)$$

We make two important observations. First we observe that the matrix M is skew-symmetric. Secondly, the system (2.13) satisfies the IPC. The all-one vector does the work, because taking $\bar{z} = e_{\bar{n}-1}$ and $\vartheta = 1$, we have

$$Mz + q = \begin{bmatrix} \bar{M} & r \\ -r^T & 0 \end{bmatrix} \begin{bmatrix} e_{\bar{n}-1} \\ 1 \end{bmatrix} + \begin{bmatrix} 0 \\ \bar{n} \end{bmatrix} = \begin{bmatrix} \bar{M}e_{\bar{n}-1} + r \\ -r^Te_{\bar{n}-1} + \bar{n} \end{bmatrix} = \begin{bmatrix} e_{\bar{n}-1} \\ 1 \end{bmatrix}. \qquad (2.14)$$

The last equality is due to the definition of r, which implies $\bar{M}e_{\bar{n}-1} + r = e_{\bar{n}-1}$ and

$$-r^Te_{\bar{n}-1} + \bar{n} = -\left(e_{\bar{n}-1} - \bar{M}e_{\bar{n}-1}\right)^T e_{\bar{n}-1} + \bar{n} = -e_{\bar{n}-1}^Te_{\bar{n}-1} + \bar{n} = 1,$$

where we used $e_{\bar{n}-1}^T \bar{M}e_{\bar{n}-1} = 0$ (cf. Exercise 3, page 20).

The usefulness of system (2.13) stems from two facts. First, it satisfies the IPC and hence can be treated by an interior-point method. What this implies will become apparent in the next chapter. Another crucial property is that there is a correspondence between the solutions of (2.8) and the solutions of (2.13) with $\vartheta = 0$. To see this it is useful to write (2.13) in terms of $\bar{z}$ and ϑ:

$$\begin{bmatrix} \bar{M} & r \\ -r^T & 0 \end{bmatrix} \begin{bmatrix} \bar{z} \\ \vartheta \end{bmatrix} + \begin{bmatrix} 0 \\ \bar{n} \end{bmatrix} \geq 0, \quad \bar{z} \geq 0, \quad \vartheta \geq 0.$$

Obviously, if $z = (\bar{z}, 0)$ satisfies (2.13), this implies $\bar{M}\bar{z} \geq 0$ and $\bar{z} \geq 0$, and hence $\bar{z}$ satisfies (2.8). On the other hand, if $\bar{z}$ satisfies (2.8) then $\bar{M}\bar{z} \geq 0$ and $\bar{z} \geq 0$; as a consequence $z = (\bar{z}, 0)$ satisfies (2.13) if and only if $-r^T\bar{z} + \bar{n} \geq 0$, i.e., if and only if

$$r^T\bar{z} \leq \bar{n}.$$

If $r^T\bar{z} \leq 0$ this certainly holds. Otherwise, if $r^T\bar{z} > 0$, the positive multiple $\bar{n}\bar{z}/r^T\bar{z}$ of $\bar{z}$ satisfies $r^T\bar{z} \leq \bar{n}$. Since a positive multiple preserves signs, this is sufficient for our goal. We summarize the above discussion in the following theorem.

Theorem I.5 *The following three statements are equivalent:*

(*i*) *Problems* (P) *and* (D) *have optimal solutions with vanishing duality gap;*
(*ii*) *If* $\bar{M}$ *and* $\bar{z}$ *are given by (2.7) then (2.8) has a solution;*
(*iii*) *If* M *and* z *are given by (2.11) then (2.13) has a solution with* $\vartheta = 0$ *and* $\kappa > 0$.

Moreover, system (2.13) satisfies the IPC.

2.5   Embedding into a self-dual LO-problem

Obviously, solving (2.8) is equivalent to finding a solution of the minimization problem

$$(SP_0) \qquad \min \left\{ 0^T \bar{z} \ : \ \bar{M}\bar{z} \geq 0, \quad \bar{z} \geq 0 \right\} \tag{2.15}$$

with $\kappa > 0$. In fact, this is the way we are going to follow: our aim will be to find out whether this problem has a(n optimal) solution with $\kappa > 0$ or not. Note that the latter condition makes our task nontrivial. Because finding an optimal solution of (SP_0) is trivial: the zero vector is feasible and hence optimal. Also note that (SP_0) is in the canonical form. However, it has a very special structure: its feasible domain is *homogeneous* and since $\bar{M}$ is skew-symmetric, the problem (SP_0) is a self-dual problem (cf. Exercise 2, page 18). We say that (SP_0) is a *self-dual embedding* of the canonical problem (P) and its dual problem (D).

If the constraints in an LO problem satisfy the IPC, then we simply say that the problem itself satisfies the IPC. As we established in the previous section, the self-dual embedding (SP_0) does not satisfy the IPC, and therefore, from an algorithmic point of view this problem is not useful.

In the previous section we reduced the problem of finding optimal solutions (P) and (D) with vanishing duality gap to finding a solution of (2.13) with $\vartheta = 0$ and $\kappa > 0$. For that purpose we consider another *self-dual embedding* of (P) and (D), namely

$$(SP) \qquad \min \left\{ q^T z \ : \ Mz \geq -q, z \geq 0 \right\}. \tag{2.16}$$

The following theorem shows that we can achieve our goal by solving this problem.

Theorem I.6 *The system (2.13) has a solution with* $\vartheta = 0$ *and* $\kappa > 0$ *if and only if the problem* (SP) *has an optimal solution with* $\kappa = z_{\bar{n}-1} > 0$.

Proof: Since $q \geq 0$ and $z \geq 0$, we have $q^T z \geq 0$, and hence the optimal value of (SP) is certainly nonnegative. On the other hand, since $q \geq 0$ the zero vector $(z = 0)$ is feasible, and yields zero as objective value, which is therefore the optimal value. Since $q^T z = \bar{n}\vartheta$, we conclude that the optimal solutions of (2.16) are precisely the vectors z satisfying (2.13) with $\vartheta = 0$. This proves the theorem. $\square$

We associate to any vector $z \in \mathbb{R}^n$ its *slack vector* $s(z)$ as follows.

$$s(z) := Mz + q. \tag{2.17}$$

Then we have

$$z \text{ is a feasible for } (SP) \quad \Longleftrightarrow \quad z \geq 0, \, s(z) \geq 0.$$

As we established in the previous section, the inequalities defining the feasible domain of (SP) satisfy the IPC. To be more specific, we found in (2.14) that the all-one vector e is feasible and its slack vector is the all-one vector. In other words,

$$s(e) = e. \tag{2.18}$$

We proceed by giving a small example.

Example I.7 By way of example we consider the case where the problems (P) and (D) are determined by the following constraint matrix A, and vectors b and c:[11]

$$A = \begin{bmatrix} 1 \\ 0 \end{bmatrix}, \quad b = \begin{bmatrix} 1 \\ -1 \end{bmatrix}, \quad c = [\, 2 \,].$$

According to (2.7) the matrix $\bar{M}$ is then equal to

$$\bar{M} = \begin{bmatrix} 0 & A & -b \\ -A^T & 0 & c \\ b^T & -c^T & 0 \end{bmatrix} = \begin{bmatrix} 0 & 0 & 1 & -1 \\ 0 & 0 & 0 & 1 \\ -1 & 0 & 0 & 2 \\ 1 & -1 & -2 & 0 \end{bmatrix}$$

and according to (2.10), the vector r becomes

$$r = e - \bar{M}e = \begin{bmatrix} 1 \\ 1 \\ 1 \\ 1 \end{bmatrix} - \begin{bmatrix} 0 \\ 1 \\ 1 \\ -2 \end{bmatrix} = \begin{bmatrix} 1 \\ 0 \\ 0 \\ 3 \end{bmatrix},$$

Thus, by (2.11) and (2.12), we obtain

$$M = \begin{bmatrix} 0 & 0 & 1 & -1 & 1 \\ 0 & 0 & 0 & 1 & 0 \\ -1 & 0 & 0 & 2 & 0 \\ 1 & -1 & -2 & 0 & 3 \\ -1 & 0 & 0 & -3 & 0 \end{bmatrix}, \quad q = \begin{bmatrix} 0 \\ 0 \\ 0 \\ 0 \\ 5 \end{bmatrix}.$$

Hence, the self-dual problem (SP), as given by (2.16), gets the form

$$\min \left\{ 5\vartheta : \begin{bmatrix} 0 & 0 & 1 & -1 & 1 \\ 0 & 0 & 0 & 1 & 0 \\ -1 & 0 & 0 & 2 & 0 \\ 1 & -1 & -2 & 0 & 3 \\ -1 & 0 & 0 & -3 & 0 \end{bmatrix} \begin{bmatrix} z_1 \\ z_2 \\ z_3 \\ z_4 \\ z_5 \end{bmatrix} + \begin{bmatrix} 0 \\ 0 \\ 0 \\ 0 \\ 5 \end{bmatrix} \geq 0, \quad \begin{bmatrix} z_1 \\ z_2 \\ z_3 \\ z_4 = \kappa \\ z_5 = \vartheta \end{bmatrix} \geq 0 \right\}. \tag{2.19}$$

Note that the all-one vector is feasible for this problem and that its surplus vector also is the all-one vector. This is in accordance with (2.18). As we shall see later on, it means that the all-one vector is the point on the central path for $\mu = 1$. $\diamond$

———————————————————————————————————————

[11] cf. Example D.5 (page 449) in Appendix D.

Remark I.8 *In the rest of this chapter, and the next chapter, we deal with the problem* (SP). *In fact, our analysis does not only apply to the case that M and q have the special form of (2.11) and (2.12). Therefore we extend the applicability of our analysis by weakening the assumptions on M and q. Unless stated otherwise below we only assume the following:*

$$M^T = -M, \quad q \geq 0, \quad s(e) = e. \tag{2.20}$$

The last two variables in the vector z play a special role. They are the homogenizing variable $\kappa = z_{\bar{n}-1}$, and $\vartheta = z_{\bar{n}}$. The variable ϑ is called the *normalizing variable*, because of the following important property.

Lemma I.9 *One has*

$$e^T z + e^T s(z) = \bar{n} + q^T z. \tag{2.21}$$

Proof: The identity in the lemma is a consequence of the *orthogonality property* (cf. Exercise 3, page 20)

$$u^T M u = 0, \quad \forall u \in \mathbb{R}^n. \tag{2.22}$$

First we deduce that for every z one has

$$q^T z = z^T \left(s(z) - Mz \right) = z^T s(z) - z^T Mz = z^T s(z). \tag{2.23}$$

Taking $u = e - z$ in (2.22) we obtain

$$(z - e)^T \left(s(z) - s(e) \right) = 0.$$

Since $s(e) = e, e^T e = \bar{n}$ and $z^T s(z) = q^T z$, the relation (2.21) follows. $\square$

It follows from Lemma I.9 that the sum of the positive coordinates in z and $s(z)$ is bounded above by $\bar{n} + q^T z$. Note that this is especially interesting if z is optimal, because then $q^T z = 0$. Hence, if z is optimal then

$$e^T z + e^T s(z) = \bar{n}. \tag{2.24}$$

Since z and $s(z)$ are nonnegative this implies that the set of optimal solutions is bounded.

Another interesting feature of the LO-problem (2.16) is that it is self-dual: the dual problem is

$$(DSP) \qquad \max \left\{ -q^T u \ : \ M^T u \leq q, \, u \geq 0 \right\};$$

since M is skew-symmetric, $M^T u \leq q$ is equivalent to $-Mu \leq q$, or $Mu \geq -q$, and maximizing $-q^T u$ is equivalent to minimizing $q^T u$, and thus the dual problem is essential the same problem as (2.16).

The rest of the chapter is devoted to our main task, namely to find an optimal solution of (2.16) with $\kappa > 0$ or to establish that such a solution does not exist.

2.6   The classes B and N

We introduce the index sets B and N according to

$$B := \{i \ : \ z_i > 0 \text{ for some optimal } z\}$$
$$N := \{i \ : \ s_i(z) > 0 \text{ for some optimal } z\}.$$

So, B contains all indices i for which an optimal solution z with positive z_i exists. We also write $z_i \in B$ if $i \in B$. Note that we certainly have $\vartheta \notin B$, because ϑ is zero in any optimal solution of (SP). The main question we have to answer is whether $\kappa \in B$ holds or not. Because if $\kappa \in B$ then there exists an optimal solution z with $\kappa > 0$, in which case (P) and (D) have optimal solutions with vanishing duality gap, and otherwise not.

The next lemma implies that the sets B and N are disjoint. In this lemma, and further on, we use the following notation. To any vector $u \in \mathbb{R}^k$, we associate the diagonal matrix U whose diagonal entries are the elements of u, in the same order. If also $v \in \mathbb{R}^k$, then Uv will be denoted shortly as uv. Thus uv is a vector whose entries are obtained by multiplying u and v componentwise.

Lemma I.10 *Let z^1 and z^2 be feasible for (SP). Then z^1 and z^2 are optimal solutions of (SP) if and only if $z^1 s(z^2) = z^2 s(z^1) = 0$.*

Proof: According to (2.23) we have for any feasible z:

$$q^T z = z^T s(z). \tag{2.25}$$

As a consequence, $z \geq 0$ is optimal if and only if $s(z) \geq 0$ and $z^T s(z) = 0$. Since, by (2.22),

$$\left(z^1 - z^2\right)^T M \left(z^1 - z^2\right) = 0,$$

we have

$$\left(z^1 - z^2\right)^T \left(s(z^1) - s(z^2)\right) = 0.$$

Expanding the product on the left and rearranging the terms we get

$$(z^1)^T s(z^2) + (z^2)^T s(z^1) = (z^1)^T s(z^1) + (z^2)^T s(z^2).$$

Now z^1 is optimal if and only if $(z^1)^T s(z^1) = 0$, by (2.25), and similarly for z^2. Hence, since $z^1, z^2, s(z^1)$ and $s(z^2)$ are all nonnegative, z^1 and z^2 are optimal if and only if

$$(z^1)^T s(z^2) + (z^2)^T s(z^1) = 0,$$

which is equivalent to

$$z^1 s(z^2) = z^2 s(z^1) = 0,$$

proving the lemma. $\qquad\square$

Corollary I.11 *The sets B and N are disjoint.*

Proof: If $i \in B \cap N$ then there exist optimal solutions z^1 and z^2 of (SP) such that $z_i^1 > 0$ and $s_i(z^2) > 0$. This would imply $z_i^1 s_i(z^2) > 0$, a contradiction with Lemma I.10. Hence $B \cap N$ is the empty set. $\qquad\square$

By way of example we determine the classes B and N for the problem considered in Example I.7.

Example I.12 Consider the self-dual problem (SP) in Example I.7, as given by (2.19):

$$\min \left\{ 5\vartheta : \begin{bmatrix} 0 & 0 & 1 & -1 & 1 \\ 0 & 0 & 0 & 1 & 0 \\ -1 & 0 & 0 & 2 & 0 \\ 1 & -1 & -2 & 0 & 3 \\ -1 & 0 & 0 & -3 & 0 \end{bmatrix} \begin{bmatrix} z_1 \\ z_2 \\ z_3 \\ z_4 \\ z_5 \end{bmatrix} + \begin{bmatrix} 0 \\ 0 \\ 0 \\ 0 \\ 5 \end{bmatrix} \geq 0, \quad \begin{bmatrix} z_1 \\ z_2 \\ z_3 \\ z_4 = \kappa \\ z_5 = \vartheta \end{bmatrix} \geq 0 \right\}.$$

For any $z \in \mathbb{R}^5$ we have

$$s(z) = \begin{bmatrix} z_3 - z_4 + z_5 \\ z_4 \\ 2z_4 - z_1 \\ z_1 - z_2 - 2z_3 + 3z_5 \\ 5 - z_1 - 3z_4 \end{bmatrix} = \begin{bmatrix} z_3 - \kappa + \vartheta \\ \kappa \\ 2\kappa - z_1 \\ z_1 - z_2 - 2z_3 + 3\vartheta \\ 5 - z_1 - 3\kappa \end{bmatrix}.$$

Now z is feasible if $z \geq 0$ and $s(z) \geq 0$, and optimal if moreover $zs(z) = 0$. So $z = (z_1, z_2, z_3, \kappa, \vartheta)$ is optimal if and only if

$$\begin{bmatrix} z_1 \\ z_2 \\ z_3 \\ \kappa \\ \vartheta \end{bmatrix} \geq 0, \quad \begin{bmatrix} z_3 - \kappa + \vartheta \\ \kappa \\ 2\kappa - z_1 \\ z_1 - z_2 - 2z_3 + 3\vartheta \\ 5 - z_1 - 3\kappa \end{bmatrix} \geq 0, \quad \begin{cases} z_1 (z_3 - \kappa + \vartheta) = 0 \\ z_2 \kappa = 0 \\ z_3 (2\kappa - z_1) = 0 \\ \kappa (z_1 - z_2 - 2z_3 + 3\vartheta) = 0 \\ \vartheta (5 - z_1 - 3\kappa) = 0 \end{cases}.$$

Adding the equalities at the right we obtain $5\vartheta = 0$, which gives $\vartheta = 0$, as it should. Substitution gives

$$\begin{bmatrix} z_1 \\ z_2 \\ z_3 \\ \kappa \\ 0 \end{bmatrix} \geq 0, \quad \begin{bmatrix} z_3 - \kappa \\ \kappa \\ 2\kappa - z_1 \\ z_1 - z_2 - 2z_3 \\ 5 - z_1 - 3\kappa \end{bmatrix} \geq 0, \quad \begin{cases} z_1 (z_3 - \kappa) = 0 \\ z_2 \kappa = 0 \\ z_3 (2\kappa - z_1) = 0 \\ \kappa (z_1 - z_2 - 2z_3) = 0 \\ \vartheta = 0 \end{cases}.$$

Note that if $\kappa = 0$ then the inequality $2\kappa - z_1 \geq 0$ implies $z_1 = 0$, and then the inequality $z_1 - z_2 - 2z_3 \geq 0$ gives also $z_2 = 0$ and $z_3 = 0$. Hence, $z = 0$ is the only optimal solution for which $\kappa = 0$. So, let us assume $\kappa > 0$. Then we deduce from the second and fourth equality that $z_2 = 0$ and $z_1 - z_2 - 2z_3 = 0$. This reduces our system to

$$\begin{bmatrix} z_1 = 2z_3 \\ 0 \\ z_3 \\ \kappa \\ 0 \end{bmatrix} \geq 0, \quad \begin{bmatrix} z_3 - \kappa \\ \kappa \\ 2\kappa - 2z_3 \\ 0 \\ 5 - 2z_3 - 3\kappa \end{bmatrix} \geq 0, \quad \begin{cases} 2z_3 (z_3 - \kappa) = 0 \\ z_2 = 0 \\ z_3 (2\kappa - 2z_3) = 0 \\ 0 = 0 \\ \vartheta = 0 \end{cases}.$$

The equations at the right make clear that either $z_3 = 0$ or $z_3 = \kappa$. However, the inequality $z_3 - \kappa \geq 0$ forces $z_3 > 0$ since $\kappa > 0$. Thus we find that any optimal solution has the form

$$z = \begin{bmatrix} 2\kappa \\ 0 \\ \kappa \\ \kappa \\ 0 \end{bmatrix}, \quad s(z) = \begin{bmatrix} 0 \\ \kappa \\ 0 \\ 0 \\ 5 - 5\kappa \end{bmatrix}, \quad 0 \le \kappa \le 1. \tag{2.26}$$

This implies that in this example the sets B and N are given by

$$B = \{1, 3, 4\}, \quad N = \{2, 5\}. \qquad \Diamond$$

In the above example the union of B and N is the full index set. This is not an incident. Our next aim is to prove that this always holds. [12,13,14,15] As a consequence these sets form a partition of the full index set $\{1, 2, \ldots, \bar{n}\}$; it is the so-called *optimal partition* of (SP). This important and nontrivial result is fundamental to our purpose but its proof requires some effort. It highly depends on properties of the *central path* of (SP), which is introduced in the next section.

2.7 The central path

2.7.1 Definition of the central path

Recall from (2.14) that $s(e) = e$, where e (as always) denotes the all-one vector of appropriate length (in this case, $\bar{n}$). As a consequence, we have a vector z such that $z_i s_i(z) = 1$ $(1 \le i \le \bar{n})$, which, using our shorthand notation can also be expressed as

$$z = e \quad \Rightarrow \quad zs(z) = e. \tag{2.27}$$

Now we come to a very fundamental notion, both from a theoretical and algorithmic point of view, namely the central path of the LO-problem at hand. The underlying

_____

[12] **Exercise 4** Following the same approach as in Example I.7 construct the embedding problem for the case where the problems (P) and (D) are determined by

$$A = \begin{bmatrix} 1 \\ 0 \end{bmatrix}, \quad b = \begin{bmatrix} 1 \\ 0 \end{bmatrix}, \quad c = \begin{bmatrix} 2 \end{bmatrix},$$

and, following the approach of Example I.12, find the set of all optimal solutions and the optimal partition.

[13] **Exercise 5** Same as in Exercise 4, but now with

$$A = \begin{bmatrix} 1 \\ 0 \end{bmatrix}, \quad b = \begin{bmatrix} 1 \\ 1 \end{bmatrix}, \quad c = \begin{bmatrix} 2 \end{bmatrix}.$$

[14] **Exercise 6** Same as in Exercise 4, but now with

$$A = \begin{bmatrix} 1 \\ 0 \end{bmatrix}, \quad b = \begin{bmatrix} 1 \\ \beta \end{bmatrix}, \quad c = \begin{bmatrix} 2 \end{bmatrix}, \quad \beta > 0.$$

[15] **Exercise 7** Same as in Exercise 4, but now with

$$A = \begin{bmatrix} 1 \\ 0 \end{bmatrix}, \quad b = \begin{bmatrix} 1 \\ \beta \end{bmatrix}, \quad c = \begin{bmatrix} 2 \end{bmatrix}, \quad \beta < 0.$$

theoretical property is that for every positive number μ there exist a nonnegative vector z such that

$$zs(z) = \mu e, \quad z \geq 0, \, s(z) \geq 0, \tag{2.28}$$

and moreover, this vector is unique. If $\mu = 1$, the existence of such a vector is guaranteed by (2.27). Also note that if we put $\mu = 0$ in (2.28) then the solutions are just the optimal solutions of (SP). As we have seen in Example I.12 there may more than one optimal solution. Therefore, if $\mu = 0$ the system (2.28) may have multiple solutions. The following lemma is of much interest. It makes clear that for $\mu > 0$ the system (2.28) has at most one solution.

Lemma I.13 *If $\mu > 0$, then there exists at most one nonnegative vector z such that (2.28) holds.*

Proof: Let z^1 and z^2 to nonnegative vectors satisfying (2.28), and let $s^1 = s(z^1)$ and $s^2 = s(z^2)$. Since $\mu > 0$, z^1, z^2, s^1, s^2 are all positive. Define $\Delta z := z^2 - z^1$, and similarly $\Delta s := s^2 - s^1$. Then we may easily verify that

$$M\Delta z = \Delta s \tag{2.29}$$
$$z^1\Delta s + s^1\Delta z + \Delta s\Delta z = 0. \tag{2.30}$$

Using that M is skew-symmetric, (2.22) implies that $\Delta z^T \Delta s = 0$, or, equivalently,

$$e^T (\Delta z \Delta s) = 0. \tag{2.31}$$

Rewriting (2.30) gives

$$(z^1 + \Delta z)\Delta s + s^1\Delta z = 0.$$

Since $z^1 + \Delta z = z^2 > 0$ and $s^1 > 0$, this implies that no two corresponding entries in Δz and Δs have the same sign. So it follows that

$$\Delta z \Delta s \leq 0. \tag{2.32}$$

Combining (2.31) and (2.32), we obtain $\Delta z \Delta s = 0$. Hence either $(\Delta z)_i = 0$ or $(\Delta s)_i = 0$, for each i. Using (2.30) once more, we conclude that $(\Delta z)_i = 0$ and $(\Delta s)_i = 0$, for each i. Hence $\Delta z = \Delta s = 0$, whence $z^1 = z^2$ and $s^1 = s^2$. This proves the lemma. $\square$

To prove the existence of a solution to (2.28) requires much more effort. We postpone this to the next section. For the moment, let us take the existence of a solution to (2.28) for granted and denote it as $z(\mu)$. We call it the μ-center of (SP). The set

$$\{z(\mu) \, : \, \mu > 0\}$$

of all μ-centers represents a parametric curve in the feasible region of (SP). This curve is called the *central path* of (SP). Note that

$$q^T z(\mu) = s(\mu)^T z(\mu) = \mu\bar{n}. \tag{2.33}$$

This proves that along the central path, when μ approaches zero, the objective value $q^T z(\mu)$ monotonically decreases to zero, at a linear rate.

2.7.2 Existence of the central path

In this section we give an algorithmic proof of the existence of a solution to (2.28). Starting at $z = e$ we construct the μ-center for any $\mu > 0$. This is done by using the so-called Newton direction as a search direction. The results in this section will also be used later when dealing with a polynomial-time method for solving (SP).

Newton direction

Assume that z is a positive solution of (SP) such that its slack vector $s = s(z)$ is positive, and let Δz denote a displacement in the z-space. Our aim is to find Δz such that $z + \Delta z$ is the μ-center. We denote

$$z^+ := z + \Delta z,$$

and the new slack vector as s^+:

$$s^+ := s(z^+) = M(z + \Delta z) + q = s + M\Delta z.$$

Thus, the displacement Δs in the s-space is simply given by

$$\Delta s = s^+ - s = M\Delta z.$$

Observe that Δz and Δs are orthogonal, since by (2.22):

$$(\Delta z)^T \Delta s = (\Delta z)^T M\Delta z = 0. \tag{2.34}$$

We want Δz to be such that z^+ becomes the μ-center, which means $(z + \Delta z)$ $(s + \Delta s) = \mu e$, or

$$zs + z\Delta s + s\Delta z + \Delta z\Delta s = \mu e.$$

This equation is nonlinear, due to the quadratic term $\Delta z\Delta s$. Applying Newton's method, we omit this nonlinear term, leaving us with the following linear system in the unknown vectors Δz and Δs:

$$M\Delta z - \Delta s = 0, \tag{2.35}$$

$$z\Delta s + s\Delta z = \mu e - zs. \tag{2.36}$$

This system has a unique solution, as easily may be verified, by using that M is skew-symmetric and $z > 0$ and $s > 0$.[16,17] The solution Δz is called the Newton direction. Since we omitted the quadratic term $\Delta z\Delta s$ in our calculation of the Newton

---

[16] **Exercise 8** The coefficient matrix of the system (2.35-2.36) of linear equations in Δz and Δs is

$$\begin{bmatrix} M & -I \\ S & Z \end{bmatrix}.$$

As usual, $Z = \text{diag}(z)$ and $S = \text{diag}(s)$, with $z > 0$ and $s > 0$, and I denotes the identity matrix. Show that this matrix is nonsingular.

[17] **Exercise 9** Let M be a skew-symmetric matrix of size $n \times n$ and Z and S positive diagonal matrices of the same size as M. Then the matrix $S + ZM$ is nonsingular. Prove this.

direction, $z + \Delta z$ will (in general) not be the μ-center, but hopefully it will be a good approximation. In fact, using (2.36), after the Newton step one has

$$z^+ s(z^+) = (z + \Delta z)(s + \Delta s) = zs + (z\Delta s + s\Delta z) + \Delta z \Delta s = \mu e + \Delta z \Delta s. \quad (2.37)$$

Comparing this with our desire, namely $z^+ s(z^+) = \mu e$, we see that the 'error' is precisely the quadratic term $\Delta z \Delta s$. Using (2.22), we deduce from (2.37) that

$$\left(z^+\right)^T s(z^+) = \mu e^T e = \mu \bar{n}, \quad (2.38)$$

showing that after the Newton step the duality gap already has the desired value.

Example I.14 Let us compute the Newton step at $z = e$ for the self-dual problem (SP) in Example I.7, as given by (2.19), with respect to some $\mu > 0$. Since $z = s(z) = e$, the equation (2.36) reduces to

$$\Delta s + \Delta z = \mu e - e = (\mu - 1)e.$$

Hence, by substitution into (2.35) we obtain

$$(M + I)\Delta z = (\mu - 1)e.$$

It suffices to know the solution of the equation $(M + I)\zeta = e$, because then $\Delta z = (\mu - 1)\zeta$. Thus we need to solve ζ from

$$\begin{bmatrix} 1 & 0 & 1 & -1 & 1 \\ 0 & 1 & 0 & 1 & 0 \\ -1 & 0 & 1 & 2 & 0 \\ 1 & -1 & -2 & 1 & 3 \\ -1 & 0 & 0 & -3 & 1 \end{bmatrix} \zeta = \begin{bmatrix} 1 \\ 1 \\ 1 \\ 1 \\ 1 \end{bmatrix},$$

which gives the unique solution

$$\zeta = \left(-\frac{1}{3}, \frac{8}{9}, \frac{4}{9}, \frac{1}{9}, 1\right)^T.$$

Hence

$$\Delta z = (\mu - 1)\left(-\frac{1}{3}, \frac{8}{9}, \frac{4}{9}, \frac{1}{9}, 1\right)^T, \quad (2.39)$$

and

$$\Delta s = M\Delta z = (\mu - 1)(e - \zeta) = (\mu - 1)\left(\frac{4}{3}, \frac{1}{9}, \frac{5}{9}, \frac{8}{9}, 0\right)^T. \quad (2.40)$$

After the Newton step we thus have

$$\begin{aligned} z^+ s^+ &= (z + \Delta z)(s + \Delta s) = zs + (\Delta z + \Delta s) + \Delta z \Delta s \\ &= e + (\mu - 1)e + \Delta z \Delta s = \mu e + \Delta z \Delta s \\ &= \mu e + \frac{(\mu - 1)^2}{81}(-36, 8, 20, 8, 0)^T. \end{aligned}$$

$\diamond$

Proximity measure

To measure the quality of any approximation z of $z(\mu)$, we introduce a *proximity measure* $\delta(z, \mu)$ that vanishes if $z = z(\mu)$ and is positive otherwise. To this end we introduce the *variance vector* of z with respect to μ as follows:

$$v := \sqrt{\frac{zs(z)}{\mu}}, \qquad (2.41)$$

where all operations are componentwise. Note that

$$zs(z) = \mu e \quad \Leftrightarrow \quad v = e.$$

The proximity measure $\delta(z, \mu)$ is now defined by[18]

$$\delta(z, \mu) := \tfrac{1}{2} \left\| v - v^{-1} \right\|. \qquad (2.42)$$

Note that if $z = z(\mu)$ then $v = e$ and hence $\delta(z, \mu) = 0$, and otherwise $\delta(z, \mu) > 0$. We show below that if $\delta(z, \mu) < 1$ then the Newton process quadratically fast converges to $z(\mu)$. For this we need the following lemma, which estimates the error term in terms of the proximity measure. In this lemma $\|.\|$ denotes the Eucledian norm (or 2-norm) and $\|.\|_\infty$ the Chebychev norm (or infinity norm) of a vector.

Lemma I.15 *If* $\delta := \delta(z, \mu)$, *then* $\|\Delta z \Delta s\|_\infty \le \mu \delta^2$ *and* $\|\Delta z \Delta s\| \le \mu \delta^2 \sqrt{2}$.

Proof: Componentwise division of (2.36) by $\sqrt{\mu}\, v = \sqrt{zs}$ yields

$$\sqrt{\frac{z}{s}}\, \Delta s + \sqrt{\frac{s}{z}}\, \Delta z = \sqrt{\mu}\, (v^{-1} - v).$$

The terms at the left represent orthogonal vectors whose componentwise product is $\Delta z \Delta s$. Applying Lemma C.4 in Appendix C to these vectors, and using that $\|v^{-1} - v\| = 2\delta$, the result immediately follows. $\square$

Quadratic convergence of the Newton process

We are now ready for the main result on the Newton direction.

Theorem I.16 *If* $\delta := \delta(z, \mu) < 1$, *then the Newton step is strictly feasible, i.e.,* $z^+ > 0$ *and* $s^+ > 0$. *Moreover,*

$$\delta(z^+, \mu) \le \frac{\delta^2}{\sqrt{2(1 - \delta^2)}}.$$

---

[18] In the analysis of interior-point methods we always need to introduce a quantity that measures the 'distance' of a feasible vector z to the central path or to the μ-center. This can be done in many ways as becomes apparent in the course of this book. In the coming chapters we make use of a variety of so-called *proximity measures*. Most of these measures are based on the simple observation that z is equal to the μ-center if and only if $v = e$ and z is on the central path if and only if the vector $zs(z)$ is a scalar multiple of the all-one vector.

Proof: Let $0 \leq \alpha \leq 1$, $z^\alpha = z + \alpha \Delta z$ and $s^\alpha = s + \alpha \Delta s$. We then have, using (2.36),

$$\begin{aligned}
z^\alpha s^\alpha &= (z + \alpha \Delta z)(s + \alpha \Delta s) = zs + \alpha (z \Delta s + s \Delta z) + \alpha^2 \Delta z \Delta s \\
&= zs + \alpha (\mu e - zs) + \alpha^2 \Delta z \Delta s = (1 - \alpha)zs + \alpha (\mu e + \alpha \Delta z \Delta s)
\end{aligned}$$

By Lemma I.15,

$$\mu e + \alpha \Delta z \Delta s \geq \mu e - \alpha \|\Delta z \Delta s\|_\infty e \geq \mu(1 - \alpha \delta^2) e > 0.$$

Hence, since $(1 - \alpha)zs \geq 0$, we have $z^\alpha s^\alpha > 0$, for all $\alpha \in [0, 1]$. Therefore, the components of z^α and s^α cannot vanish when $\alpha \in [0, 1]$. Hence, since $z > 0$ and $s > 0$, by continuity, z^α and s^α must be positive for any such α, especially for $\alpha = 1$. This proves the first statement in the lemma.

Now let us turn to the proof of the second statement. Let $\delta^+ := \delta(z^+, \mu)$ and let v^+ be the variance vector of z^+ with respect to μ:

$$v^+ = \sqrt{\frac{z^+ s^+}{\mu}}.$$

Then, by definition,

$$2\delta^+ = \left\| (v^+)^{-1} - v^+ \right\| = \left\| (v^+)^{-1} \left(e - (v^+)^2 \right) \right\|. \tag{2.43}$$

Recall from (2.37) that $z^+ s^+ = \mu e + \Delta z \Delta s$. In other words,

$$(v^+)^2 = e + \frac{\Delta z \Delta s}{\mu}.$$

Substitution into (2.43) gives

$$2\delta^+ = \left\| \frac{-\frac{\Delta z \Delta s}{\mu}}{\sqrt{e + \frac{\Delta z \Delta s}{\mu}}} \right\| \leq \frac{\left\| \frac{\Delta z \Delta s}{\mu} \right\|}{\sqrt{1 - \left\| \frac{\Delta z \Delta s}{\mu} \right\|_\infty}} \leq \frac{\delta^2 \sqrt{2}}{\sqrt{1 - \delta^2}}.$$

The last inequality follows by using Lemma I.15 twice. Thus the proof is complete. $\square$

Theorem I.16 implies that when $\delta \leq 1/\sqrt{2}$, then after a Newton step the proximity to the μ-center satisfies $\delta(z^+, \mu) \leq \delta^2$. In other words, Newton's method is quadratically convergent.

Example I.17 Using the self-dual problem (SP) in Example I.7 again, we consider in this example feasibility of the Newton step, and the proximity measure before and after Newton step at $z = e$ for several values of μ, to be specified below. We will see that the Newton step performs much better than Theorem I.16 predicts! In Example I.14 we found the values of Δz and Δs. Using these values we find for the new iterate:

$$z^+ = e + (\mu - 1) \left(-\frac{1}{3}, \frac{8}{9}, \frac{4}{9}, \frac{1}{9}, 1 \right)^T,$$

and since $s = s(e) = e$,

$$s^+ = e + (\mu - 1) \left(\frac{4}{3}, \frac{1}{9}, \frac{5}{9}, \frac{8}{9}, 0 \right)^T.$$

Hence the Newton step is feasible, i.e., z^+ and s^+ are nonnegative, if and only if

$$0.25 \leq \mu \leq 4,$$

as easily may be verified. For any such μ we have

$$\delta(z, \mu) = \frac{1}{2} \left\| \sqrt{\mu} e - \frac{e}{\sqrt{\mu}} \right\| = \frac{1}{2} \left| \sqrt{\mu} - \frac{1}{\sqrt{\mu}} \right| \|e\| = \frac{\sqrt{5}}{2} \left| \sqrt{\mu} - \frac{1}{\sqrt{\mu}} \right|.$$

Note that Theorem I.16 guarantees feasibility only if $\delta(z, \mu) \leq 1$. This holds if $5\mu^2 - 14\mu + 5 \leq 0$, which is equivalent to

$$0.4202 \approx \frac{1}{5} \left(7 - 2\sqrt{6} \right) \leq \mu \leq \frac{1}{5} \left(7 + 2\sqrt{6} \right) \approx 2.3798.$$

The same theorem guarantees quadratically convergence if $\delta(z, \mu) \leq 1/\sqrt{2}$, which holds if and only if

$$0.5367 \approx \frac{1}{5} \left(6 - \sqrt{11} \right) \leq \mu \leq \frac{1}{5} \left(6 + \sqrt{11} \right) \approx 1.8633.$$

By way of example, consider the case where $\mu = 0.5$. Then we have $\delta(z, \mu) = \frac{1}{4}\sqrt{10} \approx 0.7906$ and, by Theorem I.16, $\delta(z^+, \mu) \leq \frac{5}{12}\sqrt{3} \approx 0.7217$. Let us compute the actual value of $\delta(z^+, \mu)$. For $\mu = 0.5$ we have

$$z^+ = e - \frac{1}{2} \left(-\frac{1}{3}, \frac{8}{9}, \frac{4}{9}, \frac{1}{9}, 1 \right)^T = \left(\frac{7}{6}, \frac{5}{9}, \frac{7}{9}, \frac{17}{18}, \frac{1}{2} \right)^T,$$

and since $s = s(e) = e$,

$$s^+ = e - \frac{1}{2} \left(\frac{4}{3}, \frac{1}{9}, \frac{5}{9}, \frac{8}{9}, 0 \right)^T = \left(\frac{1}{3}, \frac{17}{18}, \frac{13}{18}, \frac{5}{9}, 1 \right)^T.$$

Therefore,

$$(v^+)^2 = \frac{z^+ s^+}{\mu} = \left(\frac{7}{9}, \frac{85}{81}, \frac{91}{81}, \frac{85}{81}, 1 \right)^T.$$

Finally, we compute $\delta(z^+, \mu)$ by using

$$4\delta(z^+, \mu)^2 = \left\| v^+ - (v^+)^{-1} \right\|^2 = \sum_{i=1}^{5} (v_i^+)^2 + \sum_{i=1}^{5} (v_i^+)^{-2} - 10.$$

Note that the first sum equals $(z^+)^T s^+/\mu = 2\bar{n}\mu = 5$. The second sum equals 5.0817. Thus we obtain $4\delta(z^+, \mu)^2 = 0.0817$, which gives $\delta(z^+, \mu) = 0.1429$. ◇

Existence of the central path

Now suppose that we know the μ-center for $\mu = \mu^0 > 0$ and let us denote $z^0 = z(\mu^0)$. Note that this is true for $\mu^0 = 1$, with $z^0 = e$, because $es(e) = e$. So e is the μ-center for $\mu = 1$.

Since $z^0 s(z^0) = \mu^0 e$, the v-vector for z^0 with respect to an arbitrary $\mu > 0$ is given by

$$v = \sqrt{\frac{z^0 s(z^0)}{\mu}} = \sqrt{\frac{\mu^0 e}{\mu}} = \sqrt{\frac{\mu^0}{\mu}} \, e.$$

Hence we have $\delta(z^0, \mu) \leq \frac{1}{\sqrt{2}}$ if and only if

$$\frac{1}{2} \left| \sqrt{\frac{\mu^0}{\mu}} - \sqrt{\frac{\mu}{\mu^0}} \right| \|e\| \leq \frac{1}{\sqrt{2}}.$$

Using $\|e\| = \sqrt{\bar{n}}$, one may easily verify that this holds if and only if

$$\frac{1}{\beta} \leq \frac{\mu}{\mu^0} \leq \beta, \qquad \beta := 1 + \frac{1}{\bar{n}} + \sqrt{\frac{1}{\bar{n}^2} + \frac{2}{\bar{n}}}. \tag{2.44}$$

Now starting the Newton process at z^0, with μ fixed, and such that μ satisfies (2.44), we can generate an infinite sequence $z^0, z^1, \cdots z^k, \cdots$ such that

$$\delta\left(z^k, \mu\right) \leq \frac{1}{2^{2^{k-1}}}.$$

Hence

$$\lim_{k \to \infty} \delta\left(z^k, \mu\right) = 0.$$

The generated sequence has at least one accumulation point z^*, since the iterates $z^1, \cdots z^k, \cdots$ lie in the compact set

$$e^T z + e^T s(z) = \bar{n}\left(1 + \mu\right), \quad z \geq 0, \, s(z) \geq 0,$$

due to (2.21) and (2.38). Since $\delta\left(z^*, \mu\right) = 0$, we obtain $z^* s\left(z^*\right) = \mu e$. Due to Lemma I.13, z^* is unique. This proves that the μ-center exists if μ satisfies (2.44) with $\mu^0 = 1$, i.e., if

$$\frac{1}{\beta} \leq \mu \leq \beta.$$

By redefining μ^0 as one of the endpoints of the above interval we can repeat the above procedure, and extend the interval where the μ-center exists to

$$\frac{1}{\beta^2} \leq \mu \leq \beta^2.$$

and so on. After applying the procedure k times the interval where the μ-center certainly exists is given by

$$\frac{1}{\beta^k} \leq \mu \leq \beta^k.$$

For arbitrary $\mu > 0$, we have to apply the above procedure at most

$$\frac{|\log \mu|}{\log \beta}$$

times, to prove the existence of the μ-center. This completes the proof of the existence of the central path.

It may be worth noting that, using $\bar{n} \geq 2$ and $\log(1+t) \geq \frac{t}{1+t}$ for $t \geq 0$,[19]

$$\log \beta = \log\left(1 + \frac{1}{\bar{n}} + \sqrt{\frac{1}{\bar{n}^2} + \frac{2}{\bar{n}}}\right) \geq \log\left(1 + \sqrt{\frac{2}{\bar{n}}}\right) \geq \frac{\sqrt{\frac{2}{\bar{n}}}}{1 + \sqrt{\frac{2}{\bar{n}}}} = \frac{\sqrt{2}}{\sqrt{\bar{n}} + \sqrt{2}} \geq \frac{1}{\sqrt{2\bar{n}}}.$$

Hence the number of times that we have to apply the above described procedure to obtain the μ-center is bounded above by

$$\sqrt{2\bar{n}}\, |\log \mu|. \tag{2.45}$$

We have just shown that the system (2.28) has a unique solution for every positive μ. The solution is called the μ-center, and denoted as $z(\mu)$. The set of all μ-centers is a curve in the interior of the feasible region of (SP). The definition of the μ-center, as given by (2.28), can be equivalently given as the unique solution of the system

$$Mz + q = s, \quad z \geq 0,\, s \geq 0$$
$$zs = \mu e, \tag{2.46}$$

with M and z as defined in (2.11), and $s = s(z)$, as in (2.17).[20,21,22]

2.8 Existence of a strictly complementary solution

Now that we have proven the existence of the central path we can use it as a guide to the optimal set, by letting the parameter μ approach to zero. As we show in this section, in this way we obtain an optimal solution z such that $z + s(z) > 0$.

Definition I.18 *Two nonnegative vectors a and b in $\mathbb{R}^n$ are said to be complementary vectors if $ab = 0$. If moreover $a + b > 0$ then a and b are called strictly complementary vectors.*

---

[19] See, e.g., Exercise 39, page 133.

[20] **Exercise 10** Using the definitions of z and q, according to (2.11) and (2.12), show that $\vartheta(\mu) = \mu$.

[21] **Exercise 11** In this exercise a skew-symmetric M and four vectors $q^{(i)}$, $i = 1, 2, 3, 4$ are given as follows:

$$M = \begin{bmatrix} 0 & 1 \\ -1 & 0 \end{bmatrix}, \quad q^{(1)} = \begin{bmatrix} 0 \\ 0 \end{bmatrix}, \quad q^{(2)} = \begin{bmatrix} 1 \\ 0 \end{bmatrix}, \quad q^{(3)} = \begin{bmatrix} 0 \\ 1 \end{bmatrix}, \quad q^{(4)} = \begin{bmatrix} 1 \\ 1 \end{bmatrix}.$$

For each of the four cases $q = q^{(i)}$, $i = 1, 2, 3, 4$, one is asked to verify (1) if the system (2.46) has a solution if $\mu > 0$ and (2) if the first equation in (2.46) satisfies the IPC, i.e., has a solution with $z > 0$ and $s > 0$.

[22] **Exercise 12** Show that $z(\mu)$ is continuous (and differentiable) at any positive μ. (Hint: Apply the implicit function theorem (cf. Proposition A.2 in Appendix A) to the system (2.46)).

Recall that optimality of z means that $zs(z) = 0$, which means that z and $s(z)$ are complementary vectors. We are going to show that there exists an optimal vector z such that z and $s(z)$ are strictly complementary vectors. Then for every index i, either $z_i > 0$ or $s_i(z) > 0$. This implies that the index sets B and N, introduced in Section 2.5 form a partition of the index set, the so-called optimal partition of (SP).

It is convenient to introduce some more notation.

Definition I.19 *If z is a nonnegative vector, we define its support, denoted by $\sigma(z)$, as the set of indices i for which $z_i > 0$:*

$$\sigma(z) = \{i \: : \: z_i > 0\}.$$

Note that if z is feasible then $zs(z) = 0$ holds if and only if $\sigma(z) \cap \sigma(s(z)) = \emptyset$. Furthermore, z is a strictly complementary optimal solution if and only if it is optimal and $\sigma(z) \cup \sigma(s) = \{1, 2, \ldots, \bar{n}\}$.

Theorem I.20 *(SP) has an optimal solution z^* with $z^* + s(z^*) > 0$.*

Proof: Let $\{\mu_k\}_{k=1}^{\infty}$ be a monotonically decreasing sequence of positive numbers μ_k such that $\mu_k \to 0$ if $k \to \infty$, and let $s(\mu_k) := s(z(\mu_k))$. Due to Lemma I.9 the set $\{(z(\mu_k), s(\mu_k))\}$ lies in a compact set, and hence it contains a subsequence converging to a point (z^*, s^*), with $s^* = s(z^*)$. Since $z(\mu_k)^T s(\mu_k) = \bar{n}\mu_k \to 0$, we have $(z^*)^T s^* = 0$. Hence, from (2.25), $q^T z^* = 0$. So z^* is an optimal solution.

We claim that (z^*, s^*) is a strictly complementary pair. To prove this, we apply the orthogonality property (2.22) to the points z^* and $z(\mu_k)$, which gives

$$(z(\mu_k) - z^*)^T (s(\mu_k) - s^*) = 0.$$

Rearranging the terms, and using $z(\mu_k)^T s(\mu_k) = \bar{n}\mu_k$ and $(z^*)^T s^* = 0$, we arrive at

$$\sum_{j \in \sigma(z^*)} z_j^* s_j(\mu_k) + \sum_{j \in \sigma(s^*)} z_j(\mu_k) s_j^* = \bar{n}\mu_k.$$

Dividing both sides by μ_k and recalling that $z_j(\mu_k) s_j(\mu_k) = \mu_k$, we obtain

$$\sum_{j \in \sigma(z^*)} \frac{z_j^*}{z_j(\mu_k)} + \sum_{j \in \sigma(s^*)} \frac{s_j^*}{s_j(\mu_k)} = \bar{n}.$$

Letting $k \to \infty$, the first sum on the left becomes equal to the number of positive coordinates in z^*. Similarly, the second sum becomes equal to the number of positive coordinates in s^*. The sum of these numbers being $\bar{n}$, we conclude that the optimal pair (z^*, s^*) is strictly complementary.[23],[24]   □

---

[23] By using a similar proof technique it can be shown that the limit of $z(\mu)$ exists if μ goes to zero. In other words, the central path converges. The limit point is (of course) a uniquely determined optimal solution of (SP), which can further be characterized as the so-called *analytic center* of the set of optimal solutions (cf. Section 2.11).

[24] Let us also mention that Theorem I.20 is a special case of an old result of Goldman and Tucker which states that every feasible linear system of equalities and inequalities has a strictly feasible solution [111].

By Theorem I.20 there exists a strictly complementary solution z of (SP). Having such a solution, the classes B and N simply follow from

$$B = \{i \; : \; z_i > 0\}, \qquad N = \{i \; : \; s_i(z) > 0\}.$$

Now recall from Theorem I.5 and Theorem I.6 that the problems (P) and (D) have optimal solutions with vanishing duality gap if and only if (SP) has an optimal solution with $\kappa > 0$. Due to Theorem I.20 this can be restated as follows.

Corollary I.21 *The problems (P) and (D) have optimal solutions with vanishing duality gap if and only if $\kappa \in B$.*

Let us consider more in detail the implications of $\kappa \in B$ for the problems (SP_0), and more importantly, for (P) and (D).

Theorem I.20 implies the existence of a strictly complementary optimal solution z of (SP). Let z be such an optimal solution. Then we have

$$zs(z) = 0, \quad z + s(z) > 0, \quad z \geq 0, \quad s(z) \geq 0.$$

Now using $s(z) = Mz + q$ and $\vartheta = 0$, and also (2.11) and (2.12), we obtain

$$z = \begin{bmatrix} y \\ x \\ \kappa \\ 0 \end{bmatrix} \geq 0, \quad s(z) = \begin{bmatrix} Ax - \kappa b \\ -A^T y + \kappa c \\ b^T y - c^T x \\ \bar{n} - [y^T, x^T, \kappa] r \end{bmatrix} \geq 0.$$

Neglecting the last entry in both vectors, it follows that

$$\bar{z} := \begin{bmatrix} y \\ x \\ \kappa \end{bmatrix} \geq 0, \quad \bar{s}(\bar{z}) := \bar{M}\bar{z} = \begin{bmatrix} Ax - \kappa b \\ -A^T y + \kappa c \\ b^T y - c^T x \end{bmatrix} \geq 0, \tag{2.47}$$

and moreover,

$$\bar{z}\bar{s}(\bar{z}) = 0, \quad \bar{z} + \bar{s}(\bar{z}) > 0, \quad \bar{z} \geq 0, \quad \bar{s}(\bar{z}) \geq 0. \tag{2.48}$$

This shows that $\bar{z}$ is a strictly complementary optimal solution of (SP_0). Hence the next theorem requires no further proof.

Theorem I.22 *(SP_0) has an optimal solution $\bar{z}$ with $\bar{z} + \bar{s}(\bar{z}) > 0$.*

Note that (2.47) and (2.48) are homogeneous in the variables x, y and κ. So, assuming $\kappa \in B$, without loss of generality we may put $\kappa = 1$. Then we come to

$$y \geq 0, \qquad Ax - b \geq 0, \qquad y(Ax - b) = 0, \qquad y + (Ax - b) > 0,$$
$$x \geq 0, \qquad c - A^T y \geq 0, \qquad x(c - A^T y) = 0, \qquad x + (c - A^T y) > 0,$$
$$1 \geq 0, \quad b^T y - c^T x \geq 0, \qquad b^T y - c^T x = 0, \qquad 1 + (b^T y - c^T x) > 0.$$

This makes clear that x is feasible for (P) and y is feasible for (D), and because $c^T x = b^T y$ these solutions are optimal with vanishing duality gap. We get a little more information from the above system, namely

$$y\,(Ax - b) = 0, \qquad y + (Ax - b) > 0,$$
$$x\,(c - A^T y) = 0, \qquad x + (c - A^T y) > 0.$$

The upper two relations show that the dual vector y and the primal slack vector $Ax - b$ are strictly complementary, whereas the lower two relations express that the primal vector x and the dual slack vector $c - A^T y$ are strictly complementary. Thus the following is also true.

Theorem I.23 *If $\kappa \in B$ then the problems (P) and (D) have optimal solutions that are strictly complementary with the slack vector of the other problem. Moreover, the optimal values of (P) and (D) are equal.*

An intriguing question is of course what can be said about the problems (P) and (D) if $\kappa \notin B$, i.e., if $\kappa \in N$. This question is completely answered in the next section.

2.9   Strong duality theorem

We start by proving the following lemma.

Lemma I.24 *If $\kappa \in N$ then there exist vectors x and y such that*

$$x \geq 0, \quad y \geq 0, \quad Ax \geq 0, \quad A^T y \leq 0, \quad b^T y - c^T x > 0.$$

Proof: Let $\kappa \in N$. Substitution of $\kappa = 0$ in (2.47) and (2.48) yields

$$
\begin{array}{llll}
y \geq 0, & Ax \geq 0, & y\,(Ax) = 0, & y + Ax > 0, \\
x \geq 0, & -A^T y \geq 0, & x\,(A^T y) = 0, & x - A^T y > 0, \\
0 \geq 0, & b^T y - c^T x \geq 0, & 0\,(b^T y - c^T x) = 0, & 0 + (b^T y - c^T x) > 0.
\end{array}
$$

It follows that the vectors x and y are as desired, thus the lemma is proved.   □

Let us call an LO-problem *solvable* if it has an optimal solution, and *unsolvable* otherwise. Note that an LO-problem can be unsolvable for two possible reasons: the domain of the problem is empty, or the domain is not empty but the objective function is unbounded on the domain. In the first case the problem is called *infeasible* and in the second case *unbounded*.

Theorem I.25 *If $\kappa \in N$ then neither (P) nor (D) has an optimal solution.*

Proof: Let $\kappa \in N$. By Lemma I.24 we then have vectors x and y such that

$$x \geq 0, \quad y \geq 0, \quad Ax \geq 0, \quad A^T y \leq 0, \quad b^T y - c^T x > 0. \tag{2.49}$$

By the last inequality we cannot have $b^T y \leq 0$ and $c^T x \geq 0$. Hence,

$$\text{either} \quad b^T y > 0 \quad \text{or} \quad c^T x < 0. \tag{2.50}$$

Suppose that (P) is not infeasible. Then there exists x^* such that

$$x^* \geq 0 \quad \text{and} \quad Ax^* \geq b.$$

Using (2.49) we find that $x^* + x \geq 0$ and $A(x^* + x) \geq b$. So $x^* + x$ is feasible for (P). We can not have $b^T y > 0$, because this would lead to the contradiction

$$0 < b^T y \leq (Ax^*)^T y = x^{*T}(A^T y) \leq 0,$$

since $x^* \geq 0$ and $A^T y \leq 0$. Hence we have $b^T y \leq 0$. By (2.50) this implies $c^T x < 0$. But then we have for any positive λ that $x^* + \lambda x$ is feasible for (P) and

$$c^T(x^* + \lambda x) = c^T x^* + \lambda c^T x,$$

showing that the objective value goes to minus infinity if λ grows to infinity. Thus we have shown that (P) is either infeasible or unbounded, and hence (P) has no optimal solution.

The other case can be handled in the same way. If (D) is feasible then there exists y^* such that $y^* \geq 0$ and $A^T y^* \leq c$. Due to (2.49) we find that $y^* + y \geq 0$ and $A^T(y^* + y) \leq c$. So $y^* + y$ is feasible for (D). We then can not have $c^T x < 0$, because this gives the contradiction

$$0 > c^T x \geq (A^T y^*)^T x = y^{*T}(Ax) \geq 0,$$

since $y^* \geq 0$ and $Ax \geq 0$. Hence $c^T x \geq 0$. By (2.50) this implies $b^T y > 0$. But then we have for any positive λ that $y^* + \lambda y$ is feasible for (D) and

$$b^T(y^* + \lambda y) = b^T y^* + \lambda b^T y.$$

If λ grows to infinity then the last expression goes to infinity as well, so (D) is an unbounded problem. Thus we have shown that (D) is either infeasible or unbounded. This completes the proof. $\qquad \square$

The following theorem summarizes the above results.

Theorem I.26 (Strong duality theorem) *For an LO problem (P) in canonical form and its dual problem (D) we have the following two alternatives:*

(i) Both (P) and (D) are solvable and there exist (strictly complementary) optimal solutions x for (P) and y for (D) such that $c^T x = b^T y$.

(ii) Neither (P) nor (D) is solvable.

This theorem is known as the *strong duality theorem*. It is the result that we announced in Section 2.2. It implies that if one of the problems (P) and (D) is solvable then the other problem is solvable as well and in that case the duality gap vanishes at optimality. So the optimal values of both problems are then equal.

If (B, N) is the optimal partition of the self-dual problem (SP) in which (P) and (D) are embedded, then case (i) occurs if $\kappa \in B$ and case (ii) if $\kappa \in N$. Also, by Theorem I.25, case (ii) occurs if and only if there exist x and y such that (2.49) holds, and then at least one of the two problems is infeasible.

Duality is a major topic in the theory of LO. At many places in the book, and in many ways, we explore duality properties. The above result concerns an LO problem (P) in canonical form and its dual problem (D). In the next section we will extend the applicability of Theorem I.26 to any LO problem.

We conclude the present section with an interesting observation.

Remark I.27 In the classical approach to LO we have so-called *theorems of the alternatives*, also known as variants of Farkas' lemma. We want to establish here that the fact that (2.47) has a strictly complementary solution for each vector $c \in \mathbb{R}^n$ implies Farkas' lemma. We show this below for the following variant of the lemma.

Lemma I.28 (Farkas' lemma [75]) *For a given $m \times n$ matrix A and a vector $b \in \mathbb{R}^m$ either the system*

$$Ax \geq b, \ x \geq 0$$

has a solution or the system

$$A^T y \leq 0, \ b^T y > 0, \ y \geq 0$$

has a solution but not both systems have a solution.

Proof: The obvious part of the lemma is that not both systems can have a solution, because this would lead to the contradiction

$$0 < b^T y \leq (Ax)^T y = x^T A^T y \leq 0.$$

Taking $c = 0$ in (2.47), it follows that there exist x and y such that the two vectors

$$z = \begin{bmatrix} y \\ x \\ \kappa \end{bmatrix} \geq 0, \quad s(z) = \begin{bmatrix} Ax - \kappa b \\ -A^T y \\ b^T y \end{bmatrix} \geq 0$$

are strictly complementary. For κ there are two possibilities: either $\kappa = 0$ or $\kappa > 0$. In the first case we obtain $A^T y \leq 0, \ b^T y > 0, \ y \geq 0$. In the second case we may assume without loss of generality that $\kappa = 1$. Then x satisfies $Ax \geq b, \ x \geq 0$, proving the claim.[25] •

2.10 The dual problem of an arbitrary LO problem

Every LO problem can be transformed into a canonical form. In fact, this can be done in many ways. In its canonical form the problem has a dual problem. In this way we can obtain a dual problem for any LO problem. Unfortunately the transformation to canonical form is not unique, and as a consequence, the dual problem obtained in this way is not uniquely determined.

_____

[25] **Exercise 13** Derive Theorem I.22 from Farkas' lemma. In other words, use Farkas' lemma to show that for any skew-symmetric matrix M there exists a vector x such that

$$x \geq 0, \quad Mx \geq 0, \quad x + Mx > 0.$$

The aim of this section is to show that we can find a dual problem for any given problem in a unique and simple way, so that when taking the dual of the dual problem we get the original problem, in its original description.

Recall that three types of variables can be distinguished: nonnegative variables, free variables and nonpositive variables. Similarly, three types of constraints can occur in an LO problem: equality constraints, inequality constraints of the $\leq$ type and inequality constraints of the $\geq$ type. For our present purpose we need to consider the LO problem in its most general form, with all types of constraint and all types of variable. Therefore, we consider the following problem as the primal problem:

$$(P) \quad \min \left\{ \begin{bmatrix} c^0 \\ c^1 \\ c^2 \end{bmatrix}^T \begin{bmatrix} x^0 \\ x^1 \\ x^2 \end{bmatrix} : \begin{array}{l} A_0 x^0 + A_1 x^1 + A_2 x^2 = b^0 \\ B_0 x^0 + B_1 x^1 + B_2 x^2 \geq b^1, \\ C_0 x^0 + C_1 x^1 + C_2 x^2 \leq b^2 \end{array} \quad x^1 \geq 0, \ x^2 \leq 0 \right\},$$

where, for each $i = 0, 1, 2$, A_i, B_i and C_i are matrices and b^i, c^i and x^i are vectors, and the sizes of these matrices and vectors, which we do not further specify, are such that all expressions in the problem are well defined.

Now let us determine the dual of this problem. We first put it into canonical form.[26] To this end we replace the equality constraint by two inequality constraints and we multiply the $\leq$ constraint by -1. Furthermore, we replace the nonpositive variable x^2 by $x^3 = -x^2$ and the free variable x^0 by $x^+ - x^-$ with x^+ and x^- nonnegative. This yields the following equivalent problem:

$$\text{minimize} \quad \begin{bmatrix} c^0 \\ -c^0 \\ c^1 \\ -c^2 \end{bmatrix}^T \begin{bmatrix} x^+ \\ x^- \\ x^1 \\ x^3 \end{bmatrix}$$

$$\text{subject to} \quad \begin{bmatrix} A_0 & -A_0 & A_1 & -A_2 \\ -A_0 & A_0 & -A_1 & A_2 \\ B_0 & -B_0 & B_1 & -B_2 \\ -C_0 & C_0 & -C_1 & C_2 \end{bmatrix} \begin{bmatrix} x^+ \\ x^- \\ x^1 \\ x^3 \end{bmatrix} \geq \begin{bmatrix} b^0 \\ -b^0 \\ b^1 \\ -b^2 \end{bmatrix}, \quad \begin{bmatrix} x^+ \\ x^- \\ x^1 \\ x^3 \end{bmatrix} \geq 0.$$

In terms of vectors z^1, z^2, z^3, z^4 that contain the appropriate nonnegative dual variables, the dual of this problem becomes

$$\text{maximize} \quad \begin{bmatrix} b^0 \\ -b^0 \\ b^1 \\ -b^2 \end{bmatrix}^T \begin{bmatrix} z^1 \\ z^2 \\ z^3 \\ z^4 \end{bmatrix}$$

---

[26] The transformations carried out below lead to an increase of the numbers of constraints and variables in the problem formulation. They are therefore 'bad' from a computational point of view. But our present purpose is purely theoretical. In Appendix D it is shown how the problem can be put in canonical form without increasing these numbers.

$$\text{subject to} \quad \begin{bmatrix} A_0^T & -A_0^T & B_0^T & -C_0^T \\ -A_0^T & A_0^T & -B_0^T & C_0^T \\ A_1^T & -A_1^T & B_1^T & -C_1^T \\ -A_2^T & A_2^T & -B_2^T & C_2^T \end{bmatrix} \begin{bmatrix} z^1 \\ z^2 \\ z^3 \\ z^4 \end{bmatrix} \le \begin{bmatrix} c^0 \\ -c^0 \\ c^1 \\ -c^2 \end{bmatrix}, \quad \begin{bmatrix} z^1 \\ z^2 \\ z^3 \\ z^4 \end{bmatrix} \ge 0.$$

We can easily check that the variables z^1 and z^2 only occur together in the combination $z^1 - z^2$. Therefore, we can replace the variables by one free variable $y^0 := z^1 - z^2$. This reduces the problem to

$$\text{maximize} \quad \begin{bmatrix} b^0 \\ b^1 \\ -b^2 \end{bmatrix}^T \begin{bmatrix} y^0 \\ z^3 \\ z^4 \end{bmatrix}$$

$$\text{subject to} \quad \begin{bmatrix} A_0^T & B_0^T & -C_0^T \\ -A_0^T & -B_0^T & C_0^T \\ A_1^T & B_1^T & -C_1^T \\ -A_2^T & -B_2^T & C_2^T \end{bmatrix} \begin{bmatrix} y^0 \\ z^3 \\ z^4 \end{bmatrix} \le \begin{bmatrix} c^0 \\ -c^0 \\ c^1 \\ -c^2 \end{bmatrix}, \quad \begin{bmatrix} z^3 \\ z^4 \end{bmatrix} \ge 0.$$

In this problem the first two blocks of constraints can be taken together into one block of equality constraints:

$$\max \left\{ \begin{bmatrix} b^0 \\ b^1 \\ -b^2 \end{bmatrix}^T \begin{bmatrix} y^0 \\ z^3 \\ z^4 \end{bmatrix} : \begin{array}{l} A_0^T y^0 + B_0^T z^3 - C_0^T z^4 = c^0 \\ A_1^T y^0 + B_1^T z^3 - C_1^T z^4 \le c^1 \\ -A_2^T y^0 - B_2^T z^3 + C_2^T z^4 \le -c^2 \end{array}, \begin{bmatrix} z^3 \\ z^4 \end{bmatrix} \ge 0 \right\}.$$

Finally we multiply the last block of constraints by -1, we replace the nonnegative variable z^3 by $y^1 = z^3$ and the nonnegative variable z^4 by the nonpositive variable $y^2 = -z^4$. This transforms the dual problem to its final form, namely

$$(D) \quad \max \left\{ \begin{bmatrix} b^0 \\ b^1 \\ b^2 \end{bmatrix}^T \begin{bmatrix} y^0 \\ y^1 \\ y^2 \end{bmatrix} : \begin{array}{l} A_0^T y^0 + B_0^T y^1 + C_0^T y^2 = c^0 \\ A_1^T y^0 + B_1^T y^1 + C_1^T y^2 \le c^1, \; y^1 \ge 0, \; y^2 \le 0 \\ A_2^T y^0 + B_2^T y^1 + C_2^T y^2 \ge c^2 \end{array} \right\}.$$

Comparison of the primal problem (P) with its dual problem (D), in its final description, reveals some simple rules for the construction of a dual problem for any given LO problem. First, the objective vector and the right-hand side vector are interchanged in the two problems, and the constraint matrix is transposed. At first sight it may not be obvious that the types of the dual variables and the dual constraints can be determined. We need to realize that the vector y^0 of dual variables relates to the first block of constraints in the primal problem, y^1 to the second block and y^2 to the third block of constraints. Then the relation becomes obvious: equality constraints in the primal problem yield free variables in the dual problem, inequality constraints in the primal problem of type $\ge$ yield nonnegative variables in the dual problem, and inequality constraints in the primal problem of type $\le$ yield nonpositive variables in the dual problem. For the types of dual constraint we have similar relations. Here the vector of primal variables x^0 relates to the first block of constraints in the dual problem, x^1 to the second block and x^2 to the third block of constraints. Free variables in the primal problem yield equality constraints in the dual problem, nonnegative variables in the primal problem yield inequality constraints of type $\leq$ in the dual problem, and nonpositive variables in the primal problem yield inequality constraints of type $\geq$ in the dual problem. Table 2.1. summarizes these observations, and as such provides a simple scheme for writing down a dual problem for any given minimization problem. To get the dual of a maximization problem, one simply has to use the table from the right to the left.

Primal problem (P)	Dual problem (D)
min $c^T x$	max $b^T y$
equality constraint	free variable
inequality constraint $\geq$	variable ≥ 0
inequality constraint $\leq$	variable ≤ 0
free variable	equality constraint
variable ≥ 0	inequality constraint $\leq$
variable ≤ 0	inequality constraint $\geq$

Table 2.1. Scheme for dualizing.

As indicated before, the dualizing scheme is such that when it is applied twice, the original problem is returned. This easily follows from Table 2.1., by inspection.[27]

2.11 Convergence of the central path

We already announced in footnote 23 (page 36) that the central path has a unique limit point in the optimal set. Because this result was not needed in the rest of this chapter, we postponed its proof to this section. We also characterize the limit point as the so-called *analytic center* of the optimal set of (SP).

As before, we assume that the central path of (SP) exists. Our aim is to investigate the behavior of the central path as μ tends to 0. From the proof of Theorem I.20 we know that the central path has a subsequence converging to an optimal solution. This was sufficient for proving the existence of a strictly complementary solution of (SP). However, as we show below, the central path itself converges. The limit point z^* and

---

[27] **Exercise 14** Using the results of this chapter prove that the following three statements are equivalent:
  (i) (SP) satisfies the interior-point condition;
  (ii) the level sets $\mathcal{L}_\gamma := \{(z, s(z)) : q^T z \leq \gamma, s(z) = Mz + q \geq 0, z \geq 0\}$ of $q^T z$ are bounded;
  (iii) the optimal set of (SP) is bounded.

its surplus vector $s^* := s(z^*)$ form a strictly complementary optimal solution pair, and hence determine the optimal partition (B, N) of (SP).

The optimal set of (SP) is given by

$$\mathcal{SP}^* = \left\{ (z, s) \ : \ Mz + q = s, \ z \ge 0, \ s \ge 0, \ q^T z = 0 \right\}.$$

This makes clear that $\mathcal{SP}^*$ is the intersection of the affine space

$$\left\{ (z, s) \ : \ Mz + q = s, \ q^T z = 0 \right\}$$

with the nonnegative orthant of $\mathbb{R}^{2n}$.

At this stage we need to define the analytic center of $\mathcal{SP}^*$. We give the definition for the more general case of an arbitrary (nonempty) set that is the intersection of an affine space in $\mathbb{R}^p$ and the nonnegative orthant of $\mathbb{R}^p$.

Definition I.29 (Analytic center) [28] *Let the nonempty and bounded set $\mathcal{T}$ be the intersection of an affine space in $\mathbb{R}^p$ with the nonnegative orthant of $\mathbb{R}^p$. We define the support $\sigma(\mathcal{T})$ of $\mathcal{T}$ as the subset of the full index set $\{1, 2, \ldots, p\}$ given by*

$$\sigma(\mathcal{T}) = \{ i \ : \ \exists x \in \mathcal{T} \ \text{such that} \ x_i > 0 \}.$$

The analytic center of $\mathcal{T}$ is defined as the zero vector if $\sigma(\mathcal{T})$ is empty; otherwise it is the vector in $\mathcal{T}$ that maximizes the product

$$\prod_{i \in \sigma(\mathcal{T})} x_i, \qquad x \in \mathcal{T}. \tag{2.51}$$

If the support of the set $\mathcal{T}$ in the above definition is nonempty then the convexity of $\mathcal{T}$ implies the existence of a vector $x \in \mathcal{T}$ such that $x_{\sigma(\mathcal{T})} > 0$. Moreover, if $\sigma(\mathcal{T})$ is nonempty then the maximum value of the product (2.51) exists since $\mathcal{T}$ is bounded. Since the logarithm of the product (2.51) is strictly concave, the maximum value (if it exists) is attained at a unique point of $\mathcal{T}$. Thus the above definition uniquely defines the analytic center for any bounded subset that is the intersection of an affine space in $\mathbb{R}^p$ with the nonnegative orthant of $\mathbb{R}^p$.

Due to Lemma I.9 any pair $(z, s) \in \mathcal{SP}^*$ satisfies

$$e^T z + e^T s(z) = \bar{n}.$$

This makes clear that the optimal set $\mathcal{SP}^*$ is bounded. Its analytic center therefore exists. We now show that the central path converges to this analytic center. The proof very much resembles that of Theorem I.20.[29]

---

[28] The notion of analytic center of a polyhedron was introduced by Sonnevend [257]. It plays a crucial role in the theory of interior-point methods.

[29] The limiting behavior of the central path as μ approaches zero has been an important subject in research on interior-point methods for a long time. In the book by Fiacco and McCormick [77] the convergence of the central path to an optimal solution is investigated for general convex optimization problems. McLinden [197] considered the limiting behavior of the path for monotone complementarity problems and introduced the idea for the proof-technique of Theorem I.20, which was later adapted by Güler and Ye [135]. Megiddo [200] extensively investigated the properties of the central path, which motivated Monteiro and Adler [218], Tanabe [261] and Kojima, Mizuno and Yoshise [178] to investigate primal-dual methods. Other relevant references for the limiting behavior of the central path are Adler and Monteiro [3], Asić, Kovačević-Vujčić and Radosavljević-Nikolić [28], Güler [131], Kojima, Mizuno and Noma [176], Monteiro and Tsuchiya [222] and Witzgall, Boggs and Domich [294], Halická [137], Wechs [290] and Zhao and Zhu [321].

Theorem I.30 *The central path converges to the analytic center of the optimal set* SP^* *of* (SP).

Proof: Let (z^*, s^*) be an accumulation point of the central path, where $s^* = s(z^*)$. The existence of such a point has been established in the proof of Theorem I.20. Let $\{\mu_k\}_{k=1}^\infty$ be a positive sequence such that $\mu_k \to 0$ and such that $(z(\mu_k), s(\mu_k))$, with $s(\mu_k) = s(z(\mu_k))$, converges to (z^*, s^*). Then z^* is optimal, which means $z^* s^* = 0$, and z^* and s^* are strictly complementary, i.e, $z^* + s^* > 0$.

Now let $\bar{z}$ be optimal in (SP) and let $\bar{s} = M\bar{z} + q$ be its surplus vector. Applying the orthogonality property (2.22) to the points $\bar{z}$ and $z(\mu)$ we obtain

$$(z(\mu_k) - \bar{z})^T (s(\mu_k) - \bar{s}) = 0.$$

Rearranging terms and using $z(\mu_k)^T s(\mu_k) = n\mu_k$ and $(\bar{z})^T \bar{s} = 0$, we get

$$\sum_{j=1}^n \bar{z}_j s_j(\mu_k) + \sum_{j=1}^n \bar{s}_j z_j(\mu_k) = n\mu_k.$$

Since the pair (z^*, s^*) is strictly complementary and $(\bar{z}, \bar{s})$ is an arbitrary optimal pair, we have for each coordinate j:

$$z_j^* = 0 \Rightarrow \bar{z}_j = 0, \quad s_j^* = 0 \Rightarrow \bar{s}_j = 0.$$

Hence, $\bar{z}_j = 0$ if $j \notin \sigma(z^*)$ and $\bar{s}_j = 0$ if $j \notin \sigma(s^*)$. Thus we may write

$$\sum_{j \in \sigma(z^*)} \bar{z}_j s_j(\mu_k) + \sum_{j \in \sigma(s^*)} \bar{s}_j z_j(\mu_k) = n\mu_k.$$

Dividing both sides by $\mu_k = z_j(\mu_k) s_j(\mu_k)$, we get

$$\sum_{j \in \sigma(z^*)} \frac{\bar{z}_j}{z_j(\mu_k)} + \sum_{j \in \sigma(s^*)} \frac{\bar{s}_j}{s_j(\mu_k)} = n.$$

Letting $k \to \infty$, it follows that

$$\sum_{j \in \sigma(z^*)} \frac{\bar{z}_j}{z_j^*} + \sum_{j \in \sigma(s^*)} \frac{\bar{s}_j}{s_j^*} = n.$$

Using the arithmetic-geometric-mean inequality we obtain

$$\left(\prod_{j \in \sigma(z^*)} \frac{\bar{z}_j}{z_j^*} \prod_{j \in \sigma(s^*)} \frac{\bar{s}_j}{s_j^*} \right)^{1/n} \leq \frac{1}{n} \left(\sum_{j \in \sigma(z^*)} \frac{\bar{z}_j}{z_j^*} + \sum_{j \in \sigma(s^*)} \frac{\bar{s}_j}{s_j^*} \right) = 1.$$

Obviously, the above inequality implies

$$\prod_{j \in \sigma(z^*)} \bar{z}_j \prod_{j \in \sigma(s^*)} \bar{s}_j \leq \prod_{j \in \sigma(z^*)} z_j^* \prod_{j \in \sigma(s^*)} s_j^*.$$

This shows that (z^*, s^*) maximizes the product $\prod_{j \in \sigma(z^*)} z_j \prod_{j \in \sigma(s^*)} s_j$ over the optimal set. Hence the central path of (SP) has only one accumulation point when μ approaches zero and this is the analytic center of SP^*. $\qquad \square$

Example I.31 Let us compute the limit point of the central path of the self-dual problem (SP) in Example I.7, as given by (2.19). Recall from (2.26) in Example I.12 that any optimal solution has the form

$$z = \begin{bmatrix} 2\kappa \\ 0 \\ \kappa \\ \kappa \\ 0 \end{bmatrix}, \quad s(z) = \begin{bmatrix} 0 \\ \kappa \\ 0 \\ 0 \\ 5 - 5\kappa \end{bmatrix}, \quad 0 \le \kappa \le 1,$$

from which the sets B and N follow:

$$B = \{1,\, 3,\, 4\}, \quad N = \{2,\, 5\}.$$

Hence we have for any optimal z,

$$\prod_{j \in B} z_j \prod_{j \in N} s_j(z) = 2\kappa^4 (5 - 5\kappa) = 10 \left(\kappa^4 - \kappa^5\right).$$

This product is maximal for $\kappa = 0.8$, so the analytical center of the optimal set is given by [30,31,32,33]

$$z = \begin{bmatrix} 1.6 \\ 0 \\ 0.8 \\ 0.8 \\ 0 \end{bmatrix}, \quad s(z) = \begin{bmatrix} 0 \\ 0.8 \\ 0 \\ 0 \\ 1 \end{bmatrix}.$$

The convergence of the central path when μ goes to zero implies the boundedness of the coordinates of $z(\mu)$ and $s(\mu)$ for any finite section of the central path. Of course, this also follows from Lemma I.9 and (2.33).[34]

---

[30] **Exercise 15** Find the analytic center of the self-dual problem considered in Exercise 4 (page 27).

[31] **Exercise 16** Find the analytic center of the self-dual problem considered in Exercise 5 (page 27).

[32] **Exercise 17** Find the analytic center of the self-dual problem considered in Exercise 6 (page 27).

[33] **Exercise 18** Find the analytic center of the self-dual problem considered in Exercise 7 (page 27).

[34] **Exercise 19** For any positive μ consider the set

$$SP_\mu := \left\{ (z, s) \; : \; Mz + q = s, \, z \ge 0, \, s \ge 0, \, q^T z = q^T z(\mu) \right\}.$$

Using the same proof-technique as for Theorem I.30, show that the pair $(z(\mu), s(\mu))$ is the analytic center of this set.

3

A Polynomial Algorithm for the Self-dual Model

3.1 Introduction

The previous chapter made clear that any (canonical) LO problem can be solved by finding a strictly complementary solution of a specific self-dual problem that satisfies the interior-point assumption. In particular, the self-dual problem has the form

$$(SP) \qquad \min \left\{ q^T z \ : \ Mz \geq -q, \ z \geq 0 \right\},$$

where M is a skew-symmetric matrix and q a nonnegative vector. Deviating from the notation in Chapter 2 we denote the order of M as n (instead of $\bar{n}$). Then, according to (2.12) the vector q has the form

$$q := \begin{bmatrix} 0_{n-1} \\ n \end{bmatrix}. \tag{3.1}$$

Note that due to the definition of the matrix M we may assume that $n \geq 5$.

Like before, we associate to any vector $z \in \mathbb{R}^n$ its *slack vector* $s(z)$:

$$s(z) := Mz + q. \tag{3.2}$$

As a consequence we have

z is a feasible for (SP) if and only if $z \geq 0$ and $s(z) \geq 0$.

Also recall that the all-one vector e is feasible for (SP) and its slack vector is the all-one vector (cf. Theorem I.5):

$$s(e) = e. \tag{3.3}$$

Assuming that the entries in M and q are integral (or rational), we show in this chapter that we can find a strictly complementary solution of (SP) *in polynomial time*. This means that we present an algorithm that yields a strictly complementary solution of (SP) after a number of arithmetic operations that is bounded by a polynomial in the *size* of (SP).

Remark I.32 The terminology is taken from *complexity theory*. For our purpose it is not necessary to have a deep understanding of this theory. Major contributions to complexity theory were given by Cook [56], Karp [166], Aho, Hopcroft and Ullman [5], and Garey and Johnson [92]. For a survey focusing on linear and combinatorial optimization problems we refer the reader to Schrijver [250]. Complexity theory distinguishes between easy and hard problems. In this theory a problem consists of a class of problem instances, so 'the' LO problem consists of all possible instances of LO problems; here we restrict ourselves to LO problems with integral input data.[1] A problem is called *solvable in polynomial time* (or simply *polynomial* or *easy*) if there exists an algorithm that solves each instance of the problem in a time that is bounded above by a polynomial in the size of the problem instance; otherwise the problem is considered to be *hard*. In general the size of an instance is defined as the length of a binary string encoding the instance. For the problem (SP) such a string consists of binary encodings of the entries in the matrix M and the vector q. Note that the binary encoding of a positive integer a requires a string of length $1 + \lceil \log_2(1 + |a|) \rceil$. (The first 1 serves to encode the sign of the number.) If the entries in M and q are integral, the total length of the string for encoding (SP) becomes

$$\sum_{i=1}^{n} (1 + \lceil \log_2 (1 + |q_i|) \rceil) + \sum_{i,j=1}^{n} (1 + \lceil \log_2 (1 + |M_{ij}|) \rceil) =$$

$$n(n+1) + \sum_{i=1}^{n} \lceil \log_2 (1 + |q_i|) \rceil + \sum_{i,j=1}^{n} \lceil \log_2 (1 + |M_{ij}|) \rceil . \tag{3.4}$$

Instead we work with the smaller number

$$L = n(n+1) + \log_2 \Pi, \tag{3.5}$$

where Π is the product of all nonzero entries in q and M. Ignoring the integrality operators, we can show that the expression in (3.4) is less than $2L$. In fact, one can easily understand that the number of operations of an algorithm is polynomial in (3.4) if and only if it is bounded by a polynomial in L. ●

We consider the number L, as given by (3.5), as the size of (SP). In fact we use this number only once. In the next section we present an algorithm that generates a positive vector z such that $z^T s(z) \leq \varepsilon$, where ε is any positive number, and we derive a bound for the number of iterations required by the algorithm. Then, in Section 3.3, we show that this algorithm can be used to find a strictly complementary solution of (SP) and we derive an iteration bound that depends on the so-called *condition number* of (SP). Finally, we show that the iteration bound can be bounded from above by a polynomial in the quantity L, which represents the size of (SP).

3.2 Finding an ε-solution

After all the theoretical results of the previous sections we are now ready to present an algorithm that finds a strictly complementary solution of (SP) in polynomial time. The working horse in the algorithm is the Newton step that was introduced in Section 2.7.2. It will be convenient to recall its definition and some of its properties.

---

[1] We could easily have included LO problems with rational input data in our considerations, because if the entries in M and q are rational numbers then after multiplication of these entries with their smallest common multiple, all entries become integral. Thus, each problem instance with rational data can easily be transformed to an equivalent problem with integral data.

Given a positive vector z such that $s = s(z) > 0$, the Newton direction Δz at z with respect to μ (or the μ-center $z(\mu)$) is uniquely determined by the linear system (cf. (2.35) – (2.36))

$$M\Delta z - \Delta s = 0, \tag{3.6}$$

$$z\Delta s + s\Delta z = \mu e - zs. \tag{3.7}$$

Substituting (3.6) into (3.7) we get [2]

$$(S + ZM)\,\Delta z = \mu e - zs.$$

Since the matrix $S + ZM$ is invertible (cf. Exercise 9, page 29), it follows that

$$\Delta z = (S + ZM)^{-1}\,(\mu e - zs) \tag{3.8}$$

$$\Delta s = M\Delta z. \tag{3.9}$$

The result of the Newton step is denoted as

$$z^+ := z + \Delta z;$$

the new slack vector is then given by

$$s^+ := s(z^+) = M(z + \Delta z) + q = s + M\Delta z.$$

The vectors Δz and Δs are orthogonal, by (2.34). After the Newton step the objective value has the desired value $n\mu$, by (2.38):

$$q^T z = s^T z = n\mu. \tag{3.10}$$

The *variance vector* of z with respect to μ is defined by (cf. (2.41))[3]:

$$v := \sqrt{\frac{zs(z)}{\mu}}. \tag{3.11}$$

This implies

$$zs(z) = \mu e \quad \Leftrightarrow \quad v = e. \tag{3.12}$$

We use $\delta(z, \mu)$ as a measure for the proximity of z to $z(\mu)$. It is defined by (cf. (2.42))

$$\delta(z, \mu) := \tfrac{1}{2} \left\| v - v^{-1} \right\|. \tag{3.13}$$

If $z = z(\mu)$ then $v = e$ and hence $\delta(z, \mu) = 0$, otherwise $\delta(z, \mu) > 0$. If $\delta(z, \mu) < 1$ then the Newton step is feasible, and if $\delta(z, \mu) < 1/\sqrt{2}$ then the Newton process quadratically fast converges to $z(\mu)$. This is a consequence of the next lemma (cf. Theorem I.16).

Lemma I.33 *If* $\delta := \delta(z, \mu) < 1$*, then the Newton step is strictly feasible, i.e.,* $z^+ > 0$ *and* $s^+ > 0$*. Moreover,*

$$\delta(z^+, \mu) \le \frac{\delta^2}{\sqrt{2(1 - \delta^2)}}.$$

---

[2] Here, as usual, $Z = \text{diag}\,(z)$ and $S = \text{diag}\,(s)$.

[3] **Exercise 20** If we define $d := \sqrt{z/s}$, where $s = s(z)$, then show that the Newton step Δz satisfies

$$(I + DMD)\,\Delta z = \frac{z}{v}\left(v^{-1} - v\right) = \mu s^{-1} - z.$$

3.2.1 Newton-step algorithm

The idea of the algorithm is quite simple. Starting at $z = e$, we choose $\mu < 1$ such that

$$\delta(z, \mu) \leq \tfrac{1}{\sqrt{2}}, \tag{3.14}$$

and perform a Newton step targeting at $z(\mu)$. After the step the new iterate z satisfies $\delta(z, \mu) \leq \tfrac{1}{2}$. Then we decrease μ such that (3.14) holds for the new values of z and μ, and repeat the procedure. Note that after each Newton step we have $q^T z = z^T s(z) = n\mu$. Thus, if μ approaches zero, then z will approach the set of optimal solutions. Formally the algorithm can be stated as follows.

Full-Newton step algorithm

Input:
    An accuracy parameter $\varepsilon > 0$;
    a barrier update parameter θ, $0 < \theta < 1$.
begin
    $z = e$; $\mu := 1$;
    while $n\mu \geq \varepsilon$ **do**
    begin
        $\mu := (1 - \theta)\mu$;
        $z := z + \Delta z$;
    end
end

Note that the reduction of the *barrier parameter* μ is realized by the multiplication with the factor $1 - \theta$. In the next section we discuss how an appropriate value of the *update parameter* θ can be obtained, so that during the course of the algorithm the iterates are kept within the region where Newton's method is quadratically convergent.

3.2.2 Complexity analysis

At the start of the algorithm we have $\mu = 1$ and $z = z(1) = e$, whence $q^T z = n$ and $\delta(z, \mu) = 0$. In each iteration μ is first reduced with the factor $1 - \theta$ and then the Newton step is made targeting the new μ-center. It will be clear that the reduction of μ has effect on the value of the proximity measure. This effect is fully described by the following lemma.

Lemma I.34 *Let $z > 0$ and $\mu > 0$ be such that $s = s(z) > 0$ and $q^T z = n\mu$. Moreover, let $\delta := \delta(z, \mu)$ and $\mu' = (1 - \theta)\mu$. Then*

$$\delta(z, \mu')^2 = (1 - \theta)\delta^2 + \frac{\theta^2 n}{4(1 - \theta)}.$$

Proof: Let $\delta^+ := \delta(z, \mu')$ and $v = \sqrt{zs/\mu}$, as in (3.11). Then, by definition,

$$4(\delta^+)^2 = \left\| \sqrt{1-\theta}\, v^{-1} - \frac{v}{\sqrt{1-\theta}} \right\|^2 = \left\| \sqrt{1-\theta}\, (v^{-1} - v) - \frac{\theta v}{\sqrt{1-\theta}} \right\|^2 .$$

From $z^T s = n\mu$ it follows that $\|v\|^2 = n$. This implies

$$v^T \left(v^{-1} - v \right) = n - \|v\|^2 = 0.$$

Hence, v is orthogonal to $v^{-1} - v$. Therefore,

$$4(\delta^+)^2 = (1-\theta) \left\| v^{-1} - v \right\|^2 + \frac{\theta^2 \|v\|^2}{1-\theta} = (1-\theta) \left\| v^{-1} - v \right\|^2 + \frac{n\theta^2}{1-\theta}.$$

Since $\|v^{-1} - v\| = 2\delta$, the result follows. $\square$

Lemma I.35 *Let $\theta = \frac{1}{\sqrt{2n}}$. Then at the start of each iteration we have*

$$q^T z = n\mu \quad \text{and} \quad \delta(z, \mu) \le \frac{1}{2} \tag{3.15}$$

Proof: At the start of the first iteration we have $\mu = 1$ and $z = e$, so $q^T z = n$ and $\delta(z, \mu) = 0$. Therefore (3.15) certainly holds at the start of the first iteration. Now suppose that (3.15) holds at the start of some iteration. We show that (3.15) then also holds at the start of the next iteration. Let $\delta = \delta(z, \mu)$. When the barrier parameter is updated to $\mu' = (1-\theta)\mu$, Lemma I.34 gives

$$\delta(z, \mu')^2 = (1-\theta)\,\delta^2 + \frac{\theta^2 n}{4(1-\theta)} \le \frac{1-\theta}{4} + \frac{1}{8(1-\theta)} \le \frac{3}{8}.$$

The last inequality can be understood as follows. Due to $n \ge 2$ we have $0 \le \theta \le 1/\sqrt{4} = 1/2$. The left hand side expression in the last inequality is a convex function of θ. Its value at $\theta = 0$ as well as at $\theta = 1/2$ equals $3/8$, hence its value does not exceed $3/8$ for $\theta \in [0, 1/2]$.

Since $3/8 \le 1/2$, it follows that after the μ-update $\delta(z, \mu') \le 1/\sqrt{2}$. Hence, by Lemma I.33, after performing the Newton step we certainly have $\delta(z^+, \mu') \le 1/2$. Finally, by (3.10), $q^T z^+ = n\mu'$. Thus the lemma has been proved. $\square$

How many iterations are needed by the algorithm? The answer is provided by the following lemma.

Lemma I.36 *After at most*

$$\left\lceil \frac{1}{\theta} \log \frac{n}{\varepsilon} \right\rceil$$

iterations we have $n\mu \le \varepsilon$.

Proof: Initially, the objective value is n and in each iteration it is reduced by the factor $1 - \theta$. Hence, after k iterations we have $\mu = (1-\theta)^k$. Therefore, the objective value, given by $q^T z(\mu) = n\mu$, is smaller than, or equal to ε if

$$(1-\theta)^k\, n \le \varepsilon.$$

Taking logarithms, this becomes

$$k \log (1 - \theta) + \log n \leq \log \varepsilon.$$

Since $- \log (1 - \theta) \geq \theta$, this certainly holds if

$$k\theta \geq \log n - \log \varepsilon = \log \frac{n}{\varepsilon}.$$

This implies the lemma. □

The above results are summarized in the next theorem which requires no further proof.

Theorem I.37 *If $\theta = \frac{1}{\sqrt{2n}}$ then the algorithm requires at most*

$$\left\lceil \sqrt{2n} \, \log \frac{n}{\varepsilon} \right\rceil$$

iterations. The output is a feasible $z > 0$ such that $q^T z = n\mu \leq \varepsilon$ and $\delta(z, \mu) \leq \frac{1}{2}$.

This theorem shows that we can get an ε-solution z of our self-dual model with ε as small as desirable.[4]

A crucial question for us is whether the variable $\kappa = z_{n-1}$ is positive or zero in the limit, when μ goes to zero. In practice, for small enough ε it is usually no serious problem to decide which of the two cases occurs. In theory, however, this means that we need to know what the optimal partition of the problem is. As we explain in the next section, the optimal partition can be found in polynomial time. This requires some further analysis of the central path.

Example I.38 In this example we demonstrate the behavior of the Full-Newton step algorithm by applying it to the problem (SP) in Example I.7, as given in (2.19) on page 23. According to Theorem I.37, with $n = 5$, the algorithm requires at most

$$\left\lceil \sqrt{10} \, \log \frac{5}{\varepsilon} \right\rceil$$

iterations. For $\varepsilon = 10^{-3}$ we have $\log (5/\varepsilon) = \log 5000 = 8.5172$, and we get 27 as an upper bound for the number of iterations. When running the algorithm with this ε the actual number of iterations is 22. The actual values of the output of the algorithm are

$$z = (1.5999, \, 0.0002, \, 0.8000, \, 0.8000, \, 0.0002)^T$$

and

$$s(z) = (0.0001, \, 0.8000, \, 0.0002, \, 0.0002, \, 1.0000)^T.$$

The left plot in Figure 3.1 shows how the coordinates of the vector $z := (z_1, z_2, z_3, z_4 = \kappa, z_5 = \vartheta)$, which contains the variables in the problem, develop in the course of the algorithm. The right plot does the same for the coordinates of the surplus vector $s(z) := (s_1, s_2, s_3, s_4, s_5)$. Observe that z and $s(z)$ converge nicely to the limit point of the central path of the sample problem as given in Example I.31. ◇

---

[4] It is worth pointing out that if we put $\varepsilon = n\mu$ in the iteration bound of Theorem I.37 we get exactly the same bound as given by (2.45).

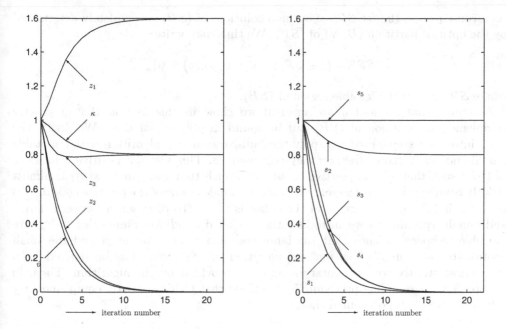

Figure 3.1   Output Full-Newton step algorithm for the problem in Example I.7.

3.3   Polynomial complexity result

3.3.1   Introduction

Having a strictly complementary solution z of (SP), we also know the optimal partition (B, N) of (SP), as defined in Section 2.6. For if z is a strictly complementary solution of (SP) then we have $zs(z) = 0$ and $z + s(z) > 0$, and the optimal partition follows from[5]

$$B = \{i \ : \ z_i > 0\}$$
$$N = \{i \ : \ s_i(z) > 0\}.$$

Definition I.39 *The restriction of a vector $z \in \mathbb{R}^n$ to the coordinates in a subset I of the full index set $\{1, 2, \ldots, n\}$ is denoted by z_I.*

Hence if $\tilde{z}$ is a strictly complementary solution of (SP) then

$$\tilde{z}_B > 0, \quad \tilde{z}_N = 0, \quad s_B(\tilde{z}) = 0, \quad s_N(\tilde{z}) > 0.$$

Now let z be any feasible solution of (SP). Then, by Lemma I.10, with $z_1 = z$, $z_2 = \tilde{z}$ we obtain that z is optimal if and only if $zs(\tilde{z}) = \tilde{z}s(z) = 0$. This gives

$$z \ \text{ is optimal for } \ (SP) \quad \Longleftrightarrow \quad z_N = 0 \ \text{ and } \ s_B(z) = 0.$$

---

[5] It may be sensible to point out that if, conversely, the optimal partition is known, then it is not obvious at all how to find a strictly complementary solution of (SP).

As a consequence, the set $\mathcal{SP}^*$ of optimal solutions of (SP) is completely determined by the optimal partition (B, N) of (SP). We thus may write

$$\mathcal{SP}^* = \{z \in \mathcal{SP} \ : \ z_N = 0, \ s_B(z) = 0\},$$

where $\mathcal{SP}$ denotes the feasible region of (SP).

Assuming that M and q are integral we show in this section that a strictly complementary solution of (SP) can be found in polynomial time. We divide the work into a few steps. First we apply the Full-Newton step algorithm with a suitable (small enough) value of the accuracy parameter ε. This yields a positive solution z of (SP) such that $s(z)$ is positive as well and such that the pair $(z, s(z))$ is almost strictly complementary in the sense that for each index i one of the positive coordinates in the pair $(z_i, s_i(z))$ is large and the other is small. To distinguish between 'large' and 'small' coordinates we introduce the so-called *condition number* of (SP). We are able to specify ε such that the large coordinates of z are in B and the small coordinates of z in N. The optimal partition of (SP) can thus be derived from the almost strictly complementary solution z provided by the algorithm. Then, in Section 3.3.6, a rounding procedure is described that yields a strictly complementary solution of (SP) in polynomial time.

3.3.2 Condition number

Below, (B, N) always denotes the optimal partition of (SP), and $\mathcal{SP}^*$ the optimal set of (SP). We first introduce the following two numbers:

$$\sigma_{SP}^z := \min_{i \in B} \max_{z \in \mathcal{SP}^*} \{z_i\}, \quad \sigma_{SP}^s := \min_{i \in N} \max_{z \in \mathcal{SP}^*} \{s_i(z)\}.$$

By convention we take $\sigma_{SP}^z = \infty$ if B is empty and $\sigma_{SP}^s = \infty$ if N is empty. Since the optimal set $\mathcal{SP}^*$ is bounded, σ_{SP}^z is finite if B is nonempty and σ_{SP}^s is finite if N is nonempty. Due to the definition of the sets B and N both numbers are positive, and since B and N cannot be both empty at least one of the two numbers is finite. As a consequence, the number

$$\sigma_{SP} := \min \{\sigma_{SP}^z, \sigma_{SP}^s\}$$

is positive and finite. This number plays a crucial role in the further analysis and is called the *condition number* of (SP).[6] Using that z and $s(z)$ are complementary if $z \in \mathcal{SP}^*$ we can easily verify that σ_{SP} can also be written as

$$\sigma_{SP} := \min_{1 \leq i \leq n} \max_{z \in \mathcal{SP}^*} \{z_i + s_i(z)\}.$$

Example I.40 Let us calculate the condition number of our sample problem (2.19) in Example I.7. We found in Example I.12 that any optimal solution z has the form

---

as given by (2.26), namely

$$
z = \begin{bmatrix} 2\kappa \\ 0 \\ \kappa \\ \kappa \\ 0 \end{bmatrix}, \quad s(z) = \begin{bmatrix} 0 \\ \kappa \\ 0 \\ 0 \\ 5 - 5\kappa \end{bmatrix}, \quad 0 \le \kappa \le 1.
$$

Hence we have for any optimal z,

$$
z + s(z) = \begin{bmatrix} 2\kappa \\ \kappa \\ \kappa \\ \kappa \\ 5 - 5\kappa \end{bmatrix}, \quad 0 \le \kappa \le 1.
$$

To find the condition number we need to find the maximum values of each of the variables in in this vector. These values are 2, 1, 1, 1 (for $\kappa = 1$) and 5 (for $\kappa = 0$), respectively. The minimum of these maximal values being 1, the condition number of our sample problem (2.19) turns out to be 1.[7,8,9,10]   ◇

In the above example we were able to calculate the condition number of a given problem. We see below that when we know the condition number of a problem we can profit from it in the solution procedure. In general, however, the calculation of the condition number is at least as hard as solving the problem. Hence, in general, we have to solve a problem without knowing its condition number. In such cases there is a cheap way to get a lower bound for the condition number. We proceed by deriving such a lower bound for σ_{SP} in terms of the data of the problem (SP). We introduce some more notation.

Definition I.41 *The submatrix of M consisting of the elements in the rows whose indices are in I and the columns whose indices are in J is denoted by M_{IJ}.*

Using this convention, we have for any vector z and its surplus vector $s = s(z)$:

$$
\begin{bmatrix} s_B \\ s_N \end{bmatrix} = \begin{bmatrix} M_{BB} & M_{BN} \\ M_{NB} & M_{NN} \end{bmatrix} \begin{bmatrix} z_B \\ z_N \end{bmatrix} + \begin{bmatrix} q_B \\ q_N \end{bmatrix}. \tag{3.16}
$$

Recall from the previous section that the vector z is optimal if and only if z and s are nonnegative, $z_N = 0$ and $s_B = 0$. Hence we have $q^T z = q_B^T z_B$. Due to the existence of

---

[7] **Exercise 21** Using the results of Exercise 4 (page 27), prove that the condition number of the self-dual problem in question equals 5/4.

[8] **Exercise 22** Using the results of Exercise 5 (page 27), prove that the condition number of the self-dual problem in question equals 5/4.

[9] **Exercise 23** Using the results of Exercise 6 (page 27), prove that the condition number of the self-dual problem in question equals $5/(1 + \beta)$ if $\beta \ge 2$ and otherwise $5\beta/(2(1 + \beta))$.

[10] **Exercise 24** Using the results of Exercise 7 (page 27), prove that the condition number of the self-dual problem in question equals $5/(4 - \beta)$ if $\beta \le -1$ and otherwise $-5\beta/(4 - \beta)$.

a strictly complementary solution z, for which z_B is positive, we conclude that

$$q_B = 0. \tag{3.17}$$

Thus it becomes clear that a vector z and its surplus vector s are optimal for (SP) if and only if $z_B \geq 0$, $z_N = 0$, $s_B = 0$, $s_N \geq 0$ and

$$\begin{bmatrix} 0 \\ s_N \end{bmatrix} = \begin{bmatrix} M_{BB} & M_{BN} \\ M_{NB} & M_{NN} \end{bmatrix} \begin{bmatrix} z_B \\ 0 \end{bmatrix} + \begin{bmatrix} 0 \\ q_N \end{bmatrix}.$$

This is equivalent to

$$\begin{bmatrix} M_{BB} & 0_{BN} \\ M_{NB} & -I_{NN} \end{bmatrix} \begin{bmatrix} z_B \\ s_N \end{bmatrix} = \begin{bmatrix} 0_B \\ -q_N \end{bmatrix}, \quad z_B \geq 0,\ z_N = 0,\ s_B = 0,\ s_N \geq 0. \tag{3.18}$$

Note that any strictly complementary solution z gives rise to a positive solution of this system. For the calculation of σ_{SP} we need to know the maximal value of each coordinate of the vector (z_B, s_N) when this vector runs through all possible solutions of (3.18). Then σ_{SP} is just the smallest of all these maximal values.

At this stage we may apply Lemma C.1 in Appendix C to (3.18), which gives us an easy to compute lower bound for σ_{SP}.

Theorem I.42 *The condition number σ_{SP} of (SP) satisfies*

$$\sigma_{SP} \geq \frac{1}{\prod_{j=1}^{n} \|M_j\|},$$

where M_j denotes the j-th column of M.

Proof: Recall that the optimal set of (SP) is determined by the equation (3.18). Also, by Lemma I.9 we have $e^T z + e^T s(z) = n$, showing that the optimal set is bounded. As we just established, the system (3.18) has a positive solution, and hence we may apply Lemma C.1 to (3.18) with

$$A = \begin{pmatrix} M_{BB} & 0_{BN} \\ M_{NB} & -I_{NN} \end{pmatrix}.$$

The columns in A made up by the two left blocks are the columns M_j of M with $j \in B$, whereas the columns made up by the two right blocks are unit vectors. Thus we obtain that the maximal value of each coordinate of the vector (z_B, s_N) is bounded below by the quantity

$$\frac{1}{\prod_{j \in B} \|M_j\|}.$$

With the definition of σ_{SP} this implies

$$\sigma_{SP} \geq \frac{1}{\prod_{j \in B} \|M_j\|} \geq \frac{1}{\prod_{j=1}^{n} \|M_j\|}.$$

The last inequality is an obvious consequence of the assumption that all columns in M are nonzero and integral. Hence the theorem has been proved. □

3.3.3 Large and small variables

It will be convenient to call the coordinates of $z(\mu)$ that are indexed by B the *large coordinates* of $z(\mu)$, and the other coordinates the *small coordinates* of $z(\mu)$. Furthermore, the coordinates of $s_N(\mu)$ are called the *large coordinates* of $s(\mu)$, and the coordinates of $s_B(\mu)$ *small coordinates* of $s(\mu)$. The next lemma provides lower bounds for the large coordinates and upper bounds for the small coordinates of $z(\mu)$ and $s(\mu)$. This lemma implies that the large coordinates of $z(\mu)$ and $s(\mu)$ are bounded away from zero along the central path, and there exists a uniform lower bound that is independent of μ. Moreover, the small coordinates are bounded above by a constant times μ, where the constant depends only on the data in the problem (SP). In other words, the order of magnitude of the small coordinates is $\mathcal{O}(\mu)$. The bounds in the lemma use the condition number σ_{SP} of (SP).

Lemma I.43 *For any positive μ we have*

$$z_i(\mu) \geq \frac{\sigma_{SP}}{n}, \; i \in B, \qquad z_i(\mu) \leq \frac{n\mu}{\sigma_{SP}}, \; i \in N,$$

$$s_i(\mu) \leq \frac{n\mu}{\sigma_{SP}}, \; i \in B, \qquad s_i(\mu) \geq \frac{\sigma_{SP}}{n}, \; i \in N.$$

Proof: First let $i \in N$ and let $\tilde{z}$ be an optimal solution such that $\tilde{s}_i := s_i(\tilde{z})$ is maximal. Then the definition of the condition number σ_{SP} implies that $\tilde{s}_i \geq \sigma_{SP}$. Applying the orthogonality property (2.22) to the points $\tilde{z}$ and $z(\mu)$ we obtain

$$(z(\mu) - \tilde{z})^T (s(\mu) - \tilde{s}) = 0,$$

which gives

$$z(\mu)^T \tilde{s} + s(\mu)^T \tilde{z} = n\mu.$$

This implies

$$z_i(\mu)\tilde{s}_i \leq z(\mu)^T \tilde{s} \leq n\mu.$$

Dividing by $\tilde{s}_i$ and using that $\tilde{s}_i \geq \sigma_{SP}$ we obtain

$$z_i(\mu) \leq \frac{n\mu}{\tilde{s}_i} \leq \frac{n\mu}{\sigma_{SP}}.$$

Since $z_i(\mu)s_i(\mu) = \mu$, it also follows that

$$s_i(\mu) \geq \frac{\sigma_{SP}}{n}.$$

This proves the second and fourth inequality in the lemma. The other inequalities are obtained in the same way. Let $i \in B$ and let $\tilde{z}$ be an optimal solution such that $\tilde{z}_i$ is maximal. Then the definition of the condition number σ_{SP} implies that $\tilde{z}_i \geq \sigma_{SP}$. Applying the orthogonality property to the points $\tilde{z}$ and $z(\mu)$ we obtain in the same way as before

$$s_i(\mu)\tilde{z}_i \leq s(\mu)^T \tilde{z} \leq n\mu.$$

From this we deduce that

$$s_i(\mu) \leq \frac{n\mu}{\tilde{z}_i} \leq \frac{n\mu}{\sigma_{SP}}.$$

Using once more $z(\mu)s(\mu) = \mu e$ we find

$$z_i(\mu) \geq \frac{\sigma_{SP}}{n},$$

completing the proof of the lemma.[11,12] □

We collect the results of the above lemma in Table 3.1..

	$i \in B$	$i \in N$
$z_i(\mu)$	$\geq \frac{\sigma_{SP}}{n}$	$\leq \frac{n\mu}{\sigma_{SP}}$
$s_i(\mu)$	$\leq \frac{n\mu}{\sigma_{SP}}$	$\geq \frac{\sigma_{SP}}{n}$

Table 3.1. Estimates for large and small variables on the central path.

The lemma has an important consequence. If μ is so small that

$$\frac{n\mu}{\sigma_{SP}} < \frac{\sigma_{SP}}{n}$$

then we have a complete separation of the small and the large variables. This means that if a point $z(\mu)$ on the central path is given so that

$$\mu < \frac{\sigma_{SP}^2}{n^2},$$

then we can determine the optimal partition (B, N) of (SP).

In the next section we show that the Full-Newton step algorithm can produce a point z in the neighborhood of the central path with this feature, namely that it gives a complete separation of the small and the large variables.

3.3.4 Finding the optimal partition

Theorem I.37 states that after at most

$$\left\lceil \sqrt{2n} \, \log \frac{n}{\varepsilon} \right\rceil \tag{3.19}$$

---

[11] **Exercise 25** Let $0 < \mu < \bar{\mu}$. Using the orthogonality property (2.22), show that for each i $(1 \leq i \leq n)$,

$$\frac{z_i(\mu)}{z_i(\bar{\mu})} + \frac{s_i(\mu)}{s_i(\bar{\mu})} \leq 2n.$$

[12] The result in Exercise 25 can be improved to

$$\frac{z_i(\mu)}{z_i(\bar{\mu})} + \frac{s_i(\mu)}{s_i(\bar{\mu})} \leq n,$$

which implies

$$z_i(\mu) \leq n z_i(\bar{\mu}), \quad s_i(\mu) \leq n s_i(\bar{\mu}).$$

For a proof we refer the reader to Vavasis and Ye [281].

iterations the Full-Newton step algorithm yields a feasible solution z such that $q^T z = n\mu \leq \varepsilon$ and $\delta(z,\mu) \leq \frac{1}{2}$. We show in this section that if μ is small enough we can recognize the optimal partition (B, N) from z, and such z can be found in a number of iterations that depends only on the dimension n and on the condition number σ_{SP} of (SP).

We need a simple measure for the distance of z to the central path. To this end, for each positive feasible vector z with $s(z) > 0$, we define the number $\delta_c(z)$ as follows:

$$\delta_c(z) := \frac{\max\left(zs(z)\right)}{\min\left(zs(z)\right)}. \tag{3.20}$$

Observe that $\delta_c(z) = 1$ if and only if $zs(z)$ is a multiple of the all-one vector e. This occurs precisely if z lies on the central path. Otherwise we have $\delta_c(z) > 1$. We consider $\delta_c(z)$ as an indicator for the 'distance' of z to the central path.[13]

Lemma I.44 If $\delta(z,\mu) \leq \frac{1}{2}$ then $\delta_c(z) \leq 4$.

Proof: Using the variance vector v of z, with respect to the given $\mu > 0$, we may write

$$\delta_c(z) = \frac{\max\left(\mu v^2\right)}{\min\left(\mu v^2\right)} = \frac{\max\left(v^2\right)}{\min\left(v^2\right)}.$$

Using (3.13), it follows from $\delta(z,\mu) \leq \frac{1}{2}$ that

$$\left\| v - v^{-1} \right\| \leq 1.$$

Without loss of generality we assume that the coordinates of v are ordered such that

$$v_1 \geq v_2 \geq \ldots \geq v_n.$$

Then $\delta_c(z) = v_1^2 / v_n^2$. Now consider the problem

$$\max \left\{ \frac{v_1^2}{v_n^2} \; : \; \left\| v - v^{-1} \right\| \leq 1 \right\}.$$

The optimal value of this problem is an upper bound for $\delta_c(z)$. One may easily verify that the optimal solution has $v_i = 1$ if $1 < i < n$, $v_1 = \sqrt{2}$ and $v_n = 1/\sqrt{2}$. Hence the optimal value is 4.[14] This proves the lemma. $\qquad\square$

---

[13] In the analysis of interior-point methods we always need to introduce a quantity that measures the 'distance' of a feasible vector x to the central path. This can be done in many ways as becomes apparent in the course of this book. In the coming chapters we make use of a variety of so-called *proximity measures*. All these measures are based on the simple observation that x is on the central path if and only if the vector $xs(x)$ is a scalar multiple of the all-one vector.

[14] **Exercise 26** Prove that

$$\max \left\{ \frac{v_1^2}{v_n^2} \; : \; \sum_{i=1}^{n} \left(v_i - \frac{1}{v_i} \right)^2 \leq 1 \right\} = 4.$$

Lemma I.45 *Let z be a feasible solution of (SP) such that $\delta_c(z) \leq \tau$. Then, with $s = s(z)$, we have*

$$z_i \geq \frac{\sigma_{SP}}{\tau n}, \, i \in B, \qquad z_i \leq \frac{z^T s}{\sigma_{SP}}, \, i \in N,$$

$$s_i \leq \frac{z^T s}{\sigma_{SP}}, \, i \in B, \qquad s_i \geq \frac{\sigma_{SP}}{\tau n}, \, i \in N.$$

Proof: The proof is basically the same as the proof of Lemma I.43. It is a little more complicated because the estimates now concern a point off the central path. From $\delta_c(z) \leq \tau$ we conclude that there exist positive numbers τ_1 and τ_2 such that $\tau \tau_1 = \tau_2$ and

$$\tau_1 \leq z_i s_i \leq \tau_2, \quad 1 \leq i \leq n. \tag{3.21}$$

When we realize that these inequalities replace the role of the identity $z_i(\mu)s_i(\mu) = \mu$ in the proof of Lemma I.43 the generalization becomes almost straightforward. First suppose that $i \in N$ and let $\tilde{z}$ be an optimal solution such that $\tilde{s}_i := s_i(\tilde{z})$ is maximal. Then, from to the definition of σ_{SP}, it follows that $\tilde{s}_i \geq \sigma_{SP}$. Applying the orthogonality property (2.22) to the points $\tilde{z}$ and z, we obtain in the same way as before

$$z_i \tilde{s}_i \leq z^T \tilde{s} \leq z^T s.$$

Hence, dividing both sides by $\tilde{s}_i$ and using that $\tilde{s}_i \geq \sigma_{SP}$ we get

$$z_i \leq \frac{z^T s}{\sigma_{SP}}.$$

From the left inequality in (3.21) we also have $z_i s_i \geq \tau_1$. Hence we must have

$$s_i \geq \frac{\tau_1 \sigma_{SP}}{z^T s}.$$

The right inequality in (3.21) gives $z^T s \leq n \tau_2$. Thus

$$s_i \geq \frac{\tau_1 \sigma_{SP}}{n \tau_2} = \frac{\sigma_{SP}}{n \tau}.$$

This proves the second and fourth inequality in the lemma. The other inequalities are obtained in the same way. If $i \in B$ and $\tilde{z}$ is an optimal solution such that $\tilde{z}_i$ is maximal, then $\tilde{z}_i \geq \sigma_{SP}$. Applying the orthogonality property (2.22) to the points $\tilde{z}$ and z we obtain

$$s_i \tilde{z}_i \leq s^T \tilde{z} \leq z^T s.$$

Thus we get

$$s_i \leq \frac{z^T s}{\tilde{z}_i} \leq \frac{z^T s}{\sigma_{SP}}.$$

Using once more that $z_i s_i \geq \tau_1$ and $z^T s \leq n \tau_2$ we obtain

$$z_i \geq \frac{\tau_1 \sigma_{SP}}{z^T s} \geq \frac{\tau_1 \sigma_{SP}}{n \tau_2} = \frac{\sigma_{SP}}{n \tau},$$

completing the proof of the lemma. $\qquad\qquad\qquad\qquad\qquad\qquad\qquad\qquad\qquad\quad \square$

	$i \in B$	$i \in N$
z_i	$\geq \frac{\sigma_{SP}}{\tau n}$	$\leq \frac{z^T s}{\sigma_{SP}}$
$s_i(z)$	$\leq \frac{z^T s}{\sigma_{SP}}$	$\geq \frac{\sigma_{SP}}{\tau n}$

Table 3.2. Estimates for large and small variables if $\delta_c(z) \leq \tau$.

The results of the above lemma are shown in Table 3.2.. We conclude that if $z^T s$ is so small that

$$\frac{z^T s}{\sigma_{SP}} < \frac{\sigma_{SP}}{\tau n}$$

then we have a complete separation of the small and the large variables. Thus we may state without further proof the following result.

Lemma I.46 *Let z be a feasible solution of (SP) such that $\delta_c(z) \leq \tau$. If*

$$z^T s(z) < \frac{\sigma_{SP}^2}{\tau n}$$

then the optimal partition of (SP) follows from

$$B = \{i \,:\, z_i > s_i(z)\} \quad and \quad N = \{i \,:\, z_i < s_i(z)\}. \tag{3.22}$$

This lemma is the basis of our next result.

Theorem I.47 *After at most*

$$\left\lceil \sqrt{2n} \log \frac{4n^2}{\sigma_{SP}^2} \right\rceil \tag{3.23}$$

iterations, the Full-Newton step algorithm yields a feasible (and positive) solution z of (SP) that reveals the optimal partition (B, N) of (SP) according to (3.22).

Proof: Let us run the Full-Newton step algorithm with $\varepsilon = \sigma_{SP}^2 / (4n)$. Then Theorem I.37 states that we obtain a feasible z with $z^T s(z) \leq \sigma_{SP}^2 / (4n)$ and $\delta(z, \mu) \leq 1/2$. Lemma I.44 implies that $\delta_c(z) \leq 4$. By Lemma I.46, with $\tau = 4$, this z gives a complete separation between the small variables and the large variables. By Theorem I.37, the required number of iterations for the given ε is at most

$$\left\lceil \sqrt{2n} \log \frac{4n^2}{\sigma_{SP}^2} \right\rceil$$

which is equal to the bound given in the theorem. Thus the proof is complete. □

Example I.48 Let us apply Theorem I.47 to the self-dual problem (2.19) in Example I.7. Then $n = 5$ and, according to Example I.40 (page 54), $\sigma_{SP} = 1$. Thus the iteration bound (3.23) in Theorem I.47 becomes

$$\left\lceil \sqrt{10} \log{(100)} \right\rceil = \lceil 14.5628 \rceil = 15.$$

With the help of Figure 3.1 (page 53) we can now determine the optimal partition and we find

$$B = \{1, 3, 5\}, \qquad N = \{2, 4\},$$

in agreement with the result of Example I.12. ◇

3.3.5 A rounding procedure for interior-point solutions

We have just established that the optimal partition of (SP) can be found after a finite number of iterations of the Full-Newton step algorithm. The required number of iterations is at most equal to the number given by (3.23). After this number of iterations the small variables and the large variables are well enough separated from each other to reveal the classes B and N that constitute the optimal partition.

The aim of this section and the next section is to show that if B has been fixed then a strictly complementary solution of (SP) can be obtained with little extra effort.[15]

First we establish that the class B is not empty.

Lemma I.49 *The class B in the optimal partition of (SP) is not empty.*

Proof: If B is the empty set then $z = 0$ is the only optimal solution. Since, by Theorem I.20, this solution must be strictly complementary we must have $s(z) > 0$. Since $s(z) = Mz + q = q$, we find $q > 0$. This contradicts that q has zero entries, by (3.1). This proves the lemma. □

Assuming that the optimal partition (B, N) has been determined, with B nonempty, we describe a rounding procedure that can be applied to any positive vector z with positive surplus vector $s(z)$ to yield a vector $\bar{z}$ such that $\bar{z}$ and its surplus vector $\bar{s} = s(\bar{z})$ are complementary (in the sense that $\bar{z}_N = \bar{s}_B = 0$) but not necessarily nonnegative. In the next section we run the algorithm an additional number of iterations to get a sharper separation between the small and the large variables and we show that the rounding procedure yields a strictly complementary solution in polynomial time.

Let us have a positive vector z with positive surplus vector $s(z)$. Recall from (3.16), page 55, that

$$\begin{bmatrix} s_B \\ s_N \end{bmatrix} = \begin{bmatrix} M_{BB} & M_{BN} \\ M_{NB} & M_{NN} \end{bmatrix} \begin{bmatrix} z_B \\ z_N \end{bmatrix} + \begin{bmatrix} q_B \\ q_N \end{bmatrix}.$$

---

[15] It is generally believed that interior-point methods for LO *never generate an exact optimal solution in polynomial time* (Andersen and Ye [11]). In fact, Ye [308] showed in 1992 that a strictly complementary solution can be found in polynomial time by all the known $\mathcal{O}(n^3 L)$ interior-point methods. See also Mehrotra and Ye [208]. The rounding procedure described in this chapter is essentially the same as the one presented in these two papers and leads to *finite termination* of the algorithm.

This implies that
$$s_B = M_{BB}z_B + M_{BN}z_N + q_B.$$

Since $q_B = 0$, by (3.17), $\xi = z_B$ satisfies the system of equations in the unknown vector ξ given by
$$M_{BB}\,\xi = s_B - M_{BN}z_N. \tag{3.24}$$

Note that z_B is a 'large' solution of (3.24), because the entries of z_B are large variables. On the other hand we can easily see that (3.24) must have more solutions. This follows from the existence of a strictly complementary solution of (SP), because for any such solution $\tilde{z}$ we derive from $\tilde{z}_N = 0$ and $s_B(\tilde{z}) = 0$ that $M_{BB}\tilde{z}_B = 0$. Since $\tilde{z}_B > 0$, it follows that the columns of M_{BB} are linearly dependent, and hence (3.24) has multiple solutions.

Now let ξ be any solution of (3.24) and consider the vector $\bar{z}$ defined by
$$\bar{z}_B = z_B - \xi, \quad \bar{z}_N = 0.$$

For the surplus vector $\bar{s} = s(\bar{z})$ of $\bar{z}$ we have
$$\bar{s}_B = M_{BB}\bar{z}_B + M_{BN}\bar{z}_N = M_{BB}\bar{z}_B = M_{BB}\,(z_B - \xi) = 0.$$

So we have $\bar{z}_N = \bar{s}_B = 0$, which means that the vectors $\bar{z}$ and $\bar{s}$ are complementary. It will be clear, however, that the vectors $\bar{z}$ and $\bar{s}$ are not necessarily nonnegative. This only holds if
$$\bar{z}_B = z_B - \xi \geq 0,$$

and
$$\bar{s}_N = M_{NB}\bar{z}_B + M_{NN}\bar{z}_N + q_N = M_{NB}\,(z_B - \xi) + q_N = s_N - M_{NN}z_N - M_{NB}\xi \geq 0.$$

We conclude that if (3.24) admits a solution ξ that satisfies the last two inequalities then $\bar{z}$ is a solution of (SP). Moreover, if ξ satisfies these inequalities strictly, so that
$$z_B - \xi > 0, \quad s_N - M_{NN}z_N - M_{NB}\xi > 0, \tag{3.25}$$

then $\bar{z}$ is a strictly complementary solution of (SP). In the next section we show that solving (3.24) by Gaussian elimination gives such a solution, provided the separation between the small and the large variables is sharp enough.

Example I.50 In this example we show that the Full-Newton step algorithm equipped with the above described rounding procedure solves the sample problem (2.19) in Example I.7 in one iteration. Recall from Example I.14 that the Newton step in the first iteration is given by (2.39) and (2.40). Since in this iteration $\mu = 1 - \theta$, substituting $\theta = 1/\sqrt{10}$, we find

$$\Delta z = -\frac{1}{\sqrt{10}}\left(-\frac{1}{3}, \frac{8}{9}, \frac{4}{9}, \frac{1}{9}, 1\right)^T = -(-0.1054, 0.2811, 0.1405, 0.0351, 0.3162)^T,$$

and

$$\Delta s = -\frac{1}{\sqrt{10}}\left(\frac{4}{3}, \frac{1}{9}, \frac{5}{9}, \frac{8}{9}, 0\right)^T = -(0.4216, 0.0351, 0.1757, 0.2811, 0.0000)^T.$$

Hence, after one iteration the new iterate is given by

$$z = (1.1054, 0.7189, 0.8595, 0.9649, 0.6838)^T,$$

and

$$s = (0.5784, 0.9649, 0.8243, 0.7189, 1.0000)^T.$$

It is interesting to observe that the sets B and N, as defined by (3.22) are already the classes of the optimal partition of the problem:

$$B = \{1, 3, 4\}, \quad N = \{2, 5\}.$$

Now we apply the rounding procedure at z with respect to the partition (B, N). The matrix M_{BB} is given by

$$M_{BB} = \begin{bmatrix} 0 & 1 & -1 \\ -1 & 0 & 2 \\ 1 & -2 & 0 \end{bmatrix}.$$

We have

$$M_{BB}z_B = \begin{bmatrix} 0 & 1 & -1 \\ -1 & 0 & 2 \\ 1 & -2 & 0 \end{bmatrix} \begin{bmatrix} 1.1054 \\ 0.8595 \\ 0.9649 \end{bmatrix} = \begin{bmatrix} -0.1054 \\ 0.8243 \\ -0.6135 \end{bmatrix}.$$

So we need to find a 'small' solution ζ of the system

$$M_{BB}z_B = \begin{bmatrix} 0 & 1 & -1 \\ -1 & 0 & 2 \\ 1 & -2 & 0 \end{bmatrix} \zeta = \begin{bmatrix} -0.1054 \\ 0.8243 \\ -0.6135 \end{bmatrix}.$$

A solution of this system is

$$\zeta = \begin{bmatrix} 0.0000 \\ 0.3067 \\ 0.4122 \end{bmatrix}.$$

The rounded solution is now defined by

$$\bar{z}_B = z_B - \zeta = \begin{bmatrix} 1.1054 \\ 0.8595 \\ 0.9649 \end{bmatrix} - \begin{bmatrix} 0.0000 \\ 0.3067 \\ 0.4122 \end{bmatrix} = \begin{bmatrix} 1.1054 \\ 0.5527 \\ 0.5527 \end{bmatrix}, \quad \bar{z}_N = 0.$$

Hence the rounded solution is

$$z = (1.1054, 0.0000, 0.5527, 0.5527, 0.0000)^T.$$

The corresponding slack vector is

$$s(z) = Mz + q = (0.0000, 0.5527, 0.0000, 0.0000, 2.2365)^T.$$

Since z and $s(z)$ are nonnegative and complementary, z is optimal. Moreover, $z + s(z) > 0$, so z is a strictly complementary solution. Hence we have solved the sample problem in one iteration. ◇

Remark I.51 In the above example we used for ξ the least norm solution of (3.24). This is the solution of the minimization problem

$$\min \left\{ \|\xi\| \ : \ M_{BB}\xi = M_{BB}z_B \right\}.$$

Formally the least norm solution can be described as

$$\xi = M_{BB}^{+} M_{BB} z_B,$$

where M_{BB}^{+} denotes the generalized inverse(cf. Appendix B) of M_{BB}. We may then write

$$z_B - \xi = \left(I_{BB} - M_{BB}^{+} M_{BB} \right) z_B,$$

where I_{BB} is the identity matrix of appropriate size.

Since we want ξ to be such that $z_B - \xi$ is positive, an alternative approach might be to use the solution of

$$\min \left\{ \left\| z_B^{-1}\xi \right\| \ : \ M_{BB}\xi = M_{BB}z_B \right\}.$$

Then ξ is given by

$$\xi = Z_B \left(M_{BB}Z_B \right)^{+} M_{BB}z_B.$$

•

Of course, we were lucky in the above example in two ways: the first iterate already determined the optimal partition and, moreover, at this iterate the rounding procedure yielded a strictly complementary solution. In general more iterations will be necessary to find the optimal partition and once the optimal partition has been found the rounding procedure may not yield a strictly complementary solution at once. But, as we see in the next section, after sufficiently many iterations we can always find an exact solution of any problem in this way, and the required number of iterations can be bounded by a (linear) polynomial of the size of the problem.

3.3.6 Finding a strictly complementary solution

In this section we assume that the optimal partition (B, N) of (SP) is known. In the previous section we argued that it may be assumed without loss of generality that the set B is not empty. In this section we show that when we run the algorithm an additional number of iterations, the rounding procedure of the previous section can be used to construct a strictly complementary solution of (SP). The additional number of iterations depends on the size of B and is aimed at creating a sufficiently large distance between the small and the large variables.

We need some more notation. First, ω will denote the infinity norm of M:

$$\omega := \|M\|_{\infty} = \max_{1 \le i \le n} \sum_{j=1}^{n} |M_{ij}|.$$

Second, B^* denotes the subset of B for which the columns in M_{BB} are nonzero, and third, the number π_B is defined by

$$\pi_B := \begin{cases} 1 & \text{if } B^* = \emptyset, \\ \prod_{j \in B^*} \|(M_{BB})_j\| & \text{otherwise.} \end{cases}$$

Lemma I.52 *Let z be a feasible solution of (SP) such that $\delta_c(z) \leq \tau \doteq 4$. If*

$$z^T s(z) \leq \frac{\sigma_{SP}^2}{4n(1+\omega)^2 \pi_B \sqrt{|B|}},$$

with ω and π_B as defined above, then a strictly complementary solution can be found in

$$\mathcal{O}(|B^*|^3)$$

arithmetical operations.

Proof: Suppose that z is positive solution of (SP) with positive surplus vector $s = s(z)$ such that $\delta_c(z) \leq 4$ and $z^T s \leq \varepsilon$, where

$$\varepsilon := \frac{\sigma_{SP}^2}{4n(1+\omega)^2 \pi_B \sqrt{|B|}}. \tag{3.26}$$

Recall that the entities $|B|$, ω and π_B are all at least 1 and also, by Lemma I.45, that the small variables in z and s are less than ε/σ_{SP} and the large variables are at least $\sigma_{SP}/(4n)$.

We now show that the system (3.24) has a solution ξ whose coordinates are small enough, so that

$$z_B - \xi > 0, \quad s_N - M_{NN} z_N - M_{NB}\xi > 0. \tag{3.27}$$

We need to distinguish between the cases where M_{BB} is zero and nonzero respectively.

We first consider the case where $M_{BB} = 0$. Then $\xi = 0$ satisfies (3.24) and for this ξ the condition (3.27) for the rounded solution $\bar{z}$ to be strictly complementary reduces to the single inequality

$$s_N - M_{NN} z_N > 0. \tag{3.28}$$

This inequality is satisfied if $M_{NN} = 0$. Otherwise, if $M_{NN} \neq 0$, since z_N is small we may write

$$\|M_{NN} z_N\|_\infty \leq \|M_{NN}\|_\infty \|z_N\|_\infty \leq \|M\|_\infty \frac{\varepsilon}{\sigma_{SP}} = \frac{\varepsilon\omega}{\sigma_{SP}}.$$

Hence, since s_N is large, (3.28) certainly holds if

$$\frac{\varepsilon\omega}{\sigma_{SP}} < \frac{\sigma_{SP}}{4n},$$

which is equivalent to

$$\varepsilon < \frac{\sigma_{SP}^2}{4n\omega}.$$

Since this inequality is implied by the hypothesis of the lemma, we conclude that the rounding procedure yields a strictly complementary solution if $M_{BB} = 0$.

Now consider the case where $M_{BB} \neq 0$. Then we solve (3.24) by Gaussian elimination. This goes as follows. Let B_1 and B_2 be two subsets of B such that $M_{B_1 B_2}$ is a nonsingular square submatrix of M_{BB} with maximal rank, and let ζ be the unique solution of the equation

$$M_{B_1 B_2} \zeta = s_{B_1} - M_{B_1 N} z_N.$$

From Cramer's rule we know that the i-th entry of ζ, with $i \in B_2$, is given by

$$\zeta_i = \frac{\det M_{B_1 B_2}^{(i)}}{\det M_{B_1 B_2}},$$

where $M_{B_1 B_2}^{(i)}$ is the matrix arising by replacing the i-th column in $M_{B_1 B_2}$ by the vector $s_{B_1} - M_{B_1 N} z_N$. Since the entries of $M_{B_1 B_2}$ are integral and this matrix is nonsingular, the absolute value of its determinant is at least 1. As a consequence we have

$$|\zeta_i| \le \left| \det M_{B_1 B_2}^{(i)} \right|.$$

The right-hand side is no larger than the product of the norms of the columns in the matrix $M_{B_1 B_2}^{(i)}$, due to the inequality of Hadamard (cf. Section 1.7.3). Thus

$$|\zeta_i| \le \|s_B - M_{BN} z_N\| \prod_{j \in B_2 \setminus \{i\}} \|(M_{B_1 B_2})_j\| \le \|s_B - M_{BN} z_N\| \pi_B. \qquad (3.29)$$

The last inequality follows because the norm of each nonzero column in M_{BB} is at least 1, and π_B is the product of these norms.

Since s_B and z_N are small variables we have

$$\|s_B\|_\infty \le \frac{\varepsilon}{\sigma_{SP}}$$

and

$$\|M_{BN} z_N\|_\infty \le \|M_{BN}\|_\infty \|z_N\|_\infty \le \|M\|_\infty \|z_N\|_\infty \le \frac{\varepsilon \omega}{\sigma_{SP}}.$$

Therefore

$$\|s_B - M_{BN} z_N\| \le \sqrt{|B|} \|s_B - M_{BN} z_N\|_\infty \le \sqrt{|B|} \frac{\varepsilon (1 + \omega)}{\sigma_{SP}}.$$

Substituting this inequality in (3.29), we obtain

$$|\zeta_i| \le \frac{\varepsilon (1 + \omega) \pi_B \sqrt{|B|}}{\sigma_{SP}}.$$

Defining ξ by
$$\xi_{B_1} = \zeta, \quad \xi_i = 0, \quad i \in B \setminus B_1,$$

the vector ξ satisfies (3.24), because $M_{B_1 B_2}$ is a nonsingular square submatrix of M_{BB} with maximal rank and because $s_B - M_{BN} z_N (= M_{BB} z_B)$ belongs to the column space of M_{BB}. Hence we have shown that Gaussian elimination yields a solution ξ of (3.24) such that

$$\|\xi\|_\infty \le \frac{\varepsilon (1 + \omega) \pi_B \sqrt{|B|}}{\sigma_{SP}}. \qquad (3.30)$$

Applying the rounding procedure of the previous section to z, using ξ, we obtain the vector $\bar{z}$ defined by

$$\bar{z}_B = z_B - \xi, \quad \bar{z}_N = 0,$$

and the surplus vector $\bar{s} = s(\bar{z})$ satisfies $\bar{s}_B = 0$. So $\bar{z}$ is complementary. We proceed by showing that $\bar{z}$ is a strictly complementary solution of (SP) by proving that ξ satisfies the condition (3.25), namely

$$\bar{z}_B = z_B - \xi > 0, \quad \bar{s}_N = s_N - M_{NN} z_N - M_{NB} \xi > 0.$$

We first establish that $\bar{z}_B$ is positive. This is now easy. The coordinates of z_B are large and the nonzero coordinates of ξ are bounded above by the right-hand side in (3.30). Therefore, $\bar{z}_B$ will be positive if

$$\frac{\varepsilon(1+\omega)\pi_B \sqrt{|B|}}{\sigma_{SP}} < \frac{\sigma_{SP}}{4n},$$

or, equivalently,

$$\varepsilon < \frac{\sigma_{SP}^2}{4n(1+\omega)\pi_B \sqrt{|B|}},$$

and this is guaranteed by the hypothesis in the lemma.

We proceed by estimating the coordinates of $\bar{s}_N$. First we write

$$\|M_{NN} z_N + M_{NB} \xi\|_\infty = \|(M_{NN}\ M_{NB})\|_\infty \left\| \begin{pmatrix} z_N \\ \xi \end{pmatrix} \right\|_\infty \leq \|M\|_\infty \left\| \begin{pmatrix} z_N \\ \xi \end{pmatrix} \right\|_\infty.$$

Using (3.30) and the fact that z_N is small we obtain

$$\|M_{NN} z_N + M_{NB} \xi\|_\infty \leq \omega \max\left(\frac{\varepsilon}{\sigma_{SP}}, \frac{\varepsilon(1+\omega)\pi_B \sqrt{|B|}}{\sigma_{SP}} \right) = \frac{\varepsilon\omega(1+\omega)\pi_B \sqrt{|B|}}{\sigma_{SP}}.$$

Here we used again that $\pi_B \geq 1$ and $|B| \geq 1$. Hence, since the coordinates of s_N are large, the coordinates of $\bar{s}_N$ will be positive if

$$\frac{\varepsilon\omega(1+\omega)\pi_B \sqrt{|B|}}{\sigma_{SP}} < \frac{\sigma_{SP}}{4n},$$

or, equivalently, if

$$\varepsilon < \frac{\sigma_{SP}^2}{4n\omega(1+\omega)\pi_B \sqrt{|B|}},$$

and this follows from the hypothesis in the lemma.

Thus we have shown that the condition for $\bar{z}$ being strictly complementary is satisfied. Finally, the calculation of ζ can be performed by Gaussian elimination and this requires $\mathcal{O}(|B^*|^3)$ arithmetic operations. Thus the proof is complete. $\square$

The next theorem now easily follows from the last lemma.

Theorem I.53 *Using the notation introduced above, the Full-Newton step algorithm yields a feasible solution z for which the rounding procedure yields a strictly complementary solution of (SP), after at most*

$$\left\lceil \sqrt{2n} \log \frac{4n^2(1+\omega)^2 \pi_B \sqrt{|B|}}{\sigma_{SP}^2} \right\rceil$$

iterations.

Proof: By Lemma I.52 the rounding procedure yields a strictly complementary solution if we run the Full-Newton step algorithm with

$$\varepsilon = \frac{\sigma_{SP}^2}{4n(1+\omega)^2 \pi_B \sqrt{|B|}}.$$

By Theorem I.37 for this value of ε the Full-Newton step algorithm requires at most

$$\left\lceil \sqrt{2n} \, \log \frac{4n^2(1+\omega)^2 \pi_B \sqrt{|B|}}{\sigma_{SP}^2} \right\rceil$$

iterations. This proves the theorem. $\qquad\qquad\qquad\qquad\qquad\qquad\qquad\qquad\qquad\qquad\square$

Remark I.54 The result in Theorem I.53 can be used to estimate the number of arithmetic operations required by the algorithm in a worst-case situation. This number can be bounded by a polynomial of the size L of (SP) (cf. Remark I.32), as we show. We thus establish that the method proposed in this chapter solves the self-dual model in polynomial time. As a consequence, by the results of the previous chapter, it also solves the canonical LO problem in polynomial time.

The iteration bound in the theorem is worst if B contains all indices. Ignoring the integrality operator, and denoting the number of iterations by K, the iteration bound becomes

$$K \leq \sqrt{2n} \, \log \frac{4n^2 \sqrt{n} \, (1+\omega)^2 \pi_M}{\sigma_{SP}^2},$$

where

$$\pi_M := \prod_{j=1}^{n} \|M_j\| \, .$$

By Theorem I.42 we have

$$\sigma_{SP} \geq \frac{1}{\prod_{j=1}^{n} \|M_j\|} = \frac{1}{\pi_M} \, .$$

Substituting this we get the upper bound

$$K \leq \sqrt{2n} \, \log \left(4n^{\frac{5}{2}} (1+\omega)^2 \pi_M^3 \right), \qquad (3.31)$$

for the number of iterations. A rather pessimistic estimate yields

$$\pi_M^2 = \prod_{j=1}^{n} \left(\sum_{i=1}^{n} M_{ij}^2 \right) \leq n^n \Pi^2 \, .$$

This follows by expanding the product in the middle, which gives n^n terms, each of which is bounded above by Π^2, where Π is defined in Remark I.32 as the product of all nonzero entries in q and M. We also have the obvious (and very pessimistic) inequality $\omega \leq \Pi$, which implies $1 + \omega \leq 2\Pi$. Substituting these pessimistic estimates in (3.31) we obtain

$$K \leq \sqrt{2n} \, \log \left(4n^{\frac{5}{2}} (2\Pi)^2 \left(n^n \Pi^2 \right)^{\frac{3}{2}} \right) = \sqrt{2n} \, \log \left(16n^{\frac{3n+5}{2}} \Pi^5 \right).$$

This can be further reduced. One has

$$
\begin{aligned}
\log\left(16\, n^{\frac{3n+5}{2}}\,\Pi^5\right) \;&=\; \log 16 + \frac{3n+5}{2}\log n + 5\log\Pi \\[2mm]
&<\; 3 + \frac{1}{2}\,(3n+5)\,(n-1) + \frac{7}{2}\log_2\Pi \\[2mm]
&=\; \frac{1}{2}\left(3n^2 + 2n + 1 + 7\log_2\Pi\right) \\[2mm]
&<\; \frac{7}{2}\,(n(n+1) + \log_2\Pi)\,.
\end{aligned}
$$

The first inequality is due to $\log 16 = 2.7726 < 3$, $\log n \le n-1$ and $\log\Pi = 0.6931\log_2\Pi$, and the second inequality holds because $7n(n+1) > 3n^2 + 2n + 1$ for all n.

Finally, using the definition (3.5) of the size $L(= n(n+1) + \log_2\Pi))$, we obtain

$$
K < \frac{7}{2}\sqrt{2n}\,L < 5\sqrt{n}\,L.
$$

Thus the claim has been proved. •

3.4 Concluding remarks

The analysis in this chapter is based on properties of the central path of (SP). To be more specific, on the property that when one moves along the central path to the optimal set, the separation between the large and small variables becomes apparent. We showed that the Full-Newton step algorithm together with a simple rounding procedure yields a polynomial algorithm for solving a canonical LP problem; the iteration bound is $5\sqrt{n}\,L$, where L is the binary input size of the problem.

In the literature many other polynomial-time interior-point algorithms have been presented. We will encounter many of these algorithms in the rest of the book. Almost all of these algorithms are based on a Newton-type search direction. At this stage we want to mention an interesting exception, which is based on an idea of Dikin and that also can be used to solve in polynomial time the self-dual problem that we considered in this and the previous chapter. In fact, an earlier version of this book used the Dikin Step Algorithm in this part of the book. The iteration bound that we could obtain for this algorithm was $7nL$. Because it leads to a better iteration bound, in this edition we preferred to use the Full-Newton step algorithm. But because the Dikin Step Algorithm is interesting in itself, and also because further on in the book we will deal with Dikins method, we decided to keep a full description and analysis of the Dikin Step Algorithm in the book. It can be found in Appendix F.[16]

---

[16] The Dikin Step Algorithm was investigated first by Jansen et al. [156]; the analysis of the algorithm used in this chapter is based on a paper of Ling [182]. By including higher-order components in the search direction, the complexity can be improved by a factor $\sqrt{n}$, thus yielding a bound of the same order as for the Full-Newton step algorithm. This has been shown by Jansen et al. [160]. See also Chapter 18.

4

Solving the Canonical Problem

4.1 Introduction

In Chapter 2 we discussed the fact that every LO problem has a canonical description
of the form

$$(P) \qquad \min \left\{ c^T x \; : \; Ax \geq b, \, x \geq 0 \right\}.$$

The matrix A is of size $m \times n$ and the vectors c and x are in $\mathbb{R}^n$ and b in $\mathbb{R}^m$. In this
chapter we further discuss how this problem, and its dual problem

$$(D) \qquad \max \left\{ b^T y \; : \; A^T y \leq c, \, y \geq 0 \right\},$$

can be solved by using the algorithm of the previous chapter for solving a self-dual
embedding of both problems. With

$$
\bar{M} := \begin{bmatrix} 0 & A & -b \\ -A^T & 0 & c \\ b^T & -c^T & 0 \end{bmatrix}, \qquad \bar{z} := \begin{bmatrix} y \\ x \\ \kappa \end{bmatrix}, \tag{4.1}
$$

as in (2.7), the embedding problem is given by (2.15). It is the self-dual homogeneous
problem

$$(SP_0) \qquad \min \left\{ 0^T \bar{z} \; : \; \bar{M} \bar{z} \geq 0, \quad \bar{z} \geq 0 \right\} \tag{4.2}$$

In Chapter 3 we showed that a strictly complementary solution $\bar{z}$ of (SP_0) can be
found in polynomial time. If a strictly complementary solution $\bar{z}$ has $\kappa > 0$ then
$\bar{x} = x/\kappa$ is an optimal solution of (P), and if $\kappa = 0$ then (P) (and also its dual (D))
must be either unbounded or infeasible. This was shown in Section 2.8, where we also
found that any strictly complementary solution of (SP_0) with $\kappa > 0$ provides a strictly
complementary pair of solutions $(\bar{x}, \bar{y})$ for (P) and (D). Thus $\bar{x}$ is primal feasible and
$\bar{y}$ dual feasible. The complementarity means that

$$\bar{x} \left(c - A^T \bar{y} \right) = 0, \qquad \bar{y} \left(A\bar{x} - b \right) = 0,$$

and the strictness of the complementarity that

$$\bar{x} + \left(c - A^T \bar{y} \right) > 0, \qquad \bar{y} + \left(A\bar{x} - b \right) > 0.$$

Obviously these results imply that every LO problem can be solved exactly in
polynomial time. The aim of this chapter is to make a more thorough investigation of

the consequences of the results in Chapter 2 and Chapter 3. We restrict ourselves to the canonical model.

The algorithm for the self-dual model, presented in Section 3.2, requires knowledge of a positive $\bar{z}$ such that the surplus vector $s(\bar{z}) = \bar{M}\bar{z}$ of $\bar{z}$ is positive. However, such $\bar{z}$ does not exist, as we argued in Section 2.4. But then, as we showed in the same section, we can embed (SP_0) in a slightly larger self-dual problem, named (SP) and given by (cf. (2.16)),

$$(SP) \qquad \min\left\{ q^T z \ : \ Mz \geq -q, \ z \geq 0 \right\}. \tag{4.3}$$

for which the constraint matrix has one extra row and one extra column, so that any strictly complementary solution of (SP) induces a strictly complementary solution of (SP_0). Hence, applying the algorithm to the larger problem (SP) yields a strictly complementary solution of (SP_0), hence also for (P) and (D) if these problems are solvable.

It should be noted that both the description of the Full-Newton Step algorithm (page 50) and its analysis apply to any problem of the form (4.3) that satisfies the IPC, provided that the matrix M is skew-symmetric and $q \geq 0$. In other words, we did not exploit the special structure of the matrix M, as given by (2.11), neither did we use the special structure of the vector q, as given by (2.12).

Also note that if the embedding problem is ill-conditioned, in the sense that the condition number σ_{SP} is small, we are forced to run the Full-Newton step algorithm with a (very) small value of the accuracy parameter. In practice, due to limitations of machine precision, it may happen that we cannot reach the state at which an exact solution of (SP) can be found. In that case the question becomes important of what conclusions can be drawn for the canonical problem (P) and its dual problem (D) when an ε-solution for the embedding self-dual problem is available.

The aim of this chapter is twofold. We want to present two other embeddings of (SP_0) that satisfy the IPC. Recall that the embedding in Chapter 2 did not require any foreknowledge about the problems (P) and (D). We present another embedding that can also be used for that case. A crucial question that we want to investigate is if we can then decide whether the given problems have optimal solutions or not without using the rounding procedure. Obviously, this amounts to deciding whether we have $\kappa > 0$ in the limit or not. This will be the subject in Section 4.3.

Our first aim, however, is to consider an embedding that applies if both (P) and (D) have a *strictly feasible solution* and such solutions are know in advance. This case is relatively easy, because we then know for sure that $\kappa > 0$ in the limit.

4.2 The case where strictly feasible solutions are known

We start with the easiest case, namely when strictly feasible solutions of (P) and (D) are given. Suppose that $x^0 \in \mathbb{R}^n$ and $y^0 \in \mathbb{R}^m$ are strictly feasible solutions of (P) and (D) respectively:

$$x^0 > 0, \ s(x^0) = Ax^0 - b > 0 \quad \text{and} \quad y^0 > 0, \ s(y^0) = c - A^T y^0 > 0.$$

4.2.1 Adapted self-dual embedding

Let

$$
M := \begin{bmatrix} 0 & A & -b & 0 \\ -A^T & 0 & c & 0 \\ b^T & -c^T & 0 & 1 \\ 0 & 0 & -1 & 0 \end{bmatrix}, \quad z := \begin{bmatrix} y \\ x \\ \kappa \\ \vartheta \end{bmatrix}, \quad q := \begin{bmatrix} 0 \\ 0 \\ 0 \\ 2 \end{bmatrix},
$$

and consider the self-dual problem

$$
(SP_1) \qquad \min \left\{ q^T z \ : \ Mz + q \geq 0, \ z \geq 0 \right\}.
$$

Note that $q \geq 0$. We proceed by showing that this problem has a positive solution with positive surplus vector. Let

$$
\vartheta^0 := 1 + c^T x^0 - b^T y^0.
$$

The weak duality property implies that $c^T x^0 - b^T y^0 \geq 0$. If $c^T x^0 - b^T y^0 = 0$ then x^0 and y^0 are optimal and we are done. Otherwise we have $\vartheta^0 > 1$. We can easily check that for

$$
z^0 := \begin{bmatrix} y^0 \\ x^0 \\ 1 \\ \vartheta^0 \end{bmatrix}
$$

we have

$$
s(z^0) := Mz^0 + q = \begin{bmatrix} Ax^0 - b \\ c - A^T y^0 \\ b^T y^0 - c^T x^0 + \vartheta^0 \\ -1 + 2 \end{bmatrix} = \begin{bmatrix} s(x^0) \\ s(y^0) \\ 1 \\ 1 \end{bmatrix},
$$

so both z^0 and its surplus vector are positive.[1] Now let $\bar{z}$ be a strictly complementary solution of (SP_1). Then we have, for suitable vectors $\bar{y}$ and $\bar{x}$ and scalars $\bar{\kappa}$ and $\bar{\vartheta}$,

$$
\bar{z} := \begin{bmatrix} \bar{y} \\ \bar{x} \\ \bar{\kappa} \\ \bar{\vartheta} \end{bmatrix} \geq 0, \quad s(\bar{z}) = \begin{bmatrix} A\bar{x} - \bar{\kappa}b \\ \bar{\kappa}c - A^T\bar{y} \\ b^T\bar{y} - c^T\bar{x} + \bar{\vartheta} \\ 2 - \bar{\kappa} \end{bmatrix} \geq 0, \quad \bar{z}s(\bar{z}) = 0, \quad \bar{z} + s(\bar{z}) > 0.
$$

Since the optimal objective value is zero, we have $\bar{\vartheta} = 0$. On the other hand, we cannot have $\bar{\kappa} = 0$, because this would imply the contradiction that either (P) or (D) is infeasible. Hence we conclude that $\bar{\kappa} > 0$. This has the consequence that $\tilde{x} = \bar{x}/\bar{\kappa}$ is feasible for (P) and $\tilde{y} = \bar{y}/\bar{\kappa}$ is feasible for (D), as follows from the feasibility of $\bar{z}$. The complementarity of $\bar{z}$ and $s(\bar{z})$ now yields that

$$
s(\bar{\kappa}) := b^T\bar{y} - c^T\bar{x} = 0.
$$

---

[1] **Exercise 27** If it happens that we have a primal feasible x^0 and a dual feasible y^0 such that $x^0 s(y^0) = \mu e_n$ and $y^0 s(x^0) = \mu e_m$ for some positive μ, find an embedding satisfying the IPC such that z^0 is on its central path.

Thus it follows that $\bar{x}/\bar{\kappa}$ is optimal for (P) and $\bar{y}/\bar{\kappa}$ is optimal for (D). Finally, the strict complementarity of $\bar{z}$ and $s(\bar{z})$ gives the strict complementarity of this solution pair.

4.2.2 Central paths of (P) and (D)

At this stage we want to point out an interesting and important consequence of the existence of strictly feasible solutions of (P) and (D). In that case we can define central paths for the problems (P) and (D). This goes as follows. Let μ be an arbitrary positive number. Then the μ-center of (SP_1) is determined as the unique solution of the system (cf. (2.46), page 35)

$$
\begin{aligned}
Mz + q &= s, \quad z \geq 0,\, s \geq 0 \\
zs &= \mu\, e_{m+n+2}.
\end{aligned}
\tag{4.4}
$$

In other words, there exist unique nonnegative x, y, κ, ϑ such that

$$
Ax - \kappa b \geq 0, \quad \kappa c - A^T y \geq 0, \quad b^T y - c^T x + \vartheta \geq 0, \quad 2 - \kappa \geq 0
$$

and, moreover

$$
\begin{aligned}
y\,(Ax - \kappa b) &= \mu e_m \\
x\,(\kappa c - A^T y) &= \mu e_n \\
\kappa\,(b^T y - c^T x + \vartheta) &= \mu \\
\vartheta\,(2 - \kappa) &= \mu.
\end{aligned}
\tag{4.5}
$$

An immediate consequence is that all the nonnegative entities mentioned above are positive. Surprisingly enough, we can compute the value of κ from (4.4). Taking the inner product of both sides in the first equation with z, while using the orthogonality property, we get $q^T z = z^T s$. The second equation in (4.4) gives $z^T s = (n + m + 2)\mu$. Due to the definition of q we obtain[2]

$$
2\vartheta = (n + m + 2)\mu.
\tag{4.6}
$$

In fact, this relation expresses that the objective value $q^T z = 2\vartheta$ along the central path equals the dimension of the matrix M times μ, already established in Section 2.7. Substitution of (4.6) in the last equation of (4.5) yields

$$
(n + m + 2)\vartheta\,(2 - \kappa) = 2\vartheta.
$$

Since $\vartheta > 0$, after dividing by ϑ it easily follows that

$$
\kappa = \frac{2(n + m + 1)}{n + m + 2}.
\tag{4.7}
$$

Substitution of the values of κ and ϑ in the third equation gives

$$
\frac{c^T x - b^T y}{\kappa} = \frac{\vartheta}{\kappa} - \frac{\mu}{\kappa^2} = \frac{\vartheta \kappa - \mu}{\kappa^2} = \frac{(n + m)\,\mu}{\kappa^2} = \frac{(n + m)\,(n + m + 2)^2}{4\,(n + m + 1)^2}\,\mu.
$$

---

[2] The relation can also be obtained by adding all the equations in (4.5).

Now, defining

$$\bar{x} = \frac{x}{\kappa}, \quad \bar{y} = \frac{y}{\kappa}, \quad \bar{\vartheta} = \frac{\vartheta}{\kappa}, \quad \bar{\mu} = \frac{\mu}{\kappa^2},$$

and using the notation

$$
\begin{aligned}
s(\bar{x}) &:= A\bar{x} - b \\
s(\bar{y}) &:= c - A^T \bar{y},
\end{aligned}
$$

we obtain that the positive vectors $\bar{x}$ and $\bar{y}$ are feasible for (P) and (D) respectively with $s(\bar{x})$ and $s(\bar{y})$ positive, and moreover,

$$
\begin{aligned}
\bar{y}\, s(\bar{x}) &= \bar{\mu}\, e_m \\
\bar{x}\, s(\bar{y}) &= \bar{\mu}\, e_n.
\end{aligned}
\tag{4.8}
$$

If μ runs through the interval $(0, \infty)$ then $\bar{\mu}$ runs through the same interval, since κ is constant. We conclude that for every positive $\bar{\mu}$ there exist positive vectors $\bar{x}$ and $\bar{y}$ that are feasible for (P) and (D) respectively and are such that $\bar{x}$, $\bar{y}$ and their associated surplus vectors $s(\bar{x})$ and $s(\bar{y})$ satisfy (4.8).

Our next aim is to show that the system (4.8) cannot have more than one solution with $\bar{x}$ and $\bar{y}$ feasible for (P) and (D). Suppose that $\bar{x}$ and $\bar{y}$ are feasible for (P) and (D) and satisfy (4.8). Then it is quite easy to derive a solution for (4.5) as follows. First we calculate κ from (4.7). Then taking $\mu = \kappa^2 \bar{\mu}$, we can find ϑ from (4.6). Finally, the values $x = \kappa \bar{x}$ and $y = \kappa \bar{y}$ satisfy (4.5). Since the solution of (4.5) is unique, it follows that the solution of (4.8) is unique as well. Thus we have shown that for each positive $\bar{\mu}$ the system (4.8) has a unique solution with $\bar{x}$ and $\bar{y}$ feasible for (P) and (D).

Denoting the solution of (4.8) by $\bar{x}(\bar{\mu})$ and $\bar{y}(\bar{\mu})$, we obtain the central paths of (P) and (D) by letting $\bar{\mu}$ run through all positive values. Summarizing the above results, we have proved the following.

Theorem I.55 *Let $(x(\mu), y(\mu), \kappa(\mu), \vartheta(\mu))$ denote the point on the central path of (SP_1) corresponding to the barrier parameter value μ. Then we have $\kappa(\mu) = \kappa$ with*

$$\kappa = \frac{2(n + m + 1)}{n + m + 2}.$$

If $\bar{\mu} = \mu/\kappa^2$, then $\bar{x}(\bar{\mu}) = x(\mu)/\kappa$ and $\bar{y}(\bar{\mu}) = y(\mu)/\kappa$ are the points on the central paths of (P) and (D) corresponding to the barrier parameter $\bar{\mu}$. As a consequence we have

$$c^T \bar{x} - b^T \bar{y} = \bar{x}^T s(\bar{y}) + \bar{y}^T s(\bar{x}) = (n + m)\bar{\mu}.$$

4.2.3 Approximate solutions of (P) and (D)

Our aim is to solve the given problem (P) by solving the embedding problem (SP_1). The Full-Newton step algorithm yields an ε-solution, i.e. a feasible solution z of (SP_1) such that $q^T z \leq \varepsilon$, where ε is some positive number. Therefore, it is of great importance to see how we can derive approximate solutions for (P) and (D) from any such solution of (SP_1). In this respect the following lemma is of interest.

Lemma I.56 *Let $z = (y, x, \kappa, \vartheta)$ be a positive solution of (SP_1). If*

$$\tilde{x} = \frac{x}{\kappa}, \quad \tilde{y} = \frac{y}{\kappa},$$

then $\tilde{x}$ is feasible for (P), $\tilde{y}$ is feasible for (D), and the duality gap at the pair $(\tilde{x}, \tilde{y})$ satisfies

$$c^T \tilde{x} - b^T \tilde{y} \leq \frac{\vartheta}{\kappa}.$$

Proof: Since z is feasible for (SP_1), we have

$$
\begin{aligned}
Ax - \kappa b &\geq 0 \\
-A^T y + \kappa c &\geq 0 \\
b^T y - c^T x + \vartheta &\geq 0 \\
-\kappa + 2 &\geq 0.
\end{aligned}
$$

With $\tilde{x}$ and $\tilde{y}$ as defined in the theorem it follows that $A\tilde{x} \geq b$, $A^T \tilde{y} \leq c$ and

$$c^T \tilde{x} - b^T \tilde{y} \leq \frac{\vartheta}{\kappa},$$

thus proving the lemma. □

The above lemma makes clear that it is important for our goal to have a solution $z = (y, x, \kappa, \vartheta)$ of (SP_1) for which the quotient ϑ/κ is small. From (4.7) in Section 4.2.2 we know that along the central path the variable κ is constant and given by

$$\kappa = \frac{2(n + m + 1)}{n + m + 2}.$$

Hence, along the central path we have the following inequality:

$$c^T \tilde{x} - b^T \tilde{y} \leq \frac{(n + m + 2)\,\vartheta}{2(n + m + 1)}.$$

For large-scale problems, where $n + m$ is large, this means that the duality gap at the feasible pair $(\tilde{x}, \tilde{y})$ is about $\vartheta/2$.

Unfortunately our algorithm for solving (SP_1) generates a feasible solution z that is not necessarily on the central path. Hence the above estimate for the duality gap at $(\tilde{x}, \tilde{y})$ is no longer valid. However, we show now that the estimate is 'almost' valid because the solution z generated by the algorithm is close to the central path. To be more precise, according to Lemma I.44 z satisfies $\delta_c(z) \leq \tau$, where $\tau = 4$, and where the proximity measure $\delta_c(z)$ is defined by

$$\delta_c(z) = \frac{\max\,(zs(z))}{\min\,(zs(z))}.$$

Recall that $\delta_c(z) = 1$ if and only if $zs(z)$ is a multiple of the all-one vector e. This occurs precisely if z lies on the central path. Otherwise we have $\delta_c(z) > 1$. Now we can prove the following generalization of Lemma I.56.

Lemma I.57 *Let* $\tau \geq 1$ *and let* $z = (y, x, \kappa, \vartheta)$ *be a feasible solution of* (SP_1) *such that* $\delta_c(z) \leq \tau$. *If*

$$\tilde{x} = \frac{x}{\kappa}, \quad \tilde{y} = \frac{y}{\kappa},$$

then $\tilde{x}$ *is feasible for* (P), $\tilde{y}$ *is feasible for* (D), *and the duality gap at the pair* $(\tilde{x}, \tilde{y})$ *satisfies*

$$c^T \tilde{x} - b^T \tilde{y} < \frac{n+m+2}{2(n+m+2-\tau)} \vartheta.$$

Proof: Recall from (2.23) that $q^T z = z^T s(z)$. Since $q^T z = 2\vartheta$, the average value of the products $z_i s_i(z)$ is equal to

$$\frac{2\vartheta}{n+m+2}.$$

From $\delta_c(z) \leq \tau$ we deduce the following bounds:[3,4]

$$\frac{2\vartheta}{\tau(n+m+2)} \leq z_i s_i(z) \leq \frac{2\tau\vartheta}{n+m+2}, \quad 1 \leq i \leq m+n+2. \qquad (4.9)$$

The lemma is obtained by applying these inequalities to the last two coordinates of z, which are κ and ϑ. Application of (4.9) to $z_i = \vartheta$ yields the inequalities

$$\frac{2\vartheta}{\tau(n+m+2)} \leq \vartheta(2-\kappa) \leq \frac{2\tau\vartheta}{n+m+2}.$$

After division by ϑ and some elementary reductions, this gives the following bounds on κ:

$$\frac{2(n+m+2-\tau)}{n+m+2} \leq \kappa \leq \frac{2(\tau(n+m+2)-1)}{\tau(n+m+2)}. \qquad (4.10)$$

Application of the left-hand side inequality in (4.9) to $z_i = \kappa$ leads to

$$\kappa(b^T y - c^T x + \vartheta) \geq \frac{2\vartheta}{\tau(n+m+2)}.$$

Using the upper bound for κ in (4.10) we obtain

$$b^T y - c^T x + \vartheta \geq \frac{2\vartheta}{\tau(n+m+2)} \frac{\tau(n+m+2)}{2(\tau(n+m+2)-1)} = \frac{\vartheta}{\tau(n+m+2)-1}.$$

Hence,

$$c^T x - b^T y \leq \vartheta - \frac{\vartheta}{\tau(n+m+2)-1} = \frac{\tau(n+m+2)-2}{\tau(n+m+2)-1}\vartheta < \vartheta.$$

Finally, dividing both sides of this inequality by κ, and using the lower bound for κ in (4.10), we obtain

$$c^T \tilde{x} - b^T \tilde{y} = \frac{c^T x - b^T y}{\kappa} < \frac{n+m+2}{2(n+m+2-\tau)}\vartheta.$$

---

[3] These bounds are sufficient for our purpose. Sharper bounds could be obtained from the next exercise.

[4] **Exercise 28** Let $x \in \mathbb{R}^n_+$ and $\tau \geq 1$. Prove that if $e^T x = n\sigma$ and $\tau \min(x) \geq \max(x)$ then

$$\frac{\sigma}{\tau} \leq \frac{n\sigma}{1+(n-1)\tau} \leq x_i \leq \frac{\tau n\sigma}{n+\tau-1} \leq \tau\sigma, \quad 1 \leq i \leq n.$$

This proves the lemma.[5]   □

For large-scale problems the above lemma implies that the duality gap at the feasible pair $(\tilde{x}, \tilde{y})$ is about $\vartheta/2$, provided that τ is small compared with $n + m$.

4.3 The general case

4.3.1 Introduction

This time we assume that there is no foreknowledge about (P) and (D). It may well be that one of the problems is infeasible, or both. This raises the question of whether the given problems have any solution at all. This question must be answered by the solution method. In fact, the method that we presented in Chapter 3 perfectly answers the question. In the next section, we present an alternative self-dual embedding. The new embedding problem can be solved in exactly the same way as the embedding problem (SP) in Chapter 3, and by using the rounding procedure described there, we can find a strictly complementary solution. Then the answer to the above question is given by the value of the homogenizing variable κ. If this variable is positive then both (P) and (D) have optimal solutions; if it is zero then at least one of the two problems is infeasible. Our aim is to develop some tools that may be helpful in deciding if κ is positive or not without using the rounding procedure.

4.3.2 Alternative embedding for the general case

Let x^0 and y^0 be arbitrary positive vectors of dimension n and m respectively. Defining positive vectors s^0 and t^0 by the relations

$$x^0 s^0 = e_n, \quad y^0 t^0 = e_m,$$

we consider the self-dual problem

$$(SP_2) \qquad \min \left\{ q^T z \ : \ Mz + q \geq 0, \ z \geq 0 \right\},$$

where M and q are given by

$$M := \begin{bmatrix} 0_{mm} & A & -b & \bar{b} \\ -A^T & 0_{nn} & c & \bar{c} \\ b^T & -c^T & 0 & \beta \\ -\bar{b}^T & -\bar{c}^T & -\beta & 0 \end{bmatrix}, \quad q := \begin{bmatrix} 0_m \\ 0_n \\ 0 \\ n+m+2 \end{bmatrix},$$

with

$$\bar{b} \quad = \quad t^0 + b - Ax^0$$

---

[5] **Exercise 29** Using the sharper bounds for $z_i s_i(z)$ obtainable from Exercise 28, and using the notation of Lemma I.57, derive the following bound for the duality gap:

$$c^T \tilde{x} - b^T \tilde{y} \leq \frac{(n+m+1+\tau)\left((n+m+1)\tau - 1\right)}{2\tau (n+m+1)^2} \vartheta.$$

$$\bar{c} = s^0 - c + A^T y^0,$$
$$\beta = 1 - b^T y^0 + c^T x^0.$$

Taking

$$z^0 := \begin{bmatrix} y^0 \\ x^0 \\ 1 \\ 1 \end{bmatrix}$$

we then have

$$Mz^0 + q = \begin{bmatrix} Ax^0 - b + \bar{b} \\ -A^T y^0 + c + \bar{c} \\ b^T y^0 - c^T x^0 + \beta \\ -\bar{b}^T y^0 - \bar{c}^T x^0 - \beta \end{bmatrix} + \begin{bmatrix} 0_m \\ 0_n \\ 0 \\ n+m+2 \end{bmatrix} = \begin{bmatrix} t^0 \\ s^0 \\ 1 \\ 1 \end{bmatrix}.$$

Except for the last entry in the last vector this is obvious. For this entry we write

$$
\begin{aligned}
-\bar{b}^T y^0 - \bar{c}^T x^0 - \beta &= -\left(t^0 + b - Ax^0\right)^T y^0 - \left(s^0 - c + A^T y^0\right)^T x^0 - \beta \\
&= -\left(t^0\right)^T y^0 - b^T y^0 + \left(x^0\right)^T A^T y^0 \\
&\quad - \left(s^0\right)^T x^0 + c^T x^0 - \left(y^0\right)^T Ax^0 - \beta \\
&= -m - b^T y^0 - n + c^T x^0 - \beta = -m - n - 1,
\end{aligned}
$$

whence

$$-\bar{b}^T y^0 - \bar{c}^T x^0 - \beta + n + m + 2 = 1.$$

We conclude that z^0 is a positive solution of (SP_2) with a positive surplus vector. Moreover, since $x^0 s^0 = e_n$ and $y^0 t^0 = e_m$, this solution lies on the central path of (SP_2) and the corresponding barrier parameter value is 1. It remains to show that if a strictly complementary solution of (SP_2) is available then we can solve problems (P) and (D). Therefore, let

$$\begin{bmatrix} \bar{y} \\ \bar{x} \\ \bar{\kappa} \\ \bar{\vartheta} \end{bmatrix}$$

be a strictly complementary solution. Then, since the optimal value of (SP_2) is zero, we have $\bar{\vartheta} = 0$. As a consequence, the vector

$$\bar{z} := \begin{bmatrix} \bar{y} \\ \bar{x} \\ \bar{\kappa} \end{bmatrix}$$

is a strictly complementary solution of

$$\min \left\{ \begin{bmatrix} 0_m \\ 0_n \\ 0 \end{bmatrix}^T \begin{bmatrix} y \\ x \\ \kappa \end{bmatrix} : \begin{bmatrix} 0_{mm} & A & -b \\ -A^T & 0_{nn} & c \\ b^T & -c^T & 0 \end{bmatrix} \begin{bmatrix} y \\ x \\ \kappa \end{bmatrix} \geq \begin{bmatrix} 0_m \\ 0_n \\ 0 \end{bmatrix}, \begin{bmatrix} y \\ x \\ \kappa \end{bmatrix} \geq 0 \right\}.$$

This is the problem (SP_0), that we introduced in Chapter 2. We can duplicate the arguments used there to conclude that if $\bar{\kappa}$ is positive then the pair $(\bar{x}/\bar{\kappa}, \bar{y}/\bar{\kappa})$ provides strictly complementary optimal solutions of (P) and (D), and if $\bar{\kappa}$ is zero then one of the two problems is feasible and the other is unbounded, or both problems are infeasible.

Thus (SP_2) provides a self-dual embedding for (P) and (D). Moreover, z^0 provides a suitable starting point for the Full-Newton step algorithm. It is the point on the central path of (SP_2) corresponding to the barrier parameter value 1.

4.3.3 The central path of (SP_2)

In this section we point out some properties of the central path of the problem (SP_2). Let μ be an arbitrary positive number. Then the μ-center of (SP_2) is determined as the unique solution of the system (cf. (2.46), page 35)

$$
\begin{aligned}
Mz + q &= s, \quad z \geq 0,\, s \geq 0 \\
zs &= \mu\, e_{m+n+2}.
\end{aligned}
\tag{4.11}
$$

This solution defines the point on the central path of (SP_2) corresponding to the barrier parameter value μ. Hence there exists unique positive x, y, κ, ϑ such that

$$
\begin{aligned}
Ax - \kappa b + \vartheta \bar{b} &> 0 \\
\kappa c - A^T y + \vartheta \bar{c} &> 0 \\
s(\kappa) := b^T y - c^T x + \vartheta \beta &> 0 \\
s(\vartheta) := n + m + 2 - \bar{b}^T x - \bar{c}^T y - \kappa \beta &> 0
\end{aligned}
\tag{4.12}
$$

and, moreover,

$$
\begin{aligned}
y\left(Ax - \kappa b + \vartheta \bar{b}\right) &= \mu e_m \\
x\left(\kappa c - A^T y + \vartheta \bar{c}\right) &= \mu e_n \\
\kappa\left(b^T y - c^T x + \vartheta \beta\right) &= \mu \\
\vartheta\left(n + m + 2 - \bar{b}^T y - \bar{c}^T x - \kappa \beta\right) &= \mu.
\end{aligned}
\tag{4.13}
$$

Just as in Section 4.2.2 we take the inner product of both sides with z in the first equation of (4.11). Using the orthogonality property, we obtain $q^T z = z^T s$. The second equation in (4.11) gives $z^T s = (n + m + 2)\mu$. Due to the definition of q we obtain

$$
(n + m + 2)\vartheta = (n + m + 2)\mu,
$$

which gives $\vartheta = \mu$. Since $\vartheta s(\vartheta) = \mu$, by the fourth equation in (4.13), we conclude that $s(\vartheta) = 1$. Since

$$
s(\vartheta) = n + m + 2 - \bar{b}^T y - \bar{c}^T x - \kappa \beta
$$

this leads to

$$
\bar{b}^T y + \bar{c}^T x + \kappa \beta = n + m + 1.
\tag{4.14}
$$

Using $\vartheta = \mu$, the third equality in (4.13) can be rewritten as

$$\kappa \left(b^T y - c^T x \right) = \mu - \mu \kappa \beta,$$

which gives

$$\kappa \beta = 1 + \frac{\kappa}{\mu} \left(c^T x - b^T y \right).$$

Substituting this in (4.14) we get

$$\bar{b}^T y + \bar{c}^T x + \frac{\kappa}{\mu} \left(c^T x - b^T y \right) = n + m,$$

which is equivalent to

$$(\kappa c + \mu \bar{c})^T x - (\kappa b - \mu \bar{b})^T y = \mu (n + m). \tag{4.15}$$

This relation admits a nice interpretation. The first two inequality in (4.12) show that x is feasible for the perturbed problem

$$\min \left\{ (\kappa c + \mu \bar{c})^T x \ : \ Ax \geq \kappa b - \mu \bar{b}, \ x \geq 0 \right\},$$

and y is feasible for the dual problem

$$\max \left\{ (\kappa b - \mu \bar{b})^T y \ : \ A^T y \leq \kappa c + \mu \bar{c}, \ y \geq 0 \right\}.$$

For these perturbed problems the duality gap at the pair (x, y) is $\mu(n + m)$, from (4.15). Now consider the behavior along the central path when μ approaches zero. Two cases can occur: either κ converges to some positive value, or κ goes to zero. In both cases the duality gap converges to zero. Roughly speaking, the limiting values of x and y are optimal solutions for the perturbed problems. In the first case, when κ converges to some positive value, asymptotically the first perturbed problem becomes equivalent to (P). We simply have to replace the variable x by κx. Also, the second problem becomes equivalent to (D): replace the variable y by κy. In the second case however, asymptotically the perturbed problems become

$$\min \left\{ 0^T x \ : \ Ax \geq 0, \ x \geq 0 \right\},$$

and

$$\max \left\{ 0^T y \ : \ A^T y \leq 0, \ y \geq 0 \right\}.$$

As we know, one of the problems (P) and (D) is then infeasible and the other unbounded, or both problems are infeasible.

When dealing with a solution method for the canonical problem, the method must decide which of these two cases occurs. In this respect we make an interesting observation. Clearly the first case occurs if and only if $\kappa \in B$ and the second case if and only if $\kappa \in N$, where (B, N) is the optimal partition of (SP_2). In other words, which of the two cases occurs depends on whether κ is a large variable or a small variable. Note that the variable ϑ is always small. In the present case we have $\vartheta(\mu) = \mu$, for each $\mu > 0$. Recall from Lemma I.43 that the large variables are bounded below by

σ_{SP}/n and the small variables above by $n\mu/\sigma_{SP}$. Hence, if κ is a large variable then $\kappa \geq \sigma_{SP}/n$ implies

$$\frac{\vartheta}{\kappa} = \frac{\mu}{\kappa} \leq \frac{n\mu}{\sigma_{SP}}.$$

This implies that the quotient ϑ/κ goes to zero if μ goes to zero. On the other hand, if κ is a small variable then

$$\frac{\kappa}{\vartheta} \leq \frac{n\mu}{\vartheta\sigma_{SP}} = \frac{n}{\sigma_{SP}},$$

proving that the quotient κ/ϑ is bounded above. Therefore, if μ goes to zero, κ^2/ϑ goes to zero as well, and hence ϑ/κ^2 goes to infinity. Thus we may state the following without further proof.

Theorem I.58 *If κ is a large variable then*

$$\lim_{\mu \downarrow 0} \frac{\vartheta}{\kappa} = \lim_{\mu \downarrow 0} \frac{\vartheta}{\kappa^2} = 0,$$

and if κ is a small variable then

$$\lim_{\mu \downarrow 0} \frac{\vartheta}{\kappa^2} = \infty.$$

The above theorem provides another theoretical tool for distinguishing between the two possible cases.

4.3.4 Approximate solutions of (P) and (D)

Assuming that an ε-solution $z = (y, x, \kappa, \vartheta)$ for the embedding problem (SP_2) is given, we proceed by investigating what information this gives on the embedded problem (P) and its dual (D). With

$$\tilde{x} := \frac{x}{\kappa}, \quad \tilde{y} := \frac{y}{\kappa},$$

the feasibility of z for (SP_2) implies the following inequalities:

$$
\begin{aligned}
A\tilde{x} &\geq & b - \frac{\vartheta}{\kappa}\bar{b} \\
A^T\tilde{y} &\leq & c + \frac{\vartheta}{\kappa}\bar{c} \\
c^T\tilde{x} - b^T\tilde{y} &\leq & \frac{\vartheta}{\kappa}\beta \\
\kappa\left(\bar{b}^T\tilde{x} + \bar{c}^T\tilde{y} + \beta\right) &\leq & n + m + 2.
\end{aligned}
\tag{4.16}
$$

Clearly we cannot conclude that $\tilde{x}$ is feasible for (P) or that $\tilde{y}$ is feasible for (D). But $\tilde{x}$ is feasible for the perturbed problem

$$(P') \quad \min\left\{ \left(c + \frac{\vartheta}{\kappa}\bar{c}\right)^T \tilde{x} \ : \ A\tilde{x} \geq b - \frac{\vartheta}{\kappa}\bar{b}, \ x \geq 0 \right\},$$

and $\tilde{y}$ is feasible for its dual problem

$$(D') \quad \max\left\{ \left(b - \frac{\vartheta}{\kappa}\bar{b}\right)^T \tilde{y} \ : \ A^T\tilde{y} \leq c + \frac{\vartheta}{\kappa}\bar{c}, \ y \geq 0 \right\}.$$

We have the following lemma.

Lemma I.59 *Let $z = (y, x, \kappa, \vartheta)$ be a feasible solution of (SP_2) with $\kappa > 0$. If*

$$\tilde{x} = \frac{x}{\kappa}, \quad \tilde{y} = \frac{y}{\kappa},$$

then $\tilde{x}$ is feasible for (P'), $\tilde{y}$ is feasible for (D'), and the duality gap at the pair $(\tilde{x}, \tilde{y})$ for this pair of perturbed problems satisfies

$$\left(c + \frac{\vartheta}{\kappa}\bar{c}\right)^T \tilde{x} - \left(b - \frac{\vartheta}{\kappa}\bar{b}\right)^T \tilde{y} \leq \frac{(n + m + 2)\vartheta}{\kappa^2}.$$

Proof: We have already established that $\tilde{x}$ is feasible for (P') and $\tilde{y}$ is feasible for (D'). We rewrite the duality gap for the perturbed problems (P') and (D') at the pair $(\tilde{x}, \tilde{y})$ as follows:

$$\left(c + \frac{\vartheta}{\kappa}\bar{c}\right)^T \tilde{x} - \left(b - \frac{\vartheta}{\kappa}\bar{b}\right)^T \tilde{y} = c^T\tilde{x} - b^T\tilde{y} + \frac{\vartheta}{\kappa}\left(\bar{c}^T\tilde{x} + \bar{b}^T\tilde{y}\right).$$

The third inequality in (4.16) gives

$$c^T\tilde{x} - b^T\tilde{y} \leq \frac{\vartheta}{\kappa}\beta$$

and the fourth inequality

$$\bar{c}^T\tilde{x} + \bar{b}^T\tilde{y} \leq \frac{n + m + 2}{\kappa} - \beta.$$

Substitution gives

$$\left(c + \frac{\vartheta}{\kappa}\bar{c}\right)^T \tilde{x} - \left(b - \frac{\vartheta}{\kappa}\bar{b}\right)^T \tilde{y} \leq \frac{\vartheta}{\kappa}\beta + \frac{\vartheta}{\kappa}\left(\frac{n + m + 2}{\kappa} - \beta\right) = \frac{(n + m + 2)\vartheta}{\kappa^2},$$

proving the lemma. □

The above lemma seems to be of interest only if κ is a large variable. For if ϑ/κ and ϑ/κ^2 are small enough then the lemma provides a pair of vectors $(\tilde{x}, \tilde{y})$ such that $\tilde{x}$ and $\tilde{y}$ are 'almost' feasible for (P) and (D) respectively and the duality gap at this pair is small. The error in feasibility for (P) is given by the vector $(\vartheta/\kappa)\bar{b}$ and the error in feasibility for (D) by the vector $(\vartheta/\kappa)\bar{c}$, whereas the duality gap with respect to (P) and (D) equals

$$\frac{\vartheta}{\kappa}\left(\bar{c}^T\tilde{x} + \bar{b}^T\tilde{y}\right).$$

Part II

The Logarithmic Barrier Approach

5

Preliminaries

5.1 Introduction

In the previous chapters we showed that every LO problem can be solved in polynomial time. This was achieved by transforming the given problem to its canonical form and then embedding it into a self-dual model. We proved that the self-dual model can be solved in polynomial time. Our proof was based on the algorithm in Chapter 3 that uses the Newton direction as search direction. As we have seen, this algorithm is conceptually simple and allows a quite elementary analysis. For the theoretical purpose of Part I of the book this algorithm therefore is an ideal choice.

From the practical point of view, however, there exist more efficient algorithms. The aim of this part of the book is to deal with a class of algorithms that has a relatively long history, going back to work of Frisch [88] in 1955. Frisch was the first to propose the use of logarithmic barrier functions in LO. The idea was worked out by Lootsma [185] and in the classical book of Fiacco and McCormick [77]. After 1984, the year when Karmarkar's paper [165] raised new interest in the interior-point approach to LO, the so-called *logarithmic barrier approach* also began a new life. It became the basis of a wide class of polynomial time algorithms. Variants of the most efficient algorithms in this class found their way into commercial optimization packages like CPLEX and OSL.[1]

The aim of this part of the book is to provide a thorough introduction to these algorithms. In the literature of the last decade these interior-point algorithms were developed for LO problems in the so-called *standard format*:

$$(P) \qquad \min \left\{ c^T x \ : \ Ax = b, \ x \geq 0 \right\},$$

where A is an $m \times n$ matrix of rank m, $c, x \in \mathbb{R}^n$, and $b \in \mathbb{R}^m$. This format also served as the standard for the literature on the Simplex Method. Because of its historical status, we adopt the standard format for this part of the book.

We want to point out, however, that all results in this part can easily be adapted to any other format, including the self-dual model of Part I. We only have to define a suitable logarithmic barrier function for the format under consideration.

A disadvantage of the change from the self-dual to the standard format is that it leads to some repetition of results. For example, we need to establish under what conditions the problem (P) in standard format has a central path, and so on. In fact,

---

[1] CPLEX is a product of CPLEX Optimization, Inc. OSL stands for Optimization Subroutine Library and is the optimization package of IBM.

we could have derived all these results from the results in Chapter 2. But, instead, to make this part of the book more accessible for readers who are better acquainted with the standard format rather than the less known self-dual format, we decided to make this part self-contained.

Readers who went through Part I may only be interested in methods for solving the self-dual problem

$$(SP) \qquad \min \left\{ q^T x \ : \ Mx \geq -q, \, x \geq 0 \right\},$$

with $q \geq 0$ and $M^T = -M$. Those readers may be advised to skip the rest of this chapter and continue with Chapters 6 and 7. The relevance of these chapters for solving (SP) is due to the fact that (SP) can easily be brought into the standard format by introducing a surplus vector s to create equality constraints. Since x and s are nonnegative, this yields (SP) in the standard format:

$$(SPS) \qquad \min \left\{ q^T x \ : \ Mx - s = -q, \, x \geq 0, \, s \geq 0 \right\}.$$

In this part of the book we take the classical duality results for the standard format of the LO problem as granted. We briefly review these results in the next section.

5.2 Duality results for the standard LO problem

The standard format problem (P) has the following dual problem:

$$(D) \qquad \max \left\{ b^T y \ : \ A^T y + s = c, \, s \geq 0 \right\},$$

where $s \in \mathbb{R}^n$ and $y \in \mathbb{R}^m$. We call (D) the standard dual problem. The feasible regions of (P) and (D) are denoted by $\mathcal{P}$ and $\mathcal{D}$, respectively:

$$\mathcal{P} := \left\{ x \ : \ Ax = b, \, x \geq 0 \right\},$$

$$\mathcal{D} := \left\{ (y, s) \ : \ A^T y + s = c, \, s \geq 0 \right\}.$$

If $\mathcal{P}$ is empty we call (P) infeasible, otherwise feasible. If (P) is feasible and the objective value $c^T x$ is unbounded below on $\mathcal{P}$, then (P) is called unbounded, otherwise bounded. We use similar terminology for the dual problem (D).

Since we assumed that A has full (row) rank m, we have a one-to-one correspondence between y and s in the pairs $(y, s) \in \mathcal{D}$. In order to facilitate the discussion we feel free to refer to any pair $(y, s) \in \mathcal{D}$ either by $y \in \mathcal{D}$ or $s \in \mathcal{D}$. The (relative) interiors of $\mathcal{P}$ and $\mathcal{D}$ are denoted by $\mathcal{P}^+$ and $\mathcal{D}^+$:

$$\mathcal{P}^+ := \left\{ x \ : \ Ax = b, \, x > 0 \right\},$$

$$\mathcal{D}^+ := \left\{ (y, s) \ : \ A^T y + s = c, \, s > 0 \right\}.$$

We recall the well known and almost trivial weak duality result for the LO problem in standard format.

Proposition II.1 (Weak duality) *Let x and s be feasible for (P) and (D), respectively. Then $c^T x - b^T y = x^T s \geq 0$. Consequently, $c^T x$ is an upper bound for the optimal value of (D), if it exists, and $b^T y$ is a lower bound for the optimal value of (P), if it exists. Moreover, if the duality gap $x^T s$ is zero then x is an optimal solution of (P) and (y, s) is an optimal solution of (D).*

Proof: The proof is straightforward. We have

$$0 \leq x^T s = x^T (c - A^T y) = c^T x - (Ax)^T y = c^T x - b^T y. \qquad (5.1)$$

This implies that $c^T x$ is an upper bound for the optimal objective value of (D), and $b^T y$ is a lower bound for the optimal objective value of (P), and, moreover, if the duality gap is zero then the pair (x, s) is optimal.   □

A direct consequence of Proposition II.1 is that if one of the problems (P) and (D) is unbounded, then the other problem is infeasible. The classical duality results for the primal and dual problems in standard format boil down to the following two results. The first result is the Duality Theorem (due to von Neumann, 1947, [227]), and the second result will be referred to as the Goldman–Tucker Theorem (Goldman and Tucker, 1956, [111]).

Theorem II.2 (Duality Theorem) *If (P) and (D) are feasible then both problems have optimal solutions. Then, if $x \in \mathcal{P}$ and $(y, s) \in \mathcal{D}$, these are optimal solutions if and only if $x^T s = 0$. Otherwise neither of the two problems has optimal solutions: either both (P) and (D) are infeasible or one of the two problems is infeasible and the other one is unbounded.*

Theorem II.3 (Goldman–Tucker Theorem) *If (P) and (D) are feasible then there exists a strictly complementary pair of optimal solutions, that is an optimal solution pair (x, s) satisfying $x + s > 0$.*

It may be noted that these two classical results follow immediately from the results in Part I.[2] For future use we also mention that (P) is infeasible if and only if there exists a vector y such that $A^T y \leq 0$ and $b^T y > 0$, and (D) is infeasible if and only if there exists a vector $x \geq 0$ such that $Ax = 0$ and $c^T x < 0$. These statements are examples of *theorems of the alternatives* and easily follow from Farkas' lemma.[3]

We denote the set of all optimal solutions of (P) by $\mathcal{P}^*$ and similarly $\mathcal{D}^*$ denotes the set of optimal solutions of (D). Of course, $\mathcal{P}^*$ is empty if and only if (P) is infeasible or unbounded, and $\mathcal{D}^*$ is empty if and only if (D) is infeasible or unbounded. Note that the Duality Theorem (II.2) implies that $\mathcal{P}^*$ is empty if and only if $\mathcal{D}^*$ is empty.

---

[2] **Exercise 30** Derive Theorem II.2 and Theorem II.3 from Theorem I.26.

[3] **Exercise 31** Using Farkas' lemma (cf. Remark I.27), prove:
   (i) either the system $Ax = b$, $x \geq 0$ or the system $A^T y \leq 0$, $b^T y > 0$ has a solution;
   (ii) either the system $A^T y \leq c$ or the system $Ax = 0$, $x \geq 0$, $c^T x < 0$ has a solution.

5.3 The primal logarithmic barrier function

We start by introducing the so-called *logarithmic barrier function* for the primal problem (P). This is the function $\tilde{g}_\mu(x)$ defined by

$$\tilde{g}_\mu(x) := c^T x - \mu \sum_{j=1}^n \log x_j, \tag{5.2}$$

where μ is a positive number called the *barrier parameter*, and x runs through all primal feasible vectors that are positive. The domain of $\tilde{g}_\mu$ is the set $\mathcal{P}^+$.

The use of logarithmic barrier functions in LO was first proposed by Frisch [88] in 1955. By minimizing $\tilde{g}_\mu(x)$, we try to realize two goals at the same time, namely to find a primal feasible vector x for which $c^T x$ is small and such that the *barrier term* $\sum_{j=1}^n \log x_j$ is large. Frisch observed that the minimization of $\tilde{g}_\mu(x)$ can be done easily by using standard techniques from nonlinear optimization. The barrier parameter can be used to put more emphasis on either the objective value $c^T x$ of the primal LO problem (P), or on the barrier term. Intuitively, by letting μ take a small (positive) value, we may expect that a minimizer of $\tilde{g}_\mu(x)$ will be a good approximation for an optimal solution of (P). It has taken approximately 40 years to make clear that this is a brilliant idea, not only from a practical but also from a theoretical point of view. In this part of the book we deal with logarithmic barrier methods for solving both the primal problem (P) and the dual problem (D), and we show that when worked out in an appropriate way, the resulting methods solve both (P) and (D) in polynomial time.

5.4 Existence of a minimizer

In the logarithmic barrier approach a major question is whether the barrier function has a minimizing point or not. This section is devoted to this question, and we present some necessary and sufficient conditions. One of these (mutually equivalent) conditions will be called the *interior-point condition*. This condition is fundamental not only for the logarithmic barrier approach, but as we shall see, for all interior-point approaches.

Note that the definition of $\tilde{g}_\mu(x)$ can be extended to the set $\mathbb{R}^n_{++}$ of all positive vectors x, and that $\tilde{g}_\mu(x)$ is differentiable on this set. We can easily verify that the gradient of $\tilde{g}_\mu$ is given by

$$\nabla \tilde{g}_\mu(x) = c - \mu x^{-1},$$

and the Hessian matrix by

$$\nabla^2 \tilde{g}_\mu(x) = \mu X^{-2}.$$

Obviously, the Hessian is positive definite for any $x \in \mathbb{R}^n_{++}$. This means that $\tilde{g}_\mu(x)$ is strictly convex on $\mathbb{R}^n_{++}$. We are interested in the behavior of $\tilde{g}_\mu$ on its domain, which is the set $\mathcal{P}^+$ of the positive vectors in the primal feasible space. Since $\mathcal{P}^+$ is the intersection of $\mathbb{R}^n_{++}$ and the affine space $\{x : Ax = b\}$, it is a relatively open subset of $\mathbb{R}^n_{++}$. Therefore, the smallest affine space containing $\mathcal{P}^+$ is the affine space

$\{x \ : \ Ax = b\}$, and the linear space parallel to it is the null space $\mathcal{N}(A)$ of A:

$$\mathcal{N}(A) = \{x \ : \ Ax = 0\}\,.$$

Taking $D = \mathbb{R}^n_{++}$ and $C = \mathcal{P}^+$, we may now apply Proposition A.1. From this we conclude that $\tilde{g}_\mu$ has a minimizer if and only if there exists an $x \in \mathcal{P}^+$ such that

$$c - \mu x^{-1} \perp \mathcal{N}(A).$$

Since the orthogonal complement of the null space of A is the row space of A, it follows that $x \in \mathcal{P}^+$ is a minimizer of $\tilde{g}_\mu$ if and only if there exists a vector $y \in \mathbb{R}^m$ such that

$$c - \mu x^{-1} = A^T y.$$

By putting $s := \mu x^{-1}$, which is equivalent to $xs = \mu e$, it follows that $\tilde{g}_\mu$ has a minimizer if and only if there exist vectors x, y and s such that

$$\begin{aligned}
Ax &= b, & x &> 0, \\
A^T y + s &= c, & s &> 0, \\
xs &= \mu e.
\end{aligned} \tag{5.3}$$

We thus have shown that this system represents the optimality conditions for the primal logarithmic barrier minimization problem, given by

$$(P_\mu) \qquad \min \left\{ \tilde{g}_\mu(x) \ : \ x \in \mathcal{P}^+ \right\}.$$

We refer to the system (5.3) as the KKT system with respect to μ.[4]
Note that the condition $x > 0$ can be relaxed to $x \geq 0$, because the third equation in (5.3) forces strict inequality. Similarly, the condition $s > 0$ can be replaced by $s \geq 0$. Thus, the first equation in (5.3) is simply the feasibility constraint for the *primal* problem (P) and the second equation is the feasibility constraint for the dual problem (D). For reasons that we shall make clear later on, the third constraint is referred to as the *centering condition* with respect to μ.

5.5 The interior-point condition

If the KKT system has a solution for some positive value of the barrier parameter μ, then the primal feasible region contains a positive vector x, and the dual feasible region contains a pair (y, s) with positive slack vector s. In short, both $\mathcal{P}$ and $\mathcal{D}$ contain a positive vector. At this stage we announce the surprising result that the converse is also true: if both $\mathcal{P}$ and $\mathcal{D}$ contain a positive vector, then the KKT system has a solution for any positive μ. This is a consequence of the following theorem.

Theorem II.4 *Let* $\mu > 0$. *Then the following statements are equivalent:*

(i) both $\mathcal{P}$ *and* $\mathcal{D}$ *contain a positive vector;*

---

[4] The reader who is familiar with the theory of nonlinear optimization will recognize in this system the first-order optimality conditions, also known as Karush–Kuhn–Tucker conditions, for (P_μ).

(ii) there exists a (unique) minimizer of $\tilde{g}_\mu$ on $\mathcal{P}^+$;
(iii) the KKT system (5.3) has a (unique) solution.

Proof: The equivalence of (ii) and (iii) has been established above. We have also observed already the implication $(iii) \Rightarrow (i)$. So the proof of the theorem will be complete if we show $(i) \Rightarrow (ii)$. The proof of this implication is more sophisticated.

Assuming (i), there exist vectors x^0 and y^0 such that x^0 is feasible for (P) and y^0 is feasible for (D), $x^0 > 0$ and $s^0 := c - A^T y^0 > 0$. Taking $K = \tilde{g}_\mu\left(x^0\right)$ and defining the level set $\mathcal{L}_K$ of $\tilde{g}_\mu$ by

$$\mathcal{L}_K := \left\{ x \in \mathcal{P}^+ \ : \ \tilde{g}_\mu(x) \le K \right\},$$

we have $x^0 \in \mathcal{L}_K$, so $\mathcal{L}_K$ is not empty. Since $\tilde{g}_\mu$ is continuous on its domain, it suffices to show that $\mathcal{L}_K$ is compact. Because then $\tilde{g}_\mu$ has a minimizer, and since $\tilde{g}_\mu$ is strictly convex this minimizer is unique. Thus to complete the proof we show below that $\mathcal{L}_K$ is compact.

Let $x \in \mathcal{L}_K$. Using Proposition II.1 we have

$$c^T x - b^T y^0 = x^T s^0,$$

so, in the definition of $\tilde{g}_\mu(x)$ we may replace $c^T x$ by $b^T y^0 + x^T s^0$:

$$\tilde{g}_\mu(x) = c^T x - \mu \sum_{j=1}^n \log x_j = b^T y^0 + x^T s^0 - \mu \sum_{j=1}^n \log x_j.$$

Since $x^T s^0 = e^T\left(x s^0\right)$ and $e^T e = n$, this can be written as

$$\tilde{g}_\mu(x) = e^T\left(x s^0 - e\right) - \mu \sum_{j=1}^n \log \frac{x_j s_j^0}{\mu} + n - n\mu \log \mu + b^T y^0 + \mu \sum_{j=1}^n \log s_j^0,$$

or, equivalently,

$$e^T\left(x s^0 - e\right) - \mu \sum_{j=1}^n \log \frac{x_j s_j^0}{\mu} = \tilde{g}_\mu(x) - n + n\mu \log \mu - b^T y^0 - \mu \sum_{j=1}^n \log s_j^0.$$

Hence, using $\tilde{g}_\mu(x) \le K$ and defining $\bar{K}$ by

$$\bar{K} := K - n + n\mu \log \mu - b^T y^0 - \mu \sum_{j=1}^n \log s_j^0,$$

we obtain

$$e^T\left(x s^0 - e\right) - \mu \sum_{j=1}^n \log \frac{x_j s_j^0}{\mu} \le \bar{K}. \tag{5.4}$$

Note that $\bar{K}$ does not depend on x.

Now let the function $\psi : (-1, \infty) \to \mathbb{R}$ be defined by

$$\psi(t) = t - \log(1 + t). \tag{5.5}$$

Then, also using $e^T e = n$, we may rewrite (5.4) as follows:

$$\mu \sum_{j=1}^{n} \psi \left(\frac{x_j s_j^0}{\mu} - 1 \right) \leq \bar{K}. \tag{5.6}$$

The rest of the proof is based on some simple properties of the function $\psi(t),$[5] namely

- $\psi(t) \geq 0$ for $t > -1$;
- ψ is strictly convex;
- $\psi(0) = 0$;
- $\lim_{t \to \infty} \psi(t) = \infty$;
- $\lim_{t \downarrow -1} \psi(t) = \infty$.

In words: $\psi(t)$ is strictly convex on its domain and minimal at $t = 0$, with $\psi(0) = 0$; moreover, $\psi(t)$ goes to infinity if t goes to one of the boundaries of the domain $(-1, \infty)$ of ψ. Figure 5.1 depicts the graph of ψ.

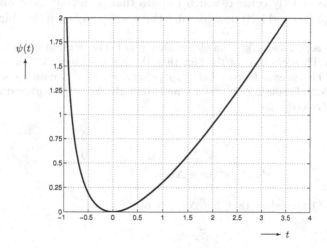

Figure 5.1 The graph of ψ.

Since ψ is nonnegative on its domain, each term in the above sum is nonnegative. Therefore,

$$\mu \psi \left(\frac{x_j s_j^0}{\mu} - 1 \right) \leq \bar{K}, \quad 1 \leq j \leq n.$$

Now using that $\psi(t)$ is strictly convex, zero at $t = 0$, and unbounded if t goes to -1 or to infinity, it follows that there must exist unique nonnegative numbers a and b,

---

[5] E. Klafszky drew our attention to the fact that this function is known in the literature. It was used in a different context for measuring discrepancy between two positive vectors in $\mathbb{R}^n$. See Csiszár [58] and Klafszky, Mayer and Terlaky [169].

with $a < 1$, such that

$$\psi(-a) = \psi(b) = \frac{\bar{K}}{\mu}.$$

We conclude that

$$-a \leq \frac{x_j s_j^0}{\mu} - 1 \leq b, \quad 1 \leq j \leq n,$$

which gives

$$\frac{\mu(1-a)}{s_j^0} \leq x_j \leq \frac{\mu(1+b)}{s_j^0}, \quad 1 \leq j \leq n.$$

Since $1 - a > 0$, this shows that each coordinate of the vector x belongs to a finite and closed interval on the set $(0, \infty)$ of positive real numbers. As a consequence, since the level set $\mathcal{L}_K$ is a closed subset of the Cartesian product of these intervals, $\mathcal{L}_K$ is compact. Thus we have shown that (ii) holds. □

The first condition in Theorem II.4 will be referred to as the *interior-point condition*. Let us point out once more that the word 'unique' in the second statement comes from the fact that $\tilde{g}_\mu$ is strictly convex, which implies that $\tilde{g}_\mu$ has at most one minimizer. The equivalence of (ii) and (iii) now justifies the word 'unique' in the third statement.

Remark II.5 It is possible to give an elementary proof (i.e., without using the equivalence of (ii) and (iii) in Theorem II.4) of the fact that the KKT system (5.3) cannot have more than one solution. This goes as follows. Let x^1, y^1, s^1 and x^2, y^2, s^2 denote two solutions of the equation system (5.3). Define $\Delta x := x^2 - x^1$, and similarly $\Delta y := y^2 - y^1$ and $\Delta s := s^2 - s^1$. Then we may easily verify that

$$
\begin{aligned}
A\Delta x &= 0 & (5.7) \\
A^T \Delta y + \Delta s &= 0 & (5.8) \\
x^1 \Delta s + s^1 \Delta x + \Delta s \Delta x &= 0. & (5.9)
\end{aligned}
$$

From (5.7) and (5.8) we deduce that $\Delta s^T \Delta x = 0$, or

$$e^T \Delta x \Delta s = 0. \tag{5.10}$$

Rewriting (5.9) gives

$$(x^1 + \Delta x)\Delta s + s^1 \Delta x = 0.$$

Since $x^1 + \Delta x = x^2 > 0$ and $s^1 > 0$, this implies that no two corresponding entries in Δx and Δs have the same sign. So it follows that

$$\Delta x \Delta s \leq 0. \tag{5.11}$$

Combining (5.10) and (5.11), we obtain $\Delta x \Delta s = 0$. Hence either $(\Delta x)_i = 0$ or $(\Delta s)_i = 0$, for each i. Using (5.9), we conclude that $(\Delta x)_i = 0$ and $(\Delta s)_i = 0$, for each i. Hence $x^1 = x^2$ and $s^1 = s^2$. Consequently, $A^T(y^1 - y^2) = 0$. Since $\text{rank}(A) = m$, the columns of A^T are linearly independent and it follows that $y^1 = y^2$. This proves the claim. •

5.6   The central path

Theorem II.4 has several important consequences. First we remark that the interior-point condition is independent of the barrier parameter. Therefore, since this condition is equivalent to the existence of a minimizer of the logarithmic barrier function $\tilde{g}_\mu$, if such a minimizer exists for some (positive) μ, then it exists for all μ. Hence, the interior point condition guarantees that the KKT system (5.3) has a unique solution for every positive value of μ. These solutions are denoted throughout as $x(\mu), y(\mu)$ and $s(\mu)$, and we call $x(\mu)$ the μ-center of (P) and $(y(\mu), s(\mu))$ the μ-center of (D). The set

$$\{x(\mu) \ : \ \mu > 0\}$$

of all primal μ-centers represents a parametric curve in the feasible region $\mathcal{P}$ of (P) and is called the *central path* of (P). Similarly, the set

$$\{(y(\mu), s(\mu)) \ : \ \mu > 0\}$$

is called the *central path* of (D).

Remark II.6 It may worthwhile to point out that along the primal central path the primal objective value $c^T x(\mu)$ is monotonically decreasing and along the dual central path the dual objective value $b^T y(\mu)$ is monotonically increasing if μ decreases. In fact, in both cases the monotonicity is strict unless the objective value is constant on the feasible region, and in the latter case the central path is just a point. Although we will not use these results we include here the proof for the primal case.[6] Recall that $x(\mu)$ is the (unique) minimizer of the primal logarithmic barrier function

$$\tilde{g}_\mu(x) = c^T x - \mu \sum_{j=1}^{n} \log x_j,$$

as given by (5.2), when x runs through the positive vectors in $\mathcal{P}$. First we deal with the case when the primal objective value is constant on $\mathcal{P}$. We have the following equivalent statements:

  (i) $c^T x$ is constant for $x \in \mathcal{P}$;
  (ii) $x(\mu)$ is constant for $\mu > 0$;
  (iii) $x(\mu_1) = x(\mu_2)$ for some μ_1 and μ_2 with $0 < \mu_1 < \mu_2$;
  (iv) there exists a $\xi \in \mathbb{R}^n$ such that $s(\mu) = \mu\xi$ for $\mu > 0$.

The proof is easy. If (i) holds then the minimizer of $\tilde{g}_\mu(x)$ is independent of μ, and hence $x(\mu)$ is constant for all $\mu > 0$, which means that (ii) holds. The implication (ii) $\Rightarrow$ (iii) is obvious. Assuming (iii), let ξ be such that $x(\mu_1) = x(\mu_2) = \xi$. Since $s(\mu_1) = \mu_1\xi^{-1}$ and $s(\mu_2) = \mu_2\xi^{-1}$ we have

$$A^T y(\mu_1) + \mu_1\xi^{-1} = c, \qquad A^T y(\mu_2) + \mu_2\xi^{-1} = c.$$

This implies

$$(\mu_2 - \mu_1)c = A^T (\mu_2 y(1) - \mu_1 y(2)),$$

---

[6]   The idea of the following proof is due to Fiacco and McCormick [77]. They deal with the more general case of a convex optimization problem and prove the monotonicity of the objective value only for the primal central path. We also refer the reader to den Hertog, Roos and Vial [146] for a different proof. The proof for the dual central path is similar to the proof for the primal central path and is left to the reader.

showing that c belongs to the row space of A. This means that (i) holds.[7] Thus we have shown the equivalence of (i) to (iii). The equivalence of (ii) and (iv) is immediate from $x(\mu)s(\mu) = \mu e$ for all $\mu > 0$.

Now consider the case where the primal objective value is not constant on $\mathcal{P}$. Letting $0 < \mu_1 < \mu_2$ and $x^1 = x(\mu_1)$ and $x^2 = x(\mu_2)$, we claim that $c^T x^1 < c^T x^2$. The above equivalence $(i) \Leftrightarrow (iii)$ makes it clear that $x^1 \neq x^2$. The rest of the proof is based on the fact that $\tilde{g}_\mu(x)$ is strictly convex. From this we deduce that $\tilde{g}_{\mu_1}(x^1) < \tilde{g}_{\mu_1}(x^2)$ and $\tilde{g}_{\mu_2}(x^2) < \tilde{g}_{\mu_2}(x^1)$. Hence

$$c^T x^1 - \mu_1 \sum_{j=1}^n \log x_j^1 < c^T x^2 - \mu_1 \sum_{j=1}^n \log x_j^2 \qquad (5.12)$$

and

$$c^T x^2 - \mu_2 \sum_{j=1}^n \log x_j^2 < c^T x^1 - \mu_2 \sum_{j=1}^n \log x_j^1. \qquad (5.13)$$

The sums in these inequalities can be eliminated by multiplying both sides of (5.12) by μ_2 and both sides of (5.13) by μ_1, and then adding the resulting inequalities. Thus we find

$$\mu_2 c^T x^1 + \mu_1 c^T x^2 < \mu_2 c^T x^2 + \mu_1 c^T x^1,$$

which is equivalent to

$$(\mu_2 - \mu_1)\left(c^T x^1 - c^T x^2\right) < 0.$$

Since $\mu_2 - \mu_1 > 0$ we obtain $c^T x^1 < c^T x^2$, proving the claim. •

It is obvious that if one of the problems (P) and (D) is infeasible, then the interior-point condition cannot be satisfied, and hence the central paths do not exist. But feasibility of both (P) and (D) is not enough for the existence of the central paths: the central paths exist if and only if both the primal and the dual feasible region contain a positive vector. In that case, when the interior-point condition is satisfied, the central path can be obtained by solving the KKT system.

Unfortunately, the KKT system is nonlinear, and hence in general it will not be possible to solve it explicitly. In order to understand better the type of nonlinearity, we show that the KKT system can be reformulated as a system of m polynomial equations of degree at most n, in the m coordinates of the vector y. This goes as follows. From the second and the third equations we derive that

$$x = \mu \left(c - A^T y\right)^{-1}.$$

Substituting this in the first equation we obtain

$$\mu A \left(c - A^T y\right)^{-1} = b. \qquad (5.14)$$

If we multiply each of the m equations in this system by the product of the n coordinates of the vector $c - A^T y$, which are linear in the m coordinates y_j, we arrive at m polynomial equations of degree at most n in the coordinates of y.

We illustrate this by a simple example.

---

[7] **Exercise 32** Assume that (P) and (D) satisfy the interior point condition. Prove that the primal objective value is constant on the primal feasible region $\mathcal{P}$ if and only if $c = A^T \lambda$ for some $\lambda \in \mathbb{R}^m$, and the dual objective value is constant on the dual feasible region $\mathcal{D}$ if and only if $b = 0$.

Example II.7 Consider the case where [8]

$$
A = \begin{bmatrix} 1 & -1 & 0 \\ 0 & 0 & 1 \end{bmatrix}, \quad c = \begin{bmatrix} 1 \\ 1 \\ 1 \end{bmatrix}.
$$

For the moment we do not further specify the vector b. The left-hand side of (5.14) becomes

$$
\mu A \left(c - A^T y \right)^{-1} = \mu \begin{bmatrix} 1 & -1 & 0 \\ 0 & 0 & 1 \end{bmatrix} \begin{bmatrix} 1 - y_1 \\ 1 + y_1 \\ 1 - y_2 \end{bmatrix}^{-1} = \begin{bmatrix} \dfrac{2\mu y_1}{1 - y_1^2} \\ \dfrac{\mu}{1 - y_2} \end{bmatrix}.
$$

This means that the KKT system (5.3) is equivalent to the system of equations

$$
\begin{bmatrix} \dfrac{2\mu y_1}{1 - y_1^2} \\ \dfrac{\mu}{1 - y_2} \end{bmatrix} = \begin{bmatrix} b_1 \\ b_2 \end{bmatrix}, \quad \begin{bmatrix} 1 - y_1 \\ 1 + y_1 \\ 1 - y_2 \end{bmatrix} \geq 0.
$$

We consider this system for special choices of the vector b. Obviously, if $b_2 \leq 0$ then the system has no solution, since $\mu > 0$ and $1 - y_2 \geq 0$. Note that the second equation in $Ax = b$ then requires that $x_3 \leq 0$, showing that the primal feasible region does not contain a positive vector in that case. Hence, the central path exists only if $b_2 > 0$. Without loss of generality we may put $b_2 = 1$. Then we find

$$
y_2 = 1 - \mu.
$$

Now consider the case where $b_1 = 0$:

$$
b = \begin{bmatrix} 0 \\ 1 \end{bmatrix}.
$$

Then we obtain $y_1 = 0$ from the first equation, and hence for each $\mu > 0$:

$$
\begin{aligned}
x(\mu) &= (\mu, \mu, 1) \\
s(\mu) &= (1, 1, \mu) \\
y(\mu) &= (0, 1 - \mu).
\end{aligned}
$$

Thus we have found a parametric representation of the central paths of (P) and (D). They are straight half lines in this case. The dual central path (in the y-space) is shown in Figure 5.2.

---

[8] Note that these data are the same as in the examples D.5, D.6 and D.7 in Appendix D. These examples differ only in the vector b.

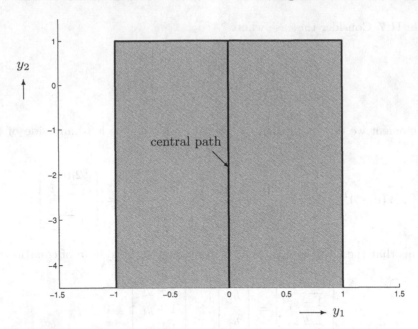

Figure 5.2 The dual central path if $b = (0, 1)$.

Let us also consider the case where $b_1 = 1$:

$$b = \begin{bmatrix} 1 \\ 1 \end{bmatrix}.$$

The first equation in the reduced KKT system then becomes

$$y_1^2 + 2\mu y_1 - 1 = 0,$$

giving

$$y_1 = -\mu \pm \sqrt{1 + \mu^2}.$$

The minus sign gives $y_1 \leq -1$, which implies $s_2 = 1 + y_1 \leq 0$. Since $1 + y_1$ must be positive, the unique solution for y_1 is determined by the plus sign:

$$y_1 = -\mu + \sqrt{1 + \mu^2}.$$

With $y(\mu)$ found, the calculation of $s(\mu)$ and $x(\mu)$ is straightforward, and yields a parametric representation of the central paths of (P) and (D). We have for each $\mu > 0$:

$$
\begin{aligned}
x(\mu) &= \left(\frac{1}{2} \left(\mu + 1 + \sqrt{1 + \mu^2} \right), \frac{1}{2} \left(-1 + \mu + \sqrt{1 + \mu^2} \right), 1 \right) \\
s(\mu) &= \left(1 + \mu - \sqrt{1 + \mu^2}, 1 - \mu + \sqrt{1 + \mu^2}, \mu \right) \\
y(\mu) &= \left(-\mu + \sqrt{1 + \mu^2}, 1 - \mu \right).
\end{aligned}
$$

The dual central path in the y-space is shown in Figure 5.3.

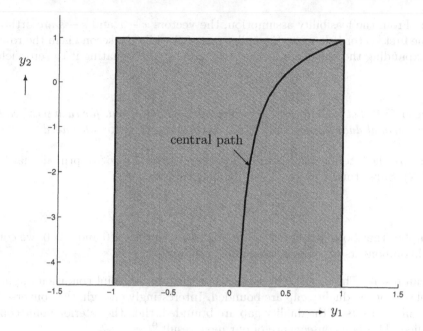

Figure 5.3 The dual central path if $b = (1, 1)$.

Note that in the above examples the limit of the central path exists if μ approaches zero, and that the limit point is an optimal solution. In fact this property of the central path is at the heart of the interior-point methods for solving the problems (P) and (D). The central path is used as a guideline to the optimal solution set. $\qquad\qquad \diamond$

5.7 Equivalent formulations of the interior-point condition

Later on we need other conditions that are equivalent to the interior-point condition. In this section we deal with one of them.

Let x be feasible for the primal problem, and (y, s) for the dual problem. Then, omitting y, we call (x, s) a *primal-dual pair*. From Proposition II.1 we recall that the duality gap for this pair is given by

$$c^T x - b^T y = x^T s.$$

We now derive an important consequence of the interior point condition on the level sets of the duality gap. In doing so, we shall use a simple relationship that we state, for further use, as a lemma. The relation in the lemma is an immediate consequence of the orthogonality of the row space and the null space of the matrix A.

Lemma II.8 *Assume $\bar{x} \in \mathcal{P}$ and $\bar{s} \in \mathcal{D}$. Then for all primal-dual feasible pairs (x, s),*

$$x^T s = \bar{s}^T x + \bar{x}^T s - \bar{x}^T \bar{s}.$$

Proof: From the feasibility assumption, the vectors $x - \bar{x}$ and $s - \bar{s}$ are orthogonal, since the first vector belongs to the null space of A while the second is in the row space of A. Expanding the scalar product $(x - \bar{x})^T(s - \bar{s})$ and equating it to zero yields the result. □

Theorem II.9 *Let the interior-point assumption hold. Then, for each positive K, the set of all primal-dual feasible pairs (x, s) such that $x^T s \leq K$ is bounded.*

Proof: By the interior-point assumption there exists a positive primal-dual feasible pair $(\bar{x}, \bar{s})$. Substituting K for $x^T s$ in Lemma II.8, we get

$$\bar{s}^T x + \bar{x}^T s \leq K + \bar{x}^T \bar{s}.$$

This implies that both $\bar{s}^T x$ and $\bar{x}^T s$ are bounded. Since $\bar{x} > 0$ and $\bar{s} > 0$, we conclude that all components of x and s must also be bounded. □

We can restate Theorem II.9 by saying that the interior-point condition implies that all level sets of the duality gap are bounded. Interestingly enough, the converse is also true. If all level sets of the duality gap are bounded, then the interior point condition is satisfied. This is a consequence of our next result.[9]

Theorem II.10 *Let the feasible regions of (P) and (D) be nonempty. Then the following statements are equivalent:*

(i) *both $\mathcal{P}$ and $\mathcal{D}$ contain a positive vector;*
(ii) *the level sets of the duality gap are bounded;*
(iii) *the optimal sets of (P) and (D) are bounded.*

Proof: The implication $(i) \Rightarrow (ii)$ is just a restatement of Theorem II.9. The implication $(ii) \Rightarrow (iii)$ is obvious, because optimal solutions of (P) and (D) are contained in any nonempty level set of the duality gap. The implication $(iii) \Rightarrow (i)$ in the theorem is nontrivial and can be proved as follows.

Since the feasible regions of (P) and (D) are nonempty we have optimal solutions x^* and (y^*, s^*) for (P) and (D). First assume that the optimal set of (P) is bounded. Since $x \in \mathcal{P}$ is optimal for (P) if and only if $x^T s^* = 0$, this set is given by

$$\mathcal{P}^* = \left\{ x \; : \; Ax = b, \; x \geq 0, \; x^T s^* = 0 \right\}.$$

The boundedness of $\mathcal{P}^*$ implies that the problem

$$\max_x \left\{ e^T x \; : \; Ax = b, \; x \geq 0, \; x^T s^* = 0 \right\}$$

is bounded, and hence it has an optimal solution. Since x and s^* are nonnegative, the problem is equivalent to

$$\max_x \left\{ e^T x \; : \; Ax = b, \; x \geq 0, \; x^T s^* \leq 0 \right\}.$$

---

[9] This result was first established by McLinden [197, 198]. See also Megiddo [200].

Hence, the dual of this problem is feasible. The dual is given by

$$\min_{y,\lambda} \left\{ b^T y \ : \ A^T y + \lambda s^* \geq e, \ \lambda \geq 0 \right\}.$$

Let $(\bar{y}, \bar{\lambda})$ be feasible for this problem. Then we have $A^T \bar{y} + \bar{\lambda} s^* \geq e$. If $\bar{\lambda} = 0$ then $A^T \bar{y} \geq e$, which implies

$$A^T \left(y^* - \bar{y} \right) = A^T y^* - A^T \bar{y} \leq c - e,$$

and hence $y^* - \bar{y}$ is dual feasible with positive slack vector. Now let $\bar{\lambda} > 0$. Then, replacing s^* by $c - A^T y^*$ in $A^T \bar{y} + \bar{\lambda} s^* \geq e$ we get

$$A^T \bar{y} + \bar{\lambda} \left(c - A^T y^* \right) \geq e.$$

Dividing by the positive number $\bar{\lambda}$ we obtain

$$A^T \left(y^* - \frac{\bar{y}}{\bar{\lambda}} \right) + \frac{e}{\bar{\lambda}} \leq c,$$

showing that $y^* - \bar{y}/\bar{\lambda}$ is feasible for (D) with a positive slack vector.

We proceed by assuming that the (nonempty!) optimal set of (D) is bounded. The same arguments apply in this case. Using that $(y, s) \in \mathcal{D}$ is optimal for (D) if and only if $s^T x^* = 0$, the dual optimal set is given by

$$\mathcal{D}^* = \left\{ (y, s) \ : \ A^T y + s = c, \ s \geq 0, \ s^T x^* = 0 \right\}.$$

The boundedness of $\mathcal{D}^*$ implies that the problem

$$\max_{y,s} \left\{ e^T s \ : \ A^T y + s = c, \ s \geq 0, \ s^T x^* = 0 \right\}$$

is bounded and hence has an optimal solution. This implies that the problem

$$\max_{y,s} \left\{ e^T s \ : \ A^T y + s = c, \ s \geq 0, \ s^T x^* \leq 0 \right\}$$

is also feasible and bounded. Hence, the dual problem, given by

$$\min_{x,\eta} \left\{ c^T x \ : \ Ax = 0, \ x + \eta x^* \geq e, \ \eta \geq 0 \right\},$$

is feasible and bounded as well. We only use the feasibility. Let $(\bar{x}, \bar{\eta})$ be a feasible solution. Then $\bar{x} + \bar{\eta} x^* \geq e$ and $A\bar{x} = 0$. If $\bar{\eta} = 0$ then we have $x^* + \bar{x} \geq e > 0$ and $A \left(x^* + \bar{x} \right) = A\bar{x} + Ax^* = b$, whence $x^* + \bar{x}$ is a positive vector in $\mathcal{P}$. If $\bar{\eta} > 0$ then we write

$$A \left(\frac{\bar{x}}{\bar{\eta}} + x^* \right) = \frac{1}{\bar{\eta}} A\bar{x} + Ax^* = b,$$

yielding that the positive vector $\bar{x}/\bar{\eta} + x^*$ is feasible for (P). Thus we have shown that (iii) implies (i), completing the proof. $\qquad \square$

Each of the three statements in Theorem II.10 deals with properties of both (P) and (D). We also have two one-sided versions of Theorem II.10 in which we have three equivalent statements where each statement involves a property of (P) or a property of (D). We state these results as corollaries, in which a primal level set means any set of the form

$$\{x \in \mathcal{P} \ : \ c^T x \leq K\}$$

and a dual level set means any set of the form

$$\{y \in \mathcal{D} \ : \ b^T y \geq K\},$$

where K may be any real number. The first corollary follows.

Corollary II.11 *Let the feasible regions of (P) and (D) be nonempty. Then the following statements are equivalent:*

  (i) $\mathcal{P}$ *contains a positive vector;*
 (ii) *the level sets of the dual objective are bounded;*
(iii) *the optimal set of (D) is bounded.*

Proof: Recall that the hypothesis in the corollary implies that the optimal sets of (P) and (D) are nonempty. The proof is cyclic, and goes as follows.

  $(i) \Rightarrow (ii)$: Letting $\bar{x} \in \mathcal{P}$, with $\bar{x} > 0$, we show that each level set of the dual objective is bounded. For any number K let $\mathcal{D}_K$ be the corresponding level set of the dual objective:

$$\mathcal{D}_K = \{(y, s) \in \mathcal{D} \ : \ b^T y \geq K\}.$$

Then $(y, s) \in \mathcal{D}_K$ implies

$$s^T \bar{x} = c^T \bar{x} - b^T y \leq c^T \bar{x} - K.$$

Since $\bar{x} > 0$, the i-th coordinate of s must be bounded above by $(c^T \bar{x} - K)/\bar{x}_i$. Therefore, $\mathcal{D}_K$ is bounded.

  $(ii) \Rightarrow (iii)$: This implication is trivial, because the optimal set of (D) is a level set of the dual objective.

  $(iii) \Rightarrow (i)$: This implication has been obtained as part of the proof of Theorem II.10.
□

The proof of the second corollary goes in the same way and is therefore omitted.

Corollary II.12 *Let the feasible regions of (P) and (D) be nonempty. Then the following statements are equivalent:*

  (i) $\mathcal{D}$ *contains a positive vector;*
 (ii) *the level sets of the primal objective are bounded;*
(iii) *the optimal set of (P) is bounded.*

We conclude this section with some interesting consequences of these corollaries. We assume that the feasible regions $\mathcal{P}$ and $\mathcal{D}$ are nonempty.

Corollary II.13 $\mathcal{D}$ *is bounded if and only if the null space of A contains a positive vector.*

Proof: The dual feasible region remains unchanged if we put $b = 0$. In that case $\mathcal{D}$ coincides with the optimal set $\mathcal{D}^*$ of (D), and this is the only nonempty dual level set. Hence, Corollary II.11 yields that $\mathcal{D}$ is bounded if and only if $\mathcal{P}$ contains a positive vector. Since $b = 0$ this gives the result. □

Corollary II.14 $\mathcal{P}$ *is bounded if and only if the row space of A contains a positive vector.*

Proof: The primal feasible region remains unchanged if we put $c = 0$. Now $\mathcal{P}$ coincides with the primal optimal set $\mathcal{P}^*$ of (P), and Corollary II.12 yields that $\mathcal{D}$ is bounded if and only if $\mathcal{D}$ contains a positive vector. Since $c = 0$ this gives the result. □

Note that the word 'positive' in the last two corollaries could be replaced by the word 'negative', because a linear space contains a positive vector if and only if it contains a negative vector. An immediate consequence of Corollary II.13 and Corollary II.14 is as follows.

Corollary II.15 *At least one of the two sets $\mathcal{P}$ and $\mathcal{D}$ is unbounded.*

Proof: If both sets are bounded then there exist a positive vector x and a vector y such that $Ax = 0$ and $A^T y > 0$. This gives the contradiction

$$0 = (Ax)^T y = x^T \left(A^T y \right) > 0.$$

The result follows. □

Remark II.16 If (P) and (D) satisfy the interior-point condition then for every positive μ we have a primal-dual pair (x, s) such that $xs = \mu e$. Letting μ go to infinity, it follows that for each index i the product $x_i s_i$ goes to infinity. Therefore, at least one of the coordinates x_i and s_i must be unbounded. It can be shown that exactly one of these two coordinates is unbounded and the other is bounded. This is an example of a *coordinatewise duality property*. We will not go further in this direction here, but refer the reader to Williams [291, 292] and to Güler et al. [134]. ●

5.8   Symmetric formulation

In this chapter we dealt with the LO problem in standard form

$$(P) \qquad \min \left\{ c^T x \; : \; Ax = b, \, x \geq 0 \right\},$$

and its dual problem

$$(D) \qquad \max \left\{ b^T y \; : \; A^T y + s = c, \, s \geq 0 \right\}.$$

Note that there is an asymmetry in problems (P) and (D). The constraints in (P) and (D) are equality constraints, but in (P) all variables are nonnegative, whereas in (D) we also have free variables, in y. Note that we could eliminate s in the formulation of

(D), leaving us with the inequality constraints $A^T y \leq c$, so this would not remove the asymmetry in the formulations.

We could have avoided the asymmetry by using a different format for problem (P), but because the chosen format is more or less standard in the literature, we decided to use the standard format in this chapter and to accept its inherent asymmetry. Note that the asymmetry is also reflected in the KKT system. This is especially true for the first two equations, because the third equation is symmetric in x and s.

In this section we make an effort to show that it is quite easy to obtain a perfect symmetry in the formulations. This has some practical value. It implies that every concept, or result, or algorithm for one of the two problems, has its natural counterpart for the other problem. It will also highlight the underlying geometry of an LO problem.

Let us define the linear space $\mathcal{L}$ as the null space of the matrix A:

$$\mathcal{L} = \{Ax = 0 \, : \, x \in \mathbb{R}^n\}, \tag{5.15}$$

and let $\mathcal{L}^\perp$ denote the orthogonal complement of $\mathcal{L}$. Then, due to a well known result in linear algebra, $\mathcal{L}^\perp$ is the row space of the matrix A, i.e.,

$$\mathcal{L}^\perp = \{A^T y \, : \, y \in \mathbb{R}^m\}. \tag{5.16}$$

Now let $\bar{x}$ be any vector satisfying $A\bar{x} = b$. Then x is primal feasible if $x \in \bar{x} + \mathcal{L}$ and $x \in \mathbb{R}^n_+$. So the primal problem can be reformulated as

$$(P') \qquad \min \left\{ c^T x \, : \, x \in (\bar{x} + \mathcal{L}) \cap \mathbb{R}^n_+ \right\}.$$

So, (P) amounts to minimizing the linear function $c^T x$ over the intersection of the affine space $\bar{x} + \mathcal{L}$ and the nonnegative orthant $\mathbb{R}^n_+$.

We can put (D) in the same format by eliminating the vector y of free variables. To this end we observe that $s \in \mathbb{R}^n$ is feasible for (D) if and only if $s \in c + \mathcal{L}^\perp$ and $s \in \mathbb{R}^n_+$. Given any vector $s \in c + \mathcal{L}^\perp$, let y be such that $A^T y + s = c$. Then

$$b^T y = (A\bar{x})^T y = \bar{x}^T A^T y = \bar{x}^T (c - s) = c^T \bar{x} - \bar{x}^T s. \tag{5.17}$$

Omitting the constant $c^T \bar{x}$, it follows that solving (D) is equivalent to solving the problem

$$(D') \qquad \min \left\{ \bar{x}^T s \, : \, s \in (c + \mathcal{L}^\perp) \cap \mathbb{R}^n_+ \right\}.$$

Thus we see that the dual problem amounts to minimizing the linear function $\bar{x}^T s$ over the intersection of the affine space $c + \mathcal{L}^\perp$ and the nonnegative orthant $\mathbb{R}^n_+$. The similarity with reformulation (P') is striking: both problems are minimization problems, the roles of the vectors $\bar{x}$ and c are interchanged, and the underlying linear spaces are each others orthogonal complement. An immediate consequence is also that the dual of the dual problem is the primal problem.[10] The KKT conditions can now be expressed in a way that is completely symmetric in x and s:

$$
\begin{aligned}
x &\in (\bar{x} + \mathcal{L}) \cap \mathbb{R}^n_+, & x &> 0, \\
s &\in (c + \mathcal{L}^\perp) \cap \mathbb{R}^n_+, & s &> 0, \\
xs &= \mu e.
\end{aligned}
\tag{5.18}
$$

---

[10] The affine spaces $c + \mathcal{L}^\perp$ and $\bar{x} + \mathcal{L}$ intersect in a unique point $\xi \in \mathbb{R}^n$. Hence, we could even take $c = \bar{x} = \xi$.

Due to (5.17), we conclude that on the dual feasible region, $b^T y$ and $\bar{x}^T s$ sum up to the constant $c^T \bar{x}$.

5.9 Dual logarithmic barrier function

We conclude this chapter by introducing the dual logarithmic barrier function, using the symmetry that has now become apparent. Recall that for any positive μ the primal μ-center $x(\mu)$ has been characterized as the minimizer of the primal logarithmic barrier function $\tilde{g}_\mu(x)$, as given by (5.2):

$$\tilde{g}_\mu(x) = c^T x - \mu \sum_{j=1}^{n} \log x_j.$$

Using the symmetry, we obtain that the dual μ-center $s(\mu)$ can be characterized as the minimizer of the function

$$\tilde{h}_\mu(s) := \bar{x}^T s - \mu \sum_{j=1}^{n} \log s_j, \tag{5.19}$$

where s runs through all positive dual feasible slack vectors. According to (5.17), we may replace $\bar{x}^T s$ by $c^T \bar{x} - b^T y$. Omitting the constant $c^T \bar{x}$, it follows that $(y(\mu), s(\mu))$ is the minimizer of the function

$$k_\mu(y, s) = -b^T y - \mu \sum_{j=1}^{n} \log s_j.$$

The last function is usually called the dual logarithmic barrier function. Recall that for any dual feasible pair (y, s), $\tilde{h}_\mu(s)$ and $k_\mu(y, s)$ differ by a constant only. It may often be preferable to use $\tilde{h}_\mu(s)$, because then we only have to deal with the nonnegative slack vectors, and not with the free variable y. It will be convenient to refer also to $\tilde{h}_\mu(s)$ as the dual logarithmic barrier function.

From now on we assume that the interior point condition is satisfied, unless stated otherwise. As a consequence, both the primal and the dual logarithmic barrier functions have a minimizer, for each $\mu > 0$. These minimizers are denoted by $x(\mu)$ and $s(\mu)$ respectively.

6

The Dual Logarithmic Barrier Method

In the previous chapter we introduced the central path of a problem as the set consisting of all μ-centers, with μ running through all positive real numbers. Using this we can now easily describe the basic idea behind the logarithmic barrier method. We do so for the dual problem in standard format:

$$(D) \qquad \max \left\{ b^T y \; : \; A^T y + s = c, \; s \geq 0 \right\}.$$

Recall that any method for the dual problem can also be used for solving the primal problem, because of the symmetry discussed in Section 5.8. The dual problem has the advantage that its feasible region—in the y-space—can be drawn if its dimension is small enough ($m = 1, 2$ or 3). This enables us to illustrate graphically some aspects of the methods to be described below.

6.1 A conceptual method

We assume that we know the μ-centers $y(\mu)$ and $s(\mu)$ for some positive $\mu = \mu^0$. Later on, in Chapter 8, we show that this assumption can be made without loss of generality. Given $s(\mu)$, the primal μ-center $x(\mu)$ follows from the relation

$$x(\mu)s(\mu) = \mu e.$$

Now the duality gap for the pair of μ-centers is given by

$$c^T x(\mu) - b^T y(\mu) = x(\mu)^T s(\mu) = n\mu.$$

The last equality follows since we have for each i that

$$x_i(\mu)s_i(\mu) = \mu.$$

It follows that if μ goes to zero, then the duality gap goes to zero as well. As a consequence we have that if μ is small enough, then the pair $(y(\mu), s(\mu))$ is 'almost' optimal for the dual problem. This can also be seen by comparing the dual objective value $b^T y(\mu)$ with the optimal value of (D). Denoting the optimal value of (P) and (D) by z^* we know from Proposition II.1 that

$$b^T y(\mu) \leq z^* \leq c^T x(\mu),$$

so we have

$$z^* - b^T y(\mu) \leq c^T x(\mu) - b^T y(\mu) = x(\mu)^T s(\mu) = n\mu,$$

and

$$c^T x(\mu) - z^* \leq c^T x(\mu) - b^T y(\mu) = x(\mu)^T s(\mu) = n\mu.$$

Thus, if μ is chosen small enough, the primal objective value $c^T x(\mu)$ and the dual objective value $b^T y(\mu)$ can simultaneously be driven arbitrarily close to the optimal value. We thus have to deal with the question of how to obtain the μ-centers for small enough values of μ.

Now let μ^* be obtained from μ by

$$\mu^* := (1 - \theta)\,\mu,$$

where θ is a positive constant smaller than 1. We may expect that if θ is not too large, the μ^*-centers will be close to the given μ-centers.[1] For the moment, let us assume that we are able to calculate the μ^*-centers, provided θ is not too large. Then the following conceptual algorithm can be used to find ε-optimal solutions of both (P) and (D).

Conceptual Logarithmic Barrier Algorithm

Input:
   An accuracy parameter $\varepsilon > 0$;
   a barrier update parameter θ, $0 < \theta < 1$;
   the center $(y(\mu^0), s(\mu^0))$ for some $\mu^0 > 0$.
begin
   $\mu := \mu^0$;
   while $n\mu \geq \varepsilon$ **do**
   begin
      $\mu := (1 - \theta)\mu$;
      $s := s(\mu)$;
   end
end

Recall that, given the dual center $s(\mu)$, the primal center $x(\mu)$ can be calculated immediately from the centering condition at μ. Hence, the output of this algorithm is a feasible primal-dual pair of solutions for (P) and (D) such that the duality gap does not exceed ε. How many iterations are needed by the algorithm? The answer is provided by the following lemma.

---

[1] This is a consequence of the fact that the μ-centers depend continuously on the barrier parameter μ, due to a result of Fiacco and McCormick [77]. See also Chapter 16.

Lemma II.17 *If the barrier parameter μ has the initial value μ^0 and is repeatedly multiplied by $1 - \theta$, with $0 < \theta < 1$, then after at most*

$$\left\lceil \frac{1}{\theta} \log \frac{n\mu^0}{\varepsilon} \right\rceil$$

iterations we have $n\mu \leq \varepsilon$.

Proof: Initially the duality gap is $n\mu^0$, and in each iteration it is reduced by the factor $1 - \theta$. Hence, after k iterations the duality gap is smaller than ε if

$$(1 - \theta)^k \, n\mu^0 \leq \varepsilon.$$

The rest of the proof goes in the same as in the proof of Lemma I.36. Taking logarithms we get

$$k \log (1 - \theta) + \log(n\mu^0) \leq \log \varepsilon.$$

Since $-\log (1 - \theta) \geq \theta$, this certainly holds if

$$k\theta \geq \log(n\mu^0) - \log \varepsilon = \log \frac{n\mu^0}{\varepsilon}.$$

This implies the lemma. $\qquad\qquad\qquad\qquad\qquad\qquad\qquad\qquad\qquad\qquad\square$

To make the algorithm more practical, we have to avoid the exact calculation of the μ-center $s(\mu)$. This is the subject of the following sections.

6.2   Using approximate centers

Recall that any μ-center is the minimizer for the corresponding logarithmic barrier function. Therefore, by minimizing the corresponding logarithmic barrier function we will find the μ-center. Since the logarithmic barrier function has a positive definite Hessian, Newton's method is a natural candidate for this purpose. If we know the μ-center, then defining μ^* by $\mu^* := (1 - \theta)\mu$, just as in the preceding section, we can move to the μ^*-center by applying Newton's method to the logarithmic barrier function corresponding to μ^*, starting at the μ-center. Having reached the μ^*-center, we can repeat this process until the barrier parameter has become small enough. In fact this would yield an implementation of the conceptual algorithm of the preceding section. Unfortunately, however, after the update of the barrier parameter to μ^*, to find the μ^*-center exactly infinitely many Newton steps are needed. To restrict the number of Newton steps between two successive updates of the barrier parameter, we do not calculate the μ^*-center exactly, but instead use an approximation of it. Our first aim is to show that this can be done in such a way that only one Newton step is taken between two successive updates of the barrier parameter. Later on we deal with a different approach where the number of Newton steps between two successive updates of the barrier parameter may be larger than one.

In the following sections we are concerned with a more detailed analysis of the use of approximate centers. In the analysis we need to measure the proximity of an approximate center to the exact center. We also have to study the behavior of

Newton's method when applied to the logarithmic barrier function. We start in the next section with the calculation of the Newton step. Then we proceed to defining a proximity measure and deal with some related properties. After this we can formulate the algorithm, and analyze it.

6.3   Definition of the Newton step

In this section we assume that we are given a dual feasible pair (y, s), and, by applying Newton's method to the dual logarithmic barrier function corresponding to the barrier parameter value μ, we try to find the minimizer of this function, which is the pair $(y(\mu), s(\mu))$. Recall that the dual logarithmic barrier function is the function $k_\mu(y, s)$ defined by

$$k_\mu(y, s) := -b^T y - \mu \sum_{i=1}^{n} \log s_i,$$

where (y, s) runs through all dual feasible pairs with positive slack vector s. Recall also that y and s are related by the dual feasibility condition

$$A^T y + s = c, \quad s \geq 0,$$

and since we assume that A has full rank, this defines a one-to-one correspondence between the components y and s in dual feasible pairs. As a consequence, we can consider $k_\mu(y, s)$ as a function of s alone. In Section 5.8 we showed that $k_\mu(y, s)$ differs only by the constant $c^T \bar{x}$ from

$$\tilde{h}_\mu(s) = \bar{x}^T s - \mu \sum_{j=1}^{n} \log s_j,$$

provided $A\bar{x} = b$.

Our present aim is to compute the minimizer $s(\mu)$ of $\tilde{h}_\mu(s)$. Assuming $s \neq s(\mu)$, we construct a search direction by applying Newton's method to $\tilde{h}_\mu(s)$. We first calculate the first and second derivatives of $\tilde{h}_\mu(s)$ with respect to s, namely

$$\nabla \tilde{h}_\mu(s) = \bar{x} - \mu s^{-1}, \qquad \nabla^2 \tilde{h}_\mu(s) = \mu S^{-2},$$

where, as usual, $S = \text{diag}(s)$. The Newton step Δs — in the s-space — is the minimizer of the second-order approximation of $\tilde{h}_\mu(s + \Delta s)$ at s, which is given by

$$t(\Delta s) := \tilde{h}_\mu(s) + \left(\bar{x} - \mu s^{-1} \right)^T \Delta s + \frac{1}{2} \Delta s^T \mu S^{-2} \Delta s,$$

subject to the condition that $s + \Delta s$ is dual feasible. The latter means that there exists Δy such that

$$A^T (y + \Delta y) + s + \Delta s = c.$$

Since $A^T y + s = c$, this is equivalent to

$$A^T \Delta y + \Delta s = 0$$

for some Δy.

We make use of an $(n-m) \times n$ matrix H whose null space is equal to the row space of A. Then the condition on Δs simply means that $H\Delta s = 0$, which is equivalent to

$$\Delta s \in \text{null space of } H.$$

Using Proposition A.1, we find that Δs minimizes $t(\Delta s)$ if and only if

$$\nabla t(\Delta s) = \bar{x} - \mu s^{-1} + \mu s^{-2}\Delta s \perp \text{null space of } H.$$

It is useful to restate these conditions in terms of the matrix HS:[2]

$$s\bar{x} - \mu e + \mu s^{-1}\Delta s \perp \text{null space of } HS,$$

and

$$\mu s^{-1}\Delta s \in \text{null space of } HS.$$

Therefore, writing

$$s\bar{x} - \mu e = -\mu s^{-1}\Delta s + \left(s\bar{x} - \mu e + \mu s^{-1}\Delta s\right),$$

we have a decomposition of the vector $s\bar{x} - \mu e$ into two components, with the first component in the null space of HS and the second component orthogonal to the null space of HS. Stated otherwise, $\mu s^{-1}\Delta s$ is the orthogonal projection of $\mu e - s\bar{x}$ into the null space of HS. Hence we have shown that

$$\mu s^{-1}\Delta s = P_{HS}\left(\mu e - s\bar{x}\right). \tag{6.1}$$

From this relation the Newton step Δs can be calculated. Since the projection matrix $P_{HS}{}^{3}$ is given by

$$P_{HS} = I - SH^{T}\left(HS^{2}H^{T}\right)^{-1}HS,$$

we obtain the following expression for Δs:

$$\Delta s = s\left(I - SH^{T}\left(HS^{2}H^{T}\right)^{-1}HS\right)\left(e - \frac{s\bar{x}}{\mu}\right).$$

Recall that $\bar{x}$ may be any vector such that $A\bar{x} = b$. It follows that the right-hand side in (6.1) must be independent of $\bar{x}$. It is left to the reader to verify that this is indeed true.[4,5,6] We are now going to explore this in a surprising way with extremely important consequences.

---

[2] **Exercise 33** Let S be a square and nonsingular matrix and H be any other matrix such that the product HS is well defined. Then $x \in$ null space of H if and only if $S^{-1}x \in$ null space of HS, and $x \perp$ null space of H if and only if $Sx \perp$ null space of HS^{T}. Prove this.

[3] For any matrix Q the matrix of the orthogonal projection onto the null space of Q is denoted as P_{Q}.

[4] **Exercise 34** Show that $P_{HS}\left(s\Delta x\right) = 0$ whenever $A\Delta x = 0$.

[5] **Exercise 35** The Newton step in the y-space is given by

$$\Delta y = \left(AS^{-2}A^{T}\right)^{-1}\left(\frac{b}{\mu} - AS^{-1}e\right).$$

Prove this. (Hint: Use that $A^{T}\Delta y + \Delta s = 0$.)

[6] Observe that the computation of Δs requires the inversion of the matrix $HS^{2}H^{T}$, and the computation of Δy the inversion of the matrix $AS^{-2}A^{T}$. It is not clear in general which of the two inversions is more attractive from a computational point of view.

If we let $\bar{x}$ run through the affine space $A\bar{x} = b$ then the vector $\mu e - s\bar{x}$ runs through another affine space that is parallel to the null space of AS^{-1}. Now using that

$$\text{null space of } AS^{-1} = \text{row space of } HS,$$

we conclude that the affine space consisting of all vectors $\mu e - s\bar{x}$, with $A\bar{x} = b$, is orthogonal to the null space of HS. This implies that these two spaces intersect in a unique point. Hence there exists a unique vector $\bar{x}$ satisfying $A\bar{x} = b$ such that $\mu e - s\bar{x}$ belongs to the null space of HS. We denote this vector as $x(s, \mu)$. From its definition we have

$$P_{HS}\left(\mu e - sx(s, \mu)\right) = \mu e - sx(s, \mu),$$

thus yielding the following expression for the Newton step:

$$\mu s^{-1} \Delta s = \mu e - sx(s, \mu). \tag{6.2}$$

Figure 6.1 depicts the situation.

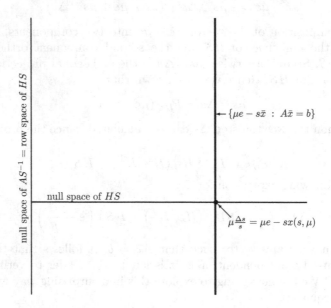

Figure 6.1   The projection yielding $s^{-1}\Delta s$.

Another important feature of the vector $x(s, \mu)$ is that it minimizes the 2-norm of $\mu e - s\bar{x}$ in the affine space $A\bar{x} = b$. Hence, $x(s, \mu)$ can be characterized by the property

$$x(s, \mu) = \text{argmin}_x \left\{ \|\mu e - sx\| \ : \ Ax = b \right\}. \tag{6.3}$$

We summarize these results in a theorem.

Theorem II.18 *Let s be any positive dual feasible slack vector. Then the Newton step Δs at s with respect to the dual logarithmic barrier function corresponding to the barrier parameter value μ satisfies (6.2), with $x(s, \mu)$ as defined in (6.3).*

6.4   Properties of the Newton step

We denote the result of the Newton step at s by s^+. Thus we may write

$$s^+ := s + \Delta s = s \left(e + s^{-1} \Delta s \right).$$

A major question is whether s^+ is feasible or not. Another important question is whether $x(s, \mu)$ is primal feasible. In this section we deal with these two questions, and we show that both questions allow a perfect answer.

We start with the feasibility of s^+. Clearly, s^+ is feasible if and only if s^+ is nonnegative, and this is true if and only if

$$e + s^{-1} \Delta s \geq 0. \tag{6.4}$$

We conclude that the (full) Newton step is feasible if (6.4) is satisfied.

Let us now consider the vector $x(s, \mu)$. By definition, it satisfies the equation $Ax = b$, so if it is nonnegative, then $x(s, \mu)$ is primal feasible. We can derive a simple condition for that. From (6.2) we obtain that

$$x(s, \mu) = \mu s^{-1} \left(e - s^{-1} \Delta s \right). \tag{6.5}$$

We conclude that $x(s, \mu)$ is primal feasible if and only if

$$e - s^{-1} \Delta s \geq 0. \tag{6.6}$$

Combining this result with (6.4) we state the following lemma.

Lemma II.19 *If the Newton step* Δs *satisfies*

$$-e \leq s^{-1} \Delta s \leq e$$

then $x(s, \mu)$ *is primal feasible, and* $s^+ = s + \Delta s$ *is dual feasible.*

Remark II.20 *We make an interesting observation. Since* s *is positive, (6.6) is equivalent to*

$$s - \Delta s \geq 0.$$

Note that $s - \Delta s$ *is obtained by moving from* s *in the opposite direction of the Newton step. Thus we conclude that* $x(s, \mu)$ *is primal feasible if and only if a backward Newton step yields a dual feasible point for the dual problem.*

We conclude this section by considering the special case where $\Delta s = 0$. From (6.2) we deduce that this occurs if and only if $sx(s, \mu) = \mu e$, i.e., if and only if s and $x(s, \mu)$ satisfy the centering condition with respect to μ. Since s and $x(s, \mu)$ are positive, they satisfy the KKT conditions. Now the unicity property gives us that $x(s, \mu) = x(\mu)$ and $s = s(\mu)$. Thus we see that the Newton step at s is equal to the zero vector if and only if $s = s(\mu)$. This could have been expected, because $s(\mu)$ is the minimizer of the dual logarithmic barrier function.

6.5   Proximity and local quadratic convergence

Lemma II.19 in the previous section states under what conditions the Newton step yields feasible solutions on both the dual and the primal side. This turned out to be the case when

$$-e \leq s^{-1}\Delta s \leq e.$$

Observe that these inequalities can be rephrased simply by saying that the infinity norm of the vector $s^{-1}\Delta s$ does not exceed 1. We refer to $s^{-1}\Delta s$ as the Newton step Δs *scaled by* s, or, in short, the scaled Newton step at s.

In the analysis of the logarithmic barrier method we need a measure for the 'distance' of s to the μ-center $s(\mu)$. The above observation might suggest that the infinity norm of the scaled Newton step could be used for that purpose. However, it turns out to be more convenient to use the 2-norm of the scaled Newton step. So we measure the proximity of s to $s(\mu)$ by the quantity[7]

$$\delta(s,\mu) := \left\| s^{-1}\Delta s \right\|. \tag{6.7}$$

At the end of the previous section we found that the Newton step Δs vanishes if and only if s is equal to $s(\mu)$. As a consequence we have

$$\delta(s,\mu) = 0 \iff s = s(\mu).$$

The obvious question that we have to deal with is about the improvement in the proximity to $s(\mu)$ after a feasible Newton step. The next theorem provides a very elegant answer to this question. In the proof of this theorem we need a different characterization of the proximity $\delta(s,\mu)$, which is an immediate consequence of Theorem II.18, namely

$$\delta(s,\mu) = \left\| e - \frac{sx(s,\mu)}{\mu} \right\| = \frac{1}{\mu}\min_x \left\{ \|\mu e - sx\| \ : \ Ax = b \right\}. \tag{6.8}$$

We have the following result.

Theorem II.21 *If $\delta(s,\mu) \leq 1$, then $x(s,\mu)$ is primal feasible, and $s^+ = s + \Delta s$ is dual feasible. Moreover,*

$$\delta(s^+,\mu) \leq \delta(s,\mu)^2.$$

Proof: The first part of the theorem is an obvious consequence of Lemma II.19, because the infinity norm of $s^{-1}\Delta s$ does not exceed its 2-norm and hence does not exceed 1. Now let us turn to the proof of the second statement. Using (6.8) we write

$$\delta(s^+,\mu) = \frac{1}{\mu}\min_x \left\{ \|\mu e - s^+x\| \ : \ Ax = b \right\}.$$

---

[7]   **Exercise 36** If $s = s(\mu)$ then we know that μs^{-1} is primal feasible. Now let $\delta = \delta(s,\mu) > 0$ and consider $x = \mu s^{-1}$. Let $Q = AS^{-2}A^T$. Then Q is positive definite, and so is its inverse. Hence Q^{-1} defines a norm that we denote as $\|.\|_{Q^{-1}}$. Thus, for any $z \in \mathbb{R}^n$:

$$\|z\|_{Q^{-1}} = \sqrt{z^T Q^{-1} z}.$$

Measuring the amount of infeasibility of x in the sense of this norm, prove that

$$\|Ax - b\|_{Q^{-1}} \leq \mu\delta.$$

Substituting for x the vector $x(s, \mu)$ we obtain the inequality

$$\delta(s^+, \mu) \le \frac{1}{\mu} \left\| \mu e - s^+ x(s, \mu) \right\|. \tag{6.9}$$

The vector $\mu e - s^+ x(s, \mu)$ can be reduced as follows:

$$\mu e - s^+ x(s, \mu) = \mu e - (s + \Delta s)\, x(s, \mu) = \mu e - s x(s, \mu) - \Delta s x(s, \mu).$$

From (6.2) this implies

$$\mu e - s^+ x(s, \mu) = \mu s^{-1} \Delta s - \Delta s x(s, \mu) = (\mu e - s x(s, \mu))\, s^{-1} \Delta s = \mu \left(s^{-1} \Delta s \right)^2. \tag{6.10}$$

Thus we obtain, by substitution of this equality in (6.9),

$$\delta(s^+, \mu) \le \left\| \left(s^{-1} \Delta s \right)^2 \right\| \le \left\| s^{-1} \Delta s \right\|_\infty \left\| s^{-1} \Delta s \right\|.$$

Now from the obvious inequality $\|z\|_\infty \le \|z\|$, with $z = s^{-1} \Delta s$, the result follows. $\square$

Theorem II.21 implies that after a Newton step the proximity to the μ-center is smaller than the square of the proximity before the Newton step. In other words, Newton's method is quadratically convergent. Moreover, the theorem defines a neighborhood of the μ-center $s(\mu)$ where the quadratic convergence occurs, namely

$$\{ s \in \mathcal{D} \; : \; \delta(s, \mu) < 1 \}. \tag{6.11}$$

This result is extremely important. It implies that when the present iterate s is close to $s(\mu)$, then only a small number of Newton steps brings us very close to $s(\mu)$. For instance, if $\delta(s, \mu) = 0.5$, then only 6 Newton steps yield an iterate with proximity less than 10^{-16}. Figure 6.2 shows a graph depicting the required number of steps to

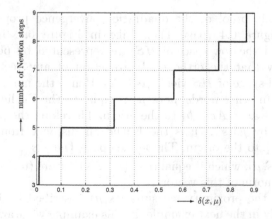

Figure 6.2 Required number of Newton steps to reach proximity 10^{-16}.

reach proximity 10^{-16} when starting at any given value of the proximity in the interval $(0, 1)$.

We can also consider it differently. If we repeatedly apply Newton steps, starting at $s^0 = s$, then after k Newton steps the resulting point, denoted by s^k, satisfies

$$\delta(s^k, \mu) \le \delta(s^0, \mu)^{2^k}.$$

Hence, taking logarithms on both sides,

$$-\log \delta(s^k, \mu) \ge -2^k \log \delta(s^0, \mu),$$

see Figure 6.3 (page 116).

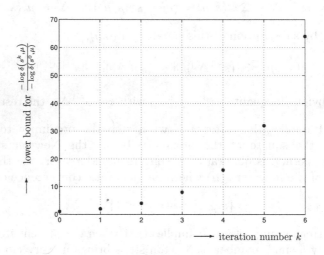

Figure 6.3   Convergence rate of the Newton process.

The above algebraic proof of the quadratic convergence property is illustrated geometrically by Figure 6.4 (page 117). Like in Figure 6.1, in Figure 6.4 the null space of HS and the row space of HS are represented by perpendicular axes. From (6.1) we know that the orthogonal projection of any vector $\mu e - sx$, with $Ax = b$, into the null space of HS yields $\mu s^{-1} \Delta s$. Hence the norm of this projection is equal to $\mu \delta(s, \mu)$. In other words, $\mu \delta(s, \mu)$ is equal to the Euclidean distance from the affine space $\{\mu e - sx : Ax = b\}$ to the origin. Therefore, the proximity after the Newton step, given by $\mu \delta(s^+, \mu)$, is the Euclidean distance from the affine space $\{\mu e - s^+ x : Ax = b\}$ to the origin. The affine space $\{\mu e - s^+ x : Ax = b\}$ contains the vector $\mu e - s^+ x(s, \mu)$, which is equal to $\mu \left(s^{-1} \Delta s\right)^2$, from (6.10). Hence, $\mu \delta(s^+, \mu)$ does not exceed the norm of this vector.

The properties of the proximity measure $\delta(s, \mu)$ described in Theorem II.21 are illustrated graphically in the next example. In this example we draw some level curves for the proximity measure for some fixed value of the barrier parameter μ, and we show how the Newton step behaves when applied at some points inside and outside the region of quadratic convergence, as given by (6.11). We do this for some simple problems.

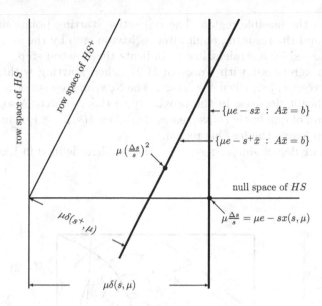

Figure 6.4 The proximity before and after a Newton step.

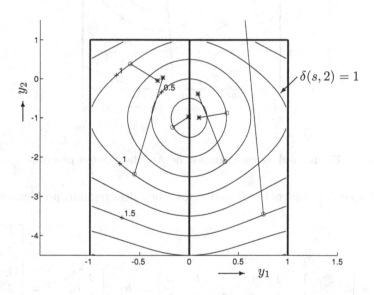

Figure 6.5 Demonstration no.1 of the Newton process.

Example II.22 First we take A and c as in Example II.7 on page 97, and $b = (0,1)^T$. Figure 5.2 (page 98) shows the feasible region and the central path. In Figure 6.5 we have added some level curves for $\delta(s,2)$. We have also depicted the Newton step at several points in the feasible region. The respective starting points are indicated by the symbol '∘', and the resulting point after a Newton step by the symbol '*'; the two points are connected by a straight line to indicate the Newton step.

Note that, in agreement with Theorem II.21, when starting within the region of quadratic convergence, i.e., when $\delta(s,\mu) < 1$, the Newton step is not only feasible, but there is a significant decrease in the proximity to the 2-center. Also, when starting outside the region of quadratic convergence, i.e., when $\delta(s,\mu) \geq 1$, it may happen that the Newton step leaves the feasible region.

In Figure 6.6 we depict similar results for the problem defined in Example II.7 with $b = (1,1)^T$.

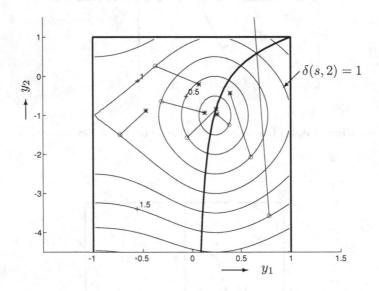

Figure 6.6 Demonstration no.2 of the Newton process.

Finally, Figure 6.7 depicts the situation for a new, less regular, problem. It is defined by

$$A = \begin{bmatrix} -2 & 1 & 1 & 0 & 1 & -1 & 0 \\ 1 & 1 & -1 & 1 & 0 & 0 & -1 \end{bmatrix}, \quad b = \begin{bmatrix} 2 \\ 1 \end{bmatrix}, \quad c = \begin{bmatrix} 1 \\ 4 \\ 1 \\ 2 \\ 2 \\ 0 \\ 0 \end{bmatrix}.$$

This figure makes clear that after a Newton step the proximity to the 2-center may increase. Concluding this example, we may state that inside the region of quadratic convergence our proximity measure provides perfect control over the Newton process,

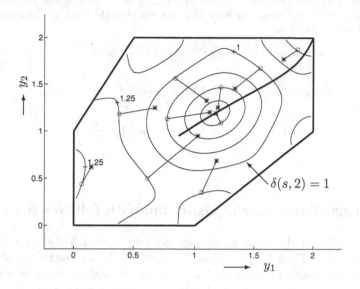

Figure 6.7   Demonstration no.3 of the Newton process.

but outside this region it has little value. ⬦

6.6   The duality gap close to the central path

A nice feature of the μ-center $s = s(\mu)$ is that the vector $x = \mu s^{-1}$ is primal feasible, and the duality gap for the primal-dual pair (x, s) is given by $n\mu$. One might ask about the situation when s is close to $s(\mu)$. The next theorem provides a satisfactory answer. It states that for small values of the proximity $\delta(s, \mu)$ the duality gap for the pair $(x(s, \mu), s)$ is close to the gap for the μ-centers.

Theorem II.23 *Let* $\delta := \delta(s, \mu) \leq 1$. *Then the duality gap for the primal-dual pair* $(x(s, \mu), s)$ *satisfies*

$$n\mu \left(1 - \delta\right) \leq s^T x(s, \mu) \leq n\mu \left(1 + \delta\right).$$

Proof: From Theorem II.21 we know that $x(s, \mu)$ is primal feasible. Hence, for the duality gap we have

$$s^T x(s, \mu) = s^T \left(\mu s^{-1} \left(e - s^{-1}\Delta s\right)\right) = \mu e^T \left(e - s^{-1}\Delta s\right).$$

Since the coordinates of the vector $e - s^{-1}\Delta s$ lie in the interval $[1 - \delta, 1 + \delta]$, the result follows. □

Remark II.24 The above estimate for the duality gap is not as sharp as it could be, but is sufficient for our goal. Nevertheless, we want to point out that the Cauchy–Schwarz inequality gives stronger bounds. We have

$$s^T x(s,\mu) = \mu e^T \left(e - s^{-1}\Delta s\right) = n\mu - \mu e^T s^{-1}\Delta s.$$

Hence

$$\left|s^T x(s,\mu) - n\mu\right| = \mu\left|e^T s^{-1}\Delta s\right| \le \mu\left\|e\right\|\left\|s^{-1}\Delta s\right\| = \mu\sqrt{n}\delta,$$

and it follows that

$$n\mu\left(1 - \frac{\delta}{\sqrt{n}}\right) \le s^T x(s,\mu) \le n\mu\left(1 + \frac{\delta}{\sqrt{n}}\right).$$

•

6.7 Dual logarithmic barrier algorithm with full Newton steps

We can now describe an algorithm using approximate centers. We assume that we are given a pair $(y^0, s^0) \in \mathcal{D}$ and a $\mu^0 > 0$ such that (y^0, s^0) is close to the μ^0-center in the sense of the proximity measure $\delta(s^0, \mu^0)$. In the algorithm the barrier parameter monotonically decreases from the initial value μ^0 to some small value determined by the desired accuracy. In the algorithm we denote by $p(s, \mu)$ the Newton step Δs at $s \in \mathcal{D}^+$ to emphasize the dependence on the barrier parameter μ.

Dual Logarithmic Barrier Algorithm with full Newton steps

Input:
    A proximity parameter τ, $0 \le \tau < 1$;
    an accuracy parameter $\varepsilon > 0$;
    $(y^0, s^0) \in \mathcal{D}$ and $\mu^0 > 0$ such that $\delta(s^0, \mu^0) \le \tau$;
    a fixed parameter θ, $0 < \theta < 1$.
begin
    $s := s^0$; $\mu := \mu^0$;
    while $n\mu \ge (1 - \theta)\varepsilon$ **do**
    begin
        $s := s + p(s, \mu)$;
        $\mu := (1 - \theta)\mu$;
    end
end

We prove the following theorem.

Theorem II.25 *If $\tau = 1/\sqrt{2}$ and $\theta = 1/(3\sqrt{n})$, then the Dual Logarithmic Barrier Algorithm with full Newton steps requires at most*

$$\left\lceil 3\sqrt{n} \log \frac{n\mu^0}{\varepsilon} \right\rceil$$

iterations. The output is a primal-dual pair (x, s) such that $x^T s \leq 2\varepsilon$.

6.7.1   Convergence analysis

The proof of Theorem II.25 depends on the following lemma. The lemma generalizes Theorem II.21 to the case where, after the Newton step corresponding to the barrier parameter value μ, the barrier parameter is updated to $\mu^+ = (1 - \theta)\mu$. Taking $\theta = 0$ in the lemma we get back the result of Theorem II.21.

Lemma II.26 [8] *Assuming $\delta(s, \mu) \leq 1$, let s^+ be obtained from s by moving along the Newton step $\Delta s = p(s, \mu)$ at s corresponding to the barrier parameter value μ, and let $\mu^+ = (1 - \theta)\mu$. Then we have*

$$\delta(s^+, \mu^+)^2 \leq \delta(s, \mu)^4 + \frac{\theta^2 n}{(1 - \theta)^2}.$$

Proof: By definition we have

$$\delta(s^+, \mu^+) = \frac{1}{\mu^+} \min_x \left\{ \|\mu^+ e - s^+ x\| \; : \; Ax = b \right\}$$

Substituting for x the vector $x(s, \mu)$ we obtain the inequality:

$$\delta(s^+, \mu^+) \leq \frac{1}{\mu^+} \|\mu^+ e - s^+ x(s, \mu)\| = \left\| e - \frac{s^+ x(s, \mu)}{\mu(1 - \theta)} \right\|.$$

From (6.10) we deduce that

$$s^+ x(s, \mu) = \mu \left(e - \left(s^{-1} \Delta s \right)^2 \right).$$

Substituting this, while simplifying the notation by using

$$h := s^{-1} \Delta s,$$

we get

$$\delta(s^+, \mu^+) \leq \left\| e - \frac{e - h^2}{1 - \theta} \right\| = \left\| h^2 - \frac{\theta}{1 - \theta} (e - h^2) \right\|. \tag{6.12}$$

To further simplify the notation we replace $\theta / (1 - \theta)$ by ρ. Then taking squares of both sides in the last inequality we obtain

$$\delta(s^+, \mu^+)^2 \leq \|h^2\|^2 - 2\rho \left(h^2 \right)^T \left(e - h^2 \right) + \rho^2 \left\| e - h^2 \right\|^2.$$

Since $\|h\| = \delta(s, \mu) \leq 1$ we have

$$0 \leq e - h^2 \leq e.$$

Hence we have

$$\left(h^2 \right)^T \left(e - h^2 \right) \geq 0, \quad \left\| e - h^2 \right\|^2 \leq \|e\|^2.$$

---

[8] This lemma and its proof are due to Ling [182]. It improves estimates used by Roos and Vial [245].

Using this, and also that $\|e\|^2 = n$, we obtain

$$\delta(s^+, \mu^+)^2 \leq \left\|h^2\right\|^2 + \rho^2 \|e\|^2 \leq \|h\|^4 + \rho^2 n = \delta(s, \mu)^4 + \rho^2 n,$$

thus proving the lemma. □

Remark II.27 It may be noted that a weaker result can be obtained in a more simple way by applying the triangle inequality to (6.12). This yields

$$\delta(s^+, \mu^+) \leq \left\|h^2\right\| + \frac{\theta}{1-\theta} \left\|e - h^2\right\| \leq \delta(s, \mu)^2 + \frac{\theta\sqrt{n}}{1-\theta}.$$

This result is strong enough to derive a polynomial iteration bound, but the resulting bound will be slightly weaker than the one in Theorem II.25. •

The proof of Theorem II.25 goes now as follows. Taking $\theta = 1/(3\sqrt{n})$, we have

$$\frac{\theta\sqrt{n}}{1-\theta} = \frac{\frac{1}{3}}{1 - \frac{1}{3\sqrt{n}}} \leq \frac{\frac{1}{3}}{\frac{2}{3}} = \frac{1}{2}.$$

Hence, applying the lemma, we obtain

$$\delta(s^+, \mu^+)^2 \leq \delta(s, \mu)^4 + \frac{1}{4}.$$

Therefore, if $\delta(s, \mu) \leq \tau = 1/\sqrt{2}$, then we obtain

$$\delta(s^+, \mu^+)^2 \leq \frac{1}{4} + \frac{1}{4} = \frac{1}{2},$$

which implies that $\delta(s^+, \mu^+) \leq 1/\sqrt{2} = \tau$. Thus it follows that after each iteration of the algorithm the property

$$\delta(s, \mu) \leq \tau$$

is maintained. The iteration bound in the theorem is an immediate consequence of Lemma I.36. Finally, if s is the dual iterate at termination of the algorithm, and μ the value of the barrier parameter, then with $x = x(s, \mu)$, Theorem II.23 yields

$$s^T x(s, \mu) \leq n\mu \left(1 + \delta(s, \mu)\right) \leq n\mu \left(1 + \tau\right) \leq 2n\mu.$$

Since upon termination we have $n\mu \leq \varepsilon$, it follows that $s^T x(s, \mu) \leq 2\varepsilon$. This completes the proof of the theorem. □

6.7.2   *Illustration of the algorithm with full Newton steps*

In this section we start with a straightforward application of the logarithmic barrier algorithm. After that we devote some sections to modifications of the algorithm that increase the practical efficiency of the algorithm without destroying the theoretical iteration bound.

As an example we solve the problem with A and c as in Example II.7, and with $b^T = (1,1)$. Written out, the (dual) problem is given by

$$\max \{y_1 + y_2 \,:\, -1 \leq y_1 \leq 1, \, y_2 \leq 1\}.$$

and the primal problem is

$$\min \{x_1 + x_2 + x_3 \,:\, x_1 - x_2 = 1, \, x_3 = 1, \, x \geq 0\}.$$

We can start the algorithm at $y = (0,0)$ and $\mu = 2$, because we then have $s = (1,1,1)$ and, since $x = (2,1,1)$ is primal feasible,

$$\delta(s,\mu) \leq \left\| \frac{sx}{\mu} - e \right\| = \left\| \begin{pmatrix} 0 \\ -\frac{1}{2} \\ -\frac{1}{2} \end{pmatrix} \right\| = \frac{1}{\sqrt{2}}.$$

With $\varepsilon = 10^{-4}$, the dual logarithmic barrier algorithm needs 53 iterations. to generate the primal feasible solution $x = (1.000015, 0.000015, 1.000000)$ and the dual feasible pair (y,s) with $y = (0.999971, 0.999971)$ and $s = (0.000029, 1.999971, 0.000029)$. The respective objective values are $c^T x = 2.000030$ and $b^T y = 1.999943$, and the duality gap is 0.000087.

Table 6.1. (page 124) shows some quantities generated in the course of the algorithm. For each iteration the table shows the values of $n\mu$, the first coordinate of $x(s,\mu)$, the coordinates of y, the first coordinate of s, the proximity $\delta = \delta(s,\mu)$ before and the proximity $\delta^+ = \delta(s^+,\mu)$ after the Newton step at y to the μ-center, and, in the last column, the barrier update parameter θ, which is constant in this example.

The columns for δ and δ^+ in Table 6.1. are of special interest. They make clear that the behavior of the algorithm differs from what might be expected. The analysis was based on the idea of maintaining the proximity of the iterates below the value $\tau = 1/\sqrt{2} = 0.7071$, so as to stay in the region where Newton's method is very efficient. Therefore we updated the barrier parameter in such a way that just before the Newton step, i.e., just after the update of the barrier parameter, the proximity should reach the value τ. The table makes clear that in reality the proximity takes much smaller values (soon after the start). Asymptotically the proximity before the Newton step is always 0.2721 and after the Newton step 0.0524.

This can also be seen from Figure 6.8, which shows the relevant part of the feasible region and the central path. The points y are indicated by small circles and the exact μ-centers as asterisks. The above observation becomes very clear in this figure: soon after the start the circles and the asterisks can hardly be distinguished. The figure also shows at each iteration the region where the proximity is smaller than τ, thus indicating the space where we are allowed to move without leaving the region of quadratic convergence. Instead of using this space the algorithm moves in a very narrow neighborhood of the central path.

6.8 A version of the algorithm with adaptive updates

The example in the previous section has been discussed in detail in the hope that the reader will now understand that there is an easy way to reduce the number of iterations

It.	$n\mu$	x_1	y_1	y_2	s_1	δ	δ^+	θ
0	6.000000	2.500000	0.000000	0.000000	1.000000	0.6124	0.2509	0.1925
1	4.845299	2.255388	0.250000	-0.500000	0.750000	0.0901	0.0053	0.1925
2	3.912821	1.969957	0.285497	-0.606897	0.714503	0.2491	0.0540	0.1925
3	3.159798	1.749168	0.342068	-0.234058	0.657932	0.2003	0.0303	0.1925
4	2.551695	1.578422	0.403015	-0.022234	0.596985	0.2334	0.0420	0.1925
5	2.060621	1.447319	0.467397	0.184083	0.532603	0.2285	0.0379	0.1925
6	1.664054	1.347011	0.532510	0.337370	0.467490	0.2406	0.0416	0.1925
7	1.343807	1.270294	0.595745	0.466322	0.404255	0.2438	0.0421	0.1925
8	1.085191	1.211482	0.654936	0.568477	0.345064	0.2500	0.0441	0.1925
9	0.876346	1.166207	0.708650	0.651736	0.291350	0.2537	0.0453	0.1925
10	0.707693	1.131170	0.756184	0.718677	0.243816	0.2574	0.0467	0.1925
11	0.571498	1.103907	0.797423	0.772849	0.202577	0.2601	0.0477	0.1925
12	0.461513	1.082581	0.832650	0.816552	0.167350	0.2624	0.0486	0.1925
13	0.372695	1.065815	0.862383	0.851862	0.137617	0.2643	0.0493	0.1925
14	0.300969	1.052577	0.887244	0.880369	0.112756	0.2658	0.0499	0.1925
15	0.243048	1.042085	0.907881	0.903393	0.092119	0.2670	0.0504	0.1925
16	0.196273	1.033741	0.924914	0.921985	0.075086	0.2680	0.0508	0.1925
17	0.158500	1.027088	0.938910	0.936999	0.061090	0.2688	0.0511	0.1925
18	0.127997	1.021771	0.950370	0.949123	0.049630	0.2695	0.0513	0.1925
19	0.103364	1.017513	0.959728	0.958915	0.040272	0.2700	0.0515	0.1925
20	0.083472	1.014098	0.967352	0.966821	0.032648	0.2704	0.0517	0.1925
21	0.067407	1.011356	0.973553	0.973207	0.026447	0.2707	0.0518	0.1925
22	0.054435	1.009152	0.978589	0.978363	0.021411	0.2710	0.0519	0.1925
23	0.043959	1.007378	0.982674	0.982527	0.017326	0.2712	0.0520	0.1925
24	0.035499	1.005950	0.985986	0.985890	0.014014	0.2714	0.0521	0.1925
25	0.028667	1.004800	0.988668	0.988605	0.011332	0.2716	0.0521	0.1925
26	0.023150	1.003873	0.990839	0.990798	0.009161	0.2717	0.0522	0.1925
27	0.018695	1.003125	0.992596	0.992569	0.007404	0.2718	0.0522	0.1925
28	0.015097	1.002522	0.994017	0.993999	0.005983	0.2718	0.0523	0.1925
29	0.012192	1.002036	0.995165	0.995154	0.004835	0.2719	0.0523	0.1925
30	0.009845	1.001643	0.996094	0.996087	0.003906	0.2720	0.0523	0.1925
31	0.007951	1.001327	0.996845	0.996840	0.003155	0.2720	0.0523	0.1925
32	0.006421	1.001071	0.997451	0.997448	0.002549	0.2720	0.0523	0.1925
33	0.005185	1.000865	0.997941	0.997939	0.002059	0.2721	0.0523	0.1925
34	0.004187	1.000698	0.998337	0.998336	0.001663	0.2721	0.0523	0.1925
35	0.003381	1.000564	0.998657	0.998656	0.001343	0.2721	0.0524	0.1925
36	0.002731	1.000455	0.998915	0.998915	0.001085	0.2721	0.0524	0.1925
37	0.002205	1.000368	0.999124	0.999124	0.000876	0.2721	0.0524	0.1925
38	0.001781	1.000297	0.999292	0.999292	0.000708	0.2721	0.0524	0.1925
39	0.001438	1.000240	0.999429	0.999428	0.000571	0.2721	0.0524	0.1925
40	0.001161	1.000194	0.999539	0.999538	0.000461	0.2721	0.0524	0.1925
41	0.000938	1.000156	0.999627	0.999627	0.000373	0.2721	0.0524	0.1925
42	0.000757	1.000126	0.999699	0.999699	0.000301	0.2721	0.0524	0.1925
43	0.000612	1.000102	0.999757	0.999757	0.000243	0.2722	0.0524	0.1925
44	0.000494	1.000082	0.999804	0.999804	0.000196	0.2722	0.0524	0.1925
45	0.000399	1.000066	0.999841	0.999841	0.000159	0.2722	0.0524	0.1925
46	0.000322	1.000054	0.999872	0.999872	0.000128	0.2722	0.0524	0.1925
47	0.000260	1.000043	0.999897	0.999897	0.000103	0.2722	0.0524	0.1925
48	0.000210	1.000035	0.999917	0.999917	0.000083	0.2722	0.0524	0.1925
49	0.000170	1.000028	0.999933	0.999933	0.000067	0.2722	0.0524	0.1925
50	0.000137	1.000023	0.999946	0.999946	0.000054	0.2722	0.0524	0.1925
51	0.000111	1.000018	0.999956	0.999956	0.000044	0.2722	0.0524	0.1925
52	0.000089	1.000015	0.999964	0.999964	0.000036	0.2722	0.0524	0.1925
53	0.000072	1.000015	0.999971	0.999971	0.000029	—	—	—

Table 6.1. Output of the dual full-step algorithm.

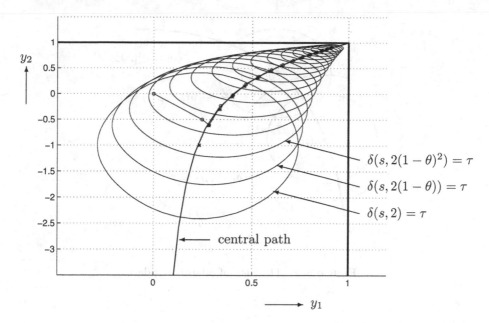

Figure 6.8   Iterates of the dual logarithmic barrier algorithm.

required by the algorithm without losing the quality of the solution guaranteed by Theorem II.25. The obvious way to reach this goal is to make larger updates of the barrier parameter while keeping the iterates in the region of quadratic convergence.

This is called the *adaptive-update strategy*,[9] which we discuss in the next section. After that we deal with a more greedy approach, using larger updates of the barrier parameter, and in which we may leave temporarily the region of quadratic convergence. This is the so-called *large-update strategy*. The analysis of the large-update strategy cannot be based on the proximity measure $\delta(y, \mu)$ alone, because outside the region of quadratic convergence this measure has no useful meaning. But, as we shall see, there exists a different way of measuring the progress of the algorithm in that case.

6.8.1   An adaptive-update variant

Observe that the iteration bound of Theorem II.25 was obtained by requiring that after each update of the barrier parameter μ the proximity satisfies

$$\delta(s, \mu) \leq \tau. \tag{6.13}$$

In order to make clear how this observation can be used to improve the performance of the algorithm without losing the iteration bound of Theorem II.25, let us briefly recall the idea behind the proof of this theorem. At the start of an iteration we are given s and μ such that (6.13) holds. We then make a Newton step to the μ-center,

---

[9] The adaptive-update strategy was first proposed by Ye [303]. See also Roos and Vial [245].

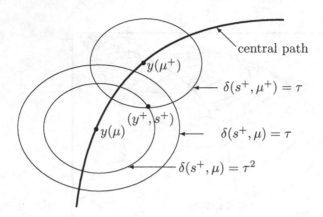

Figure 6.9 The idea of adaptive updating.

which yields s^+, and we have

$$\delta(s^+, \mu) \leq \tau^2. \tag{6.14}$$

Then we update μ to a smaller value $\mu^+ = (1 - \theta)\mu$ such that

$$\delta(s^+, \mu^+) \leq \tau, \tag{6.15}$$

and we start the next iteration. Our estimates in the proof of Theorem II.25 were such that it has become clear that the value $\theta = 1/(3\sqrt{n})$ guarantees that (6.15) will hold. But from the example in the previous section we know that actually the new proximity may be much smaller than τ. In other words, it may well happen that using the given value of θ we start the next iteration with an s^+ and a μ^+ such that $\delta(s^+, \mu^+)$ is (much) smaller than τ.

It will be clear that this opens a way to speed up the algorithm without degrading the iteration bound. For if we take θ larger than the value $\theta = 1/(3\sqrt{n})$ used in Theorem II.25, thus enforcing a deeper update of the barrier parameter in such a way that (6.15) still holds, then the analysis in the proof of Theorem II.25 remains valid but the number of iterations decreases. The question arises of how deep the update might be. In other words, we have to deal with the problem that we are given s^+ and μ such that (6.14) holds, and we ask how large we can take θ in $\mu^+ = (1 - \theta)\mu$ so that (6.15) holds with equality:

$$\delta(s^+, \mu^+) = \tau.$$

See Figure 6.9. Note that we know beforehand that this value of θ is at least $\theta = 1/(3\sqrt{n})$.

To answer the above question we need to introduce the so-called *affine-scaling direction* and the *centering direction* at s.

6.8.2 The affine-scaling direction and the centering direction

From (6.1) we recall that the Newton step at s to the μ-center is given by

$$\mu s^{-1} \Delta s = P_{HS} \left(\mu e - s\bar{x} \right),$$

so we may write

$$\Delta s = S P_{HS} \left(e - \frac{s\bar{x}}{\mu} \right) = S P_{HS} \left(e \right) - \frac{1}{\mu} S P_{HS} s\bar{x}.$$

The directions

$$\Delta^c s := S P_{HS} \left(e \right) \tag{6.16}$$

and

$$\Delta^a s := -S P_{HS} \left(s\bar{x} \right) \tag{6.17}$$

are called the *centering direction* and the *affine-scaling direction* respectively. Note that these two directions depend only on the iterate s and not on the barrier parameter μ. Now the Newton step at s to the μ-center can be written as

$$\Delta s = \Delta^c s + \frac{1}{\mu} \Delta^a s,$$

and the definition (6.7) of the proximity $\delta(s, \mu)$ implies

$$\delta(s, \mu) = \left\| d^c + \frac{1}{\mu} d^a \right\|,$$

where

$$d^c = s^{-1} \Delta^c s, \quad d^a = s^{-1} \Delta^a s$$

are the scaled centering and affine-scaling directions respectively.

6.8.3 Calculation of the adaptive update

Now that we know how the proximity depends on the barrier parameter we are able to solve the problem posed above. We assume that we have an iterate s such that for some $\mu > 0$ and $0 < \tau < 1/\sqrt{2}$,

$$\delta := \delta(s, \mu) \leq \tau^2,$$

and we ask for the smallest value μ^+ of the barrier parameter such that

$$\delta(s, \mu^+) = \tau.$$

Clearly, μ^+ is the smallest positive root of the equation

$$\delta(s, \mu) = \left\| d^c + \frac{1}{\mu} d^a \right\| = \tau. \tag{6.18}$$

Note that in the case where $b = 0$, the dual objective value is constant on the dual feasible region and hence s is optimal.[10,11] We assume that $d^a \neq 0$. This is true if and only if $b \neq 0$. It then follows from (6.18) that $\delta(s, \mu)$ depends continuously on μ and goes to infinity if μ approaches zero. Hence, since $\tau > \tau^2$, equation (6.18) has at least one positive solution.

Squaring both sides of (6.18), we arrive at the following quadratic equation in $1/\mu$:

$$\frac{1}{(\mu)^2} \|d^a\|^2 + \frac{2}{\mu}(d^a)^T d^c + \|d^c\|^2 - \tau^2 = 0. \tag{6.19}$$

The two roots of (6.19) are given by

$$\frac{-(d^a)^T d^c \pm \sqrt{((d^a)^T d^c)^2 - \|d^a\|^2 \left(\|d^c\|^2 - \tau^2\right)}}{\|d^a\|^2}.$$

We already know that at least one of the roots is positive. Hence, although we do not know the sign of the second root, we may conclude that $1/\mu^*$, where μ^* is the value of the barrier parameter we are looking for, is equal to the larger of the two roots. This gives, after some elementary calculations,

$$\mu^* = \frac{\|d^c\|^2 - \tau^2}{(d^a)^T d^c + \sqrt{((d^a)^T d^c)^2 - \|d^a\|^2 \left(\|d^c\|^2 - \tau^2\right)}}.$$

It is interesting to observe that it is easy to characterize the situation that both roots of (6.18) are positive. By considering the constant term in the quadratic equation (6.19) we see that both roots are positive if and only if $\|d^c\|^2 - \tau^2 > 0$. From (6.18) it follows that $\|d^c\| = \delta(s, \infty)$. Thus, both roots are positive if and only if

$$\delta(s, \infty) > \tau.$$

Obviously this situation occurs only if

$$(d^a)^T d^c < 0.$$

Thus we find the interesting result

$$\delta(s, \infty) > \tau \quad \Rightarrow \quad (d^a)^T d^c < 0.$$

At the central path, when $\delta(s, \mu) = 0$, we have $d^a = -\mu d^c$, so in that case the above implication is obvious.

---

[10] **Exercise 37** Show that $d^a = 0$ if and only if $b = 0$.

[11] **Exercise 38** Consider the case $b = 0$. Then the primal feasibility condition is $Ax = 0$, $x \geq 0$, which is homogeneous in x. Show that $x(s, \mu) = \mu x(s, 1)$ for each $\mu > 0$, and that $\delta(s, \mu)$ is independent of μ. Taking $s = s(1)$, it now easily follows that $s(\mu) = s(1)$ for each $\mu > 0$. This means that the dual central path is a point in this case, whereas the primal central path is a straight half line. If s and $\mu > 0$ are such that $\delta(s, \mu) < 1$ then the Newton process converges quadratically to $s(1)$, which is the analytic center of the dual feasible region. See also Roos and Vial [243] and Ye [310].

6.8.4 Illustration of the use of adaptive updates

By way of example we solve the same problem as in Section 6.7.2 with the dual logarithmic barrier algorithm, now using adaptive updates. As before, we start the algorithm at $y = (0,0)$ and $\mu = 2$. With $\varepsilon = 10^{-4}$ and adaptive updates, the dual full-step algorithm needs 20 iterations to generate the primal feasible solution $x = (1.000013, 0.000013, 1.000000)$ and the dual feasible pair (y, s) with $y = (0.999973, 0.999986)$ and $s = (0.000027, 1.999973, 0.000014)$. The respective objective values are $c^T x = 2.000027$ and $b^T y = 1.999960$, and the duality gap is 0.000067. Table 6.2. (page 129) gives some information on how the algorithm progresses. From the seventh column in this table (with the heading δ) it is clear that we have reached our goal: after each update of the barrier parameter the proximity equals τ. Moreover, the adaptive barrier parameter update strategy reduced the number of iterations, from 53 to 20.

Figure 6.10 (page 130) provides a graphical illustration of the adaptive strategy. It shows the relevant part of the feasible region and the central path, as well as the first four points generated by the algorithm and their regions of quadratic convergence. After each update the iterate lies on the boundary of the region of quadratic convergence for the next value of the barrier parameter.

It.	$n\mu$	x_1	y_1	y_2	s_1	δ	δ^+	θ
0	3.000000	1.500000	0.000000	0.000000	1.000000	0.7071	0.1581	0.5000
1	1.778382	1.374235	0.500000	0.000000	0.500000	0.7071	0.4725	0.4072
2	0.863937	1.149409	0.579559	0.686927	0.420441	0.7071	0.4563	0.5142
3	0.505477	1.091171	0.864662	0.714208	0.135338	0.7071	0.4849	0.4149
4	0.280994	1.047762	0.847943	0.913169	0.152057	0.7071	0.4912	0.4441
5	0.165317	1.028293	0.954529	0.906834	0.045471	0.7071	0.4776	0.4117
6	0.093710	1.015735	0.947640	0.971182	0.052360	0.7071	0.4937	0.4332
7	0.055038	1.009255	0.984428	0.968951	0.015572	0.7071	0.4799	0.4127
8	0.031566	1.005275	0.982196	0.990450	0.017804	0.7071	0.4915	0.4265
9	0.018469	1.003087	0.994677	0.989568	0.005323	0.7071	0.4827	0.4149
10	0.010662	1.001779	0.993971	0.996813	0.006029	0.7071	0.4894	0.4227
11	0.006220	1.001038	0.998188	0.996484	0.001812	0.7071	0.4846	0.4167
12	0.003603	1.000601	0.997961	0.998931	0.002039	0.7071	0.4881	0.4207
13	0.002098	1.000350	0.999385	0.998814	0.000615	0.7071	0.4856	0.4177
14	0.001217	1.000203	0.999311	0.999640	0.000689	0.7071	0.4874	0.4198
15	0.000708	1.000118	0.999792	0.999599	0.000208	0.7071	0.4862	0.4183
16	0.000411	1.000069	0.999767	0.999879	0.000233	0.7071	0.4870	0.4193
17	0.000239	1.000040	0.999930	0.999865	0.000070	0.7071	0.4864	0.4186
18	0.000139	1.000023	0.999921	0.999959	0.000079	0.7071	0.4868	0.4191
19	0.000081	1.000013	0.999976	0.999954	0.000024	0.7071	0.4865	0.4187
20	0.000081	1.000013	0.999973	0.999986	0.000027	—	—	—

Table 6.2. Output of the dual full-step algorithm with adaptive updates.

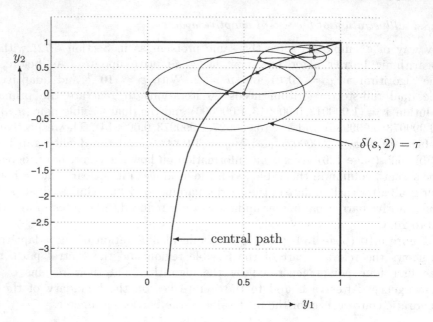

Figure 6.10 The iterates when using adaptive updates.

6.9 A version of the algorithm with large updates

In this section we consider a more greedy approach than the adaptive strategy, using larger updates of the barrier parameter. As before, we assume that we have an iterate s and a $\mu > 0$ such that s belongs to the region of quadratic convergence around the μ-center. In fact we assume that[12]

$$\delta(s, \mu) \leq \tau = \frac{1}{\sqrt{2}}.$$

Starting at s we want to reach the region of quadratic convergence around the μ^+-center, with

$$\mu^+ = (1 - \theta)\mu,$$

and we assume that θ is so large that s lies outside the region of quadratic convergence around the μ^+-center. In fact, it may well happen that $\delta(s, \mu^+)$ is much larger than 1. It is clear that the analysis of the previous sections, where we always took full Newton steps for the target value of the barrier parameter, is then no longer useful: this analysis was based on the nice behavior of Newton's method in a close neighborhood of the μ^+-center. Being outside this region, we can no longer profit from this nice behavior and we need an alternative approach.

Now remember that the target center $s(\mu^+)$ can be characterized as the (unique)

---

[12] We could have taken a different value for τ, for example $\tau = 1/2$, but the choice $\tau = 1/\sqrt{2}$ seems to be natural. The analysis below supports our choice. In the literature the choice $\tau = 1/2$ is very popular (see, e.g., [140]). It is easy to adapt the analysis below to this value.

minimizer of the dual logarithmic barrier function

$$k_{\mu^+}(y, s) = -b^T y - \mu^+ \sum_{j=1}^{n} \log s_j$$

and that this function is strictly convex on the interior of the dual feasible region. Hence, the difference

$$k_{\mu^+}(y, s) - k_{\mu^+}(y(\mu^+), s(\mu^+))$$

vanishes if and only if $s = s(\mu^+)$ and is positive elsewhere. The difference can therefore be used as another indicator for the 'distance' from s to $s(\mu^+)$. That is exactly what we plan to do. Outside the region of quadratic convergence the barrier function value will act as a measure for proximity to the μ-center. We show that when moving in the direction of the Newton step at s the barrier function decreases, and that by choosing an appropriate step-size we can guarantee a sufficient decrease of the barrier function value. In principle, the step-size can be obtained from a one-dimensional line search in the Newton direction so as to minimize the barrier function in this direction. Based on these ideas we derive an upper bound for the required number of *damped Newton steps* to reach the vicinity of $s(\mu^+)$; the upper bound will be a function of θ.

The algorithm is described on page 131. We refer to the first **while**-loop in the

---

Dual Logarithmic Barrier Algorithm with Large Updates

Input:
    A proximity parameter $\tau = 1/\sqrt{2}$;
    an accuracy parameter $\varepsilon > 0$;
    a variable damping factor α;
    an update parameter θ, $0 < \theta < 1$;
    $(y^0, s^0) \in \mathcal{D}$ and $\mu^0 > 0$ such that $\delta(s^0, \mu^0) \leq \tau$.

begin
    $s := s^0$; $\mu := \mu^0$;
    while $n\mu \geq \varepsilon$ **do**
    begin
        $\mu := (1 - \theta)\mu$;
        while $\delta(s, \mu) \geq \tau$ **do**
        begin
            $s := s + \alpha p(s, \mu)$;
            (The damping factor α must be such that $k_\mu(y, s)$ decreases
            sufficiently. The default value is $1/(1 + \delta(s, \mu))$.)
        end
    end
end

---

algorithm as the *outer loop* and to the second **while**-loop as the *inner loop*. Each execution of the outer loop is called an *outer iteration* and each execution of the inner loop an *inner iteration*. The required number of outer iterations depends only on the dimension n of the problem, on μ^0 and ε, and on the (fixed) barrier update parameter θ. This number immediately follows from Lemma I.36. The main task in the analysis of the algorithm is to derive an upper bound for the number of iterations in the inner loop. For that purpose we need some lemmas that estimate barrier function values and objective values in the region of quadratic convergence around the μ-center. Since these estimates are interesting in themselves, and also because their importance goes beyond the analysis of the present algorithm with line searches alone, we discuss them in separate sections.

6.9.1 Estimates of barrier function values

We start with the barrier function values. Our goal is to estimate dual barrier function values in the region of quadratic convergence around the μ-center. It will be convenient not to deal with the barrier function itself, but to scale it by the barrier parameter. Therefore we introduce

$$h_\mu(s) := \frac{1}{\mu} k_\mu(y, s) = \frac{-b^T y}{\mu} - \sum_{j=1}^n \log s_j.$$

Let us point out once more that y is omitted in the argument of $h_\mu(s)$ because of the one-to-one correspondence between y and s in dual feasible pairs (y, s). We also use the primal barrier function scaled by μ:

$$g_\mu(x) := \frac{1}{\mu} \tilde{g}_\mu(x) = \frac{c^T x}{\mu} - \sum_{j=1}^n \log x_j.$$

Recall that both barrier functions are strictly convex on their domain and that $s(\mu)$ and $x(\mu)$ are their respective minimizers. Therefore, defining

$$\phi_\mu^p(x) := g_\mu(x) - g_\mu(x(\mu)), \quad \phi_\mu^d(s) := h_\mu(s) - h_\mu(s(\mu)),$$

we have $\phi_\mu^d(s) \geq 0$, with equality if and only if $s = s(\mu)$, and also $\phi_\mu^p(x) \geq 0$, with equality if and only if $x = x(\mu)$. As a consequence, defining

$$\phi_\mu(x, s) := \phi_\mu^p(x) + \phi_\mu^d(s), \tag{6.20}$$

where (x, s) is any pair of positive primal and dual feasible solutions, we have $\phi_\mu(x, s) \geq 0$, and the equality holds if and only if $x = x(\mu)$ and $s = s(\mu)$. The function $\phi_\mu : \mathcal{P}^+ \times \mathcal{D}^+ \to \mathbb{R}^+$ is called the *primal-dual logarithmic barrier function with barrier parameter* μ. Now the following lemma is almost obvious.

Lemma II.28 *Let $x > 0$ be primal feasible and $s > 0$ dual feasible. Then*

$$\phi_\mu^p(x) = \phi_\mu(x, s(\mu)) \leq \phi_\mu(x, s) \quad \text{and} \quad \phi_\mu^d(s) = \phi_\mu(x(\mu), s) \leq \phi_\mu(x, s).$$

Proof: The inequalities in the lemma are immediate from (6.20) since $\phi_\mu^p(x)$ and $\phi_\mu^d(s)$ are nonnegative. Similarly, the equalities follow since $\phi_\mu^p(x(\mu)) = \phi_\mu^d(s(\mu)) = 0$. Thus the lemma has been proved. $\qquad\square$

In the sequel, properties of the function ϕ_μ form the basis of many of our estimates. These estimates follow from properties of the univariate function

$$\psi(t) = t - \log(1 + t), \quad t > -1, \tag{6.21}$$

as defined in (5.5).[13] The definition of ψ is extended to any vector $z = (z_1, z_2, \ldots, z_n)$ satisfying $z + e > 0$ according to

$$Psi(z) = \sum_{j=1}^n \psi(z_j) = \sum_{j=1}^n (z_j - \log(1 + z_j)) = e^T z - \sum_{j=1}^n \log(1 + z_j). \tag{6.22}$$

We now make a crucial observation, namely that the barrier functions $\phi_\mu(x, s)$, $\phi_\mu^p(x)$ and $\phi_\mu^d(s)$ can be nicely expressed in terms of the function Ψ.

Lemma II.29 *Let $x > 0$ be primal feasible and $s > 0$ dual feasible. Then*

(i) $\phi_\mu(x, s) = \Psi\left(\frac{xs}{\mu} - e\right)$;

(ii) $\phi_\mu^p(x) = \Psi\left(\frac{xs(\mu)}{\mu} - e\right)$;

(iii) $\phi_\mu^d(s) = \Psi\left(\frac{x(\mu)s}{\mu} - e\right)$.

Proof: [14] First we consider item (i). We use that $c^T x - b^T y = x^T s$ and $c^T x(\mu) - b^T y(\mu) = x(\mu)^T s(\mu) = n\mu$. Now $\phi_\mu(x, s)$ can be reduced as follows:

$$
\begin{aligned}
\phi_\mu(x, s) \quad &= \quad h_\mu(s) + g_\mu(x) - (h_\mu(s(\mu)) + g_\mu(x(\mu))) \\
&= \quad \frac{x^T s}{\mu} - \sum_{j=1}^n \log x_j s_j - \frac{x(\mu)^T s(\mu)}{\mu} + \sum_{j=1}^n \log z_j(\mu) s_j(\mu) \\
&= \quad \frac{x^T s}{\mu} - \sum_{j=1}^n \log x_j s_j - n + n \log \mu.
\end{aligned}
$$

Since $x^T s = e^T(xs)$ and $e^T e = n$, we find the following expression for $\phi_\mu(x, s)$:[15]

$$\phi_\mu(x, s) = e^T\left(\frac{sx}{\mu} - e\right) - \sum_{j=1}^n \log \frac{x_j s_j}{\mu} = \Psi\left(\frac{sx}{\mu} - e\right). \tag{6.23}$$

This proves the first statement in the lemma. The second statement follows by substituting $s = s(\mu)$ in the first statement, and using Lemma II.28. Similarly, the third statement follows by substituting $x = x(\mu)$ in the first statement. $\square$

_____

[13] **Exercise 39** Let $t > -1$. Prove that

$$\psi\left(\frac{-t}{1+t}\right) + \psi(t) = \frac{t^2}{1+t}.$$

[14] Note that the dependence of $\phi_\mu(x, s)$ on x and s is such that it depends only on the coordinatewise product xs of x and s.

[15] **Exercise 40** Considering (6.23) as the definition of $\phi_\mu(x, s)$, and without using the properties of ψ, show that $\phi_\mu(x, s)$ is nonnegative, and zero if and only if $xs = \mu e$. (Hint: Use the arithmetic-geometric-mean inequality.)

Now we are ready to derive lower and upper bounds for the value of the dual logarithmic barrier function in the region of quadratic convergence around the μ-center. These bounds heavily depend on the following two inequalities:

$$\psi\left(\|z\|\right) \leq \Psi(z) \leq \psi\left(-\|z\|\right), \quad z > -e. \tag{6.24}$$

The second inequality is valid only if $\|z\| < 1$. The inequalities in (6.24) are fundamental for our purpose and are immediate consequences of Lemma C.2 in Appendix C.[16,17]

Lemma II.30 [18] *Let* $\delta := \delta(s, \mu)$. *Then*

$$\phi_\mu^d(s) \geq \delta - \log(1 + \delta) = \psi(\delta).$$

Moreover, if $\delta < 1$, *then*

$$\phi_\mu^d(s) \leq \phi_\mu(x(s,\mu), s) \leq -\delta - \log(1 - \delta) = \psi(-\delta).$$

Proof: By applying the inequalities in (6.24) to (6.23) we obtain for any positive primal feasible x:

$$\psi\left(\left\|\frac{sx}{\mu} - e\right\|\right) \leq \phi_\mu(x,s) \leq \psi\left(-\left\|\frac{sx}{\mu} - e\right\|\right), \tag{6.25}$$

where the second inequality is valid only if the norm of $xs/\mu - e$ does not exceed 1. Using (6.8) we write

$$\delta = \delta(s,\mu) = \left\|e - \frac{sx(s,\mu)}{\mu}\right\| \leq \left\|e - \frac{sx}{\mu}\right\|.$$

Hence, by the monotonicity of $\psi(t)$ for $t \geq 0$,

$$\psi(\delta) \leq \psi\left(\left\|\frac{sx}{\mu} - e\right\|\right),$$

---

[16] At least one of the inequalities in (6.24) shows up in almost every paper on interior-point methods. As far as we know, all usual proofs use the power series expansion of $\log(1 + x)$, $-1 < x < 1$ and do not characterize the case of equality, at least not explicitly. We give an elementary proof in Appendix C (page 435).

[17] **Exercise 41** Let $z \in \mathbb{R}^n$. Prove that

$$z \geq 0 \quad \Rightarrow \quad \Psi(z) \leq n\psi\left(\frac{\|z\|}{\sqrt{n}}\right) \leq \frac{\|z\|^2}{2}$$

$$-e < z \leq 0 \quad \Rightarrow \quad \Psi(z) \geq n\psi\left(\frac{-\|z\|}{\sqrt{n}}\right) \geq \frac{\|z\|^2}{2}.$$

[18] This lemma improves a similar result of den Hertog et al. [146] and den Hertog [140]. The improvement is due to a suggestion made by Osman Güler [130] during a six month stay at Delft in 1992, namely to use the primal logarithmic barrier function in the analysis of the dual logarithmic barrier method. This approach not only simplifies the analysis significantly, but also leads to sharper estimates. It may be appropriate to mention that even stronger bounds for $\phi_\mu(x, s)$ will be derived in Lemma II.69, but there we use a different proximity measure.

for any positive primal feasible x. Taking $x = x(\mu)$ and using the left inequality in (6.25) and the third statement in Lemma II.29, we get

$$\psi(\delta) \leq \psi\left(\left\|\frac{sx(\mu)}{\mu} - e\right\|\right) \leq \phi_\mu(x(\mu), s) = \phi_\mu^d(s),$$

proving the first inequality in the lemma. For the proof of the second inequality in the lemma we assume $\delta < 1$ and put $x = x(s, \mu)$ in the right inequality in (6.25). This gives

$$\phi_\mu(x(s, \mu), s) \leq \psi\left(-\left\|\frac{sx(s, \mu)}{\mu} - e\right\|\right) = \psi(-\delta).$$

By Lemma II.28 we also have $\phi_\mu^d(s) \leq \phi_\mu(x(s, \mu), s)$. Thus the lemma follows. $\square$

The functions $\psi(\delta)$ and $\psi(-\delta)$, for $0 \leq \delta < 1$, play a dominant role in many of the estimates below. Figure 6.11 shows their graphs.

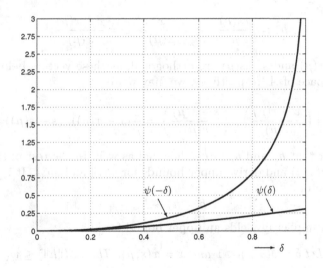

Figure 6.11 The functions $\psi(\delta)$ and $\psi(-\delta)$ for $0 \leq \delta < 1$.

6.9.2 Estimates of objective values

We proceed by considering the dual objective value $b^T y$ in the region of quadratic convergence around the μ-center. Using that $x(\mu)s(\mu) = \mu e$ and $c^T x(\mu) - b^T y(\mu) = x(\mu)^T s(\mu) = n\mu$, we write

$$
\begin{aligned}
b^T y(\mu) - b^T y &= c^T x(\mu) - n\mu - b^T y = s^T x(\mu) - n\mu = e^T (sx(\mu) - \mu e) \\
&= \mu e^T \left(\frac{sx(\mu)}{\mu} - e\right) = \mu e^T \left(\frac{s}{s(\mu)} - e\right).
\end{aligned} \tag{6.26}
$$

Applying the Cauchy–Schwarz inequality to the expression for $b^T y(\mu) - b^T y$ in (6.26), we obtain

$$\left| b^T y(\mu) - b^T y \right| \leq \mu \sqrt{n} \left\| \frac{s}{s(\mu)} - e \right\|. \tag{6.27}$$

We assume $\delta := \delta(s, \mu) < 1/\sqrt{2}$. It seems reasonable then to expect that the norm of the vector

$$h_s := \frac{s}{s(\mu)} - e = \frac{s x(\mu)}{\mu} - e$$

will not differ too much from δ. In any case, that is what we are going to show. It will then follow that the absolute value of $b^T y(\mu) - b^T y$ is of order $\mu \delta \sqrt{n}$.

Note that h_s can be written as

$$h_s = \frac{s - s(\mu)}{s(\mu)},$$

and hence $\|h_s\|$ measures the relative difference between s and $s(\mu)$. We also introduce a similar vector for any primal feasible $x > 0$:

$$h_x := \frac{x s(\mu)}{\mu} - e = \frac{x}{x(\mu)} - e = \frac{x - x(\mu)}{x(\mu)}.$$

Using that $x - x(\mu)$ and $s - s(\mu)$ are orthogonal, as these vectors belong to the null space and row space of A, respectively, we may write

$$h_x^T h_s = \left(\frac{x - x(\mu)}{x(\mu)} \right)^T \left(\frac{s - s(\mu)}{s(\mu)} \right) = \frac{1}{\mu} (x - x(\mu))^T (s - s(\mu)) = 0.$$

This makes clear that h_x and h_s are orthogonal as well. In the rest of this section we work with $x = x(s, \mu)$ and derive upper bounds for $\|h_x\|$ and $\|h_s\|$. It is convenient to introduce the vector

$$h = h_x + h_s.$$

The next lemma implicitly yields an upper bound for $\|h\|$.

Lemma II.31 *Let $\delta = \delta(s, \mu) < 1$ and $x = x(s, \mu)$. Then $\psi(\|h\|) \leq \psi(-\delta)$.*

Proof: Using Lemma II.29 we may rewrite (6.20) as

$$\phi_\mu(x, s) = \Psi(h_x) + \Psi(h_s).$$

By the first inequality in (6.24) we have

$$\Psi(h_x) \geq \psi(\|h_x\|) \quad \text{and} \quad \Psi(h_s) \geq \psi(\|h_s\|).$$

Applying the first inequality in (6.24) to the 2-dimensional vector $(\|h_x\|, \|h_s\|)$, we obtain

$$\psi(\|h_x\|) + \psi(\|h_s\|) \geq \psi(\|h\|).$$

Here we used that h_x and h_s are orthogonal. Substitution gives

$$\phi_\mu(x, s) \geq \psi(\|h\|).$$

On the other hand, by Lemma II.30 we have $\phi_\mu(x, s) \leq \psi(-\delta)$, thus completing the proof. $\qquad\qquad\qquad\qquad\qquad\qquad\qquad\qquad\qquad\qquad\qquad\qquad\qquad\qquad$ $\square$

Let us point out that we can easily deduce from Lemma II.31 an interesting upper bound for $\|h\|$ if $\delta < 1$. It can then be shown that $\psi(\|h\|) \leq \psi(-\delta)$ implies $\|h\| \leq \delta/(1 - \delta)$.[19,20] This implies that $\|h\| \leq 1$ if $\delta \leq 1/2$. However, for our purpose this bound is not strong enough. We prove a stronger result that implies that $\|h\| \leq 1$ if $\delta < 1/\sqrt{2}$.

Lemma II.32 *Let $\delta = \delta(s, \mu) \leq 1/\sqrt{2}$. Then $\|h\| < \sqrt{2}$.*

Proof: By Lemma II.31 we have $\psi(\|h\|) \leq \psi(-\delta)$. Since $\psi(-\delta)$ is monotonically increasing in δ, this implies

$$\psi(\|h\|) \leq \psi(-1/\sqrt{2}) = 0.52084.$$

Since

$$\psi(\sqrt{2}) = 0.53284 > 0.52084,$$

and $\psi(t)$ is monotonically increasing for $t \geq 0$, we conclude that $\|h\| < \sqrt{2}$. $\qquad$ $\square$

We now have the following result.

Lemma II.33 [21] *Let $\delta := \delta(s, \mu) \leq 1/\sqrt{2}$. Then*

$$\|h_s\| = \left\| \frac{s}{s(\mu)} - e \right\| \leq \sqrt{1 - \sqrt{1 - 2\delta^2}}.$$

Moreover, if $x = x(s, \mu)$ then also

$$\|h_x\| = \left\| \frac{x}{x(\mu)} - e \right\| \leq \sqrt{1 - \sqrt{1 - 2\delta^2}}.$$

Proof: Lemma II.32 implies that

$$\|h_x + h_s\| = \|h\| < \sqrt{2}.$$

On the other hand, since

$$\frac{xs}{\mu} - e = \frac{xs}{x(\mu)s(\mu)} - e = (e + h_x)(e + h_s) - e = h_x + h_s + h_x h_s,$$

with $x = x(s, \mu)$, and using (6.8), it follows that

$$\|h_x + h_s + h_x h_s\| = \delta \leq \frac{1}{\sqrt{2}}.$$

---

[19] **Exercise 42** Let $0 \leq t < 1$. Prove that

$$\psi\left(\frac{-t}{1+t}\right) \leq \frac{t^2}{2(1+t)} \leq \psi(t) \leq \frac{t^2}{2} \leq \psi(-t) \leq \frac{t^2}{2(1-t)} \leq \psi\left(\frac{t}{1-t}\right).$$

Also show that the first two inequalities are valid for any $t > 0$.
[20] **Exercise 43** Let $0 \leq \delta < 1$ and $r \geq 0$ be such that $\psi(r) \leq \psi(-\delta)$. Prove that $r \leq \delta/(1-\delta)$.
[21] For $\delta \leq 1/2$ this lemma was first shown by Gonzaga (private communication, Delft, 1994).

At this stage we may apply the fourth uv-lemma (Lemma C.8 in Appendix C) with $u = h_x$ and $v = h_s$, to obtain the lemma. □

We are now ready for the main result of this section.

Theorem II.34 *If $\delta = \delta(s, \mu) \le 1/\sqrt{2}$ then*

$$\left| b^T y(\mu) - b^T y \right| \le \mu \sqrt{n} \sqrt{1 - \sqrt{1 - 2\delta^2}}.$$

Proof: Recall from (6.27) that

$$\left| b^T y(\mu) - b^T y \right| \le \mu \sqrt{n} \, \|h_s\| \, .$$

Substituting the bound of Lemma II.33 on $\|h_s\|$, the theorem follows. □

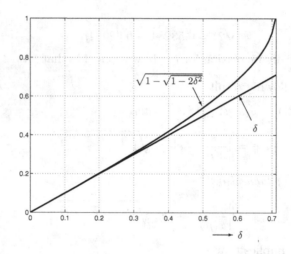

Figure 6.12 The graphs of δ and $\sqrt{1 - \sqrt{1 - 2\delta^2}}$ for $0 \le \delta \le 1/\sqrt{2}$.

Figure 6.12 (page 138) shows the graphs of δ and $\sqrt{1 - \sqrt{1 - 2\delta^2}}$. It is clear that for small values of δ ($\delta \le 0.3$ say) the functions can hardly be distinguished.

6.9.3   Effect of large update on barrier function value

We start by considering the effect of an update of the barrier parameter on the difference between the dual barrier function value and its minimal value. More precisely, we assume that for given dual feasible s and $\mu > 0$ we have $\delta = \delta(s, \mu) \le 1/\sqrt{2}$, and we want to estimate

$$\phi_{\mu^+}^d(s) = h_{\mu^+}(s) - h_{\mu^+}(s(\mu^+)),$$

where $\mu^+ = \mu(1 - \theta)$ for $0 \le \theta < 1$. Note that Lemma II.30 gives the answer if $\theta = 0$:

$$\phi_\mu^d(s) \le \psi(-\delta).$$

For the general case, where $\theta > 0$, we write

$$
\begin{aligned}
\phi_{\mu^+}^d(s) &= h_{\mu^+}(s) - h_{\mu^+}(s(\mu^+)) \\
&= h_{\mu^+}(s) - h_{\mu^+}(s(\mu)) + h_{\mu^+}(s(\mu)) - h_{\mu^+}(s(\mu^+)) \\
&= h_{\mu^+}(s) - h_{\mu^+}(s(\mu)) + \phi_{\mu^+}^d(s(\mu)), \qquad (6.28)
\end{aligned}
$$

and we treat the first two terms and the last term in the last expression separately.

Lemma II.35 *In the above notation,*

$$
h_{\mu^+}(s) - h_{\mu^+}(s(\mu)) \le \psi(-\delta) + \frac{\theta\sqrt{n}}{1-\theta}\sqrt{1 - \sqrt{1 - 2\delta^2}}.
$$

Proof: Just using definitions we write

$$
\begin{aligned}
h_{\mu^+}(s) - h_{\mu^+}(s(\mu)) &= \frac{-b^T y}{\mu^+} - \sum_{j=1}^n \log s_j + \frac{b^T y(\mu)}{\mu^+} + \sum_{j=1}^n \log s_j(\mu) \\
&= -\sum_{j=1}^n \log s_j + \sum_{j=1}^n \log s_j(\mu) + \frac{b^T y(\mu) - b^T y}{\mu^+} \\
&= h_\mu(s) - h_\mu(s(\mu)) + \frac{b^T y(\mu) - b^T y}{\mu^+} - \frac{b^T y(\mu) - b^T y}{\mu} \\
&= \phi_\mu^d(s) + \frac{\theta}{1-\theta}\frac{b^T y(\mu) - b^T y}{\mu}.
\end{aligned}
$$

Applying Lemma II.30 to the first term in the last expression, and Theorem II.34 to the second term gives the lemma. $\qquad\square$

Lemma II.36 *In the above notation,*

$$
\phi_{\mu^+}^d(s(\mu)) \le \phi_{\mu^+}(x(\mu), s(\mu)) = n\psi\left(\frac{\theta}{1-\theta}\right).
$$

Proof: The inequality follows from Lemma II.28. The equality is obtained as follows. From (6.23),

$$
\phi_{\mu^+}(x(\mu), s(\mu)) = e^T\left(\frac{s(\mu)x(\mu)}{\mu^+} - e\right) - \sum_{j=1}^n \log \frac{x_j(\mu)s_j(\mu)}{\mu^+}.
$$

Since $x(\mu)s(\mu) = \mu e$ and $\mu^+ = (1-\theta)\mu$, this can be simplified to

$$
\begin{aligned}
\phi_{\mu^+}(x(\mu), s(\mu)) &= e^T\left(\frac{\mu e}{\mu^+} - e\right) - \sum_{j=1}^n \log \frac{\mu}{\mu^+} \\
&= e^T\left(\frac{e}{1-\theta} - e\right) - \sum_{j=1}^n \log \frac{1}{1-\theta} \\
&= n\left(\frac{\theta}{1-\theta} - \log\left(1 + \frac{\theta}{1-\theta}\right)\right) \\
&= n\psi\left(\frac{\theta}{1-\theta}\right).
\end{aligned}
$$

This completes the proof. $\qquad\square$

Combining the results of the last two lemmas we find the next lemma.

Lemma II.37 *Let $\delta(s,\mu) \leq 1/\sqrt{2}$ for some dual feasible s and $\mu > 0$. Then, if $\mu^+ = \mu(1-\theta)$ with $0 \leq \theta < 1$, we have*

$$\phi^d_{\mu^+}(s) \leq \psi\left(\frac{-1}{\sqrt{2}}\right) + \frac{\theta\sqrt{n}}{1-\theta} + n\psi\left(\frac{\theta}{1-\theta}\right).$$

Proof: The lemma follows from (6.28) and the bounds provided by the previous lemmas, by substitution of $\delta = 1/\sqrt{2}$. $\qquad\square$

With s, μ and μ^+ as in the last lemma, our aim is to estimate the number of damped Newton steps required to reach the vicinity of the μ^+-center when starting at s. To this end we proceed by estimating the decrease in the barrier function value during a damped Newton step.

6.9.4 Decrease of the barrier function value

In this section we consider a damped Newton step to the μ-center at an arbitrary positive dual feasible s and we estimate its effect on the barrier function value. The analysis also yields a suitable value for the damping factor α. The result of the damped Newton step is denoted by s^+, so

$$s^+ = s + \alpha\Delta s, \qquad (6.29)$$

where Δs denotes the full Newton step.

Lemma II.38 *Let $\delta = \delta(s,\mu)$. If $\alpha = 1/(1+\delta)$ then the damped Newton step (6.29) is feasible and it reduces the barrier function value by at least $\delta - \log(1+\delta)$. In other words,*

$$\phi^d_\mu(s) - \phi^d_\mu(s^+) \geq \delta - \log(1+\delta) = \psi(\delta).$$

Proof: First recall from (6.5) in Section 6.5 that the Newton step Δs is determined by

$$x(s,\mu) = \mu s^{-1}\left(e - s^{-1}\Delta s\right).$$

We denote $x(s,\mu)$ briefly as x. With

$$z := \frac{\Delta s}{s} = e - \frac{xs}{\mu},$$

the damped Newton step can be described as follows:

$$s^+ = s + \alpha\Delta s = s(e + \alpha z).$$

Since s^+ is feasible if and only if it is nonnegative, the step is certainly feasible if $\alpha\|z\| < 1$. Since $\delta = \|z\|$, the value for α specified by the lemma satisfies this condition, and hence the feasibility of s^+ follows. Now we consider the decrease in the dual barrier function value during the step. We may write

$$
\begin{aligned}
\phi_\mu^d(s) - \phi_\mu^d(s^+) &= h_\mu(s) - h_\mu(s^+) \\
&= \frac{-b^T y}{\mu} - \sum_{j=1}^n \log s_j - \frac{-b^T y^+}{\mu} + \sum_{j=1}^n \log s_j^+ . \\
&= \frac{b^T y^+ - b^T y}{\mu} + \sum_{j=1}^n \log\left(1 + \alpha z_j\right) .
\end{aligned}
$$

The difference $b^T y^+ - b^T y$ can be written as follows:

$$
\begin{aligned}
b^T y^+ - b^T y &= c^T x - b^T y - \left(c^T x - b^T y^+\right) = x^T s - x^T s^+ \\
&= -\alpha x^T (sz) = -\alpha e^T (xs) z = \alpha \mu (z - e) z.
\end{aligned}
$$

Thus we obtain

$$
\begin{aligned}
\phi_\mu^d(s) - \phi_\mu^d(s^+) &= \alpha e^T (z - e) z + \sum_{j=1}^n \log\left(1 + \alpha z_j\right) \\
&= \alpha e^T z^2 - \left(e^T (\alpha z) - \sum_{j=1}^n \log\left(1 + \alpha z_j\right) \right) \\
&= \alpha \delta^2 - \Psi(\alpha z).
\end{aligned}
$$

Since $\|\alpha z\| < 1$ we may apply the right-hand side inequality in (6.24), which gives $\Psi(\alpha z) \le \psi\left(-\alpha \|z\|\right) = \psi\left(-\alpha\delta\right)$, whence

$$
\phi_\mu^d(s) - \phi_\mu^d(s^+) \ge \alpha \delta^2 - \psi(-\alpha\delta) = \alpha\delta^2 + \alpha\delta + \log(1 - \alpha\delta).
$$

As a function of α, the right-hand side expression is increasing for $0 \le \alpha \le 1/(1+\delta)$, as can be easily verified, and it attains its maximal value at $\alpha = 1/(1+\delta)$, which is the value specified in the lemma. Substitution of this value yields the bound in the lemma. Thus the proof is complete.[22,23] □

We are now ready to estimate the number of (inner) iterations between two successive updates of the barrier parameter.

6.9.5 Number of inner iterations

Lemma II.39 *The number of (inner) iterations between two successive updates of the barrier parameter is no larger than*

$$\left\lceil 3\left(\frac{\theta\sqrt{n}}{1-\theta}+1\right)^2\right\rceil.$$

Proof: From Lemma II.37 we know that after the update of μ we have

$$\phi_{\mu^+}^d(s) \leq \psi(-\tau) + \frac{\theta\sqrt{n}}{1-\theta} + n\psi\left(\frac{\theta}{1-\theta}\right),$$

where $\tau = 1/\sqrt{2}$. The algorithm repeats damped Newton steps as long the iterate s satisfies $\delta = \delta(s,\mu^+) > \tau$. In that case the step decreases the barrier function value by at least $\psi(\delta)$, by Lemma II.38. Since $\delta > \tau$, the decrease is at least

$$\psi(\tau) = 0.172307.$$

As soon as the barrier function value has reached $\psi(\tau)$ we are sure that $\delta(s,\mu^+) \leq \tau$, from Lemma II.30. Hence, the number of inner iterations is no larger than

$$\left\lceil \frac{1}{\psi(\tau)}\left(\psi(-\tau) - \psi(\tau) + \frac{\theta\sqrt{n}}{1-\theta} + n\psi\left(\frac{\theta}{1-\theta}\right)\right)\right\rceil.$$

The rest of the proof consists in reducing this expression to the one in the lemma. First, using that $\psi(-\tau) = 0.52084$, we obtain

$$\frac{\psi(-\tau) - \psi(\tau)}{\psi(\tau)} = \frac{0.34853}{0.172307} \leq 3.$$

---

[22] **Exercise 44** In the proof of Lemma II.38 we found the following expression for the decrease in the dual barrier function value:

$$\phi_\mu^d(s) - \phi_\mu^d(s^+) = \alpha e^T z^2 - \Psi(\alpha z),$$

where α denotes the size of the damped Newton step. Show that the decrease is maximal for the unique step-size $\bar{\alpha}$ determined by the equation

$$e^T z^2 = \sum_{j=1}^{n}\frac{\alpha z_j^2}{1+\alpha z_j}$$

and that for this value the decrease is given by

$$\Psi\left(\frac{\bar{\alpha}z}{e+\bar{\alpha}z}\right).$$

[23] It is interesting to observe that Lemma II.38 provides a second proof of the first statement in Lemma II.30, namely

$$\phi_\mu^d(s) \geq \psi(\delta),$$

where $\delta := \delta(s,\mu)$. This follows from Lemma II.38, since $\phi_\mu^d(s^+) \geq 0$.

Furthermore, using $\psi(t) \leq t^2/2$ for $t \geq 0$ we get[24]

$$n\psi\left(\frac{\theta}{1-\theta}\right) \leq \frac{n\theta^2}{2(1-\theta)^2}. \tag{6.30}$$

Finally, using that $1/\psi(\tau) < 1/6$ we obtain the following upper bound for the number of inner iterations:

$$\left\lceil 3 + \frac{60\sqrt{n}}{1-\theta} + \frac{3n\theta^2}{(1-\theta)^2} \right\rceil = \left\lceil 3\left(\frac{\theta\sqrt{n}}{1-\theta} + 1\right)^2 \right\rceil.$$

This proves the lemma. □

Remark II.40 It is tempting to apply Lemma II.39 to the case where $\theta = 1/(3\sqrt{n})$. We know that for that value of θ one full Newton step keeps the iterate in the region of quadratic convergence around the μ^+-center. Substitution of this value in the bound of Lemma II.39 however yields that at least 6 damped Newton steps are required for the same purpose. This disappointing result reveals a weakness of the above analysis. The weakness probably stems from the fact that the estimate of the number of inner iterations in one outer iteration is based on the assumption that the decrease in the barrier function value is given by the constant $\psi(\tau)$. Actually the decrease is at least $\psi(\delta)$. Since in many inner iterations, in particular in the iterations immediately after the update of the barrier parameter, the proximity δ may be much larger than τ, the actual number of iterations may be much smaller than the pessimistic estimate of Lemma II.39. This is the reason why for the algorithm with large updates there exists a gap between theory and practice. In practice the number of inner iterations is much smaller than the upper bound given by the lemma. Hopefully future research will close this gap.[25] ●

6.9.6 Total number of iterations

We proceed by estimating the total number of iterations required by the algorithm.

Theorem II.41 *To obtain a primal-dual pair* (x, s), *with* $x = x(s, \mu)$, *such that* $x^T s \leq 2\varepsilon$, *at most*

$$\left\lceil \frac{1}{\theta} \left\lceil 3\left(\frac{\theta\sqrt{n}}{1-\theta} + 1\right)^2 \right\rceil \log \frac{n\mu^0}{\varepsilon} \right\rceil$$

iterations are required by the logarithmic barrier algorithm with large updates.

---

[24] A different estimate arises by using Exercise 39, which implies $\psi(t) \leq t^2/(1+t)$ for $t > -1$. Hence

$$n\psi\left(\frac{\theta}{1-\theta}\right) \leq \frac{n\theta^2}{1-\theta},$$

which is sharper than (6.30) if $\theta > \frac{1}{2}$. The use of (6.30) however does not deteriorate the order of our estimates below.

[25] **Exercise 45** Let $\delta = \delta(s, \mu) > 0$ and $x = x(s, \mu)$. Then the vector $z = (xs/\mu) - e$ has at least one positive coordinate. Prove this. Hence, if z has only one nonzero coordinate then this coordinate equals $\|z\|$. Show that in that case the single damped Newton step with step-size $\alpha = 1/(1 + \delta)$ yields $s^+ = s(\mu)$.

Proof: The number of outer iterations follows from Lemma I.36. The bound in the theorem is obtained by multiplying this number by the bound of Lemma II.39 for the number of inner iterations per outer iteration and rounding the product, if not integral, to the smallest integer above it. □

We end this section by drawing two conclusions. If we take θ to be a fixed constant (independent of n), for example $\theta = 1/2$, the iteration bound of Theorem II.41 becomes

$$\mathcal{O}\left(n \, \log \, \frac{n\mu^0}{\varepsilon}\right).$$

For such values of θ we say that the algorithm uses *large updates*. The number of inner iterations per outer iteration is then $\mathcal{O}(n)$.

If we take $\theta = \nu/\sqrt{n}$ for some fixed constant ν (independent of n), the iteration bound of Theorem II.41 becomes

$$\mathcal{O}\left(\sqrt{n} \, \log \, \frac{n\mu^0}{\varepsilon}\right),$$

provided that n is large enough ($n \geq \nu^2$ say). It has become common to say that the algorithm uses *medium updates*. The number of inner iterations per outer iteration is then bounded by a constant, depending on ν.

In the next section we give an illustration.

6.9.7 Illustration of the algorithm with large updates

We use the same sample problem as before (see Sections 6.7.2 and 6.8.4) and solve it using the dual logarithmic barrier algorithm with large updates. We do this for several values of the barrier update parameter θ. As before, we start the algorithm at $y = (0,0)$ and $\mu = 2$, and the accuracy parameter is set to $\varepsilon = 10^{-4}$. For $\theta = 0.5$, Table 6.3. (page 145) lists the algorithm's progress.

The table needs some explanation. The first two columns contain counters for the outer and inner iterations, respectively. The algorithm requires 16 outer and 16 inner iterations. The table shows the effect of each outer iteration, which involves an update of the barrier parameter, and also the effect of each inner iteration, which involves a move in the dual space. During a barrier parameter update the dual variables y and s remain unchanged, but, because of the change in μ, the primal variable $x(s,\mu)$ and the proximity attain new values. After each update, damped Newton steps are taken until the proximity reaches the value τ. In this example the number of inner iterations per outer iteration is never more than one. Note that we can guarantee the primal feasibility of x only if the proximity is at most one. Since the table shows only the second coordinate of x (and also of s), infeasibility of x can only be detected from the table if x_2 is negative. In this example this does not occur, but it occurs in the next example, where we solve the same problem with $\theta = 0.9$.

With $\theta = 0.9$, Table 6.4. (page 146) shows that in some iterations x is infeasible indeed. Moreover, although the number of outer iterations is much smaller than in the previous case (5 instead of 16), the total number of iterations is almost the same (14 instead of 16). Clearly, and understandably, the deeper updates make it harder to reach the new target region.

Outer	Inner	$n\mu$	x_2	y_1	y_2	s_2	δ
0	0	6.000000	1.500000	0.000000	0.000000	1.000000	0.6124
1		3.000000	0.500000	0.000000	0.000000	1.000000	0.7071
	1	3.000000	0.690744	0.292893	0.000000	1.292893	0.2229
2		1.500000	0.230248	0.292893	0.000000	1.292893	1.3081
	2	1.500000	0.302838	0.519549	0.433260	1.519549	0.2960
3		0.750000	0.105977	0.519549	0.433260	1.519549	1.7316
	3	0.750000	0.138696	0.717503	0.696121	1.717503	0.3618
4		0.375000	0.056177	0.717503	0.696121	1.717503	2.0059
	4	0.375000	0.065989	0.847850	0.840792	1.847850	0.4050
5		0.187500	0.029627	0.847850	0.840792	1.847850	2.1632
	5	0.187500	0.032120	0.920315	0.918672	1.920315	0.4367
6		0.093750	0.015201	0.920315	0.918672	1.920315	2.2575
	6	0.093750	0.015842	0.959178	0.958681	1.959178	0.4591
7		0.046875	0.007704	0.959178	0.958681	1.959178	2.3176
	7	0.046875	0.007866	0.979268	0.979161	1.979268	0.4744
8		0.023438	0.003878	0.979268	0.979161	1.979268	2.3556
	8	0.023438	0.003920	0.989548	0.989516	1.989548	0.4844
9		0.011719	0.001946	0.989548	0.989516	1.989548	2.3793
	9	0.011719	0.001956	0.994747	0.994740	1.994747	0.4907
10		0.005859	0.000975	0.994747	0.994740	1.994747	2.3937
	10	0.005859	0.000977	0.997366	0.997364	1.997366	0.4945
11		0.002930	0.000488	0.997366	0.997364	1.997366	2.4023
	11	0.002930	0.000488	0.998681	0.998680	1.998681	0.4968
12		0.001465	0.000244	0.998681	0.998680	1.998681	2.4074
	12	0.001465	0.000244	0.999340	0.999340	1.999340	0.4982
13		0.000732	0.000122	0.999340	0.999340	1.999340	2.4103
	13	0.000732	0.000122	0.999670	0.999670	1.999670	0.4990
14		0.000366	0.000061	0.999670	0.999670	1.999670	2.4120
	14	0.000366	0.000061	0.999835	0.999835	1.999835	0.4994
15		0.000183	0.000031	0.999835	0.999835	1.999835	2.4130
	15	0.000183	0.000031	0.999917	0.999917	1.999917	0.4997
16		0.000092	0.000015	0.999917	0.999917	1.999917	2.4135
	16	0.000092	0.000015	0.999959	0.999959	1.999959	0.4998

Table 6.3. Progress of the dual algorithm with large updates, $\theta = 0.5$.

This is even more true in the last example where we take $\theta = 0.99$. Table 6.5. (page 146) shows the result. The number of outer iterations is only 3, but the total number of iterations is still 14. This leads us to the important observation that the deep update strategy has its limits. On the other hand, the number of iterations is competing with the methods using full Newton steps, and is significantly less than the iteration bound of Theorem II.41.

Outer	Inner	$n\mu$	x_2	y_1	y_2	s_2	δ
0	0	6.000000	1.500000	0.000000	0.000000	1.000000	0.6124
1		0.600000	−0.300000	0.000000	0.000000	1.000000	5.3385
	1	0.600000	0.014393	0.394413	0.631060	1.394413	2.4112
	2	0.600000	0.108620	0.762163	0.722418	1.762163	0.5037
2		0.060000	−0.005240	0.762163	0.722418	1.762163	16.8904
	3	0.060000	0.008563	0.906132	0.922246	1.906132	4.7236
	4	0.060000	0.010057	0.967364	0.961475	1.967364	1.1306
	5	0.060000	0.010098	0.977293	0.978223	1.977293	0.1716
3		0.006000	0.000891	0.977293	0.978223	1.977293	14.3247
	6	0.006000	0.000994	0.992649	0.992275	1.992649	3.9208
	7	0.006000	0.001001	0.996651	0.996769	1.996651	0.9143
	8	0.006000	0.001001	0.997834	0.997808	1.997834	0.1277
4		0.000600	0.000099	0.997834	0.997808	1.997834	13.9956
	9	0.000600	0.000100	0.999254	0.999264	1.999254	3.8257
	10	0.000600	0.000100	0.999676	0.999673	1.999676	0.8883
	11	0.000600	0.000100	0.999782	0.999783	1.999782	0.1224
5		0.000060	0.000010	0.999782	0.999783	1.999782	13.9508
	12	0.000060	0.000010	0.999926	0.999926	1.999926	3.8128
	13	0.000060	0.000010	0.999967	0.999968	1.999967	0.8847
	14	0.000060	0.000010	0.999978	0.999978	1.999978	0.1216

Table 6.4. Progress of the dual algorithm with large updates, $\theta = 0.9$.

Outer	Inner	$n\mu$	x_2	y_1	y_2	s_2	δ
0	0	6.000000	1.500000	0.000000	0.000000	1.000000	0.6124
1		0.060000	−0.480000	0.000000	0.000000	1.000000	60.4235
	1	0.060000	−0.133674	0.407010	0.797740	1.407010	28.2966
	2	0.060000	0.008587	0.906680	0.860655	1.906680	7.0268
	3	0.060000	0.009852	0.949767	0.964246	1.949767	1.7270
	4	0.060000	0.010099	0.978068	0.974574	1.978068	0.2919
2		0.000600	−0.000021	0.978068	0.974574	1.978068	166.4835
	5	0.000600	0.000086	0.992297	0.993722	1.992297	48.2832
	6	0.000600	0.000099	0.998161	0.997593	1.998161	13.7438
	7	0.000600	0.000100	0.999183	0.999394	1.999183	3.6913
	8	0.000600	0.000100	0.999720	0.999656	1.999720	0.8224
	9	0.000600	0.000100	0.999781	0.999792	1.999781	0.1013
3		0.000006	0.000001	0.999781	0.999792	1.999781	149.4817
	10	0.000006	0.000001	0.999939	0.999934	1.999939	43.4727
	11	0.000006	0.000001	0.999980	0.999981	1.999980	12.4359
	12	0.000006	0.000001	0.999994	0.999993	1.999994	3.3655
	13	0.000006	0.000001	0.999997	0.999997	1.999997	0.7573
	14	0.000006	0.000001	0.999998	0.999998	1.999998	0.0949

Table 6.5. Progress of the dual algorithm with large updates, $\theta = 0.99$.

We conclude this section with a graphical illustration of the algorithm, with $\theta = 0.9$. Figure 6.13 shows the first outer iteration, which consists of 2 inner iterations.

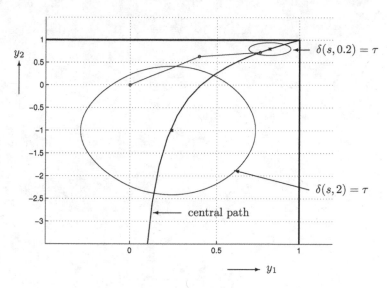

Figure 6.13   The first iterates for a large update with $\theta = 0.9$.

7

The Primal-Dual Logarithmic Barrier Method

7.1 Introduction

In the previous chapter we dealt extensively with the dual logarithmic barrier approach to the LO problem. It has become clear that Newton's method, when applied to find the minimizer of the dual logarithmic barrier function, yields a search direction Δs in the dual space that allows us to follow the dual central path (approximately) to the dual optimal set. We were able to show that an ε-solution of (D) can be obtained in a number of iterations that is proportional to the product of the logarithm of the initial duality gap divided by the desired accuracy, and $\sqrt{n}$ (for the full-step method and the medium-update method) or n (for the large-update method). Although the driving force in the dual logarithmic barrier approach is the desire to solve the dual problem (D), it also yields an ε-solution of the primal problem (P). The problem (P) also plays a crucial role in the analysis of the method. For example, the Newton step Δs at (y, s) for the barrier parameter value μ is described by the primal variable $x(s, \mu)$. Moreover, the convergence proof of the method uses the duality gap $c^T x(s, \mu) - b^T y$. Finally, the analysis of the medium-update and large-update versions of the dual method strongly depend on the properties of the primal-dual logarithmic barrier function $\phi_\mu(x, s)$.

The aim of this chapter is to show that we can benefit from the primal problem not only in the analysis but also in the design of the algorithm. The idea is to solve both the dual and the primal problem simultaneously, by taking in each iteration a step Δs in the dual space and a step Δx in the primal space. Here, the search directions Δs and Δx still have to be defined. This is done in the next section. Again, Newton's name is given to the search directions, but now the search directions arise from an iterative method — also due to Newton — for solving the system of equations defining the μ-centers of (P) and (D).

In the following paragraphs we follow the same program as for the dual algorithms: we first introduce a proximity measure, then we deal with full-step methods, with both fixed and adaptive updates of the barrier parameter, and finally we consider methods that use deep (but fixed) updates and damped Newton steps.

For the sake of clarity, it might be useful to emphasize that it is not our aim to take for Δs the dual Newton and for Δx its counterpart, the primal Newton step. For this would mean that we were executing two algorithms simultaneously, namely the dual logarithmic barrier algorithm and the primal logarithmic barrier algorithm.

Apart from the fact that this makes no sense, it doubles the computational work (roughly speaking). Instead, we define the search directions Δs and Δx in a new way and we show that the resulting algorithms, called *primal-dual algorithms*, allow similar theoretical iteration bounds to their dual (or primal) counterparts. In practice, however, primal-dual methods have a very good reputation. Many computational studies give support to this reputation. This is especially true for the so-called *predictor-corrector method*, which is discussed in Section 7.7.

7.2 Definition of the Newton step

In this section we are given a positive primal-dual feasible pair $(x, (y, s))$, and some $\mu > 0$. Our aim is to define search directions $\Delta x, \Delta y, \Delta s$ that move in the direction of the μ-center $x(\mu), y(\mu), s(\mu)$. In fact, we want the new iterates $x + \Delta x, y + \Delta y, s + \Delta s$ to satisfy the KKT system (5.3) with respect to μ. After substitution this yields the following conditions on $\Delta x, \Delta y, \Delta s$:

$$
\begin{aligned}
A(x + \Delta x) &= b, & x + \Delta x > 0, \\
A^T(y + \Delta y) + s + \Delta s &= c, & s + \Delta s > 0, \\
(x + \Delta x)(s + \Delta s) &= \mu e.
\end{aligned}
$$

If we neglect for the moment the inequality constraints, then, since $Ax = b$ and $A^T y + s = c$, this system can be rewritten as follows:

$$
\begin{aligned}
A\Delta x &= 0, \\
A^T \Delta y + \Delta s &= 0, \\
s\Delta x + x\Delta s + \Delta x \Delta s &= \mu e - xs.
\end{aligned}
\tag{7.1}
$$

Unfortunately, this system of equations in $\Delta x, \Delta y$ and Δs is nonlinear, because of the term $\Delta x \Delta s$ in the third equation. To overcome this difficulty we simply neglect this quadratic term, according to Newton's method for solving nonlinear equations, and we obtain the linear system

$$
\begin{aligned}
A\Delta x &= 0, \\
A^T \Delta y + \Delta s &= 0, \\
s\Delta x + x\Delta s &= \mu e - xs.
\end{aligned}
\tag{7.2}
$$

Below we show that this system determines the displacements $\Delta x, \Delta y$ and Δs uniquely. We call them the primal-dual Newton directions and these are the directions we are going to use.

Theorem II.42 *The system (7.2) has a unique solution, namely*

$$
\begin{aligned}
\Delta y &= \left(AXS^{-1}A^T\right)^{-1}\left(b - \mu As^{-1}\right) \\
\Delta s &= -A^T \Delta y \\
\Delta x &= \mu s^{-1} - x - xs^{-1}\Delta s.
\end{aligned}
$$

Proof: We divide the third equation in (7.2) coordinatewise by s, and obtain

$$\Delta x + xs^{-1}\Delta s = \mu s^{-1} - x. \tag{7.3}$$

Multiplying this equation from the left by A, and using that $A\Delta x = 0$ and $Ax = b$, we get

$$AXS^{-1}\Delta s = \mu As^{-1} - Ax = \mu As^{-1} - b.$$

The second equation gives $\Delta s = -A^T\Delta y$. Substituting this we find

$$AXS^{-1}A^T\Delta y = b - \mu As^{-1}.$$

Since A is an $m \times n$ matrix of rank m, the matrix $AXS^{-1}A^T$ has size $m \times m$ and is nonsingular, so the last equation determines Δy uniquely as specified in the theorem. Now Δs follows uniquely from $\Delta s = -A^T\Delta y$. Finally, (7.3) yields the expression for Δx.[1]   □

Remark II.43 In the analysis below we do not use the expressions just found for the search directions in the primal and the dual space. But it is important to see that their computation requires the solution of a linear system of equations with $AXS^{-1}A^T$ as coefficient matrix. We refer the reader to Chapter 20 for a discussion of computational issues related to efficient solution methods for such systems.   ●

Remark II.44 We can easily deduce from Theorem II.42 that the primal-dual directions for the y- and the s-space differ from the dual search directions used in the previous chapter. For example, the dual direction for y was given by

$$\left(AS^{-2}A^T\right)^{-1}\left(\frac{b}{\mu} - As^{-1}\right)$$

whereas the primal-dual direction is given by

$$\left(AXS^{-1}A^T\right)^{-1}\left(b - \mu As^{-1}\right).$$

The difference is that the *scaling matrix* S^{-2} in the dual case is replaced by the scaling matrix XS^{-1}/μ in the primal-dual case. Note that the two scaling matrices coincide if and only if $XS = \mu I$, which happens if and only if $x = x(\mu)$ and $s = s(\mu)$. In that case both expressions vanish, since then $\mu As^{-1} = Ax = b$. We conclude that if $s \neq s(\mu)$ then the dual directions in the y- and in the s-space differ from the corresponding primal-dual directions. A similar result holds for the search direction in the primal space. It may be worthwhile to point out that the dual search direction at y depends only on y itself and the slack vector $s = c - A^Ty$, whereas the primal-dual direction at y also depends on the given primal variable x.   ●

---

[1] **Exercise 46** An alternative proof of the unicity property in Theorem II.42 can be obtained by showing that the matrix in the linear system (7.2) is nonsingular. This matrix is given by

$$\begin{bmatrix} A & 0 & 0 \\ 0 & A^T & I \\ S & 0 & X \end{bmatrix}.$$

Prove that this matrix in nonsingular.

7.3 Properties of the Newton step

We denote the result of the (full) Newton step at (x, y, s) by (x^+, y^+, s^+):

$$x^+ = x + \Delta x, \quad y^+ = y + \Delta y, \quad s^+ = s + \Delta s.$$

Note that the new iterates satisfy the affine equations $Ax^+ = b$ and $A^T y^+ + s^+ = c$, since $A\Delta x = 0$ and $A^T \Delta y + \Delta s = 0$, so we only have to concentrate on the sign of the vectors x^+ and s^+. We call the Newton step feasible if x^+ and s^+ are nonnegative and strictly feasible if x^+ and s^+ are positive. The main aim of this section is to find conditions for feasibility and strict feasibility of the (full) Newton step.

First we deal with two simple lemmas.[2]

Lemma II.45 Δx and Δs are orthogonal.

Proof: Since $A\Delta x = 0$, Δx belongs to the null space of A, and since $\Delta s = -A^T \Delta y$, Δs belongs to the row space of A. Since these spaces are orthogonal, the lemma follows. □

If x^+ and s^+ are nonnegative (positive), then their product is nonnegative (positive) as well. We may write

$$x^+ s^+ = (x + \Delta x)(s + \Delta s) = xs + (s\Delta x + x\Delta s) + \Delta x \Delta s.$$

Since $s\Delta x + x\Delta s = \mu e - xs$ this leads to

$$x^+ s^+ = \mu e + \Delta x \Delta s. \tag{7.4}$$

Thus it follows that x^+ and s^+ are feasible only if $\mu e + \Delta x \Delta s$ is nonnegative. Surprisingly enough, the converse is also true. This is the content of our next lemma.

Lemma II.46 *The primal-dual Newton step is feasible if and only if $\mu e + \Delta x \Delta s \geq 0$ and strictly feasible if and only if $\mu e + \Delta x \Delta s > 0$.*

Proof: The 'only if' part of both statements in the lemma follows immediately from (7.4). For the proof of the converse part we introduce a step length $\alpha, 0 \leq \alpha \leq 1$, and we define

$$x^\alpha = x + \alpha \Delta x, \quad y^\alpha = y + \alpha \Delta y, \quad s^\alpha = s + \alpha \Delta s.$$

We then have $x^0 = x, x^1 = x^+$ and similar relations for the dual variables. Hence we have $x^0 s^0 = xs > 0$. The proof uses a continuity argument, namely that x^1 and s^1 are nonnegative if $x^\alpha s^\alpha$ is positive for all α in the open interval $(0, 1)$. This argument has a simple geometric interpretation: x^1 and s^1 are feasible if and only if the open segment connecting x^0 and x^1 lies in the interior of the primal feasible region, and the open segment connecting s^0 and s^1 lies in the interior of the dual feasible region. Now we write

$$x^\alpha s^\alpha = (x + \alpha \Delta x)(s + \alpha \Delta s) = xs + \alpha (s\Delta x + x\Delta s) + \alpha^2 \Delta x \Delta s.$$

---

[2] One might observe that some of the results in this and the next section are quite similar to analogous results in Section 2.7.2 in Part I for the Newton step for the self-dual model. To keep the treatment here self-supporting we do not invoke these results, however.

Using $s\Delta x + x\Delta s = \mu e - xs$ gives

$$x^\alpha s^\alpha = xs + \alpha(\mu e - xs) + \alpha^2 \Delta x \Delta s.$$

Now suppose $\mu e + \Delta x \Delta s \geq 0$. Then it follows that

$$x^\alpha s^\alpha \geq xs + \alpha(\mu e - xs) - \alpha^2 \mu e = (1 - \alpha)(xs + \alpha \mu e).$$

Since xs and e are positive it follows that $x^\alpha s^\alpha > 0$ for $0 \leq \alpha < 1$. Hence, none of the entries of x^α and s^α vanish for $0 \leq \alpha < 1$. Since x^0 and s^0 are positive, this implies that $x^\alpha > 0$ and $s^\alpha > 0$ for $0 \leq \alpha < 1$. Therefore, by continuity, the vectors x^1 and s^1 cannot have negative entries. This completes the proof of the first statement in the lemma. Assuming $\mu e + \Delta x \Delta s > 0$, we derive in the same way

$$x^\alpha s^\alpha > xs + \alpha(\mu e - xs) - \alpha^2 \mu e = (1 - \alpha)(xs + \alpha \mu e).$$

This implies that $x^1 s^1 > 0$. Hence, by continuity, x^1 and s^1 must be positive, proving the second statement in the lemma. □

We proceed with a discussion of the vector $\Delta x \Delta s$. From (7.4) it is clear that the error made by neglecting the second-order term in the nonlinear system (7.1) is given by this vector. It represents the so-called second-order effect in the Newton step. Therefore it will not be surprising that the vector $\Delta x \Delta s$ plays a crucial role in the analysis of primal-dual methods.

It is worth considering the ideal case where the second-order term vanishes. If $\Delta x \Delta s = 0$, then Δx and Δs solve the nonlinear system (7.1). By Lemma II.46 the Newton iterates x^+ and s^+ are feasible in this case. Hence they satisfy the KKT conditions. Now the unicity property gives us that $x^+ = x(\mu)$ and $s^+ = s(\mu)$. Thus we see that the Newton process is exact in this case: it produces the μ-centers in one iteration.[3]

In general the second-order term is nonzero and the new iterates do not coincide with the μ-centers. But we have the surprising property that the duality gap always assumes the same value as at the μ-centers, where the duality gap equals $n\mu$.

Lemma II.47 *If the primal-dual Newton step is feasible then* $(x^+)^T s^+ = n\mu$.

Proof: Using (7.4) and the fact that the vectors Δx and Δs are orthogonal, the duality gap after the Newton step can be written as follows:

$$(x^+)^T s^+ = e^T(x^+ s^+) = e^T(\mu e + \Delta x \Delta s) = \mu e^T e = n\mu.$$

This proves the lemma. □

In the general case we need some quantity for measuring the progress of the Newton iterates on the way to the μ-centers. As in the case of the dual logarithmic barrier method we start by considering a 'full-step method'. We then deal with

---

[3] **Exercise 47** Let (x, s) be a positive primal-dual feasible pair with $x = x(\mu)$. Show that the Newton process is exact in this case, with $\Delta x = 0$ and $\Delta s = s(\mu) - s$. (A similar results holds if $s = s(\mu)$, and follows in the same way.)

an 'adaptive method', in which the barrier parameter is updated 'adaptively', and then turn to the 'large-update method', which uses large fixed updates and damped Newton steps. For the large-update method we already have an excellent candidate for measuring proximity to the μ-centers, namely the primal-dual logarithmic barrier function $\phi_\mu(x,s)$. For the full-step method and the adaptive method we need a new measure that is introduced in the next section.

7.4   Proximity and local quadratic convergence

Recall that for the dual method we have used the Euclidean norm of the Newton step Δs scaled by s as a proximity measure. It is not at all obvious how this successful approach can be generalized to the primal-dual case. However, there is a natural way of doing this, but we first have to reformulate the linear system (7.2) that defines the Newton directions in the primal-dual case. To this end we introduce the vectors

$$d := \sqrt{\frac{x}{s}}, \quad u := \sqrt{\frac{xs}{\mu}}.$$

Using d we can rescale x and s to the same vector, namely u:

$$\frac{d^{-1}x}{\sqrt{\mu}} = u, \quad \frac{ds}{\sqrt{\mu}} = u.$$

Now we scale Δx and Δs similarly to d_x and d_s:

$$\frac{d^{-1}\Delta x}{\sqrt{\mu}} =: d_x, \quad \frac{d\Delta s}{\sqrt{\mu}} =: d_s. \tag{7.5}$$

For easy reference in the future we write

$$\begin{aligned} x^+ = x + \Delta x &= \sqrt{\mu}\, d\,(u + d_x) & (7.6) \\ s^+ = s + \Delta s &= \sqrt{\mu}\, d^{-1}\,(u + d_s) & (7.7) \end{aligned}$$

and, using (7.4),

$$x^+ s^+ = \mu e + \Delta x \Delta s = \mu\,(e + d_x d_s). \tag{7.8}$$

Thus we may restate Lemma II.46 without further proof as follows.

Lemma II.48 *The primal-dual Newton step is feasible if and only if*

$$e + d_x d_s \geq 0 \tag{7.9}$$

and strictly feasible if and only if

$$e + d_x d_s > 0. \tag{7.10}$$

Since

$$\Delta x \Delta s = \mu d_x d_s, \tag{7.11}$$

the orthogonality of Δx and Δs implies that the scaled displacements d_x and d_s are orthogonal as well. Now we may reformulate the left-hand side in the third equation of the KKT system as follows:

$$s\Delta x + x\Delta s = \sqrt{\mu}\left(sdd_x + xd^{-1}d_s\right) = \mu u\left(d_x + d_s\right),$$

and the right-hand side can be rewritten as

$$\mu e - xs = \mu e - \mu u^2 = \mu u\left(u^{-1} - u\right).$$

The third equation can then be restated simply as

$$d_x + d_s = u^{-1} - u.$$

On the other hand, the first and the second equations can be reformulated as $ADd_x = 0$ and $(AD)^T d_y + d_s = 0$, where

$$d_y = \frac{\Delta y}{\sqrt{\mu}}.$$

We conclude that the scaled displacements d_x, d_y and d_s satisfy

$$\begin{aligned} ADd_x &= 0 \\ (AD)^T d_y + d_s &= 0 \\ d_x + d_s &= u^{-1} - u. \end{aligned} \tag{7.12}$$

The first two equations show that the vectors d_x and d_s belong to the null space and the row space of the matrix AD respectively. These two spaces are orthogonal and the row space of AD is equal to the null space of the matrix HD^{-1}, where H is any matrix whose null space is equal to the row space of A, as defined in Section 6.3 (page 111). The last equation makes clear that d_x and d_s form the orthogonal components of the vector $u^{-1} - u$ in these complementary subspaces. Therefore, we find[4,5]

$$\begin{aligned} d_x &= P_{AD}(u^{-1} - u) & (7.13) \\ d_s &= P_{HD^{-1}}(u^{-1} - u). & (7.14) \end{aligned}$$

The orthogonality of d_x and d_s also implies

$$\|d_x\|^2 + \|d_s\|^2 = \|u^{-1} - u\|^2. \tag{7.15}$$

Note that the displacements d_x, d_s (and also d_y) are zero if and only if $u^{-1} - u = 0$. In this case x, y and s coincide with the respective μ-centers. It will be clear that the quantity $\|u^{-1} - u\|$ is a natural candidate for measuring closeness to the pair of

---

[4] **Exercise 48** Verify that the expressions for the scaled displacements d_x and d_s in (7.13) and (7.14) are in accordance with Theorem II.42.

[5] **Exercise 49** Show that
$$P_{AD} + P_{HD^{-1}} = I,$$
where I denotes the identity matrix in $\mathbb{R}^n$. Also show that

$$P_{AD} = D^{-1}H^T\left(HD^{-2}H^T\right)^{-1}HD^{-1}, \qquad P_{HD^{-1}} = DA^T\left(AD^2A^T\right)^{-1}AD.$$

μ-centers. It turns out that it is more convenient not to use the norm of $u^{-1} - u$ itself, but to divide it by 2. Therefore, we define

$$\delta(x, s; \mu) := \frac{1}{2} \left\| u^{-1} - u \right\| = \frac{1}{2} \left\| \sqrt{\frac{xs}{\mu}} - \sqrt{\frac{\mu}{xs}} \right\|. \qquad (7.16)$$

By (7.15), $\delta(x, s; \mu)$ is simply half of the Euclidean norm of the concatenation of the search direction vectors Δx and Δs after some appropriate scaling.[6,7,8]

In the previous section we discussed that the quality of the Newton step greatly depends on the second-order term $\Delta x \Delta s$. Recall that this term, when expressed in the scaled displacements, equals $\mu d_x d_s$. We proceed by showing that the vector $d_x d_s$ can be nicely bounded in terms of the proximity measure.

Lemma II.49 *Let (x, s) be any positive primal-dual pair and suppose $\mu > 0$. If $\delta := \delta(x, s; \mu)$, then $\|d_x d_s\|_\infty \leq \delta^2$ and $\|d_x d_s\| \leq \delta^2 \sqrt{2}$.*

Proof: Since the vectors d_x and d_s are orthogonal, the lemma follows immediately from the first uv−lemma (Lemma C.4 in Appendix C) by noting that $d_x + d_s = u^{-1} - u$ and $\|u^{-1} - u\| = 2\delta$. $\qquad \square$

We are now ready for the main result of this section (Theorem II.50 below), which is the primal-dual analogue of Theorem II.21 for the dual logarithmic barrier method.

Theorem II.50 *If $\delta := \delta(x, s; \mu) \leq 1$, then the primal-dual Newton step is feasible, i.e., x^+ and s^+ are nonnegative. Moreover, if $\delta < 1$, then x^+ and s^+ are positive and*

$$\delta(x^+, s^+; \mu) \leq \frac{\delta^2}{\sqrt{2(1 - \delta^2)}}.$$

---

[6] This proximity measure was introduced by Jansen et al. [157]. In the context of primal-dual methods, most authors used a different but closely related proximity measure. See Section 7.5.3. Because of the analogy with the proximity measure in the dual case, and also because of its natural interpretation as the norm of the scaled Newton direction, we prefer the proximity measure as defined by (7.16). Another motivation for the use of this measure is that it allows sharper estimates in the analysis of the primal-dual methods. This will become clear later.

[7] **Exercise 50** Let $\delta = \delta(x, s; \mu)$. In general the vector $\bar{x} = \mu s^{-1}$ is not primal feasible, and the vector $\bar{s} = \mu x^{-1}$ not dual feasible. The aim of this exercise is to show that the deviation from feasibility can be measured in a natural way. Defining

$$G_p = AD^2 A^T, \qquad G_d = HD^{-2}H^T,$$

we have

$$\|A\bar{x} - b\|_{G_p^{-1}} = \sqrt{\mu} \|d_s\|, \qquad \|H\bar{s} - Hc\|_{G_d^{-1}} = \sqrt{\mu} \|d_x\|.$$

As a consequence, prove that

$$\|A\bar{x} - b\|_{G_p^{-1}}^2 + \|H\bar{s} - Hc\|_{G_d^{-1}}^2 = 4\mu\delta^2.$$

[8] **Exercise 51** Prove that

$$\delta(x, s; \mu) = \sqrt{\frac{1}{2} \sum_{i=1}^n \left(\cosh \log \frac{x_i s_i}{\mu} - 1 \right)} = \sqrt{\sum_{i=1}^n \sinh^2 \left(\frac{1}{2} \log \frac{x_i s_i}{\mu} \right)}.$$

Proof: The first part of the theorem is a direct consequence of Lemma II.49 and Lemma II.48. The second lemma yields that $\|d_x d_s\|_\infty \leq 1$ and the first lemma that the primal-dual Newton step is feasible in this case. Now let us turn to the proof of the second statement. Let $\delta^+ := \delta(x^+, s^+; \mu)$ and

$$u^+ := \sqrt{\frac{x^+ s^+}{\mu}}.$$

Then we have, by definition,

$$2\delta^+ = \left\|(u^+)^{-1} - u^+\right\| = \left\|(u^+)^{-1}\left(e - (u^+)^2\right)\right\|.$$

Recall from (7.8) that

$$x^+ s^+ = \mu\left(e + d_x d_s\right).$$

Hence,

$$u^+ = \sqrt{e + d_x d_s}.$$

Substitution gives

$$2\delta^+ = \left\|\frac{d_x d_s}{\sqrt{e + d_x d_s}}\right\| \leq \frac{\|d_x d_s\|}{\sqrt{1 - \|d_x d_s\|_\infty}}.$$

Now using the bounds in Lemma II.49 we obtain

$$2\delta^+ \leq \frac{\delta^2 \sqrt{2}}{\sqrt{1 - \delta^2}}.$$

Dividing both sides by 2 we arrive at the result in the theorem. □

Theorem II.50 makes clear that the primal-dual Newton method is quadratically convergent in the region

$$\left\{(x, s) \in \mathcal{P} \times \mathcal{D} \ : \ \delta(x, s; \mu) \leq \frac{1}{\sqrt{2}} = 0.7071\right\}, \tag{7.17}$$

where we have $\delta^+ \leq \delta^2$. It is clear that Theorem II.50 has no value if the upper bound for $\delta(x^+, s^+; \mu)$ is not smaller than δ, which is the case if $\delta \geq \sqrt{2/3} = 0.8165$.

As for the dual Newton method, we provide a graphical example to illustrate how the primal-dual Newton process behaves.

Example II.51 We use the same problem as in Example II.7 with $b = (1, 1)^T$. So A, b and c are given by

$$A = \begin{bmatrix} 1 & -1 & 0 \\ 0 & 0 & 1 \end{bmatrix}, \quad c = \begin{bmatrix} 1 \\ 1 \\ 1 \end{bmatrix}, \quad b = \begin{bmatrix} 1 \\ 1 \end{bmatrix}.$$

Instead of drawing a graph in the dual (or primal) space we take another approach. We associate with each primal-dual pair (x, s) the positive vector $w = xs$, and represent this vector by a point in the so-called w-space, which is the interior of the nonnegative

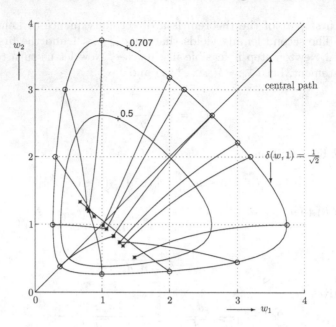

Figure 7.1 Quadratic convergence of primal-dual Newton process ($\mu = 1$).

orthant of $\mathbb{R}^n$, with $n = 3$. Note that $\delta(x, s; \mu) = 0$ if and only if $x = x(\mu)$ and $s = s(\mu)$, and that in that case $xs = \mu e$. Hence, in the w-space the central path is represented by the half-line μe, $\mu > 0$. Figure 7.1 (page 158) shows the level curves (in the w-space) for the proximity values $\tau = 1/\sqrt{2}$ and τ^2 with respect to $\mu = 1$, and also how the Newton step behaves when applied at some points on the boundary of the region of quadratic convergence. This figure depicts the w-space projected onto its first two coordinates. The starting point for a Newton step is always indicated by the symbol '$\circ$', and the point resulting from the step by the symbol '$*$'.[9] The curve connecting the two points shows the intermediate values of xs on the way from the starting point to the point after the full Newton step. The points on these curves represent

$$x^\alpha s^\alpha = (x + \alpha \Delta x)(s + \alpha \Delta s) = xs + \alpha \left(x \Delta s + s \Delta x \right) + \alpha^2 \Delta x \Delta s, \ 0 \le \alpha \le 1,$$

where (x^0, s^0) is the starting point of the iteration and (x^1, s^1) the result of the full Newton step. If there were no second-order effects (i.e., if $\Delta x \Delta s = 0$) then this curve would be a straight line. So the curvature of the line connecting the point before and after a step is an indication of the second-order effect. Note that after the Newton step the new proximity value is always smaller than $\tau^2 = 1/2$, in agreement with Theorem II.50. In fact, one may observe that often the decrease in the proximity to the 1-center is much more significant.

---

[9] The starting points in this example were obtained by using theory that will be developed later in the book, in Part III. There we show that for any positive vector $w \in \mathbb{R}^n$ there exists a primal-dual pair (x, s) such that $xs = w$ and we also deal with methods that yield such a pair. For each starting point the first two entries of w can be read from the figure; for the third coordinate of w we used the value 1, which is the value of w_3 at the 1-center, since $x(1)s(1) = e$.

When starting outside the region of quadratic convergence the behavior of the Newton process is quite unpredictable. Note that the feasibility of the (full) Newton step is then not guaranteed by the theory.

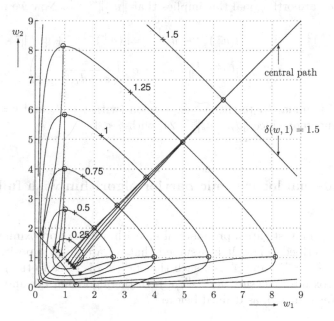

Figure 7.2 Demonstration of the primal-dual Newton process.

In Figure 7.2 we consider the behavior of the Newton process outside this region, even for proximity values larger than 1. The behavior (in this simple example) is surprisingly good if we start on (or close to) the central path. When starting closer to the boundary of the w-space the second-order effect becomes more evident and this may result in infeasibility of the Newton step, as Figure 7.2 demonstrates (for example if $w_1 = 8$ and $w_2 = 1$). This observation, that Newton's method performs better when the starting point is on or close to the central path than when we start close to the boundary of the nonnegative orthant, is not supported by the theory, but is in agreement with common computational practice. ◇

7.4.1 A sharper local quadratic convergence result

In this section we show that Theorem II.50 can be slightly improved. By using the third $uv-$lemma (Lemma C.7 in Appendix C) we obtain the following.

Theorem II.52 *If* $\delta = \delta(x, s; \mu) < 1$ *then*

$$\delta(x^+, s^+; \mu) \leq \frac{\delta^2}{\sqrt{2(1 - \delta^4)}}.$$

Proof: From the proof of Theorem II.50 we recall the definitions of δ^+ and u^+, and the relation

$$u^+ = \sqrt{e + d_x d_s}.$$

Since d_x and d_s are orthogonal this implies that $\|u^+\|^2 = n$. Now we may write

$$
\begin{aligned}
4(\delta^+)^2 &= \|(u^+)^{-1} - u^+\|^2 = \|(u^+)^{-1}\|^2 + \|u^+\|^2 - 2n \\
&= \|(u^+)^{-1}\|^2 - n = e^T \left(\frac{e}{e + d_x d_s} - e \right).
\end{aligned}
$$

Application of Lemma C.7 to the last expression (with $u = d_x$ and $v = d_s$) yields the result of the theorem, since $\|d_x + d_s\| = 2\delta$, with $\delta < 1$. □

7.5 Primal-dual logarithmic barrier algorithm with full Newton steps

In this section we investigate a primal-dual algorithm using approximate centers. The algorithm is described below. It is assumed that we are given a positive primal-dual pair $(x^0, s^0) \in \mathcal{P}^+ \times \mathcal{D}^+$ and $\mu^0 > 0$ such that (x^0, s^0) is close to the μ^0-center in the sense of the proximity measure $\delta(x^0, s^0; \mu^0)$. In the algorithm Δx and Δs denote the primal-dual Newton step, as defined before.

Primal-Dual Logarithmic Barrier Algorithm with full Newton steps

---

Input:
    A proximity parameter τ, $0 \leq \tau < 1$;
    an accuracy parameter $\varepsilon > 0$;
    $(x^0, s^0) \in \mathcal{P}^+ \times \mathcal{D}^+$ and $\mu^0 > 0$ such that $(x^0)^T s^0 = n\mu^0$ and
    $\delta(x^0, s^0; \mu^0) \leq \tau$;
    a barrier update parameter θ, $0 < \theta < 1$.
begin
    $x := x^0$; $s := s^0$; $\mu := \mu^0$;
    while $n\mu \geq (1 - \theta)\varepsilon$ **do**
    begin
        $x := x + \Delta x$;
        $s := s + \Delta s$;
        $\mu := (1 - \theta)\mu$;
    end
end

We have the following theorem. The proof will follow below.

Theorem II.53 *If* $\tau = 1/\sqrt{2}$ *and* $\theta = 1/\sqrt{2n}$, *then the Primal-Dual Logarithmic Barrier Algorithm with full Newton steps requires at most*

$$\left\lceil \sqrt{2n} \log \frac{n\mu^0}{\varepsilon} \right\rceil$$

iterations. The output is a primal-dual pair (x, s) *such that* $x^T s \leq \varepsilon$.

7.5.1 Convergence analysis

Just as in the dual case the proof depends on a lemma that quantifies the effect on the proximity measure of an update of the barrier parameter to $\mu^+ = (1 - \theta)\mu$.

Lemma II.54 *Let* (x, s) *be a positive primal-dual pair and* $\mu > 0$ *such that* $x^T s = n\mu$. *Moreover, let* $\delta := \delta(x, s; \mu)$ *and let* $\mu^+ = (1 - \theta)\mu$. *Then*

$$\delta(x, s; \mu^+)^2 = (1 - \theta)\delta^2 + \frac{\theta^2 n}{4(1 - \theta)}.$$

Proof: Let $\delta^+ := \delta(x, s; \mu^+)$ and $u = \sqrt{xs/\mu}$. Then, by definition,

$$4(\delta^+)^2 = \left\| \sqrt{1 - \theta}\, u^{-1} - \frac{u}{\sqrt{1 - \theta}} \right\|^2 = \left\| \sqrt{1 - \theta}\, (u^{-1} - u) + \frac{\theta u}{\sqrt{1 - \theta}} \right\|^2.$$

From $x^T s = n\mu$ it follows that $\|u\|^2 = n$. Hence, u is orthogonal to $u^{-1} - u$:

$$u^T \left(u^{-1} - u \right) = n - \|u\|^2 = 0.$$

Therefore,

$$4(\delta^+)^2 = (1 - \theta) \left\| u^{-1} - u \right\|^2 + \frac{\theta^2 \|u\|^2}{1 - \theta}.$$

Finally, since $\left\| u^{-1} - u \right\| = 2\delta$ and $\|u\|^2 = n$ the result follows.   □

   The proof of Theorem II.53 now goes as follows. At the start of the algorithm we have $\delta(x, s; \mu) \leq \tau = 1/\sqrt{2}$. After the primal-dual Newton step to the μ-center we have, by Theorem II.50, $\delta(x^+, s^+; \mu) \leq 1/2$. Also, from Lemma II.47, $(x^+)^T s^+ = n\mu$. Then, after the barrier parameter is updated to $\mu^+ = (1 - \theta)\mu$, with $\theta = 1/\sqrt{2n}$, Lemma II.54 yields the following upper bound for $\delta(x^+, s^+; \mu^+)$:

$$\delta(x^+, s^+; \mu^+)^2 \leq \frac{1 - \theta}{4} + \frac{1}{8(1 - \theta)} \leq \frac{3}{8}.$$

Assuming $n \geq 2$, the last inequality follows since its left hand side is a convex function of θ, whose value is $3/8$ both in $\theta = 0$ and $\theta = 1/2$. Since $\theta \in [0, 1/2]$, the left hand side does not exceed $3/8$. Since $3/8 < 1/2$, we obtain $\delta(x^+, s^+; \mu^+) \leq 1/\sqrt{2} = \tau$. Thus, after each iteration of the algorithm the property

$$\delta(x, s; \mu) \leq \tau$$

is maintained, and hence the algorithm is well defined. The iteration bound in the theorem follows from Lemma I.36. Finally, since after each full Newton step the duality gap attains its target value, by Lemma II.47, the duality gap for the pair (x, s) generated by the algorithm is at most ε. This completes the proof of the theorem. $\square$

Remark II.55 It is worthwhile to discuss the quality of the iteration bound in Theorem II.53. For that purpose we consider the hypothetical situation where the Newton step in each iteration is exact. Then, putting $\delta^+ = \delta(x^+, s^+, \mu^+)$, after the update of the barrier parameter we have

$$4(\delta^+)^2 = \frac{\theta^2 n}{1 - \theta},$$

and hence we have $\delta^+ \leq 1/\sqrt{2}$ only if $\theta^2 n \leq 2(1 - \theta)$. This occurs only if $\theta < \sqrt{2/n}$. Hence, if we maintain the property $\delta(x, s; \mu) \leq 1/\sqrt{2}$ after the update of the barrier parameter, then the iteration bound will never be smaller than

$$\left\lceil \sqrt{\frac{n}{2}} \log \frac{n\mu^0}{\varepsilon} \right\rceil. \tag{7.18}$$

Note that the iteration bound of Theorem II.53 is only a factor 2 worse than the 'ideal' iteration bound (7.18). Recall that the bound (7.18) assumes that the Newton step is exact in each iteration. In this respect it is interesting to indicate that for larger values of n the result of Theorem II.53 can be improved so that it becomes closer to the 'ideal' iteration bound. But then we need to use the stronger quadratic convergence result of Theorem II.52. If we take $\theta = 1/\sqrt{n}$, then by using Lemma II.54 and Theorem II.52, we may easily verify that the property $\delta(x, s; \mu) \leq \tau = 1/\sqrt{2}$ is maintained if

$$\frac{1}{4(1 - \theta)} + \frac{1 - \theta}{6} \leq \frac{1}{2}.$$

This holds if $\theta \leq 0.36602$, which corresponds to $n \geq 8$. Thus, for $n \geq 8$ the iteration bound of Theorem II.53 can be improved to

$$\left\lceil \sqrt{n} \log \frac{n\mu^0}{\varepsilon} \right\rceil. \tag{7.19}$$

This iteration bound is the best among all known iteration bounds for interior-point methods. It differs by only a factor $\sqrt{2}$ from the ideal bound (7.18). $\bullet$

7.5.2 Illustration of the algorithm with full Newton steps

We use the same sample problem as before (see Sections 6.7.2, 6.8.4 and 6.9.7). As starting point we use the vectors $x = (2, 1, 1)$, $y = (0, 0)$ and $s = (1, 1, 1)$, and since $x^T s = 4$, we take the initial value of the barrier parameter μ equal to $4/3$. We can easily check that $\delta(x, s; \mu) = 0.2887$. So these data can indeed be used to initialize the algorithm. With $\varepsilon = 10^{-4}$, the algorithm generates the data collected in Table 7.1.. As before, Table 7.1. contains one entry (the first) of the vectors x and s. The seventh column contains the values of the proximity $\delta = \delta(x, s; \mu)$ before the Newton step, and the eighth column the proximity $\delta^+ = \delta(x^+, s^+; \mu)$ after the Newton step at (x, s) to the current μ-center.

It.	$n\mu$	x_1	y_1	y_2	s_1	δ	δ^+	θ
0	4.000000	2.000000	0.000000	0.000000	1.000000	0.2887	0.0000	0.4082
1	2.367007	2.000000	0.333333	−0.333333	0.666667	0.4596	0.0479	0.4082
2	1.400680	1.510102	0.442200	0.210998	0.557800	0.4611	0.0586	0.4082
3	0.828855	1.267497	0.601207	0.533107	0.398793	0.4618	0.0437	0.4082
4	0.490476	1.148591	0.744612	0.723715	0.255388	0.4608	0.0271	0.4082
5	0.290240	1.085283	0.843582	0.836508	0.156418	0.4601	0.0162	0.4082
6	0.171750	1.049603	0.905713	0.903253	0.094287	0.4598	0.0096	0.4082
7	0.101633	1.029055	0.943610	0.942750	0.056390	0.4597	0.0057	0.4082
8	0.060142	1.017089	0.966423	0.966122	0.033577	0.4596	0.0034	0.4082
9	0.035589	1.010076	0.980058	0.979953	0.019942	0.4596	0.0020	0.4082
10	0.021060	1.005950	0.988174	0.988137	0.011826	0.4596	0.0012	0.4082
11	0.012462	1.003516	0.992993	0.992980	0.007007	0.4596	0.0007	0.4082
12	0.007375	1.002079	0.995850	0.995846	0.004150	0.4596	0.0004	0.4082
13	0.004364	1.001230	0.997543	0.997542	0.002457	0.4596	0.0002	0.4082
14	0.002582	1.000728	0.998546	0.998545	0.001454	0.4596	0.0001	0.4082
15	0.001528	1.000430	0.999139	0.999139	0.000861	0.4596	0.0001	0.4082
16	0.000904	1.000255	0.999491	0.999491	0.000509	0.4596	0.0001	0.4082
17	0.000535	1.000151	0.999699	0.999699	0.000301	0.4596	0.0000	0.4082
18	0.000317	1.000089	0.999822	0.999822	0.000178	0.4596	0.0000	0.4082
19	0.000187	1.000053	0.999894	0.999894	0.000106	0.4596	0.0000	0.4082
20	0.000111	1.000031	0.999938	0.999938	0.000062	0.4596	0.0000	0.4082
21	0.000066	1.000018	0.999963	0.999963	0.000037	0.4596	0.0000	0.4082
22	0.000039	1.000011	0.999978	0.999978	0.000022	−	−	−

Table 7.1.   Output of the primal-dual full-step algorithm.

Comparing the results in Table 7.1. with those in the corresponding table for the dual algorithm with full steps (Table 6.1., page 124), the most striking differences are the number of iterations and the behavior of the proximity measure. In the primal-dual case the number of iterations is 22 (instead of 53). This can be easily understood from the fact that we could use the larger barrier update parameter $\theta = 1/\sqrt{2n}$ (instead of $\theta = 1/(3\sqrt{n})$).

The second difference is probably more important. In the primal-dual case Newton's method is much more efficient than in the dual case. This is especially evident in the final iterations where both methods show very stable behavior. In the dual case the proximity takes in these iterations the values 0.2722 (before) and 0.0524 (after the Newton step), whereas in the primal-dual case these values are respectively 0.4596 and 0.0000. Note that in the dual case the effect of the Newton step is slightly better than the quadratic convergence result of Theorem II.21. In the primal-dual case, however, the effect of the Newton step is much better than predicted by Theorem II.50, and even much better than the improved quadratic convergence result of Theorem II.52.

The figures in Table 7.1. justify the statement (at least for this sample problem, but we observed the same phenomenon in other experiments) that asymptotically the primal-dual Newton method is almost exact.

Remark II.56 It is of interest to have a closer (and more accurate) look at the proximity values in the final iterations. They are given in Table 7.2. (page 164). These figures show that

It.	δ	δ^+
11	0.45960642869434	0.00069902816289
12	0.45960584496214	0.00041365328341
13	0.45960564054812	0.00024478048789
14	0.45960556896741	0.00014484936548
15	0.45960554390189	0.00008571487895
16	0.45960553512461	0.00005072193012
17	0.45960553205110	0.00003001478966
18	0.45960553097480	0.00001776130347
19	0.45960553059816	0.00001051028182
20	0.45960553046642	0.00000621947704
21	0.45960553041942	0.00000368038542

Table 7.2. Proximity values in the final iterations.

in the final iterations, where Newton's method is almost exact, the quality of the method gradually improves. After the step the proximity decreases monotonically. In fact, surprisingly enough, the rate of decrease of subsequent values of the proximity after the step is almost constant (0.59175). Remember that the barrier parameter μ also decreases at a linear rate by a factor $1 - \theta$, where $\theta = 1/\sqrt{2n}$. In our case we have $n = 3$. This gives $\theta = 0.4082$ and $1 - \theta = 0.59175$, precisely the rate of decrease in δ^+. Before the Newton step the proximity is almost constant (0.4596). Not surprisingly, this is precisely the value of $\theta\sqrt{n}/(2(1 - \theta))$. Thus, our numerical experiment gives rise to a conjecture:

Conjecture II.57 *Asymptotically the quality of the primal-dual Newton step gradually improves. The proximity before the step converges to some constant and the proximity after the step decreases monotonically to zero with a linear convergence rate. The rate of convergence is equal to $1 - \theta$.*

This observed behavior of the primal-dual Newton method has no theoretical justification at the moment.   •

We conclude this section with a graphical illustration. Figure 7.3 shows on two graphs the progress of the algorithm in the w-space (cf. Example II.51 on page 157). In both figures the w-space is projected onto its first two coordinates. The difference between the two graphs is due to the scaling of the axes. On the left graph the scale is linear and on the right graph it is logarithmic. As in Example II.51, the curves connecting the subsequent iterates show the intermediate values of xs on the way to the next iterate. The graphs show that after the first iteration the iterates follow the central path quite accurately.

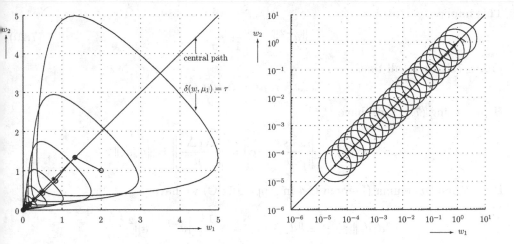

Figure 7.3 The iterates of the primal-dual algorithm with full steps.

7.5.3 The classical analysis of the algorithm

In this section we give a different analysis of the primal-dual logarithmic barrier algorithm with full Newton steps. The analysis uses the proximity measure

$$\sigma(x, s; \mu) := \left\| \frac{xs}{\mu} - e \right\|,$$

which is very common in the literature on primal-dual methods.[10]

Because of its widespread use, it seems useful to show in this section how the analysis can be easily adapted to the use of the classical proximity measure. In fact, the only thing we have to do is find suitable analogues of the quadratic convergence result in Theorem II.50 and the barrier update result of Lemma II.54.

Theorem II.58 [11] *If* $\sigma := \sigma(x, s; \mu) \leq 2 / \left(1 + \sqrt{1 + \sqrt{2}} \right) = 0.783155$, *then the primal-dual Newton step is feasible. Moreover, in that case*

$$\sigma(x^+, s^+; \mu) \leq \frac{\sigma^2}{2\sqrt{2}(1 - \sigma)}.$$

Proof: First we derive from $\|u^2 - e\| = \sigma$ the obvious inequality

$$1 - \sigma \leq u_i^2 \leq 1 + \sigma, \quad 1 \leq i \leq n.$$

---

[10] It was introduced by Kojima, Mizuno and Yoshise [178] and used in many other papers. See, e.g., Gonzaga [124], den Hertog [140], Marsten Shanno and Simantiraki [196], McShane, Monma and Shanno [199], Mehrotra and Sun [205], Mizuno [215], Monteiro and Adler [218], Todd [262, 264], Zhang and Tapia [319].

[11] This result is due to Mizuno [212].

This implies

$$\left\| u^{-2} \right\|_\infty \le \frac{1}{1-\sigma}. \tag{7.20}$$

From (7.4) we recall that

$$x^+ s^+ = \mu e + \Delta x \Delta s.$$

Hence, using (7.11), we have [12,13]

$$\sigma(x^+, s^+; \mu) := \left\| \frac{x^+ s^+}{\mu} - e \right\| = \left\| \frac{\Delta x \Delta s}{\mu} \right\| = \left\| d_x d_s \right\|.$$

By the first uv-lemma (Lemma C.4 in Appendix C) we have

$$\left\| d_x d_s \right\| \le \frac{1}{2\sqrt{2}} \left\| d_x + d_s \right\|^2 = \frac{1}{2\sqrt{2}} \left\| u^{-1} - u \right\|^2.$$

Using (7.20) we write

$$\left\| u^{-1} - u \right\|^2 = \left\| u^{-1}(e - u^2) \right\|^2 \le \left\| u^{-2} \right\|_\infty \left\| e - u^2 \right\|^2 \le \frac{\sigma^2}{1-\sigma}.$$

Hence we get

$$\sigma(x^+, s^+; \mu) \le \frac{\sigma^2}{2\sqrt{2}(1-\sigma)}.$$

Since $\sigma(x^+, s^+; \mu) = \left\| d_x d_s \right\|$, feasibility of the new iterates is certainly guaranteed if $\sigma(x^+, s^+; \mu) \le 1$, from (7.9). This condition is certainly satisfied if

$$\frac{\sigma^2}{2\sqrt{2}(1-\sigma)} \le 1,$$

and this inequality holds if and only if $\sigma \le 2/\left(1 + \sqrt{1 + \sqrt{2}}\right)$, as can easily be verified. The theorem follows. $\qquad\qquad\qquad\qquad\qquad\qquad\qquad\qquad\qquad\quad\Box$

---

[12] **Exercise 52** This exercise provides an alternative proof of the first inequality in Lemma C.4. Let u and v denote vectors in $\mathbb{R}^n$ and $\delta > 0$ ($\delta \in \mathbb{R}$). First prove that

$$\min_{u,v} \left\{ u_1 v_1 \; : \; \sum_{i=1}^n u_i v_i = 0, \; \sum_{i=1}^n \left(u_i^2 + v_i^2 \right) = 4\delta^2 \right\} = \delta^2.$$

Using this, show that if u and v are orthogonal and $\left\| u + v \right\| = 2\delta$ then $\left\| uv \right\|_\infty \le \delta^2$.

[13] **Exercise 53** This exercise provides tighter version of the second inequality in Lemma C.4. Let u and v denote vectors in $\mathbb{R}^n$ and $\delta > 0$ ($\delta \in \mathbb{R}$). First prove that

$$\max_{u,v} \left\{ \sum_{i=1}^n u_i^2 v_i^2 \; : \; \sum_{i=1}^n u_i v_i = 0, \; \sum_{i=1}^n \left(u_i^2 + v_i^2 \right) = 4\delta^2 \right\} = \frac{n\delta^4}{n-1}.$$

Using this show that if u and v are orthogonal and $\left\| u + v \right\| = 2\delta$ then $\left\| uv \right\| \le \delta^2 \sqrt{2}$.

Lemma II.59 *Let (x, s) be a positive primal-dual pair and $\mu > 0$ such that $x^T s = n\mu$. Moreover, let $\sigma := \sigma(x, s; \mu)$ and let $\mu^+ = (1 - \theta)\mu$. Then we have*

$$\sigma(x, s; \mu^+) = \frac{\sqrt{\sigma^2 + \theta^2 n}}{1 - \theta}.$$

Proof: Let $\sigma^+ := \sigma(x, s; \mu^+)$, with $x^T s = n\mu$. Then, by definition,

$$(\sigma^+)^2 = \left\| \frac{xs}{(1 - \theta)\mu} - e \right\|^2 = \frac{1}{(1 - \theta)^2} \left\| \frac{xs}{\mu} - e + \theta e \right\|^2.$$

The vectors e and $xs/\mu - e$ are orthogonal, as easily follows. Hence

$$\left\| \frac{xs}{\mu} - e + \theta e \right\|^2 = \left\| \frac{xs}{\mu} - e \right\|^2 + \|\theta e\|^2 = \sigma^2 + \theta^2 n.$$

The lemma follows. $\square$

From the above results, it is clear that maintaining the property $\sigma(x, s : \mu) \leq \tau$ during the course of the algorithm amounts to the following condition on θ:

$$\frac{1}{1 - \theta} \sqrt{\frac{\tau^4}{8(1 - \tau)^2} + n\theta^2} \leq \tau. \tag{7.21}$$

For any given τ this inequality determines how deep the updates of the barrier parameter are allowed to be. Since the full Newton step must be feasible we may assume that

$$\tau \leq \frac{2}{1 + \sqrt{1 + \sqrt{2}}} = 0.783155.$$

Squaring both sides of (7.21) gives

$$\frac{\tau^4}{8(1 - \tau)^2} + n\theta^2 \leq \tau^2(1 - \theta)^2.$$

This implies $n\theta^2 \leq \tau^2$, and hence the parameter θ must satisfy $\theta \leq \tau/\sqrt{n}$.

The iteration bound of Lemma I.36 becomes smaller for larger values of θ. Our aim here is to show that for the best possible choice of θ the iteration bound resulting from the classical analysis cannot be better than the bound of Theorem II.53. For that purpose we may assume that n is so large that $1 - \theta \approx 1$. Then the condition on θ becomes

$$\frac{\tau^4}{8(1 - \tau)^2} + n\theta^2 \leq \tau^2,$$

or equivalently,

$$n\theta^2 \leq \tau^2 - \frac{\tau^4}{8(1 - \tau)^2}. \tag{7.22}$$

Note that the right-hand side expression must be nonnegative, which holds only if

$$\tau \leq \frac{2\sqrt{2}}{1 + 2\sqrt{2}} = 0.738796.$$

We can easily verify that the right-hand side expression in (7.22) is maximal if

$$7\tau^3 - 22\tau^2 + 24\tau - 8 = 0,$$

which occurs for $\tau = 0.60155$. Substituting this value in (7.22) we obtain

$$n\theta^2 \leq 0.258765,$$

which amounts to

$$\theta \leq \frac{0.508689}{\sqrt{n}} \approx \frac{1}{2\sqrt{n}}.$$

Obviously, this upper bound for θ is too optimistic. The above argument makes clear that by using the 'classical' proximity measure $\sigma(x, s; \mu)$ in the analysis of the primal-dual method with full Newton steps, the iteration bound obtained with the proximity measure $\delta(x, s; \mu)$ cannot be improved.

7.6 A version of the algorithm with adaptive updates

7.6.1 Adaptive updating

We have seen in Section 7.5 that when the property

$$\delta(x, s; \mu) \leq \tau = \frac{1}{\sqrt{2}} \tag{7.23}$$

is maintained after the update of the barrier parameter, the values of the barrier update parameter θ are limited by the upper bound $\theta < \sqrt{2/n}$, and therefore, the iteration bound cannot be better than the 'ideal' bound

$$\left\lceil \sqrt{\frac{n}{2}} \, \log \frac{n\mu^0}{\varepsilon} \right\rceil.$$

Thus, larger updates of the barrier parameter are possible only when abandoning the idea that property (7.23) must hold after each update of the barrier parameter.

To make clear how this can be done without losing the iteration bound of Theorem II.53, we briefly recall the idea behind the proof of this theorem. After each Newton step we have a primal-dual pair (x, s) and $\mu > 0$ such that

$$\delta(x, s; \mu) \leq \bar{\tau} = \frac{\tau^2}{\sqrt{2(1 - \tau^2)}}. \tag{7.24}$$

Then we update μ to a smaller value $\mu^+ = (1 - \theta)\mu$ such that

$$\delta(x, s; \mu^+) \leq \tau, \tag{7.25}$$

and we perform a Newton step to the μ^+-center, yielding a primal-dual pair (x^+, s^+) such that $\delta(x^+, s^+; \mu^+) \leq \bar{\tau}$. Figure 7.4 illustrates this.

Why does this scheme work? It works because every time we perform a Newton step the iterates x and s are such that xs is in the region around the μ-center where

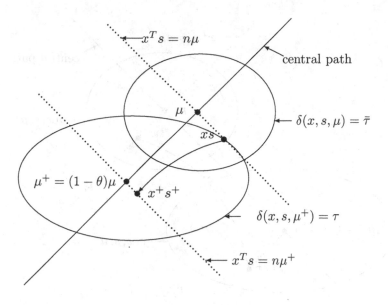

Figure 7.4 The primal-dual full-step approach.

Newton's method behaves well. The theory guarantees that if the proximity does not exceed the parameter $\tau = 1/\sqrt{2}$ then we stay within this region. However, in practice the region where Newton's method behaves well may be much larger.

Thus we can adapt our strategy to this phenomenon and choose the smallest barrier parameter $\mu^+ = (1-\theta)\mu$ so that after the Newton step to the μ^+-center the iterates satisfy $\delta(x^+, s^+; \mu^+) \leq \bar{\tau}$. Therefore, let us consider the following problem:

> Given a primal-dual pair (x, s) and $\mu > 0$ such that $\delta := \delta(x, s; \mu) \leq \bar{\tau}$, find the largest θ such that after the Newton step at (x, s) with barrier parameter value $\mu^+ = (1-\theta)\mu$ we have $\delta^+ = \delta(x^+, s^+; \mu^+) \leq \bar{\tau}$.

Here we use the parameter $\bar{\tau}$ instead of τ, because until now τ referred to the proximity before the Newton step, whereas $\bar{\tau}$ is an upper bound for the proximity just after the Newton step. It is natural to take for $\bar{\tau}$ the value $1/2$, because this is an upper bound for the proximity after the Newton step when the proximity before the step is $1/\sqrt{2}$. Our aim in this section is to investigate how deep the updates can be taken, so as to enhance the performance of the algorithm as much as possible. See Figure 7.5.[14] Just as in the case of the dual method with adaptive updates, we need to introduce the so-called *primal-dual affine-scaling* and *primal-dual centering* directions at (x, s).

---

[14] The idea of using adaptive updates of the barrier parameter in a primal-dual method can be found in, e.g., Jarre and Saunders [163].

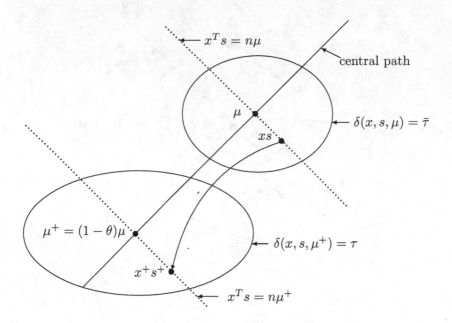

Figure 7.5 The full-step method with an adaptive barrier update.

7.6.2 The primal-dual affine-scaling and centering direction

We first recall some definitions and properties from Section 7.4. With

$$d = \sqrt{\frac{x}{s}}, \quad u = \sqrt{\frac{xs}{\mu}},$$

the vectors x and s can be scaled by d to the vector u as follows:

$$\frac{d^{-1}x}{\sqrt{\mu}} = \frac{ds}{\sqrt{\mu}} = u.$$

The same scaling applied to the Newton steps Δx and Δs yields the scaled Newton steps d_x and d_s:

$$d_x = \frac{d^{-1}\Delta x}{\sqrt{\mu}}, \quad d_s = \frac{d\Delta s}{\sqrt{\mu}},$$

and these satisfy

$$d_x + d_s = u^{-1} - u.$$

Moreover, the vectors d_x and d_s are orthogonal. They are the components of the vector $u^{-1} - u$ in the null space of AD and the null space of HD^{-1} respectively:

$$d_x = P_{AD}(u^{-1} - u) \tag{7.26}$$

$$d_s = P_{HD^{-1}}(u^{-1} - u). \tag{7.27}$$

In this section we work mainly with the scaled Newton steps d_x and d_s. The last expressions yield a natural way of separating these directions into a so-called *affine-scaling component* and a *centering component*. The (scaled) centering directions are defined by

$$d_x^c = P_{AD}(u^{-1}), \quad d_s^c = P_{HD^{-1}}(u^{-1}), \tag{7.28}$$

and the (scaled) affine directions by

$$d_x^a = -P_{AD}(u), \quad d_s^a = -P_{HD^{-1}}(u). \tag{7.29}$$

Now we have the obvious relations

$$
\begin{aligned}
d_x &= d_x^c + d_x^a \\
d_s &= d_s^c + d_s^a
\end{aligned}
$$

and

$$
\begin{aligned}
d_x^c + d_s^c &= u^{-1} \\
d_x^a + d_s^a &= -u.
\end{aligned}
$$

The unscaled centering and affine-scaling directions are defined in the obvious way: $\Delta^a x := \sqrt{\mu} d d_x^a$, etc. For the sake of completeness we list these definitions below and we also give some alternative expressions which can straightforwardly be verified.

$$\Delta^a x := \sqrt{\mu} d d_x^a = -\sqrt{\mu} D P_{AD}(u) = -D P_{AD}(\sqrt{xs})$$

$$\Delta^a s := \sqrt{\mu} d^{-1} d_s^a = -\sqrt{\mu} D^{-1} P_{HD^{-1}}(u) = -D^{-1} P_{HD^{-1}}(\sqrt{xs})$$

$$\Delta^c x := \sqrt{\mu} d d_x^c = \sqrt{\mu} D P_{AD}(u^{-1}) = \mu D P_{AD}(\tfrac{e}{\sqrt{xs}})$$

$$\Delta^c s := \sqrt{\mu} d^{-1} d_s^c = \sqrt{\mu} D^{-1} P_{HD^{-1}}(u^{-1}) = \mu D^{-1} P_{HD^{-1}}(\tfrac{e}{\sqrt{xs}}).$$

Note that the affine-scaling directions $\Delta^a x$ and $\Delta^a s$ depend only on the iterates x and s and not on the barrier parameter μ. For the centering directions we have that $\Delta^c x/\mu$ and $\Delta^c s/\mu$ depend only on the iterates x and s and not on the barrier parameter μ. Also note that if we are on the central path, i.e., if $x = x(\mu)$ and $s = s(\mu)$, then we have $u = e$. This implies $u^{-1} - u = 0$, whence $d_x = d_s = 0$. Hence, on the central path we have $d_x^a = -d_x^c$ and $d_s^a = -d_s^c$.

For future reference we observe that the above definitions imply the obvious relations

$$
\begin{aligned}
\Delta x &= \Delta^a x + \Delta^c x \\
\Delta s &= \Delta^a s + \Delta^c s,
\end{aligned}
\tag{7.30}
$$

which show that the (unscaled) full Newton step $(\Delta x, \Delta s)$ — at (x, s) and for the barrier parameter value μ — can be nicely decomposed in its affine scaling and its centering component.

7.6.3 Condition for adaptive updates

In this section we start to deal with the problem stated before. Let (x, s) be a positive primal-dual pair and $\mu > 0$ such that $\delta = \delta(x, s; \mu) \leq \bar{\tau}$. We want to investigate how large θ can be so that after the Newton step at (x, s) with barrier parameter value $\mu^+ = (1 - \theta)\mu$ we have $\delta^+ = \delta(x^+, s^+; \mu^+) \leq \bar{\tau}$. We derive a condition for the barrier update parameter θ that guarantees the desired behavior.

The vector u, the scaled search directions d_x and d_s and their (scaled) centering components d_x^c, d_s^c and (scaled) affine components d_x^a, d_s^a, have the same meaning as in the previous section; the entities u, d_x^a and d_s^a depend on the given value μ of the barrier parameter. The scaled search directions at (x, s) with barrier parameter value μ^+ are denoted by d_x^+ and d_s^+. Letting Δx and Δs denote the (unscaled) Newton directions with respect to μ^+, we have

$$s \Delta x + x \Delta s = \mu^+ e - xs,$$

and therefore, also using (7.11),

$$x^+ s^+ = \mu^+ e + \Delta x \Delta s = \mu^+ \left(e + d_x^+ d_s^+ \right).$$

By Lemma II.48, the step is feasible if $e + d_x^+ d_s^+ \geq 0$, and this certainly holds if

$$\left\| d_x^+ d_s^+ \right\|_\infty \leq 1.$$

Moreover, from the proof of Theorem II.50 we recall that the proximity $\delta^+ := \delta(x^+, s^+; \mu^+)$ of the new pair (x^+, s^+) with respect to the μ^+-center is given by

$$2\delta^+ = \left\| \frac{d_x^+ d_s^+}{\sqrt{e + d_x^+ d_s^+}} \right\|.$$

This implies that we have $\delta^+ \leq \bar{\tau}$ if and only if

$$\left\| \frac{d_x^+ d_s^+}{\sqrt{e + d_x^+ d_s^+}} \right\|^2 \leq 4\bar{\tau}^2.$$

In the sequel we use the weaker condition

$$\left\| d_x^+ d_s^+ \right\|^2 \leq 4\bar{\tau}^2 \left(1 - \left\| d_x^+ d_s^+ \right\|_\infty \right), \tag{7.31}$$

which we refer to as the *condition for adaptive updating*. A very important observation is that when this condition is satisfied, the Newton step is feasible. Because, if (7.31) holds, since the left-hand side expression is nonnegative, the right-hand side expression must be nonnegative as well, and hence $\left\| d_x^+ d_s^+ \right\|_\infty \leq 1$. Thus, in the further analysis we may concentrate on the condition for adaptive updating (7.31).

7.6.4 Calculation of the adaptive update

We proceed by deriving upper bounds for the 2-norm and the infinity norm of the vector $d_x^+ d_s^+$. It is convenient to introduce the vector

$$\bar{u} := \sqrt{\frac{xs}{\mu^+}}.$$

We then have

$$\bar{u} = \sqrt{\frac{xs}{\mu^+}} = \sqrt{\frac{xs}{(1-\theta)\mu}} = \frac{u}{\sqrt{1-\theta}}.$$

Hence, using this and (7.26),

$$d_x^+ = P_{AD}\left(\bar{u}^{-1} - \bar{u}\right) = \sqrt{1-\theta}P_{AD}\left(u^{-1}\right) - \frac{1}{\sqrt{1-\theta}}P_{AD}\left(u\right)$$

and

$$d_s^+ = P_{HD^{-1}}\left(\bar{u}^{-1} - \bar{u}\right) = \sqrt{1-\theta}P_{HD^{-1}}\left(u^{-1}\right) - \frac{1}{\sqrt{1-\theta}}P_{HD^{-1}}\left(u\right).$$

Now using (7.28) and (7.29) we obtain

$$d_x^+ = \sqrt{1-\theta}\,d_x^c + \frac{d_x^a}{\sqrt{1-\theta}} \tag{7.32}$$

$$d_s^+ = \sqrt{1-\theta}\,d_s^c + \frac{d_s^a}{\sqrt{1-\theta}}. \tag{7.33}$$

Note that d_x^+ can be rewritten in the following way:

$$\sqrt{1-\theta}d_x^c + \frac{d_x^a}{\sqrt{1-\theta}} = \sqrt{1-\theta}\left(d_x^c + d_x^a\right) + \left(\frac{1}{\sqrt{1-\theta}} - \sqrt{1-\theta}\right)d_x^a$$

$$= \sqrt{1-\theta}d_x + \frac{\theta}{\sqrt{1-\theta}}d_x^a.$$

Since d_s^+ can be reformulated in exactly the same way we find

$$d_x^+ = \sqrt{1-\theta}d_x + \frac{\theta}{\sqrt{1-\theta}}d_x^a$$

$$d_s^+ = \sqrt{1-\theta}d_s + \frac{\theta}{\sqrt{1-\theta}}d_s^a.$$

Multiplication of both expressions gives

$$d_x^+ d_s^+ = (1-\theta)d_x d_s + \theta\left(d_x d_s^a + d_s d_x^a\right) + \frac{\theta^2}{1-\theta}d_x^a d_s^a. \tag{7.34}$$

At this stage we see how the coordinates of the vector $d_x^+ d_s^+$ depend on θ. The coordinates of $(1-\theta)d_x^+ d_s^+$ are quadratic functions of θ:

$$(1-\theta)d_x^+ d_s^+ = (1-\theta)^2 d_x d_s + \theta(1-\theta)\left(d_x d_s^a + d_s d_x^a\right) + \theta^2 d_x^a d_s^a.$$

When multiplying the condition (7.31) for adaptive updating by $(1-\theta)^2$, this condition can be rewritten as

$$4\bar{\tau}^2(1-\theta)^2 - \left\|(1-\theta)d_x^+ d_s^+\right\|^2 \geq 4\bar{\tau}^2(1-\theta)\left\|(1-\theta)d_x^+ d_s^+\right\|_\infty. \tag{7.35}$$

Now denoting the left-hand side member by $p(\theta)$ and the i-th coordinate of the vector $(1-\theta)d_x^+ d_s^+$ by $q_i(\theta)$, with $\bar{\tau}$ given, we need to find the largest positive θ that satisfies the following inequalities:

$$p(\theta) \geq 4\bar{\tau}^2(1-\theta)q_i(\theta), \quad 1 \leq i \leq n$$

$$p(\theta) \geq -4\bar{\tau}^2(1-\theta)q_i(\theta), \quad 1 \leq i \leq n.$$

Since $p(\theta)$ is a polynomial of degree 4 in θ, and each $q_i(\theta)$ is a polynomial of degree 2 in θ, the largest positive θ satisfying each single one of these $2n$ inequalities can be found straightforwardly by solving a polynomial equation of degree 4. The smallest of the $2n$ positive numbers obtained in this way (some of them may be infinite, but not all of them!) is the value of θ determined by the condition of adaptive updating. Thus we have shown that the largest θ satisfying the condition for adaptive updating can be found by solving $2n$ polynomial equations of degree 4.[15]

Below we deal with a second approach. We consider a further relaxation of the condition for adaptive updating that requires the solution of only one quadratic equation. Of course, this approach yields a smaller value of θ than the above procedure, which gives the exact solution of the condition (7.31) for adaptive updating. Before proceeding it is of interest to investigate the special case where we start at the μ-centers $x = x(\mu)$ and $s = s(\mu)$.

7.6.5 Special case: adaptive update at the μ-center

When we start with $x = x(\mu)$ and $s = s(\mu)$, we established earlier that $u = e$, $d_x = d_s = 0$, $d_x^a = -d_x^c$ and $d_s^a = -d_s^c$. Substituting this in (7.34) we obtain

$$d_x^+ d_s^+ = \frac{\theta^2}{1 - \theta} d_x^a d_s^a.$$

Now we can use the first uv-lemma (Lemma C.4 in Appendix C) to estimate the 2-norm and the infinity norm of $d_x^a d_s^a$. Since $d_x^a + d_s^a = -u = -e$, we obtain

$$\|d_x^a d_s^a\| \leq \frac{n}{2\sqrt{2}}, \quad \|d_x^a d_s^a\|_\infty \leq \frac{n}{4}.$$

Substitution in (7.31) gives

$$\frac{\theta^4 n^2}{8(1 - \theta)^2} \leq 4\bar{\tau}^2 \left(1 - \frac{\theta^2 n}{4(1 - \theta)}\right).$$

This can be rewritten as

$$\frac{\theta^4 n^2}{8(1 - \theta)^2} + \frac{\bar{\tau}^2 \theta^2 n}{1 - \theta} \leq 4\bar{\tau}^2,$$

which is equivalent to

$$\left(\frac{\theta^2 n}{2\sqrt{2}(1 - \theta)} + \bar{\tau}^2 \sqrt{2}\right)^2 \leq 4\bar{\tau}^2 + 2\bar{\tau}^4,$$

or

$$\frac{\theta^2 n}{1 - \theta} \leq 2\sqrt{2} \left(\sqrt{4\bar{\tau}^2 + 2\bar{\tau}^4} - \bar{\tau}^2 \sqrt{2}\right).$$

Substituting $\bar{\tau} = 1/2$ gives

$$\frac{\theta^2 n}{1 - \theta} \leq 2.$$

---

[15] In fact, more efficient procedures exist for solving the condition for adaptive updating, but here our only aim has been to show that there exists an efficient procedure for finding the maximal value of the parameter θ satisfying the condition for adaptive updating.

This result has its own interest. The bound obtained is exactly the 'ideal' bound for θ derived in Section 7.5 for the hypothetical situation where the Newton step is exact. Here we obtained a better bound without this assumption, but under the more realistic assumption that we start at the μ-centers $x = x(\mu)$ and $s = s(\mu)$.

7.6.6   A simple version of the condition for adaptive updating

We return to the general case, and show how a weakened version of the condition for adaptive updating

$$\left\| d_x^+ d_s^+ \right\|^2 \leq 4\bar{\tau}^2 \left(1 - \left\| d_x^+ d_s^+ \right\|_\infty \right)$$

can be reduced to a quadratic inequality in θ. With

$$d^+ := d_x^+ + d_s^+,$$

the first uv-lemma (Lemma C.4 in Appendix C) implies that

$$\left\| d_x^+ d_s^+ \right\| \leq \frac{\left\| d^+ \right\|^2}{2\sqrt{2}}, \quad \left\| d_x^+ d_s^+ \right\|_\infty \leq \frac{\left\| d^+ \right\|^2}{4}.$$

Substituting these bounds in the condition for adaptive updating we obtain the weaker condition

$$\frac{\left\| d^+ \right\|^4}{8} \leq 4\bar{\tau}^2 \left(1 - \frac{\left\| d^+ \right\|^2}{4} \right).$$

Rewriting this as

$$\left(\left\| d^+ \right\|^2 + 4\bar{\tau}^2 \right)^2 \leq 32\bar{\tau}^2 + 16\bar{\tau}^4,$$

we obtain

$$\left\| d^+ \right\|^2 \leq \sqrt{32\bar{\tau}^2 + 16\bar{\tau}^4} - 4\bar{\tau}^2.$$

Substituting $\bar{\tau} = 1/2$ leads to the condition

$$\left\| d^+ \right\|^2 \leq 2. \tag{7.36}$$

From the expressions (7.32) and (7.33) for d_x^+ and d_s^+, and also using that $d_x^c + d_s^c = u^{-1}$ and $d_x^a + d_s^a = -u$, we find

$$d^+ = \sqrt{1 - \theta}\, u^{-1} - \frac{u}{\sqrt{1 - \theta}}.$$

From this expression we can calculate the norm of d^+:

$$\left\| d^+ \right\|^2 = (1 - \theta) \left\| u^{-1} \right\|^2 + \frac{\left\| u \right\|^2}{1 - \theta} - 2n.$$

Since $\left\| u \right\|^2 = n$ and $\left\| u^{-1} \right\|^2 = n + 4\delta^2$, where $\delta = \delta(x, s; \mu)$, we obtain

$$\left\| d^+ \right\|^2 = (1 - \theta) \left(n + 4\delta^2 \right) + \frac{n}{1 - \theta} - 2n = 4(1 - \theta)\delta^2 + \frac{\theta^2 n}{1 - \theta}. \tag{16}$$

---

[16] Since $\left\| d^+ \right\| = 2\delta(x, s : \mu^+)$ this analysis yields in a different way the same result as in Lemma II.54, namely

$$\delta(x, s; \mu^+)^2 = (1 - \theta)\delta^2 + \frac{\theta^2 n}{4(1 - \theta)}.$$

Putting this in (7.36) we obtain the following condition on θ:

$$4(1-\theta)\delta^2 + \frac{\theta^2 n}{1-\theta} \leq 2.$$

The largest θ satisfying this inequality is given by

$$\theta = \frac{\sqrt{2n+1-4n\delta^2}+4\delta^2-1}{n+4\delta^2}. \tag{7.37}$$

With this value of θ we are sure that when starting with $\delta(x,s;\mu)=\delta$, after the Newton step with barrier parameter value $\mu^+ = (1-\theta)\mu$ we have $\delta(x^+,s^+;\mu^+) \leq 1/2$. If $\delta = 0$, the above expression reduces to

$$\theta = \frac{2}{1+\sqrt{2n+1}} \leq \frac{1}{\sqrt{2n}}$$

and if $\delta = 1/2$ to

$$\theta = \frac{1}{\sqrt{n+1}},\,^{17}$$

as easily may be verified. Hence, when using cheap adaptive updates the actual value of θ varies from iteration to iteration but it *always* lies between the above two extreme values. The ratio between these extreme values is about $\sqrt{2}$. As a consequence, the speedup factor is bounded above by (approximately) $\sqrt{2}$.

7.6.7 Illustration of the algorithm with adaptive updates

With the same example as in the previous illustrations, and the same initialization of the algorithm as in Section 7.5.2, we experiment in this section with two adaptive-update strategies. First we consider the most expensive strategy, and calculate the barrier update parameter θ from (7.35). In this case we need to solve $2n$ polynomial inequalities of degree four. The algorithm, with $\varepsilon = 10^{-4}$, then runs as shown in Table 7.3.. As before, Table 7.3. contains one entry (the first) of the vectors x and s. A new column shows the value of the barrier update parameter in each iteration. The fast increase of this parameter to almost 1 is surprising. It results in very fast convergence of the method: only 5 iterations yield the desired accuracy.

When we calculate θ according to (7.37), the performance of the algorithm is as shown in Table 7.4.. Now 15 iterations are needed instead of 6. In this example in the final iterations θ seems to stabilize around the value 0.58486. This implies that the convergence rate for the duality gap is linear. This is in contrast with the other approach, where the convergence rate for the duality gap appears to be quadratic.

Unfortunately, at this time no theoretical justification for a quadratic convergence rate of the adaptive version of the full-step method exists. For the moment we leave

---

^{17} We could have used this value of θ in Theorem II.53, leading to the iteration bound

$$\left\lceil \sqrt{n+1}\,\log\frac{n\mu^0}{\varepsilon}\right\rceil$$

for the Primal-Dual Logarithmic Barrier Algorithm with full Newton steps.

It.	$n\mu$	x_1	y_1	y_2	s_1	δ	δ^+	θ
0	4.000000	2.000000	0.000000	0.000000	1.000000	0.2887	0.7071	0.679623
1	1.281509	1.093836	0.333333	0.572830	0.666667	0.7071	0.7071	0.846142
2	0.197170	1.010191	0.888935	0.934277	0.111065	0.7071	0.7071	0.976740
3	0.004586	1.000224	0.997391	0.998471	0.002609	0.7071	0.7071	0.999460
4	0.000002	1.000000	0.999999	0.999999	0.000001	0.7071	0.1472	0.999999
5	0.000000	1.000000	1.000000	1.000000	0.000000	–	–	–

Table 7.3. The primal-dual full-step algorithm with expensive adaptive updates.

It.	$n\mu$	x_1	y_1	y_2	s_1	δ	δ^+	θ
0	4.000000	2.000000	0.000000	0.000000	1.000000	0.2887	0.2934	0.534847
1	1.860612	1.286871	0.333333	0.379796	0.666667	0.2934	0.1355	0.534357
2	0.866381	1.138033	0.698479	0.711206	0.301521	0.1355	0.0670	0.545715
3	0.393584	1.063707	0.865026	0.868805	0.134974	0.0670	0.0308	0.547890
4	0.177943	1.029247	0.939865	0.940686	0.060135	0.0308	0.0140	0.548438
5	0.080352	1.013307	0.973046	0.973216	0.026954	0.0140	0.0063	0.548554
6	0.036275	1.006028	0.987874	0.987908	0.012126	0.0063	0.0028	0.548578
7	0.016375	1.002726	0.994534	0.994542	0.005466	0.0028	0.0013	0.548583
8	0.007392	1.001231	0.997535	0.997536	0.002465	0.0013	0.0006	0.548584
9	0.003337	1.000556	0.998887	0.998888	0.001113	0.0006	0.0003	0.548584
10	0.001506	1.000251	0.999498	0.999498	0.000502	0.0003	0.0001	0.548584
11	0.000680	1.000113	0.999773	0.999773	0.000227	0.0001	0.0001	0.548584
12	0.000307	1.000051	0.999898	0.999898	0.000102	0.0001	0.0000	0.548584
13	0.000139	1.000023	0.999954	0.999954	0.000046	0.0000	0.0000	0.548584
14	0.000063	1.000010	0.999979	0.999979	0.000021	0.0000	0.0000	0.548584
15	0.000028	1.000005	0.999991	0.999991	0.000009	–	–	–

Table 7.4. The primal-dual full-step algorithm with cheap adaptive updates.

this topic with the conclusion that the above comparison between the 'expensive' and the 'cheap' adaptive update full-step method suggests that it is worth spending extra effort in finding as large values for θ as possible.

We conclude the section with a graphical illustration of the adaptive updating strategy. Figure 7.6 shows on two graphs the progress of the algorithm with the expensive update. The graphs show the first two coordinates of the iterates in the w-space. The left graph has a linear scale and the right graph a logarithmic scale. Figure 7.7 concerns the case when cheap updates are used.

7.7 The predictor-corrector method

In the previous section it became clear that the Newton step can be decomposed into an affine-scaling component and a centering component. Using the notations introduced

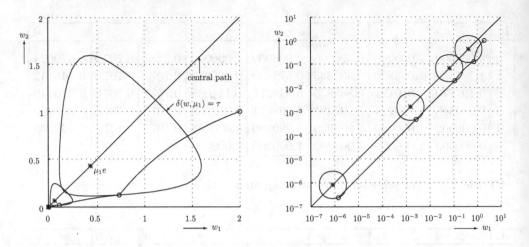

Figure 7.6 Iterates of the primal-dual algorithm with adaptive updates.

there, we recall from Section 7.6.2 that the (scaled) centering components are given by

$$d_x^c = P_{AD}(u^{-1}), \quad d_s^c = P_{HD^{-1}}(u^{-1}),$$

and the (scaled) affine components by

$$d_x^a = -P_{AD}(u), \quad d_s^a = -P_{HD^{-1}}(u),$$

where

$$d = \sqrt{\frac{x}{s}}, \quad u = \sqrt{\frac{xs}{\mu}}.$$

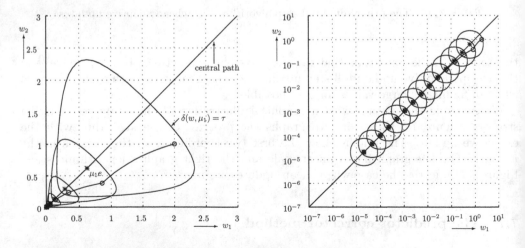

Figure 7.7 Iterates of the primal-dual algorithm with cheap adaptive updates.

We also recall the relations

$$
\begin{aligned}
d_x &= d_x^c + d_x^a \\
d_s &= d_s^c + d_s^a
\end{aligned}
$$

and

$$
\begin{aligned}
d_x^c + d_s^c &= u^{-1} \\
d_x^a + d_s^a &= -u.
\end{aligned}
$$

The unscaled centering and affine-scaling components are given by

$$
\Delta^a x = \sqrt{\mu} d d_x^a, \quad \Delta^a s = \sqrt{\mu} d^{-1} d_s^a
$$

and

$$
\Delta^c x = \sqrt{\mu} d d_x^c, \quad \Delta^c s = \sqrt{\mu} d^{-1} d_s^c;
$$

as a consequence we have

$$
\Delta^a x \Delta^a s = \mu d_x^a d_s^a, \quad \Delta^c x \Delta^c s = \mu d_x^c d_s^c.
$$

It is interesting to consider the effect of moving along these directions. Let us define

$$
\begin{aligned}
x^a(\theta) &= x + \theta \Delta^a x & s^a(\theta) &= s + \theta \Delta^a s \\
x^c &= x + \Delta^c x & s^c &= s + \Delta^a s.
\end{aligned}
$$

We say that $x^a(\theta)$ and $s^a(\theta)$ result from an *affine-scaling step* of size θ at (x, s). In preparation for the next lemma we first establish the following two relations:

$$
\begin{aligned}
x\Delta^a s + s\Delta^a x &= -xs & (7.38) \\
x\Delta^c s + s\Delta^c x &= \mu e. & (7.39)
\end{aligned}
$$

These relations easily follow from the previous ones. We show this for the first of the two relations. We first write

$$
x\Delta^a s = \sqrt{\mu} x d^{-1} d_s^a = \mu u d_s^a,
$$

and

$$
s\Delta^a x = \sqrt{\mu} s d d_x^a = \mu u d_x^a.
$$

Adding the last two equalities we get (7.38):

$$
x\Delta^a s + s\Delta^a x = \mu u \left(d_x^a + d_s^a\right) = -\mu u^2 = -xs.
$$

Now we can prove

Lemma II.60 *Let $x^T s = n\mu$. Assuming feasibility of the steps, the affine-scaling step reduces the duality gap by a factor $1 - \theta$ and the step along the centering components doubles the duality gap.*

Proof: We have

$$x^a(\theta)s^a(\theta) = (x + \theta\Delta^a x)(s + \theta\Delta^a s) = xs + \theta(x\Delta^a s + s\Delta^a x) + \theta^2\Delta^a x\Delta^a s.$$

Using (7.38) we find

$$x^a(\theta)s^a(\theta) = (1 - \theta)xs + \theta^2\Delta^a x\Delta^a s.$$

Using that $\Delta^a x$ and $\Delta^a s$ are orthogonal we obtain

$$(x^a(\theta))^T s^a(\theta) = e^T\left((1 - \theta)xs + \theta^2\Delta^a x\Delta^a s\right) = (1 - \theta)x^T s,$$

proving the first statement. For the second statement we write

$$x^c s^c = (x + \Delta^c x)(s + \Delta^c s) = xs + (x\Delta^c s + s\Delta^c x) + \Delta^c x\Delta^c s.$$

Substitution of (7.39) gives

$$x^c s^c = xs + \mu e + \theta^2\Delta^c x\Delta^c s.$$

Thus we obtain

$$(x^c)^T s^c = e^T\left(xs + \mu e + \theta^2\Delta^c x\Delta^c s\right) = x^T s + \mu e^T e = 2n\mu.$$

This completes the proof. □

Recall from (7.30) that the (unscaled) full Newton step $(\Delta x, \Delta s)$ at (x, s) — with barrier parameter value μ — can be decomposed in its affine scaling and its centering component. The above lemma makes clear that in the algorithms we dealt with before, the reduction in the duality gap during a (full) Newton step is delivered by the affine-scaling component in the Newton step. The centering component in the Newton step forces the iterates to stay close to the central path.

When solving a given LO problem, we wish to find a primal-dual pair with a duality gap close to zero. We want to reduce the duality gap as fast as possible to zero. Therefore, it becomes natural to consider algorithms that put more emphasis on the affine-scaling component. That is the underlying idea of the *predictor-corrector method* which is the subject of this section. Note that when the full affine-scaling step (with step-size 1) is feasible, it produces a feasible pair with duality gap zero, and hence it yields an optimal solution pair in a single step. This makes clear that the full affine step will be infeasible in general.

In the predictor-corrector method, instead of combining the two directions in a single Newton step, we decompose the Newton step into two steps, an affine-scaling step first and, next, a so-called pure centering step.[18] Since a full affine-scaling step is infeasible, we use a *damping parameter* θ. By taking θ small enough we enforce feasibility of the step, and at the same time gain control over the loss of proximity to the central path. The aim of the centering step is to restore the proximity to the central path. This is obtained by using a Newton step with barrier parameter value μ, where $n\mu$ is equal to the present duality gap. Such a step leaves the duality gap unchanged, by Lemma II.47.

7.7.1 The predictor-corrector algorithm

In the description of the predictor-corrector algorithm below (page 182), Δx and Δs denote the full Newton step at (x, s) with the current value of the barrier parameter μ, and $\Delta^a x$ and $\Delta^a s$ denote the full affine-scaling step at the current iterate (x, s). Observe that according to Lemma II.60 the damping factor θ for the affine-scaling step can also be interpreted as an updating parameter for the barrier parameter μ.

We have the following theorem.

Theorem II.61 *If $\tau = 1/2$ and $\theta = 1/(2\sqrt{n})$, then the Predictor-Corrector Algorithm requires at most*

$$\left\lceil 2\sqrt{n} \log \frac{n\mu^0}{\varepsilon} \right\rceil$$

iterations. The output is a primal-dual pair (x, s) such that $x^T s \leq \varepsilon$.

The proof of this result is postponed to Section 7.7.3. It requires a careful analysis of the affine-scaling step, which is the subject of the next section. Let us note now that the iteration bound is a factor $\sqrt{2}$ worse than the bound in Theorem II.53 for the algorithm with full Newton steps. Moreover, each major iteration in the predictor-corrector algorithm consists of two steps: the centering step (also called the *corrector step*) and the affine-scaling step (also called the *predictor step*).

7.7.2 Properties of the affine-scaling step

The purpose of this section is to analyze the effect of an affine-scaling step with size θ on the proximity measure. As before, (x, s) denotes a positive primal-dual pair. We

---

[18] The idea of breaking down the Newton direction into its affine-scaling and its centering component seems to be due to Mehrotra [205]. The method considered in this chapter was proposed first by Mizuno, Todd and Ye [217]; they were the first to use the name predictor-corrector method. The analysis in this chapter closely resembles their analysis. Like them we alternate (single) primal-dual affine-scaling steps and (single) primal-dual centering steps. An earlier paper of Sonnevend, Stoer and Zhao [258] is based on similar ideas, except that they use multiple centering steps. It soon appeared that one could prove that the method asymptotically has a quadratic convergence rate (see, e.g., Mehrotra [206, 205], Ye et al. [317], Gonzaga and Tapia [126, 127], Ye [309] and Luo and Ye [188].). Quadratic convergence of the primal-dual predictor-corrector method is the subject in Section 7.7.6. A dual version of the predictor-corrector method was considered by Barnes, Chopra and Jensen [36]; they showed polynomial-time convergence with an $\mathcal{O}(nL)$ iteration bound. Mehrotra's variant of the primal-dual predictor-corrector method will be discussed in Chapter 20. It significantly cuts down the computational effort to achieve the greatest practical efficiency among all interior-point methods. See, e.g., Lustig, Marsten and Shanno [192]. As a consequence the method has become very popular.

Predictor-Corrector Algorithm

Input:
  A proximity parameter τ, $0 \le \tau < 1$;
  an accuracy parameter $\varepsilon > 0$;
  $(x^0, s^0) \in \mathcal{P} \times \mathcal{D}$, $\mu^0 > 0$ with $(x^0)^T s^0 = n\mu^0$, $\delta(x^0, s^0; \mu^0) \le \tau$;
  a barrier update parameter θ, $0 < \theta < 1$.

begin
  $x := x^0;\ s := s^0;\ \mu := \mu^0;$
  while $n\mu \ge (1 - \theta)\varepsilon$ **do**
  begin
    $x := x + \Delta x;$
    $s := s + \Delta s;$
    $x := x + \theta \Delta^a x;$
    $s := s + \theta \Delta^a s;$
    $\mu := (1 - \theta)\mu;$
  end
end

---

assume that $\mu > 0$ is such that $x^T s = n\mu$, and $\delta := \delta(x, s; \mu)$. Recall from (7.16) that

$$\delta = \frac{1}{2} \left\| u^{-1} - u \right\| = \frac{1}{2} \left\| \frac{e - u^2}{u} \right\|,$$

where

$$u = \sqrt{\frac{xs}{\mu}}.$$

We need a simple bound on the coordinates of the vector u.

Lemma II.62 *Let $\rho(\delta) := \delta + \sqrt{1 + \delta^2}$. Then*

$$\frac{1}{\rho(\delta)} \le u_i \le \rho(\delta), \quad 1 \le i \le n.$$

Proof: Since u_i is positive for each i, we have

$$-2\delta u_i \le 1 - u_i^2 \le 2\delta u_i.$$

This implies

$$u_i^2 - 2\delta u_i - 1 \le 0 \le u_i^2 + 2\delta u_i - 1.$$

Rewriting this as

$$(u_i - \delta)^2 - 1 - \delta^2 \le 0 \le (u_i + \delta)^2 - 1 - \delta^2$$

we obtain

$$(u_i - \delta)^2 \leq 1 + \delta^2 \leq (u_i + \delta)^2,$$

which implies

$$u_i - \delta \leq |u_i - \delta| \leq \sqrt{1 + \delta^2} \leq u_i + \delta.$$

Thus we arrive at

$$-\delta + \sqrt{1 + \delta^2} \leq u_i \leq \delta + \sqrt{1 + \delta^2} = \rho(\delta).$$

For the left-hand expression we write

$$-\delta + \sqrt{1 + \delta^2} = \frac{1}{\delta + \sqrt{1 + \delta^2}} = \frac{1}{\rho(\delta)}.$$

This proves the lemma. $\qquad\square$

Now we can prove the following.

Lemma II.63 *Let the pair* (x^+, s^+) *result from an affine-scaling step at* (x, s) *with step-size* θ. *If* $x^T s = n\mu$ *and* $\delta := \delta(x, s; \mu) < \tau$, *then we have* $\delta^+ := \delta(x^+, s^+; (1 - \theta)\mu) \leq \tau$ *if* θ *satisfies the inequality*

$$\frac{\theta^2 n}{1 - \theta} \leq 2\sqrt{2} \left(\tau \sqrt{\frac{4}{\rho(\delta)^2} + 4\delta\rho(\delta)\sqrt{2} + 2\tau^2} - 2\delta\rho(\delta) - \tau^2\sqrt{2} \right). \qquad (7.40)$$

For fixed τ, *the right-hand side expression in (7.40) is a monotonically decreasing function of* δ.

Proof: From the proof of Lemma II.60 we recall that

$$x^+ s^+ = (1 - \theta)xs + \theta^2 \Delta^a x \Delta^a s.$$

This can be rewritten as

$$x^+ s^+ = \mu \left((1 - \theta)u^2 + \theta^2 d_x^a d_s^a \right).$$

Defining

$$u^+ := \sqrt{\frac{x^+ s^+}{(1 - \theta)\mu}},$$

we thus have

$$(u^+)^2 = u^2 + \frac{\theta^2 d_x^a d_s^a}{1 - \theta}.$$

The proximity after the affine-scaling step satisfies

$$\delta^+ = \frac{1}{2} \left\| (u^+)^{-1} \left(e - (u^+)^2 \right) \right\| \leq \frac{1}{2} \left\| (u^+)^{-1} \right\|_\infty \left\| e - (u^+)^2 \right\|.$$

We proceed by deriving bounds for the last two norms. First we consider the second norm:

$$\left\| e - (u^+)^2 \right\| = \left\| e - u^2 - \frac{\theta^2 d_x^a d_s^a}{1-\theta} \right\| \leq \left\| e - u^2 \right\| + \frac{\theta^2}{1-\theta} \left\| d_x^a d_s^a \right\|$$

$$\leq \left\| e - u^2 \right\| + \frac{\theta^2 n}{2\sqrt{2}(1-\theta)}.$$

For the last inequality we applied the first uv-lemma (Lemma C.4 in Appendix C) to the vectors d_x^a and d_s^a and further utilized $\|u\|^2 = n$. From Lemma II.62, we further obtain

$$\left\| e - u^2 \right\| = \left\| u \frac{e-u^2}{u} \right\| \leq \|u\|_\infty \left\| \frac{e-u^2}{u} \right\| \leq 2\delta\rho(\delta).$$

For the estimate of $\left\| (u^+)^{-1} \right\|_\infty$ we write, using Lemma II.62 and the first uv-lemma once more,

$$(u_i^+)^2 \geq u_i^2 - \frac{\theta^2}{1-\theta} \left\| d_x^a d_s^a \right\|_\infty \geq \frac{1}{\rho(\delta)^2} - \frac{\theta^2 n}{4(1-\theta)}.$$

We conclude, by substitution of these estimates, that

$$\delta^+ \leq \frac{2\delta\rho(\delta) + \frac{\theta^2 n}{2\sqrt{2}(1-\theta)}}{2\sqrt{\frac{1}{\rho(\delta)^2} - \frac{\theta^2 n}{4(1-\theta)}}}.$$

Hence, $\delta^+ \leq \tau$ holds if

$$\left(2\delta\rho(\delta) + \frac{\theta^2 n}{2\sqrt{2}(1-\theta)} \right)^2 \leq \frac{4\tau^2}{\rho(\delta)^2} - \frac{\theta^2 n \tau^2}{1-\theta}. \tag{7.41}$$

This can be rewritten as

$$\left(2\delta\rho(\delta) + \frac{\theta^2 n}{2\sqrt{2}(1-\theta)} \right)^2 + 2\tau^2\sqrt{2} \left(2\delta\rho(\delta) + \frac{\theta^2 n}{2\sqrt{2}(1-\theta)} \right) \leq \frac{4\tau^2}{\rho(\delta)^2} + 4\tau^2\delta\rho(\delta)\sqrt{2},$$

or equivalently,

$$\left(2\delta\rho(\delta) + \frac{\theta^2 n}{2\sqrt{2}(1-\theta)} + \tau^2\sqrt{2} \right)^2 \leq \frac{4\tau^2}{\rho(\delta)^2} + 4\tau^2\delta\rho(\delta)\sqrt{2} + 2\tau^4.$$

By taking the square root we get

$$2\delta\rho(\delta) + \frac{\theta^2 n}{2\sqrt{2}(1-\theta)} + \tau^2\sqrt{2} \leq \tau\sqrt{\frac{4}{\rho(\delta)^2} + 4\delta\rho(\delta)\sqrt{2} + 2\tau^2}.$$

By rearranging terms this can be rewritten as

$$\frac{\theta^2 n}{2\sqrt{2}(1-\theta)} \leq \tau\sqrt{\frac{4}{\rho(\delta)^2} + 4\delta\rho(\delta)\sqrt{2} + 2\tau^2} - 2\delta\rho(\delta) - \tau^2\sqrt{2}.$$

This implies the first statement in the lemma. For the proof of the second statement we observe that the inequality (7.40) in the lemma is equivalent to the inequality (7.41). We can easily verify that the left-hand side expression in (7.41) is increasing in both δ and θ and the right-hand side expression is decreasing in both δ and θ. Hence, if θ satisfies (7.41) for some value of δ, then the same value of θ satisfies (7.41) also for smaller values of δ. Since the inequalities (7.40) and (7.41) are equivalent, the last inequality has the same property: if θ satisfies (7.40) for some value of δ, then the same value of θ satisfies (7.40) also for smaller values of δ. This implies the second statement in the lemma and completes the proof. □

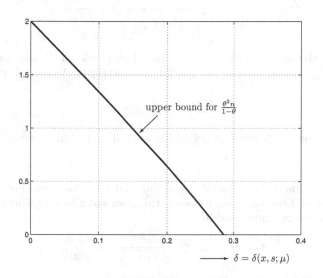

Figure 7.8 The right-hand side of (7.40) for $\tau = 1/2$.

Figure 7.8 shows the graph of the right-hand side of (7.40) as a function of δ.

With the above lemma the analysis of the predictor-corrector algorithm can easily be accomplished. We do this in the next section. At the end of this section we apply the lemma to the special case where we start the affine-scaling step at the μ-centers. Then $\delta = 0$ and $\rho(\delta) = 1$. Substitution of these values in the lemma yields that the proximity after the step does not exceed τ if

$$\frac{\theta^2 n}{1 - \theta} \leq 2\sqrt{2}\left(\tau\sqrt{4 + 2\tau^2} - \tau^2\sqrt{2}\right).$$

Note that this bound coincides with the corresponding bound obtained in Section 7.6.5 for an adaptive update at the μ-center with the full-step method.

7.7.3 Analysis of the predictor-corrector algorithm

In this section we provide the proof of Theorem II.61. Taking $\tau = 1/2$ and $\theta = 1/(2\sqrt{n})$ we show that each iteration starts with x, s and μ such that $\delta(x, s; \mu) \leq \tau$. This makes the algorithm well defined, and implies the result of the theorem.

The corrector step is simply a Newton step to the μ-center. By Theorem II.50 (on page 156) the result is a pair (x, s) such that

$$\delta := \delta(x, s; \mu) \leq \frac{\frac{1}{4}}{\sqrt{2(1 - \frac{1}{4})}} = \frac{1}{\sqrt{24}}.$$

Now we apply Lemma II.63 to this pair (x, s). This lemma states that the affine step with step-size θ leaves the proximity with respect to the barrier parameter $(1 - \theta)\mu$ smaller than (or equal to) τ if θ satisfies (7.40) and, moreover, that for fixed τ the right-hand side of (7.40) is monotonically decreasing in δ. For $\delta = 1/\sqrt{24}$ we have

$$\rho(\delta) = \frac{1}{\sqrt{24}} + \sqrt{1 + \frac{1}{24}} = \sqrt{\frac{3}{2}}.$$

Substitution of the given values in the right-hand side of (7.40) yields the value 0.612626 (cf. Figure 7.8, with $\delta = 1/\sqrt{24} = 0.204124$). Hence (7.40) is certainly satisfied if

$$\frac{\theta^2 n}{1 - \theta} \leq 0.612626.$$

If $\theta = 1/(2\sqrt{n})$ this condition is satisfied for each $n \geq 1$. This proves Theorem II.61. □

Remark II.64 In the above analysis we could also have used the improved quadratic convergence result of Theorem II.52. However, this does not give a significant change. After the centering step the proximity satisfies

$$\delta := \delta(x, s; \mu) \leq \frac{\frac{1}{4}}{\sqrt{2(1 - \frac{1}{16})}} = \frac{1}{\sqrt{30}},$$

and the condition on θ becomes a little weaker, namely:

$$\frac{\theta^2 n}{1 - \theta} \leq 0.768349. \qquad \bullet$$

7.7.4 An adaptive version of the predictor-corrector algorithm

As stated before, the predictor-corrector method is the most popular interior-point method for solving LO problems in practice. But this is not true for the version we dealt with in the previous section. When we update the barrier parameter each time by the factor $1 - \theta$, with $\theta = 1/(2\sqrt{n})$, as in that algorithm, the required number of iterations will be as predicted by Theorem II.61. That is, each iteration reduces the duality gap by the constant factor $1 - \theta$ and hence the duality gap reaches the desired accuracy in a number of iterations that is proportional to $\sqrt{n}$. The obvious way to reduce the number of iterations is to use adaptive updates of the barrier parameter. The following lemma is crucial.

Lemma II.65 Let the pair (x^+, s^+) result from an affine-scaling step at (x, s) with step-size θ. If $x^T s = n\mu$ and $\delta := \delta(x, s; \mu) < \tau$, then $\delta^+ := \delta(x^+, s^+; \mu(1 - \theta)) \leq \tau$ if

$$\frac{\theta^2}{1 - \theta} \|d_x^a d_s^a\| \leq 2\tau \left(\sqrt{\frac{1}{\rho(\delta)^2} + 2\delta\rho(\delta) + \tau^2} - \tau \right) - 2\delta\rho(\delta). \qquad (7.42)$$

Proof: The proof is a slight modification of the proof of Lemma II.63. We recall from that proof that the proximity after the affine-scaling step satisfies

$$\delta^+ = \frac{1}{2}\left\|(u^+)^{-1}\left(e - (u^+)^2\right)\right\| \leq \frac{1}{2}\left\|(u^+)^{-1}\right\|_\infty \left\|e - (u^+)^2\right\|,$$

where, as before,

$$(u^+)^2 = u^2 + \frac{\theta^2 d_x^a d_s^a}{1 - \theta},$$

d_x^a and d_s^a denote the scaled affine-scaling components, and $u = \sqrt{xs/\mu}$. We also recall some estimates:

$$\left\|e - (u^+)^2\right\| \leq \left\|e - u^2\right\| + \frac{\theta^2}{1 - \theta}\left\|d_x^a d_s^a\right\|,$$

and

$$\left\|e - u^2\right\| \leq 2\delta\rho(\delta).$$

Moreover,

$$(u_i^+)^2 \geq u_i^2 - \frac{\theta^2}{1 - \theta}\left\|d_x^a d_s^a\right\|_\infty \geq \frac{1}{\rho(\delta)^2} - \frac{\theta^2}{1 - \theta}\left\|d_x^a d_s^a\right\|.$$

By substitution of these estimates we obtain

$$\delta^+ \leq \frac{2\delta\rho(\delta) + \frac{\theta^2}{1-\theta}\left\|d_x^a d_s^a\right\|}{2\sqrt{\frac{1}{\rho(\delta)^2} - \frac{\theta^2}{1-\theta}\left\|d_x^a d_s^a\right\|}}.$$

Hence, $\delta^+ \leq \tau$ holds if

$$\left(2\delta\rho(\delta) + \frac{\theta^2}{1 - \theta}\left\|d_x^a d_s^a\right\|\right)^2 \leq \frac{4\tau^2}{\rho(\delta)^2} - \frac{4\theta^2\tau^2}{1 - \theta}\left\|d_x^a d_s^a\right\|.$$

This can be rewritten as

$$\left(2\delta\rho(\delta) + \frac{\theta^2}{1 - \theta}\left\|d_x^a d_s^a\right\|\right)^2 + 4\tau^2\left(2\delta\rho(\delta) + \frac{\theta^2}{1 - \theta}\left\|d_x^a d_s^a\right\|\right) \leq \frac{4\tau^2}{\rho(\delta)^2} + 8\tau^2\delta\rho(\delta),$$

or equivalently,

$$\left(2\delta\rho(\delta) + \frac{\theta^2}{1 - \theta}\left\|d_x^a d_s^a\right\| + 2\tau^2\right)^2 \leq \frac{4\tau^2}{\rho(\delta)^2} + 8\tau^2\delta\rho(\delta) + 4\tau^4.$$

By taking the square root we get

$$2\delta\rho(\delta) + \frac{\theta^2}{1 - \theta}\left\|d_x^a d_s^a\right\| + 2\tau^2 \leq 2\tau\sqrt{\frac{1}{\rho(\delta)^2} + 2\delta\rho(\delta) + \tau^2},$$

which reduces to

$$\frac{\theta^2}{1 - \theta}\left\|d_x^a d_s^a\right\| \leq 2\tau\left(\sqrt{\frac{1}{\rho(\delta)^2} + 2\delta\rho(\delta) + \tau^2} - \tau\right) - 2\delta\rho(\delta).$$

This completes the proof. □

From this lemma we derive the next theorem.

Theorem II.66 *If* $\tau = 1/3$ *then the property* $\delta(x, s; \mu) \leq \tau$ *is maintained in each iteration if* θ *is taken equal to*

$$\theta = \frac{2}{1 + \sqrt{1 + 13 \|d_x^a d_s^a\|}}.$$

Proof: We only need to show that when we start some iteration with x, s and μ such that $\delta(x, s; \mu) \leq \tau$, then after this iteration the property $\delta(x, s; \mu) \leq \tau$ is maintained.

By Theorem II.50 (on page 156) the result of the corrector step is a pair (x, s) such that

$$\delta := \delta(x, s; \mu) \leq \frac{\frac{1}{9}}{\sqrt{2(1 - \frac{1}{9})}} = \frac{1}{12}.$$

Now we apply Lemma II.65 to (x, s). By this lemma the affine step with step-size θ leaves the proximity with respect to the barrier parameter $(1 - \theta)\mu$ smaller than (or equal to) τ if θ satisfies (7.42). For $\delta = 1/12$ we have $\rho(\delta) = 1.0868$. Substitution of the given values in the right-hand side expression yields 0.308103, which is greater than 4/13. The right-hand side is monotonic in δ, as can be verified by elementary means, so smaller values of δ yield larger values than 4/13. Thus the proximity after the affine-scaling step does not exceed τ if θ satisfies

$$\frac{\theta^2}{1 - \theta} \|d_x^a d_s^a\| \leq \frac{4}{13}.$$

We may easily verify that the value in the theorem satisfies this condition with equality. Hence the proof is complete. □

7.7.5 Illustration of adaptive predictor-corrector algorithm

With the same example as in the previous illustrations, and the same initialization, the adaptive predictor-corrector algorithm, with $\varepsilon = 10^{-4}$, runs as shown in Table 7.5. (page 189). Each iteration consists of two steps: the corrector step (with $\theta = 0$) and the affine-scaling step (with θ as given by Theorem II.66). Table 7.5. shows that only 7 iterations yield the desired accuracy. After the corrector step the proximity is always very small, especially in the final iterations. This is the same phenomenon as observed previously, namely that the Newton process is almost exact. For the affine-scaling steps we see the same behavior as in the full-step method with adaptive updates. The value of the barrier update parameter increases very quickly to 1. As a result the duality gap goes very quickly to zero. This is not accidental. It is a property of the predictor-corrector method with adaptive updates, as shown in the next section. Figure 7.9 (page 190) shows on two graphs the progress of the algorithm in the w-space.

7.7.6 Quadratic convergence of the predictor-corrector algorithm

It is clear that the rate of convergence in the predictor-corrector method depends on the values taken by the barrier update parameter θ. We show in this section that the

It.	$n\mu$	x_1	y_1	y_2	s_1	δ	θ
1	4.000000	2.000000	0.000000	0.000000	1.000000	0.2887	0.000000
1	4.000000	2.000000	0.333333	−0.333333	0.666667	0.0000	0.601242
2	1.595030	1.278509	0.493665	0.468323	0.506335	0.1576	0.000000
2	1.595030	1.334918	0.606483	0.468323	0.393517	0.0085	0.628030
3	0.593303	1.088217	0.780899	0.802232	0.219101	0.1486	0.000000
3	0.593303	1.108991	0.822447	0.802232	0.177553	0.0031	0.752648
4	0.146755	1.019821	0.941805	0.951082	0.058195	0.1543	0.000000
4	0.146755	1.025085	0.952333	0.951082	0.047667	0.0008	0.907623
5	0.013557	1.001775	0.994513	0.995481	0.005487	0.1568	0.000000
5	0.013557	1.002265	0.995492	0.995481	0.004508	0.0001	0.989826
6	0.000138	1.000018	0.999944	0.999954	0.000056	0.1575	0.000000
6	0.000138	1.000023	0.999954	0.999954	0.000046	0.0000	0.999894
7	0.000000	1.000000	1.000000	1.000000	0.000000	0.1576	0.000000
7	0.000000	1.000000	1.000000	1.000000	0.000000	0.0000	1.000000

Table 7.5. The adaptive predictor-corrector algorithm.

rate of convergence eventually becomes quadratic. To achieve a quadratic convergence rate it must be true that in the limit, $(1-\theta)\mu$ is of the order $\mathcal{O}(\mu^2)$, so that $1-\theta = \mathcal{O}(\mu)$. In this section we show that the value of θ in Theorem II.66 has this property. The following lemma makes clear that for our purpose it is sufficient to concentrate on the magnitude of the norm of the vector $d_x^a d_s^a$.

Lemma II.67 *The value of the barrier update parameter θ in Theorem II.66 satisfies*

$$1 - \theta \le \frac{13}{4} \left\| d_x^a d_s^a \right\|.$$

Hence, the rate of convergence for the adaptive predictor-corrector method is quadratic if $\left\| d_x^a d_s^a \right\| = \mathcal{O}(\mu)$.

Proof: The lemma is an easy consequence of properties of the function $f : [0, \infty) \to \mathbb{R}_+$ defined by

$$f(x) = 1 - \frac{2}{1 + \sqrt{1 + 13x}}.$$

The derivative is given by

$$f'(x) = \frac{13}{\sqrt{1 + 13x}\left(1 + \sqrt{1 + 13x}\right)^2}$$

and the second derivative by

$$f''(x) = \frac{-169\left(1 + 3\sqrt{1 + 13x}\right)}{(1 + 13x)^{\frac{3}{2}}\left(1 + \sqrt{1 + 13x}\right)^3}.$$

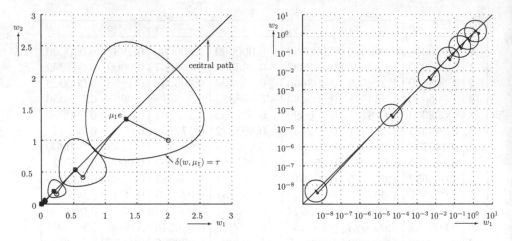

Figure 7.9 The iterates of the adaptive predictor-corrector algorithm.

This implies that f is monotonically increasing and concave. Since $f'(0) = 13/4$, it follows that $f(x) \leq 13x/4$ for each $x \geq 0$. Putting $x = \|d_x^a d_s^a\|$ gives the lemma. $\qquad\square$

We need one more basic fact in the analysis below. This concerns the optimal sets $\mathcal{P}^*$ and $\mathcal{D}^*$ of the primal and dual problems. Defining the index sets

$$B := \{i \; : \; x_i > 0 \text{ for some } x \in \mathcal{P}^*\}$$

and

$$N := \{i \; : \; s_i > 0 \text{ for some } s \in \mathcal{D}^*\},$$

we know that these sets are disjoint, because $x^T s = 0$ whenever $x \in \mathcal{P}^*$ and $s \in \mathcal{D}^*$. We need the far from obvious fact that each index i, $1 \leq i \leq n$, belongs either to B or N.[19] As a consequence, the sets B and N form a partition of the index set $\{i \; : \; 1 \leq i \leq n\}$. This partition is called the *optimal partition* of the problems (P) and (D).

The behavior of the components of the vectors d_x^a and d_s^a strongly depends on whether a component belongs to one set of the optimal partition or to the complementary set. Table 7.6. summarizes some facts concerning the order of magnitude of the components of various vectors of interest.   From this table we read, for example, that $x_B = \Theta(1)$ and $\Delta^a x_N = \mathcal{O}(\mu)$. According to the definition of the symbols Θ and $\mathcal{O}$ this means that there exist positive constants c_1, c_2, c_3 such that $c_1 e \leq x_B \leq c_2 e$ and $\Delta^a x_N \leq c_3 \mu$.[20] In our case it is important to stress that

---

[19] This is the content of the Goldman–Tucker Theorem (Theorem II.3), an early result in the theory of Linear Optimization that has often been considered exotic. The original proof was based on Farkas' lemma (see, e.g., Schrijver [250], pp. 95–96). In Part I of this book we have shown that the corresponding result for the self-dual model is a natural byproduct of the limiting behavior of the central path. We also refer the reader to Güler et al. [134], who derived the Goldman–Tucker Theorem from the limiting behavior of the central path for the standard format. Güler and Ye [135] showed that interior-point algorithms — in a wide class — keep the iterates so close to the central path that these algorithms yield the optimal partition of the problem.

[20] See Section 1.7.4 for definitions of the order symbols $\mathcal{O}$ and Θ.

	Vector	B	N
1	x	$\Theta(1)^*$	$\Theta(\mu)$
2	s	$\Theta(\mu)$	$\Theta(1)^*$
3	u	$\Theta(1)$	$\Theta(1)$
4	d	$\Theta(\frac{1}{\sqrt{\mu}})$	$\Theta(\sqrt{\mu})$
5	d_x^a	$\mathcal{O}(\mu)^*$	$\mathcal{O}(1)$
6	d_s^a	$\mathcal{O}(1)$	$\mathcal{O}(\mu)^*$
7	$\Delta^a x$	$\mathcal{O}(\mu)^*$	$\mathcal{O}(\mu)$
8	$\Delta^a s$	$\mathcal{O}(\mu)$	$\mathcal{O}(\mu)^*$

Table 7.6. Asymptotic orders of magnitude of some relevant vectors.

these constants are independent of the iterates x, s and of the value μ of the barrier parameter. They depend only on the problem data A, b and c. Some of the statements in the table are almost trivial; the more difficult ones are indicated by an asterisk. Below we present the relevant proofs.

Let us temporarily postpone the proof of the statements in Table 7.6. and show that the order estimates given in the table immediately imply quadratic convergence of the adaptive predictor-corrector method.

Theorem II.68 *The adaptive predictor-corrector method is asymptotically quadratically convergent.*

Proof: From Table 7.6. we deduce that each component of the vector $d_x^a d_s^a$ is bounded by $\mathcal{O}(\mu)$. From our conventions this implies that $d_x^a d_s^a = \mathcal{O}(\mu)$. Hence the result follows from Lemma II.67. $\square$

The rest of this section is devoted to proving the estimates in Table 7.6.. Note that at the start of an affine-scaling step we have $\delta = \delta(x, s; \mu) \leq 1/12$, from the proof of Theorem II.66. This property will be used several times in the sequel. We start with line 3 in the table.

Line 3: With $\delta \leq 1/12$, Lemma II.62 implies that each component u_i of u satisfies

$$0.92013 \leq \frac{1}{\rho(\delta)} \leq u_i \leq \rho(\delta) \leq 1.0868.$$

This proves that $u = \Theta(1)$.

Lines 1 and 2: We start with the estimates for x_B and s_B. We need the following two positive numbers:[21]

$$\sigma_p \quad := \quad \min_{i \in B} \max\{x_i \; : \; x \in \mathcal{P}^*\}$$

$$\sigma_d \quad := \quad \min_{i \in N} \max\{s_i \; : \; s \in \mathcal{D}^*\}.$$

Note that these numbers depend only on the data of the problem and not on the iterates. Moreover, due to the existence of a strictly complementary optimal solution pair, both numbers are positive. Now let $i \in B$ and let $\bar{x} \in \mathcal{P}^*$ be such that $\bar{x}_i$ is maximal. Then, using that $\bar{x}_i \geq \sigma_p > 0$, we may write

$$s_i = \frac{s_i \bar{x}_i}{\bar{x}_i} \leq \frac{s^T \bar{x}}{\bar{x}_i} \leq \frac{s^T \bar{x}}{\sigma_p}.$$

Since $\bar{x}$ is optimal, $c^T \bar{x} \leq c^T x$. Hence, with y such that $s = c - A^T y$, we have

$$s^T \bar{x} = c^T \bar{x} - b^T y \leq c^T x - b^T y = s^T x = n\mu,$$

so that

$$s_i \leq \frac{n\mu}{\sigma_p}, \quad \forall i \in B.$$

This implies that $s_B = \mathcal{O}(\mu)$. From the third line in Table 7.6. we derive that $x_B s_B = \mu u_B^2 = \Theta(\mu)$. The last two estimates imply that

$$(x_B)^{-1} = \frac{s_B}{\Theta(\mu)} = \frac{\mathcal{O}(\mu)}{\Theta(\mu)} = \mathcal{O}(1).$$

This implies that x_B is bounded away from zero. On the other hand, since the pair (x, s) has duality gap $n\mu$ and hence, by Theorem II.9 (on page 100), belongs to a bounded set, we have $x_B = \mathcal{O}(1)$. Thus we may conclude that $x_B = \Theta(1)$. Since we also have $x_B s_B = \Theta(\mu)$, it follows that $s_B = \Theta(\mu)$. In exactly the same way we derive that $s_N = \Theta(1)$ and $x_N = \Theta(\mu)$.

Line 4: The estimates in the fourth line follow directly from the definition of d and the estimates for x and s in the first two lines.

Line 5 and 6: We obtain an order estimate for $(d_x^a)_N$ and $(d_s^a)_B$ by the following simple argument. By its definition d_x^a is the component of the vector $-u$ in the null space of the matrix AD. Hence we have $\|d_x^a\| \leq \|u\| = \sqrt{n}$. Therefore, $d_x^a = \mathcal{O}(1)$. Since $(d_x^a)_N$ is a subvector of d_x^a, we must also have $(d_x^a)_N = \mathcal{O}(1)$. A similar argument applies to $(d_s^a)_B$.

The estimates for $(d_x^a)_B$ and $(d_s^a)_N$ are much more difficult to obtain. We only deal with the estimate for $(d_x^a)_B$; the result for $(d_s^a)_N$ can be obtained in a similar way.

---

[21] These quantities were introduced by Ye [311]. See also Vavasis and Ye [280]. The numbers σ_p and σ_d closely resemble the numbers σ_{SP}^x and σ_{SP}^s for the self-dual model, as introduced in Section 3.3.2 of Part I. According to the definition of the condition number σ_{SP} for the self-dual model, the smallest of the two numbers σ_p and σ_d is a natural candidate for a condition number for the standard problems (P) and (D). We refer the reader to the above-mentioned papers for a discussion of other condition numbers and their mutual relations.

The main force in the derivation below is the observation that d_x^a can be written as the projection on the null space of AD of a vector that vanishes on the index set B.[22]

This can be seen as follows. We may write

$$d_x^a = -P_{AD}(u) = -\frac{1}{\sqrt{\mu}}P_{AD}(\sqrt{xs}) = -\frac{1}{\sqrt{\mu}}P_{AD}(ds).$$

Now let $(\tilde{y}, \tilde{s})$ be any dual optimal pair. Then

$$s = c - A^T y = A^T \tilde{y} + \tilde{s} - A^T y = \tilde{s} + A^T(\tilde{y} - y),$$

so we have

$$ds = d\tilde{s} + (AD)^T(\tilde{y} - y).$$

This means that $ds - d\tilde{s}$ belongs to the row space of AD. The row space being orthogonal to the null space of AD, it follows that

$$P_{AD}(ds) = P_{AD}(d\tilde{s}).$$

Thus we obtain

$$d_x^a = -\frac{1}{\sqrt{\mu}}P_{AD}(d\tilde{s}). \tag{7.43}$$

Since $\tilde{s}$ is dual optimal, all its positive coordinates belong to the index set N, and hence we have $\tilde{s}_B = 0$. Now we can rewrite (7.43) in the following way:

$$-\sqrt{\mu}\,d_x^a = \operatorname{argmin}_h \left(\|d\tilde{s} - h\|^2 \ : \ ADh = 0 \right),$$

or equivalently,

$$-\sqrt{\mu}\,d_x^a = \operatorname{argmin}_h \left(\|d_B \tilde{s}_B - h_B\|^2 + \|d_N \tilde{s}_N - h_N\|^2 \ : \ A_B D_B h_B + A_N D_N h_N = 0 \right).$$

This means that the solution of the last minimization problem is given by $h_B = -\sqrt{\mu}(d_x^a)_B$ and $h_N = -\sqrt{\mu}(d_x^a)_N$. Hence, substituting the optimal value for h_N as above, and also using that $\tilde{s}_B = 0$, we obtain

$$-\sqrt{\mu}(d_x^a)_B = \operatorname{argmin}_{h_B} \left(\|h_B\|^2 \ : \ A_B D_B h_B = \sqrt{\mu}\,A_N D_N (d_x^a)_N \right).$$

Stated otherwise, $-\sqrt{\mu}(d_x^a)_B$ can be characterized as the vector of smallest norm in the affine space

$$S = \{\xi \ : \ A_B D_B \xi = \sqrt{\mu}\,A_N D_N (d_x^a)_N\}.$$

Now consider the least norm solution of the equation $A_B z = \sqrt{\mu}\,A_N D_N (d_x^a)_N$. This solution is given by

$$z^* = \sqrt{\mu}\,A_B^+ A_N D_N (d_x^a)_N,$$

---

[22] We kindly acknowledge that the basic idea of the analysis below was communicated privately to us by our colleague Gonzaga. We also refer the reader to Gonzaga and Tapia [127] and Ye et al. [317]; these papers deal with the asymptotically quadratic convergence rate of the predictor-corrector method.

where A_B^+ denotes the pseudo-inverse[23] of the matrix A_B. It is obvious that $D_B^{-1} z^*$ belongs to the affine space $\mathcal{S}$. Hence, $-\sqrt{\mu}(d_x^a)_B$ being the vector of smallest norm in $\mathcal{S}$, we obtain

$$\left\| \sqrt{\mu}\,(d_x^a)_B \right\| \leq \left\| D_B^{-1} z^* \right\| = \sqrt{\mu} \left\| D_B^{-1} A_B^+ A_N D_N (d_x^a)_N \right\|,$$

or, dividing both sides by $\sqrt{\mu}$,

$$\left\| (d_x^a)_B \right\| \leq \left\| D_B^{-1} A_B^+ A_N D_N (d_x^a)_N \right\|.$$

This implies

$$\left\| (d_x^a)_B \right\| \leq \left\| D_B^{-1} \right\| \left\| A_B^+ \right\| \left\| A_N \right\| \left\| D_N \right\| \left\| (d_x^a)_N \right\|.$$

Since, by convention, $\left\| A_B^+ \right\|$ and $\left\| A_N \right\|$ are bounded by $\mathcal{O}(1)$, and the order of magnitudes of the other norms on the right-hand side multiply to $\mathcal{O}(\mu)$, we obtain that $\left\| (d_x^a)_B \right\| = \mathcal{O}(\mu)$. This implies the entry $(d_x^a)_B = \mathcal{O}(\mu)$ in the table.

Line 7 and 8: These lines are not necessary for the proof of Theorem II.68. We only add them because of their own interest. They immediately follow from the previous lines in the table and the relations

$$\Delta x = \sqrt{\mu}\, d d_x, \quad \Delta s = \sqrt{\mu}\, d^{-1} d_s.$$

This completes the proof of all the entries in Table 7.6..

7.8 A version of the algorithm with large updates

The primal-dual methods considered so far share the property that the iterates stay close to the central path. More precisely, each generated primal-dual pair (x, s) belongs to the region of quadratic convergence around some μ-center. In this section we consider an algorithm in which the iterates may temporarily get quite far from the central path, because of a large, but fixed, update of the barrier parameter. Then, by using damped Newton steps, we return to the neighborhood of the point of the central path corresponding to the new value of the barrier parameter. The algorithm is the natural primal-dual analogue of the dual algorithm with large updates in Section 6.9. Just as in the dual case, when the iterates leave the neighborhood of the central path the proximity measure for the full-step method, $\delta(x, s; \mu)$, becomes less relevant as a measure for closeness to the central path. It will be of no surprise that in the primal-dual case the primal-dual logarithmic barrier function $\phi_\mu(x, s)$ is a perfect tool for this job. Recall from (6.23), on page 133, that $\phi_\mu(x, s)$ is given by

$$\phi_\mu(x, s) = \Psi\left(\frac{xs}{\mu} - e\right) = e^T\left(\frac{xs}{\mu} - e\right) - \sum_{j=1}^n \log \frac{x_j s_j}{\mu}, \tag{7.44}$$

and from Section 6.9 (page 130) that $\phi_\mu(x, s)$ is nonnegative on its domain (the set of all positive primal-dual pairs), is strictly convex, has a (unique) minimizer, namely

---

[23] See Appendix B.

$(x, s) = (x(\mu), s(\mu))$ and, finally that $\phi_\mu(x(\mu), s(\mu)) = 0.$[24]
The algorithm is described below (page 195). As usual, Δx and Δs denote the Newton step at the current pair (x, s) with the barrier parameter equal to its current value μ. The first **while**-loop in the algorithm is called the *outer loop* and the second

---

Primal-Dual Logarithmic Barrier Algorithm with Large Updates

Input:
    A proximity parameter τ;
    an accuracy parameter $\varepsilon > 0$;
    a variable damping factor α;
    a fixed barrier update parameter θ, $0 < \theta < 1$;
    $(x^0, s^0) \in \mathcal{P} \times \mathcal{D}$ and $\mu^0 > 0$ such that $\delta(x^0, s^0; \mu^0) \leq \tau$.
begin
    $x := x^0$; $s := s^0$; $\mu := \mu^0$;
    while $n\mu \geq \varepsilon$ **do**
    begin
        $\mu := (1 - \theta)\mu$;
        while $\delta(x, s; \mu) \geq \tau$ **do**
        begin
            $x := x + \alpha \Delta x$;
            $s := s + \alpha \Delta s$;
            (The damping factor α must be such that $\phi_\mu(x, s)$ decreases
            sufficiently. Lemma II.72 gives a default value for α.)
        end
    end
end

---

while-loop the *inner loop*. Each execution of the outer loop is called an *outer iteration* and each execution of the inner loop an *inner iteration*. The required number of outer iterations depends only on the dimension n of the problem, on μ^0 and ε, and on the (fixed) barrier update parameter θ. This number immediately follows from Lemma I.36 and is given by

$$\left\lceil \frac{1}{\theta} \log \frac{n\mu^0}{\varepsilon} \right\rceil.$$

Just as in the dual case, the main task in the analysis of the algorithm is the estimation of the number of iterations between two successive updates of the barrier parameter.

---

[24] **Exercise 54** Let the positive primal-dual pair (x, s) be given. We want to find $\mu > 0$ such that $\phi_\mu(x, s)$ is minimal. Show that this happens if $\mu = x^T s/n$ and verify that for this value of μ we have

$$\phi_\mu(x, s) = \Psi\left(\frac{nxs}{x^T s} - e\right) = n \log \frac{x^T s}{n} - \sum_{j=1}^{n} \log x_j s_j.$$

This is the purpose of the next sections. We first derive some estimates of $\phi_\mu(x, s)$ in terms of the proximity measure $\delta(x, s; \mu)$.

7.8.1 Estimates of barrier function values

The estimates in this section are of the same type as the estimates in Section 6.9.1 for the dual case.[25] Many of these estimates there were given in terms of the function $\psi : (-1, \infty) \to \mathbb{R}$ determined by (5.5):

$$\psi(t) = t - \log(1 + t),$$

which is nonnegative on its domain, strictly convex and zero at $t = 0$. For $z \in \mathbb{R}^n$, with $z + e > 0$, we defined in (6.22), page 133,

$$\Psi(z) = \sum_{j=1}^n \psi(z_j). \tag{7.45}$$

The estimates in Section 6.9.1 were given in terms of the dual proximity measure $\delta(y, \mu)$. Our aim is to derive similar estimates, but now in terms of the primal-dual proximity measure $\delta(x, s; \mu)$.

Let (x, s) be any positive primal-dual pair and $\mu > 0$. Then, with u as usual:

$$u = \sqrt{\frac{xs}{\mu}},$$

we may write

$$\phi_\mu(x, s) = e^T \left(u^2 - e \right) - \sum_{j=1}^n \log u_j^2 = \Psi \left(u^2 - e \right).$$

Using this we prove the next lemma.

Lemma II.69 *Let $\delta := \delta(x, s; \mu)$ and $\rho(\delta) := \delta + \sqrt{1 + \delta^2}$. Then*

$$\psi \left(\frac{-2\delta}{\rho(\delta)} \right) \le \phi_\mu(x, s) \le \psi \left(2\delta\rho(\delta) \right).$$

The first (second) inequality holds with equality if and only if one of the coordinates of u attains the value $\rho(\delta)$ $(1/\rho(\delta))$ and all other coordinates are equal to 1.

Proof: Fixing δ, we consider the behavior of $\Psi \left(u^2 - e \right)$ on the set

$$\mathcal{T} := \left\{ u \in \mathbb{R}^n \; : \; \left\| u^{-1} - u \right\| = 2\delta, \, u \ge 0 \right\}.$$

Note that this set is invariant under inverting coordinates of u. Because of the inequality

$$\psi(t - 1) > \psi \left(\frac{1}{t} - 1 \right), \, t > 1, \tag{7.46}$$

---

[25] The estimates in this section are new and dramatically improve existing estimates from the literature. See, e.g., Monteiro and Adler [218], Mizuno and Todd [216], Jansen et al. [157] and den Hertog [140].

whose elementary proof is left as an exercise [26], this implies that $u \geq e$ if u maximizes $\Psi(u^2 - e)$ on $\mathcal{T}$ and $u \leq e$ if u minimizes $\Psi(u^2 - e)$ on $\mathcal{T}$.

Consider first the case where u is a maximizer of Ψ on the set $\mathcal{T}$. The first-order optimality conditions are

$$\frac{u^2 - e}{u^2} 2u = 2\lambda \left(u - \frac{e}{u^3} \right), \tag{7.47}$$

where $\lambda \in \mathbb{R}$. This can be rewritten as

$$u^2 \left(u^2 - e \right) = \lambda \left(u^2 - e \right) \left(u^2 + e \right).$$

It follows that each coordinate of u satisfies

$$u_i = 1 \quad \text{or} \quad u_i^2 = \lambda \left(u_i^2 + 1 \right).$$

Since $u > 0$, we may conclude from this that the coordinates of u that differ from 1 are mutually equal. Suppose that u has k such coordinates, and that their common value is ν. Note that $k > 0$, unless $\delta = 0$, in which case the lemma is trivial. Therefore, we may assume that $k \geq 1$. Now, since $u \in \mathcal{T}$,

$$k \left(\frac{1}{\nu} - \nu \right)^2 = 4\delta^2,$$

which gives

$$\left| \frac{1}{\nu} - \nu \right| = \frac{2\delta}{\sqrt{k}}.$$

Since u is a maximizer, we have $\nu \geq 1$, and hence

$$\nu = \rho \left(\frac{\delta}{\sqrt{k}} \right).$$

Therefore, using that

$$\rho(t)^2 - 1 = 2t\rho(t), \quad t \in \mathbb{R}, \tag{7.48}$$

we obtain

$$\Psi \left(u^2 - e \right) = k\psi \left(\nu^2 - 1 \right) = k\psi \left(\frac{2\delta}{\sqrt{k}} \rho \left(\frac{\delta}{\sqrt{k}} \right) \right).$$

The expression on the right-hand side is decreasing as a function of k.[27] Hence the maximal value is attained if $k = 1$, and this value equals $\psi \left(2\delta \rho \left(\delta \right) \right)$. The second inequality in the lemma follows.

The first inequality is obtained in the same way. If u is a minimizer of Ψ on the set $\mathcal{T}$, then the first-order optimality conditions (7.47) imply in the same way as before

---

[26] **Exercise 55** Derive (7.46) from the inequalities in Exercise 42 (page 137).

[27] **Exercise 56** Let δ and $\rho(\delta)$ be as defined in Lemma II.69, and let $k \geq 1$. Prove that

$$k\psi \left(\frac{2\delta}{\sqrt{k}} \rho \left(\frac{\delta}{\sqrt{k}} \right) \right) = k\psi \left(\frac{2\delta^2 + 2\delta\sqrt{\delta^2 + k}}{k} \right)$$

and that this expression is maximal if $k = 1$.

that the coordinates of u that differ from 1 are mutually equal. Assuming that u has k such coordinates, and that their common value is ν again, we now have $\nu \leq 1$, and hence

$$\nu = \frac{1}{\rho\left(\frac{\delta}{\sqrt{k}}\right)}.$$

Using (7.48), it follows that

$$\frac{1}{\rho(t)^2} - 1 = \frac{1 - \rho(t)^2}{\rho(t)^2} = \frac{-2t\rho(t)}{\rho(t)^2} = \frac{-2t}{\rho(t)}.$$

Hence we may write

$$\Psi\left(u^2 - e\right) = k\psi\left(\nu^2 - 1\right) = k\psi\left(\frac{-2\delta}{\sqrt{k}\,\rho\left(\frac{\delta}{\sqrt{k}}\right)}\right).$$

The expression on the right-hand side is increasing as a function of k.[28] Hence the minimal value is attained if $k = 1$, and this value equals $\psi\left(-2\delta/\rho\left(\delta\right)\right)$. Thus the proof of the lemma is complete. $\qquad\square$

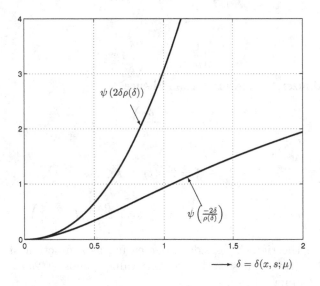

Figure 7.10   Bounds for $\psi_\mu(x, s)$.

---

[28] **Exercise 57** Let δ and $\rho(\delta)$ be as defined in Lemma II.69, and let $k \geq 1$. Prove that

$$k\psi\left(\frac{-2\delta}{\sqrt{k}\,\rho\left(\frac{\delta}{\sqrt{k}}\right)}\right) = k\psi\left(\frac{-2\delta}{\delta + \sqrt{\delta^2 + k}}\right)$$

and that this expression is minimal if $k = 1$.

Figure 7.10 shows the graphs of the bounds in Lemma II.69 for $\phi_\mu(x, s)$ as a function of the proximity δ.

Remark II.70 It may be worthwhile to discuss the quality of these bounds. Both bounds are valid for all (nonnegative) values of the proximity. Especially for the upper bound this is worth noting. Proximity measures known in the literature do not have this feature. For example, with the popular measure

$$\left\| \frac{xs}{\mu} - e \right\|$$

all known upper bounds grow to infinity if the measure approaches 1. The upper bound of Lemma II.69 goes to infinity only if our proximity measure goes to infinity.

The lower bound goes to infinity as well if if our proximity measure goes to infinity, due to the fact that $-2\delta/\rho(\delta)$ converges to -1 if δ goes to infinity. This is a new feature, which will be used below in the analysis of the large-update method. On the other hand, it must be noted that the lower bound grows very slowly if δ increases. For example, if $\delta = 1,000,000$ then the lower bound is only 28.0168.   ●

7.8.2 Decrease of barrier function value

Suppose again that (x, s) is any positive primal-dual pair and $\mu > 0$. In this section we analyze the effect on the barrier function value of a damped Newton step at (x, s) to the μ-center. With u as defined before, the Newton displacements Δx and Δs satisfy

$$x\Delta s + s\Delta x = \mu e - xs.$$

Let x^+ and s^+ result from a damped Newton step of size α at (x, s). Then we have

$$x^+ = x + \alpha\Delta x, \quad s^+ = s + \alpha\Delta s.$$

Using the scaled displacements d_x and d_s, defined in (7.5), page 154, we can also write

$$x^+ = \sqrt{\mu}\, d\, (u + \alpha d_x), \quad s^+ = \sqrt{\mu}\, d^{-1}\, (u + \alpha d_s).$$

As a consequence,

$$x^+ s^+ = \mu\, (u + \alpha d_x)\, (u + \alpha d_s) = \mu\left(u^2 + \alpha\left(e - u^2\right) + \alpha^2 d_x d_s\right).$$

Here we used that $u\, (d_x + d_s) = e - u^2$, which follows from

$$d_x + d_s = u^{-1} - u. \tag{7.49}$$

Now, defining

$$u^+ := \sqrt{\frac{x^+ s^+}{\mu}},$$

it follows that

$$\left(u^+\right)^2 = (u + \alpha d_x)\, (u + \alpha d_s) = u^2 + \alpha\left(e - u^2\right) + \alpha^2 d_x d_s. \tag{7.50}$$

Subtracting e we get

$$\left(u^+\right)^2 - e = (1 - \alpha)\left(u^2 - e\right) + \alpha^2 d_x d_s.$$

Note that the orthogonality of d_x and d_s implies that $e^T d_x d_s = 0$. Using this we find the following expression for $\phi_\mu(x^+, s^+)$:

$$
\phi_\mu(x^+, s^+) \quad = \quad e^T \left((u^+)^2 - e \right) - \sum_{j=1}^{n} \log \left(u_j^+ \right)^2
$$

$$
= \quad (1 - \alpha) e^T \left(u^2 - e \right) - \sum_{j=1}^{n} \log \left(u_j^+ \right)^2.
$$

The next lemma provides an expression for the decrease of the barrier function value during a damped Newton step.

Lemma II.71 *Let $\delta = \delta(x, s; \mu)$ and let α be such that the pair (x^+, s^+) resulting from the damped Newton step of size α is feasible. Then we have*

$$
\phi_\mu(x, s) - \phi_\mu(x^+, s^+) = 4\alpha\delta^2 - \Psi \left(\frac{\alpha d_x}{u} \right) - \Psi \left(\frac{\alpha d_s}{u} \right).
$$

Proof: For the moment let us denote $\Delta := \phi_\mu(x, s) - \phi_\mu(x^+, s^+)$. Then we have

$$
\Delta \quad = \quad e^T \left(u^2 - e \right) - \sum_{j=1}^{n} \log u_j^2 - (1-\alpha) e^T \left(u^2 - e \right) + \sum_{j=1}^{n} \log \left(u_j^+ \right)^2
$$

$$
= \quad \alpha e^T \left(u^2 - e \right) + \sum_{j=1}^{n} \log \left(\frac{u_j^+}{u_j} \right)^2.
$$

Since

$$
\left(u^+ \right)^2 = (u + \alpha d_x)(u + \alpha d_s)
$$

we may write

$$
\left(\frac{u^+}{u} \right)^2 = \left(e + \alpha \frac{d_x}{u} \right) \left(e + \alpha \frac{d_s}{u} \right).
$$

Substituting this we obtain

$$
\Delta \quad = \quad \alpha e^T \left(u^2 - e \right) + \sum_{j=1}^{n} \log \left(\frac{u_j^+}{u_j} \right)^2
$$

$$
= \quad \alpha e^T \left(u^2 - e \right) + \sum_{j=1}^{n} \log \left(1 + \alpha \frac{(d_x)_j}{u_j} \right) + \sum_{j=1}^{n} \log \left(1 + \alpha \frac{(d_s)_j}{u_j} \right).
$$

Observe that, by the definition of Ψ,

$$
\sum_{j=1}^{n} \log \left(1 + \alpha \frac{(d_x)_j}{u_j} \right) = \alpha e^T \left(\frac{d_x}{u} \right) - \Psi \left(\frac{\alpha d_x}{u} \right)
$$

and, similarly,

$$
\sum_{j=1}^{n} \log \left(1 + \alpha \frac{(d_s)_j}{u_j} \right) = \alpha e^T \left(\frac{d_s}{u} \right) - \Psi \left(\frac{\alpha d_s}{u} \right).
$$

Substituting this in the last expression for Δ we arrive at

$$\Delta = \alpha e^T \left(u^2 - e \right) + \alpha e^T \left(\frac{d_x}{u} \right) + \alpha e^T \left(\frac{d_s}{u} \right) - \Psi \left(\frac{\alpha d_x}{u} \right) - \Psi \left(\frac{\alpha d_s}{u} \right).$$

Using (7.49) once more, the coefficients of α in the first three terms can be taken together as follows:

$$e^T \left(u^2 - e + \frac{d_x + d_s}{u} \right) = e^T \left(u^2 - e + \left(u^{-2} - e \right) \right) = e^T \left(u^{-1} - u \right)^2.$$

Thus we obtain

$$\Delta = \alpha \left\| u^{-1} - u \right\|^2 - \Psi \left(\frac{\alpha d_x}{u} \right) - \Psi \left(\frac{\alpha d_s}{u} \right).$$

Since $\left\| u^{-1} - u \right\| = 2\delta$, the lemma follows.[29,30] □

We proceed by deriving a lower bound for the expression in the above lemma. The next lemma also specifies a value of the damping parameter α for which the decrease in the barrier function value attains the lower bound.

Lemma II.72 *Let* $\delta = \delta(x, s; \mu)$ *and let* $\alpha = 1/\omega - 1/(\omega + 4\delta^2)$, *where*

$$\omega := \sqrt{\left\| \frac{\Delta x}{x} \right\|^2 + \left\| \frac{\Delta s}{s} \right\|^2} = \sqrt{\left\| \frac{d_x}{u} \right\|^2 + \left\| \frac{d_s}{u} \right\|^2}.$$

Then the pair (x^+, s^+) *resulting from the damped Newton step of size* α *is feasible. Moreover, the barrier function value decreases by at least* $\psi(2\delta/\rho(\delta))$. *In other words,*

$$\phi_\mu(x, s) - \phi_\mu(x^+, s^+) \geq \psi \left(\frac{2\delta}{\rho(\delta)} \right).$$

---

[29] **Exercise 58** Verify that

$$\frac{\Delta x}{x} = \frac{d_x}{u}, \quad \frac{\Delta s}{s} = \frac{d_s}{u}.$$

[30] **Exercise 59** Using Lemma II.71, show that the decrease in the primal-dual barrier function value after a damped step of size α can be written as:

$$\Delta := \phi_\mu(x, s) - \phi_\mu(x^+, s^+) = \alpha \left\| d_x \right\|^2 + \alpha \left\| d_s \right\|^2 - \Psi \left(\frac{\alpha d_x}{u} \right) - \Psi \left(\frac{\alpha d_s}{u} \right).$$

Now let z be the concatenation of the vectors d_x and d_s. Then we may write

$$\Delta = \alpha \left\| z \right\|^2 - \Psi \left(\frac{\alpha z}{u} \right).$$

Using this, show that the decrease is maximal for the unique step-size $\bar{\alpha}$ determined by the equation

$$e^T z^2 = e^T \left(\frac{\alpha \left(\frac{z}{u} \right)^2}{e + \alpha \frac{z}{u}} \right),$$

and that for this value the decrease is given by

$$\Psi \left(\frac{-\bar{\alpha} z}{u + \bar{\alpha} z} \right) = \Psi \left(\frac{-\bar{\alpha} d_x}{u + \bar{\alpha} d_x} \right) + \Psi \left(\frac{-\bar{\alpha} d_s}{u + \bar{\alpha} d_s} \right).$$

Proof: Assuming feasibility of the damped step with size α, we know from Lemma II.71 that the decrease in the barrier function value is given by

$$\Delta := 4\alpha\delta^2 - \Psi\left(\frac{\alpha d_x}{u}\right) - \Psi\left(\frac{\alpha d_s}{u}\right).$$

We now apply the right-hand side inequality in (6.24), page 134, to the vector in $\mathbb{R}^{2n}$ obtained by concatenating the vectors $\alpha d_x/u$ and $\alpha d_s/u$. Note that the norm of this vector is given by $\alpha\omega$, with ω as defined in the lemma, and that $\alpha\omega < 1$ for the value of α specified in the lemma. Then we obtain

$$\Delta \geq 4\alpha\delta^2 - \psi\left(-\alpha\omega\right) = 4\alpha\delta^2 + \alpha\omega + \log\left(1 - \alpha\omega\right). \qquad (7.51)$$

As a function of α, the derivative of the right-hand side expression is given by

$$4\delta^2 + \omega - \frac{\omega}{1 - \alpha\omega} = \frac{4\delta^2(1 - \alpha\omega) - \alpha\omega^2}{1 - \alpha\omega}.$$

From this we see that the right-hand side expression in (7.51) is increasing for

$$0 \leq \alpha \leq \bar{\alpha} := \frac{4\delta^2}{\omega\left(\omega + 4\delta^2\right)} = \frac{1}{\omega} - \frac{1}{\omega + 4\delta^2},$$

and decreasing for larger values of α. Hence it attains its maximal value at $\alpha = \bar{\alpha}$, which is the value specified in the lemma. Moreover, since the barrier function is finite for $0 \leq \alpha \leq \bar{\alpha}$, the damped Newton step of size $\bar{\alpha}$ is feasible. Substitution of $\alpha = \bar{\alpha}$ in (7.51) yields the following bound for Δ:

$$\Delta \geq \frac{4\delta^2}{\omega} + \log\frac{\omega}{\omega + 4\delta^2} = \frac{4\delta^2}{\omega} - \log\left(1 + \frac{4\delta^2}{\omega}\right) = \psi\left(\frac{4\delta^2}{\omega}\right).$$

In this bound we may replace ω by a larger value, since $\psi(t)$ is monotonically increasing for t nonnegative. An upper bound for ω can be obtained as follows:

$$\omega = \sqrt{\left\|\frac{d_x}{u}\right\|^2 + \left\|\frac{d_s}{u}\right\|^2} \leq \left\|u^{-1}\right\|_\infty \sqrt{\|d_x\|^2 + \|d_s\|^2} = \left\|u^{-1}\right\|_\infty \left\|u^{-1} - u\right\|.$$

Since $\left\|u^{-1}\right\|_\infty \leq \rho(\delta)$, by Lemma II.62, page 182, and $\left\|u^{-1} - u\right\| = 2\delta$ we obtain

$$\omega \leq 2\delta\rho(\delta). \qquad (7.52)$$

Substitution of this bound yields

$$\Delta \geq \psi\left(\frac{2\delta}{\rho(\delta)}\right),$$

completing the proof.[31] □

---

[31] **Exercise 60** With ω as defined in Lemma II.72, show that

$$\omega \geq \frac{2\delta}{\rho(\delta)}.$$

Using this and (7.52), prove that the step-size α specified in Lemma II.72 satisfies

$$\frac{1}{2\rho(\delta)\left(2\rho(\delta) + \delta\right)} \leq \alpha = \frac{\delta^2}{\omega\left(\omega + \delta^2\right)} \leq \frac{\rho(\delta)^2}{2\left(2 + \delta\rho(\delta)\right)}.$$

Remark II.73 The same analysis as in Lemma II.72 can be applied to the case where different step-sizes are taken for the x-space and the s-space. Let $x^+ = x + \alpha \Delta x$ and $s^+ = s + \beta \Delta s$, with α and β such that both steps are feasible. Then the decrease in the primal-dual barrier function value is given by

$$\Delta := \phi_\mu(x, s) - \phi_\mu(x^+, s^+) = \alpha \|d_x\|^2 - \Psi \left(\frac{\alpha d_x}{u} \right) + \beta \|d_s\|^2 - \Psi \left(\frac{\beta d_s}{u} \right).$$

Defining $\omega_1 := \|d_x/u\|$, the x-part of the right-hand side can be bounded by

$$\Delta_1 := \alpha \|d_x\|^2 - \Psi \left(\frac{\alpha d_x}{u} \right) \geq \psi \left(\frac{\|d_x\|^2}{\omega_1} \right),$$

and this bound holds with equality if

$$\alpha = \bar{\alpha} := \frac{1}{\omega_1} - \frac{1}{\omega_1 + \|d_x\|^2}.$$

Similarly, defining $\omega_2 := \|d_s/u\|$, the s-part of the right-hand side can be bounded by

$$\Delta_2 := \beta \|d_s\|^2 - \Psi \left(\frac{\beta d_s}{u} \right) \geq \psi \left(\frac{\|d_s\|^2}{\omega_2} \right),$$

and this bound holds with equality if

$$\beta = \bar{\beta} := \frac{1}{\omega_2} - \frac{1}{\omega_2 + \|d_s\|^2}.$$

Hence,

$$\Delta = \Delta_1 + \Delta_2 \geq \psi \left(\frac{\|d_x\|^2}{\omega_1} \right) + \psi \left(\frac{\|d_s\|^2}{\omega_2} \right).$$

We can easily verify that

$$\omega_1 \leq \rho(\delta) \|d_x\|, \quad \omega_2 \leq \rho(\delta) \|d_s\|.$$

Using the monotonicity of ψ, it follows that

$$\Delta_1 \geq \psi \left(\frac{\|d_x\|}{\rho(\delta)} \right), \quad \Delta_2 \geq \psi \left(\frac{\|d_s\|}{\rho(\delta)} \right).$$

We obtain in this way

$$\Delta = \Delta_1 + \Delta_2 \geq \psi \left(\frac{\|d_x\|}{\rho(\delta)} \right) + \psi \left(\frac{\|d_s\|}{\rho(\delta)} \right).$$

Finally, applying the left inequality in (6.24) to the right-hand side expression, we can easily derive that

$$\Delta \geq \psi \left(\sqrt{\frac{\|d_x\|^2 + \|d_s\|^2}{\rho(\delta)^2}} \right) = \psi \left(\frac{2\delta}{\rho(\delta)} \right).$$

Note that this is exactly the same bound as obtained in Lemma II.72. Thus, different step-sizes in the x-space and s-space give in this analysis no advantage over equal step-sizes in both spaces. This contradicts an earlier (and incorrect) result of Roos and Vial in [246].[32] •

For our goal it is of interest to derive the following two conclusions from the above lemma. First, if $\delta(x, s; \mu) = 1/\sqrt{2}$ then a damped Newton step reduces the barrier function by at least 0.182745, which is larger than 1/6. On the other hand for larger values of $\delta(x, s; \mu)$ the lower bound for the reduction in the barrier function value seems to be rather poor. It seems reasonable to expect that the reduction grows to infinity if δ goes to infinity. However, if δ goes to infinity then $2\delta/\rho(\delta)$ goes to 1, and hence the lower bound in the lemma is bounded by the rather small constant $\psi(1) = 1 - \log 2$.[33]

7.8.3 A bound for the number of inner iterations

As before, we assume that we have an iterate (x, s) and $\mu > 0$ such that (x, s) belongs to the region around the μ-center determined by

$$\delta = \delta(x, s; \mu) \leq \tau,$$

for some positive τ. Starting at (x, s) we count the number of inner iterations needed to reach the corresponding region around the μ^+-center, with

$$\mu^+ = (1 - \theta)\mu.$$

Implicitly it is assumed that θ is so large that (x, s) lies outside the region of quadratic convergence around the μ^+-center, but this is not essential for the analysis below. Recall that the target centers $x(\mu^+)$ and $s(\mu^+)$ are the (unique) minimizers of the primal-dual logarithmic barrier function $\phi_{\mu^+}(x, s)$, and that the value of this function is an indicator for the 'distance' from (x, s) to $(x(\mu^+), s(\mu^+))$.

We start by considering the effect of an update of the barrier parameter to $\mu^+ = (1 - \theta)\mu$ with $0 \leq \theta < 1$, on the barrier function value. Note that Lemma II.69 gives the answer if $\theta = 0$:

$$\phi_\mu(x, s) \leq \psi(2\delta\rho(\delta)).$$

---

[32] **Exercise 61** In this exercise we consider the case where different step-sizes are taken for the x-space and the s-space. Let $x^+ = x + \alpha\Delta x$ and $s^+ = s + \beta\Delta s$, with α and β such that both steps are feasible. Prove that the decrease in the primal-dual barrier function value is given by

$$\Delta := \phi_\mu(x, s) - \phi_\mu(x^+, s^+) = \alpha \|d_x\|^2 + \beta \|d_s\|^2 - \Psi\left(\frac{\alpha d_x}{u}\right) - \Psi\left(\frac{\beta d_s}{u}\right).$$

Using this, show that the decrease is maximal for the unique step-sizes $\bar{\alpha}$ and $\bar{\beta}$ determined by the equations

$$e^T (d_x)^2 = e^T\left(\frac{\alpha\left(\frac{d_x}{u}\right)^2}{e + \alpha\frac{d_x}{u}}\right), \quad e^T (d_s)^2 = e^T\left(\frac{\beta\left(\frac{d_s}{u}\right)^2}{e + \beta\frac{d_s}{u}}\right),$$

and that for these values of α and β the decrease is given by

$$\Psi\left(\frac{-\bar{\alpha}d_x}{u + \bar{\alpha}d_x}\right) + \Psi\left(\frac{-\bar{\beta}d_s}{u + \bar{\beta}d_s}\right).$$

[33] We want to explicitly show the inherent weakness of the lower bound in Lemma II.72 in the hope that it will stimulate the reader to look for a stronger result.

For the general case, with $\theta > 0$, we have the following lemma.

Lemma II.74 *Using the above notation, we have*

$$\phi_{\mu^+}(x, s) \leq \phi_\mu(x, s) + \frac{2\delta\rho(\delta)\theta\sqrt{n}}{1 - \theta} + n\psi\left(\frac{\theta}{1-\theta}\right).$$

Proof: The proof is more or less straightforward. The vector u is defined as usual.

$$
\begin{aligned}
\phi_{\mu^+}(x, s) \quad &= \quad e^T\left(\frac{u^2}{1-\theta} - e\right) - \sum_{j=1}^n \log\frac{u_j^2}{1-\theta} \\
&= \quad e^T\left(u^2 - e\right) - \sum_{j=1}^n \log u_j^2 + e^T\left(\frac{u^2}{1-\theta} - u^2\right) + n\log(1-\theta) \\
&= \quad \phi_\mu(x, s) + \frac{\theta e^T u^2}{1 - \theta} + n\log(1-\theta) \\
&= \quad \phi_\mu(x, s) + \frac{\theta}{1-\theta}u^T\left(u - u^{-1}\right) + \frac{\theta n}{1-\theta} + n\log(1-\theta).
\end{aligned}
$$

The second term in the last expression can be bounded by using

$$u^T\left(u - u^{-1}\right) \leq \|u\|\,\|u - u^{-1}\| \leq 2\delta\rho(\delta)\sqrt{n}.$$

The first inequality is simply the Cauchy–Schwarz inequality and the second inequality follows from $\|u^{-1} - u\| = 2\delta$ and $\|u\| \leq \sqrt{n}\,\|u\|_\infty \leq \sqrt{n}\rho(\delta)$, where we used Lemma II.62, page 182. We also have

$$\frac{\theta n}{1-\theta} + n\log(1-\theta) = n\left(\frac{\theta}{1-\theta} - \log\left(1 + \frac{\theta}{1-\theta}\right)\right) = n\psi\left(\frac{\theta}{1-\theta}\right).$$

Substitution yields

$$\phi_{\mu^+}(x, s) \leq \phi_\mu(x, s) + \frac{2\delta\rho(\delta)\theta\sqrt{n}}{1 - \theta} + n\psi\left(\frac{\theta}{1-\theta}\right),$$

and hence the lemma has been proved. $\qquad\square$

Now we are ready to estimate the number of (inner) iterations between two successive updates of the barrier parameter.

Lemma II.75 *For given θ $(0 < \theta < 1)$, let*

$$R := \frac{\theta\sqrt{n}}{1 - \theta}.$$

Then, when

$$\tau = \frac{\sqrt{R}}{2\sqrt{1 + \sqrt{R}}},$$

the number of (inner) iterations between two successive updates of the barrier parameter is not larger than

$$\left\lceil 2\left(1 + \sqrt{\frac{\theta\sqrt{n}}{1-\theta}}\right)^4 \right\rceil.$$

Proof: Suppose that $\delta = \delta(x, s; \mu) \leq \tau$. Then it follows from Lemma II.74 that after the update of the barrier parameter to $\mu^+ = (1 - \theta)\mu$ we have

$$\phi_{\mu^+}(x, s) \leq \phi_\mu(x, s) + \frac{2\delta\rho(\delta)\theta\sqrt{n}}{1 - \theta} + n\psi\left(\frac{\theta}{1 - \theta}\right).$$

By Lemma II.69 we have $\phi_\mu(x, s) \leq \psi\left(2\delta\rho(\delta)\right)$. Using the monotonicity of ψ and, since $\delta \leq \tau$, $2\delta\rho(\delta) \leq 2\tau\rho(\tau)$ we obtain

$$\phi_{\mu^+}(x, s) \leq \psi\left(2\tau\rho(\tau)\right) + \frac{2\tau\rho(\tau)\theta\sqrt{n}}{1 - \theta} + n\psi\left(\frac{\theta}{1 - \theta}\right).$$

Application of the inequality $\psi(t) \leq t^2/2$ for $t \geq 0$ to the first and the third terms yields

$$\phi_{\mu^+}(x, s) \leq 2\tau^2\rho(\tau)^2 + \frac{2\tau\rho(\tau)\theta\sqrt{n}}{1 - \theta} + \frac{n\theta^2}{2(1 - \theta)^2} = \left(\tau\rho(\tau)\sqrt{2} + \frac{\theta\sqrt{n}}{\sqrt{2}(1 - \theta)}\right)^2.$$

The algorithm repeats damped Newton steps until the iterate (x, s) satisfies $\delta = \delta(x, s; \mu^+) \leq \tau$. Each damped step decreases the barrier function value by at least $\psi\left(2\tau/\rho(\tau)\right)$. Hence, after

$$\left[\frac{1}{\psi\left(\frac{2\tau}{\rho(\tau)}\right)}\left(\tau\rho(\tau)\sqrt{2} + \frac{\theta\sqrt{n}}{\sqrt{2}(1 - \theta)}\right)^2\right] \tag{7.53}$$

iterations the value of the barrier function will have reached (or bypassed) the value $\psi\left(2\tau/\rho(\tau)\right)$. From Lemma II.69, using that $\psi\left(2\tau/\rho(\tau)\right) < \psi\left(-2\tau/\rho(\tau)\right)$, the iterate (x, s) then certainly satisfies $\delta(x, s; \mu^+) \leq \tau$, and hence (7.53) provides an upper bound for the number of inner iterations between two successive updates of the barrier parameter.

The rest of the proof consists in manipulating this expression. First, using $\psi(t) \geq t^2/(2(1 + t))$ and $0 \leq 2\tau/\rho(\tau) \leq 1$, we obtain

$$\psi\left(\frac{2\tau}{\rho(\tau)}\right) \geq \frac{\frac{4\tau^2}{\rho(\tau)^2}}{2\left(1 + \frac{2\tau}{\rho(\tau)}\right)} = \frac{2}{1 + \frac{2\tau}{\rho(\tau)}} \frac{\tau^2}{\rho(\tau)^2} \geq \frac{\tau^2}{\rho(\tau)^2}.$$

Substitution reduces the upper bound (7.53) to

$$\left[\frac{\rho(\tau)^2}{\tau^2}\left(\tau\rho(\tau)\sqrt{2} + \frac{\theta\sqrt{n}}{\sqrt{2}(1 - \theta)}\right)^2\right] = \left[2\left(\rho(\tau)^2 + \frac{\theta\rho(\tau)\sqrt{n}}{2\tau(1 - \theta)}\right)^2\right].$$

For fixed θ the number of inner iterations is a function of τ. Note that this function goes to infinity if τ goes to zero or to infinity. Our aim is to determine τ such that this function is minimized. To this end we consider

$$T(\tau) := \rho(\tau)^2 + \frac{\rho(\tau)R}{2\tau},$$

with R as given in the lemma. The derivative of $T(\tau)$ with respect to τ can be simplified to

$$T'(\tau) = \frac{4\tau^2 \rho(\tau)^2 - R}{2\tau^2\sqrt{1+\tau^2}}.$$

Hence $T(\tau)$ is minimal if

$$2\tau\rho(\tau) = \sqrt{R}.$$

We can solve this equation for τ. It can be rewritten as

$$\rho(\tau)^2 - 1 = \sqrt{R},$$

which gives

$$\rho(\tau) = \sqrt{1 + \sqrt{R}}.$$

Hence,

$$\tau = \frac{1}{2}\left(\rho(\tau) - \frac{1}{\rho(\tau)}\right) = \frac{1}{2}\left(\sqrt{1+\sqrt{R}} - \frac{1}{\sqrt{1+\sqrt{R}}}\right) = \frac{\sqrt{R}}{2\sqrt{1+\sqrt{R}}}. \qquad (7.54)$$

Substitution of this value in $T(\tau)$ gives

$$T(\tau) = \rho(\tau)^2 + \frac{\rho(\tau)R}{2\tau} = 1 + \sqrt{R} + \frac{R\left(1+\sqrt{R}\right)}{\sqrt{R}} = \left(1+\sqrt{R}\right)^2.$$

For the value of τ given by (7.54) the number of inner iterations between two successive updates of the barrier parameter will not be larger than

$$\left\lceil 2\left(1+\sqrt{R}\right)^4 \right\rceil = \left\lceil 2\left(1 + \sqrt{\frac{\theta\sqrt{n}}{1-\theta}}\right)^4 \right\rceil,$$

which proves the lemma. $\qquad\qquad\qquad\qquad\qquad\qquad\qquad\qquad\qquad\qquad\qquad\quad$ $\square$

Remark II.76 Note that for small values of θ, so that $\theta\sqrt{n}$ is bounded by a constant, the above lemma implies that the number of inner iterations between two successive updates of the barrier parameter is bounded by a constant. For example, with $\theta = 1/\sqrt{2n}$, which gives (for large values of n) $\tau = 0.309883$, this number is given by

$$\left\lceil 2\left(1 + \sqrt{\frac{1}{\sqrt{2}}}\right)^4 \right\rceil = 23.$$

Unfortunately the constant is rather large. Because, if $\tau = 0.309883$ then we know that after an update with $\theta = 1/\sqrt{2n}$ one full Newton step will be sufficient to reach the vicinity of the new target. In fact, it turns out that the bound has the same weakness as the bound in Theorem II.41 for the dual case. As discussed earlier, this weak result is due to the poor analysis. $\qquad\qquad\qquad\qquad\qquad\qquad\qquad\qquad\qquad\qquad\qquad\qquad\qquad\qquad$ $\bullet$

In practice the number of inner iterations is much smaller than the number predicted by the lemma. This is illustrated by some examples in the next section. But first we formulate the main conclusion of this section, namely that the primal-dual logarithmic barrier method with large updates is polynomial. This is the content of our final result in this section.

Theorem II.77 *The following expression is an upper bound for the total number of iterations required by the logarithmic barrier algorithm with line searches:*

$$\left\lceil \frac{1}{\theta} \left[2 \left(1 + \sqrt{\frac{\theta \sqrt{n}}{1 - \theta}} \right)^4 \right] \log \frac{n \mu^0}{\varepsilon} \right\rceil.$$

Here it is assumed that τ is chosen as in Lemma II.75:

$$\tau = \frac{\sqrt{R}}{2\sqrt{1 + \sqrt{R}}}, \quad \text{where } R = \frac{\theta \sqrt{n}}{1 - \theta}.$$

If $\theta \leq n/(n + \sqrt{n})$ the output is a primal-dual pair (x, s) such that $x^T s \leq 2\varepsilon$.

Proof: The number of outer iterations follows from Lemma I.36. The bound in the theorem is obtained by multiplying this number by the bound of Lemma II.75 for the number of inner iterations per outer iteration and rounding the product, if not integral, to the smallest integer above it. Finally, for the last statement we use the inequality

$$x^T s \leq \left(1 + \frac{2\delta \rho(\delta)}{\sqrt{n}} \right) n\mu,$$

where $\delta = \delta(x, s; \mu)$; the elementary proof of this inequality is left as an exercise.[34,35] For the output pair (x, s) we may apply this inequality with $\delta \leq \tau$. Since

$$\tau = \frac{\sqrt{R}}{2\sqrt{1 + \sqrt{R}}}, \quad \rho(\tau) = \sqrt{1 + \sqrt{R}},$$

we have $2\tau\rho(\tau) = \sqrt{R}$. Now $\theta \leq n/(n + \sqrt{n})$ implies that $R \leq n$, and hence we obtain that $x^T s \leq 2n\mu \leq 2\varepsilon$. $\qquad \square$

Just as in the dual case, we draw two conclusions from the last theorem. If we take for θ a fixed constant (independent of n), for example $\theta = 1/2$, the algorithm is called a *large-update algorithm* and the iteration bound of Theorem II.77 becomes

$$\mathcal{O}\left(n \, \log \frac{n \mu^0}{\varepsilon} \right).$$

---

[34] **Exercise 62** Let (x, s) be a positive primal-dual pair and $\mu > 0$. If $\delta = \delta(x, s; \mu)$, prove that

$$\left| x^T s - n\mu \right| = \mu \left| u^T \left(u - u^{-1} \right) \right| \leq \frac{2\delta \rho(\delta)}{\sqrt{n}} n\mu.$$

[35] **Exercise 63** The bound in Exercise 62 is based on the estimate $\|u\| \leq \rho(\delta)\sqrt{n}$. Prove the sharper estimate

$$\|u\| \leq \rho\left(\frac{\delta}{\sqrt{n}} \right) \sqrt{n}.$$

If we take $\theta = \nu/\sqrt{n}$ for some fixed constant ν (independent of n), the algorithm is called a *medium-update algorithm* and the iteration bound of Theorem II.77 becomes

$$\mathcal{O}\left(\sqrt{n}\,\log\frac{n\mu^0}{\varepsilon}\right),$$

provided that n is large enough ($n \geq \nu^2$ say).

7.8.4 Illustration of the algorithm with large updates

We use the same sample problem as in the numerical examples given before, and solve this problem using the primal-dual logarithmic barrier algorithm with large updates. We use the same initialization as before, namely $x = (2,1,1)$, $y = (0,0)$, $s = (1,1,1)$ and $\mu = 4/3$. We do this for the values $0.5, 0.9, 0.99$ and 0.999 of the barrier update parameter θ. It may be interesting to mention the values of the parameter τ, as given by Lemma II.75, for these values of θ. With $n = 3$, these values are respectively $0.43239, 0.88746, 1.74397$ and 3.18671. The progress of the algorithm for the three successive values of θ is shown in Tables 7.7. (page 210), 7.8., 7.9. and 7.10. (page 211). The tables need some explanation. They show only the first coordinates of x and of s. As in the corresponding tables for the dual case, the tables not only have lines for the inner iterations, but also for the outer iterations, which multiply the value of the barrier parameter by the fixed factor $1 - \theta$. The last column shows the proximity to the current μ-center. The proximity value δ increases in the outer iterations and decreases in the inner iterations.

The tables clearly demonstrate the advantages of the large-update strategy. The number of inner iterations between two successive updates of the barrier parameter is never more than two.

In the last table (with $\theta = 0.999$) the sample problem is solved in only 3 iterations, which is the best result obtained so far. The practical behavior is significantly better than the theoretical analysis justifies. This is typical, and the same phenomenon occurs for larger problems than the small sample problem.

We conclude this section with a graphical illustration of the algorithm, in Figure 7.11 (page 212).

Outer	Inner	$n\mu$	x_1	y_1	y_2	s_1	δ
0	0	4.000000	2.000000	0.000000	0.000000	1.000000	0.2887
1		2.000000	2.000000	0.000000	0.000000	1.000000	0.6455
	1	2.000000	1.372070	0.313965	0.313965	0.686035	0.2334
2		1.000000	1.372070	0.313965	0.313965	0.686035	0.6838
	2	1.000000	1.158784	0.649743	0.666131	0.350257	0.1559
3		0.500000	1.158784	0.649743	0.666131	0.350257	0.6237
	3	0.500000	1.082488	0.835475	0.835249	0.164525	0.0587
4		0.250000	1.082488	0.835475	0.835249	0.164525	0.6031
	4	0.250000	1.041691	0.916934	0.916776	0.083066	0.0281
5		0.125000	1.041691	0.916934	0.916776	0.083066	0.6115
	5	0.125000	1.020805	0.958399	0.958395	0.041601	0.0147
6		0.062500	1.020805	0.958399	0.958395	0.041601	0.6111
	6	0.062500	1.010423	0.979157	0.979156	0.020843	0.0073
7		0.031250	1.010423	0.979157	0.979156	0.020843	0.6129
	7	0.031250	1.005201	0.989597	0.989598	0.010403	0.0039
8		0.015625	1.005201	0.989597	0.989598	0.010403	0.6111
	8	0.015625	1.002606	0.994789	0.994789	0.005211	0.0019
9		0.007812	1.002606	0.994789	0.994789	0.005211	0.6129
	9	0.007812	1.001300	0.997399	0.997399	0.002601	0.0015
10		0.003906	1.001300	0.997399	0.997399	0.002601	0.6111
	10	0.003906	1.000651	0.998697	0.998697	0.001303	0.0007
11		0.001953	1.000651	0.998697	0.998697	0.001303	0.6129
	11	0.001953	1.000325	0.999350	0.999350	0.000650	0.0012
12		0.000977	1.000325	0.999350	0.999350	0.000650	0.6112
	12	0.000977	1.000163	0.999674	0.999674	0.000326	0.0006
13		0.000488	1.000163	0.999674	0.999674	0.000326	0.6129
	13	0.000488	1.000081	0.999837	0.999837	0.000163	0.0011
14		0.000244	1.000081	0.999837	0.999837	0.000163	0.6112
	14	0.000244	1.000041	0.999919	0.999919	0.000081	0.0005
15		0.000122	1.000041	0.999919	0.999919	0.000081	0.6129
	15	0.000122	1.000020	0.999959	0.999959	0.000041	0.0012
16		0.000061	1.000020	0.999959	0.999959	0.000041	0.6112
	16	0.000061	1.000010	0.999980	0.999980	0.000020	0.0005

Table 7.7. Progress of the primal-dual algorithm with large updates, $\theta = 0.5$.

Outer	Inner	$n\mu$	x_1	y_1	y_2	s_1	δ
0	0	4.000000	2.000000	0.000000	0.000000	1.000000	0.2887
1		0.400000	2.000000	0.000000	0.000000	1.000000	2.4664
	1	0.400000	1.051758	0.263401	0.684842	0.736599	1.1510
	2	0.400000	1.078981	0.875555	0.861676	0.124445	0.0559
2		0.040000	1.078981	0.875555	0.861676	0.124445	2.5417
	3	0.040000	1.004551	0.976424	0.983729	0.023576	0.3661
3		0.004000	1.004551	0.976424	0.983729	0.023576	2.7838
	4	0.004000	1.000621	0.998596	0.998677	0.001404	0.0447
4		0.000400	1.000621	0.998596	0.998677	0.001404	2.4533
	5	0.000400	1.000066	0.999867	0.999868	0.000133	0.0070
5		0.000040	1.000066	0.999867	0.999868	0.000133	2.4543
	6	0.000040	1.000007	0.999987	0.999987	0.000013	0.0027

Table 7.8. Progress of the primal-dual algorithm with large updates, $\theta = 0.9$.

Outer	Inner	$n\mu$	x_1	y_1	y_2	s_1	δ
0	0	4.000000	2.000000	0.000000	0.000000	1.000000	0.2887
1		0.040000	2.000000	0.000000	0.000000	1.000000	8.5737
	1	0.040000	2.000000	0.000000	0.000000	1.000000	4.2530
	2	0.040000	1.004883	0.251292	0.743825	0.748708	0.0816
2		0.000400	1.004883	0.251292	0.743825	0.748708	8.7620
	3	0.000400	1.007772	0.987570	0.986233	0.012430	0.4532
3		0.000004	1.007772	0.987570	0.986233	0.012430	9.5961
	4	0.000004	1.000038	0.999743	0.999834	0.000257	0.0392

Table 7.9. Progress of the primal-dual algorithm with large updates, $\theta = 0.99$.

Outer	Inner	$n\mu$	x_1	y_1	y_2	s_1	δ
0	0	4.000000	2.000000	0.000000	0.000000	1.000000	0.2887
1		0.004000	2.000000	0.000000	0.000000	1.000000	27.3587
	1	0.004000	1.000977	0.250006	0.749018	0.749994	13.6684
	2	0.004000	1.000481	0.999268	0.998990	0.000732	0.3722
2		0.000004	1.000481	0.999268	0.998990	0.000732	22.4872
	3	0.000004	1.000000	0.999998	0.999999	0.000002	0.2066

Table 7.10. Progress of the primal-dual algorithm with large updates, $\theta = 0.999$.

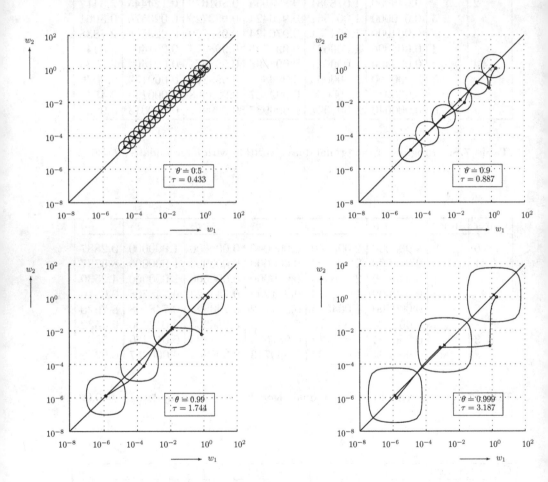

Figure 7.11 The iterates when using large updates with $\theta = 0.5, 0.9, 0.99$ and 0.999.

8

Initialization

All the methods of this part of the book assume the availability of a starting point on or close to the central path of the problem. Sometimes such a point is known, but more often we have no foreknowledge of the problem under consideration. For these cases we provide in this chapter a transformation of the problem yielding an equivalent problem for which a point on the central path is available. This transformation is based on results in Part I and is described below in detail.[1]

Suppose that we want to solve the problem (P) in standard format:

$$(P) \qquad \min \left\{ c^T x \ : \ Ax = b, \, x \geq 0 \right\},$$

where A is an $m \times n$ matrix of rank m, $c, x \in \mathbb{R}^n$, and $b \in \mathbb{R}^m$. Let I be a subset of the full index set $\{1, 2, \ldots, n\}$ such that the submatrix A_I of A has size $m \times m$ and is nonsingular. Thus, A_I is a *basis* for (P). After reordering the columns of A, we may write

$$A = (A_I \ A_J),$$

where J denotes the complement of I. Now $Ax = b$ can be rewritten as

$$A_I x_I + A_J x_J = b,$$

which is equivalent to

$$x_I = A_I^{-1} (b - A_J x_J).$$

As a consequence we have

$$c^T x = c_I^T x_I + c_J^T x_J = c_I^T A_I^{-1} (b - A_J x_J) + c_J^T x_J = c_I^T A_I^{-1} b + \left(c_J - A_J^T A_I^{-T} c_I \right)^T x_J.$$

Hence, omitting the constant $c_I^T A_I^{-1} b$ we can reformulate (P) as

$$(P^c) \qquad \min \left\{ \left(c_J - A_J^T A_I^{-T} c_I \right)^T x_J \ : \ A_I^{-1} (b - A_J x_J) \geq 0, \, x_J \geq 0 \right\},$$

or equivalently,

$$(P^c) \qquad \min \left\{ \left(c_J - A_J^T A_I^{-T} c_I \right)^T x_J \ : \ -A_I^{-1} A_J x_J \geq -A_I^{-1} b, \, x_J \geq 0 \right\}.$$

---

[1] We want to point out an advantage of the approach in this chapter over the approach in the existing literature. The technique of embedding a given standard form problem in a homogeneous and self–dual problem was introduced by Ye, Todd and Mizuno [316] in 1994. See also Wu, Wu and Ye [299]; their final model contains free variables. In our approach the occurrence of free variables is avoided by first reducing the given standard problem to a canonical problem. For a different approach to the initialization problem we refer to, e.g., Lustig [189, 190].

Thus we have transformed (P) to the equivalent problem (P^c), which is in canonical format. Chapter 4 describes how we can embed any canonical model in a self-dual model so that a strictly complementary solution of the latter model either yields a strictly complementary solution of the canonical problem or makes clear that the canonical problem is infeasible or unbounded. Moreover, for the embedding problem we have a point on its central path available. If we apply such an embedding to (P^c), the resulting self-dual model may be given by

$$(SP^c) \quad \min \left\{ q^T \xi \ : \ M\xi \geq -q, \ \xi \geq 0 \right\},$$

where M is skew-symmetric and $q \geq 0$. Let $\xi(\mu)$ be a given point on the central path of (SP^c) for some positive μ. Now (SP^c) can be written in the standard form by associating the surplus vector $\sigma(\xi) := M\xi + q$ with any ξ. We then may rewrite (SP^c) as

$$(SSP^c) \quad \min \left\{ q^T \xi \ : \ M\xi - \sigma = -q, \ \xi \geq 0, \ \sigma \geq 0 \right\},$$

and we have

$$\xi(\mu)\sigma(\xi(\mu)) = \mu e,$$

where e is an all-one vector of appropriate size. Note that (SSP^c) is in the standard format. We can rewrite it as

$$(\bar{P}) \quad \min \left\{ \bar{c}^T \bar{x} \ : \ \bar{A}\bar{x} = \bar{b}, \ \bar{x} \geq 0 \right\},$$

with

$$\bar{A} = \begin{bmatrix} M & -I \end{bmatrix}, \quad \bar{c} = \begin{bmatrix} q \\ 0 \end{bmatrix}, \quad \bar{b} = -q.$$

The problem $(\bar{P})$ is in the standard format and hence the methods of this chapter can be used to yield an ε-solution of $(\bar{P})$ provided that we have a solution on or close to its central path. We now show that this condition is satisfied by showing that the μ-center of $(\bar{P})$ is known. To this end we need to consider also the dual problem of $(\bar{P})$, namely

$$(\bar{D}) \quad \max \left\{ \bar{b}^T \bar{y} \ : \ \bar{A}^T \bar{y} + \bar{s} = \bar{c}, \ \bar{s} \geq 0 \right\}.$$

For the slack vector $\bar{s}$ we have

$$\bar{s} = \bar{c} - \bar{A}^T \bar{y} = \begin{bmatrix} q - M^T \bar{y} \\ \bar{y} \end{bmatrix} = \begin{bmatrix} q + M\bar{y} \\ \bar{y} \end{bmatrix}.$$

Here we used that $M^T = -M$. Now with the definition

$$\bar{x} := \begin{bmatrix} \xi(\mu) \\ \sigma(\xi(\mu)) \end{bmatrix}, \quad \bar{y} =: \xi(\mu),$$

$\bar{x}$ is feasible for $(\bar{P})$ and $\bar{y}$ is feasible for $(\bar{D})$. The feasibility of $\bar{y}$ follows by considering its slack vector:

$$\bar{s} = \begin{bmatrix} q + M\bar{y} \\ \bar{y} \end{bmatrix} = \begin{bmatrix} q + M\xi(\mu) \\ \xi(\mu) \end{bmatrix} = \begin{bmatrix} \sigma(\xi(\mu)) \\ \xi(\mu) \end{bmatrix}.$$

For the product of $\bar{x}$ and $\bar{s}$ we have

$$\bar{x}\bar{s} = \begin{bmatrix} \xi(\mu) \\ \sigma(\xi(\mu)) \end{bmatrix} \begin{bmatrix} \sigma(\xi(\mu)) \\ \xi(\mu) \end{bmatrix} = \begin{bmatrix} \xi(\mu)\sigma(\xi(\mu)) \\ \sigma(\xi(\mu))\xi(\mu) \end{bmatrix} = \begin{bmatrix} \mu e \\ \mu e \end{bmatrix}.$$

This proves that $\bar{x}$ is the μ-center of $(\bar{P})$, as required.

By way of example we apply the above transformation to the sample problem used throughout this part of the book.

Example II.78 Taking A and c as in Example II.7 (page 97), and $b = (1,1)^T$, we have

$$A = \begin{bmatrix} 1 & -1 & 0 \\ 0 & 0 & 1 \end{bmatrix}, \quad b = \begin{bmatrix} 1 \\ 1 \end{bmatrix}, \quad c = \begin{bmatrix} 1 \\ 1 \\ 1 \end{bmatrix},$$

and (P) is the problem

$$(P) \qquad \min\{x_1 + x_2 + x_3 \; : \; x_1 - x_2 = 1, \; x_3 = 1, \; x \geq 0\}.$$

The first and the third column of A form a basis. With the index set I defined accordingly, the matrix A_I is the identity matrix. Then we express x_1 and x_3 in terms of x_2:

$$\begin{bmatrix} x_1 \\ x_3 \end{bmatrix} = \begin{bmatrix} 1 + x_2 \\ 1 \end{bmatrix}.$$

Using this we eliminate x_1 and x_3 from (P) and we obtain the canonical problem (P^c):

$$(P^c) \qquad \min\left\{2x_2 + 2 \; : \; \begin{bmatrix} 1 \\ 0 \end{bmatrix} x_2 \geq \begin{bmatrix} -1 \\ -1 \end{bmatrix}, \; x_2 \geq 0\right\}.$$

Being unrealistic, but just to demonstrate the transformation process for this simple case, we do not assume any foreknowledge and embed this problem in a self-dual problem as described in Section 4.3.[2] Taking 1 for x^0 and s^0, and for y^0 and t^0 the all-one vector of length 2, the self-dual embedding problem is given by (SP^c) with

$$M = \begin{bmatrix} 0 & 0 & 1 & 1 & -1 \\ 0 & 0 & 0 & 1 & 0 \\ -1 & 0 & 0 & 2 & 0 \\ -1 & -1 & -2 & 0 & 5 \\ 1 & 0 & 0 & -5 & 0 \end{bmatrix}, \quad q = \begin{bmatrix} 0 \\ 0 \\ 0 \\ 0 \\ 5 \end{bmatrix}.$$

Now the all-one vector is feasible for (SP^c) and its surplus vector is also the all-one vector, as easily can be verified. It follows that the all-one vector is the point on the central path for $\mu = 1$. Adding surplus variables to this problem we get a problem in the standard format with 5 equality constraints and 10 variables. Solving this problem

_____

[2] **Exercise 64** The canonical problem (P^c) contains an empty row. Remove this row and then perform the embedding. Show that this leads to the same solution of (P^c).

with the large-update logarithmic barrier method (with $\theta = 0.999$ and $\varepsilon = 10^{-4}$), we find in 4 iterations the strictly complementary solution

$$\xi = (0, 0, 0, \frac{4}{5}, 0, \frac{4}{5}, \frac{4}{5}, \frac{8}{5}, 0, 1).$$

The slack vector is

$$\sigma(\xi) = (\frac{4}{5}, \frac{4}{5}, \frac{8}{5}, 0, 1, 0, 0, 0, \frac{4}{5}, 0).$$

Note that the first five coordinates of ξ are equal to the last five coordinates of $\sigma(\xi)$ and vice versa. In fact, the first five coordinates of ξ form a solution of the self-dual embedding (SP^c) of (P^c). The homogenizing variable, the fourth entry in ξ, is positive. Therefore, we have found an optimal solution of (P^c). The optimal value of x_2 in (P^c), the third coordinate in the vector ξ, is given by $x_2 = 0$. Hence $x = (1, 0, 1)$ is optimal for the original problem (P). $\diamond$

A clear disadvantage of the above embedding procedure seems to be that it increases the size of the problem. If the constraint matrix A of (P) has size $m \times n$ then the final standard form problem that we have to solve has size $(n + 2) \times 2(n + 2)$. However, when the procedure is implemented efficiently the amount of extra computation can be reduced significantly. In fact, the computation of the search direction for the larger problem can be organized in such a way that it requires the solution of three linear systems with the same matrix of size $(m+2) \times (m+2)$. This is explained in Chapter 20.

Part III

The Target-following Approach

9

Preliminaries

9.1 Introduction

In this part we deal again with the problems (P) and (D) in the standard form:

$$(P) \qquad \min \left\{ c^T x \, : \, Ax = b, \, x \geq 0 \right\},$$

$$(D) \qquad \max \left\{ b^T y \, : \, A^T y \leq c \right\}.$$

As before, the matrix A is of size $m \times n$ with full row rank and the vectors c and x are in $\mathbb{R}^n$ and b in $\mathbb{R}^m$. Assuming that the interior-point condition is satisfied we recall from Theorem II.4 that the KKT system (5.3)

$$
\begin{aligned}
Ax &= b, & x \geq 0, \\
A^T y + s &= c, & s \geq 0, \\
xs &= \mu e
\end{aligned}
\tag{9.1}
$$

has a unique solution for every positive value of μ. These solutions are called the μ-centers of (P) and (D). The above result is fundamental for the algorithms analyzed in Part II. When μ runs through the positive real line then the solutions of the KKT system run through the central paths of (P) and (D); the methods in Part II just approximately follow the central path to the optimal sets of (P) and (D). These methods were called *logarithmic barrier methods* because the points on the central path are minimizers of the logarithmic barrier functions for (P) and (D). For obvious reasons they have also become known under the name *central-path-following methods*. In each (outer) iteration of such a method the value of the parameter μ is fixed and starting at a given feasible solution of (P) and/or (D) a good approximation is constructed of the μ-centers of (P) and (D). Numerically the approximate solutions are obtained either by using Newton's method for solving the KKT system or by using Newton's method for minimizing the logarithmic barrier function of (P) and (D). In the first case Newton's method provides displacements for both (P) and (D); then we speak of a *primal-dual method*. In the second case Newton's method provides a displacement for either (P) or (D), depending on whether the logarithmic barrier function of (P) or (D) is used in Newton's method. This gives the so-called *primal methods* and *dual methods* respectively. In all cases the result of an (outer) iteration is a primal-dual pair approximating the μ-centers and such that the duality gap is approximately $n\mu$.

In this part we present a generalization of the above results. The starting point is the observation that if the vector μe on the right-hand side of the KKT system (9.1) is replaced by any positive vector w then the resulting system still has a (unique) solution. Thus, for any positive vector w the system

$$
\begin{aligned}
Ax &= b, & x \geq 0, \\
A^T y + s &= c, & s \geq 0, \\
xs &= w
\end{aligned}
\tag{9.2}
$$

has a unique solution, denoted by $x(w), y(w), s(w)$.[1] This result is interesting in itself. It means that we can associate with each positive vector w the primal-dual pair $(x(w), s(w))$.[2] The map Φ_{PD} associating with any $w > 0$ the pair $(x(w), s(w))$ will be called the *target map* associated with (P) and (D). In the next section we discuss its existence and also some interesting properties.

In the present context, it is convenient to refer to the interior of the nonnegative orthant in $\mathbb{R}^n$ as the *w-space*. Any (possibly infinite) sequence of positive vectors w^k $(k = 1, 2, \ldots)$ in the w-space is called a *target sequence*. If a target sequence converges to the origin, then the duality gap $e^T w^k$ for the corresponding pair in the sequence $\Phi_{PD}(w^k)$ converges to zero. We are especially interested in target sequences of this type for which the sequence $\Phi_{PD}(w^k)$ is convergent as well, and for which the limiting primal-dual pair is strictly complementary. In Section 9.3 we derive a sufficient condition on target sequences (converging to the origin) that yields this property. We also give a condition such that the limiting pair consists of so-called *weighted-analytic centers* of the optimal sets of (P) and (D).

With any central-path-following method we can associate a target sequence on the central path by specifying the values of the barrier parameter μ used in the successive (outer) iterations. The central-path-following method can be interpreted as a method that takes the points on the central path as intermediate targets on the way to the origin. Thus it becomes apparent how the notion of central-path-following methods can be generalized to *target-following methods*, which (approximately) follow arbitrary target sequences. To develop this idea further we need numerical procedures that can be used to obtain a good approximation of the primal-dual pair corresponding to some specified positive target vector. Chapters 10, 12 and 13 are devoted to such procedures. The basic principle is again Newton's method. Chapter 10 describes a primal-dual method, Chapter 12 a dual method, and Chapter 13 a primal method.

The target-following approach offers a very general framework for the analysis of almost all known interior-point methods. In Chapter 11 we analyze some of the methods of Part II in this framework. We also deal with some other applications, including a target-following method that is based on the Dikin direction, as introduced in Appendix E. Finally, in Chapter 14 we deal with the so-called *method of centers*. This method will be described and after putting it into the target-following framework we provide a new and relatively easy analysis of the method.

---

[1] This result, which establishes a one-to-one correspondence between primal-dual pairs (x, s) and positive vectors in $\mathbb{R}^n$, was proved first in Kojima et al. [175]. Below we present a simple alternative proof. Mizuno [212, 214] was the first to use this property in the design of an algorithm.

[2] Here, as before, we use that any dual feasible pair (y, s) can be uniquely represented by either y or s. This is due to the assumption that A has full row rank.

9.2 The target map and its inverse

Our first aim in this section is to establish that the target map Φ_{PD} is well defined. That is, we need to show that for any positive vector $w \in \mathbb{R}^n$ the system (9.2) has a unique solution. To this end we use a modification of the primal-dual logarithmic barrier as given by (6.23). Replacing the role of the vector μe in this function by the vector w, we consider the modified primal-dual logarithmic barrier function defined by

$$\phi_w(x,s) = \frac{1}{\max(w)} \sum_{j=1}^{n} w_j \psi\left(\frac{x_j s_j}{w_j} - 1\right). \tag{9.3}$$

Here the function ψ has its usual meaning (cf. (5.5), page 92). The scaling factor $1/\max(w)$ serves to scale $\phi_w(x,s)$ in such a way that $\phi_w(x,s)$ coincides with the primal-dual logarithmic barrier function (7.44) in Section 7.8 (page 194) if w is on the central path.[3]

Note that $\phi_w(x,s)$ is defined for all positive primal-dual pairs (x,s). Moreover, $\phi_w(x,s) \geq 0$ and the equality holds if and only if $xs = w$. Hence, the weighted KKT system (9.2) has a solution if and only if the minimal value of ϕ_w is 0.

By expanding $\phi_w(x,s)$ we get

$$
\begin{aligned}
\max(w)\,\phi_w(x,s) &= \sum_{j=1}^{n} w_j \left(\frac{x_j s_j}{w_j} - 1 - \log\frac{x_j s_j}{w_j}\right) \\
&= \sum_{j=1}^{n} x_j s_j - \sum_{j=1}^{n} w_j - \sum_{j=1}^{n} w_j \log x_j s_j + \sum_{j=1}^{n} w_j \log w_j \\
&= x^T s - \sum_{j=1}^{n} w_j \log x_j s_j - e^T w + \sum_{j=1}^{n} w_j \log w_j. \tag{9.4}
\end{aligned}
$$

Neglecting for the moment the constant part, that is the part that does not depend on x and s, we are left with the function

$$x^T s - \sum_{j=1}^{n} w_j \log x_j s_j. \tag{9.5}$$

This function is usually called a *weighted primal-dual logarithmic barrier function* with the coefficients of the vector w as *weighting coefficients*. Since $x^T s = c^T x - b^T y$, the first term in (9.5) is linear on the domain of $\phi_w(x,s)$. The second term, called the *barrier term*, is strictly convex and hence it follows that $\phi_w(x,s)$ is strictly convex on its domain.

---

[3] If $w = \mu e$ then $\max(w) = \mu$ and hence

$$\phi_w(x,s) = \frac{1}{\mu}\sum_{j=1}^{n} \mu\psi\left(\frac{x_j s_j}{\mu} - 1\right) = \sum_{j=1}^{n} \psi\left(\frac{x_j s_j}{\mu} - 1\right) = \Psi\left(\frac{xs}{\mu} - e\right);$$

this is precisely the primal-dual logarithmic barrier function $\phi_\mu(x,s)$ as given by (6.23) and (7.44), and that was used in the analysis of the large-update central-path-following logarithmic barrier method.

In the sequel we need a quantity to measure the distance from a positive vector w to the central path of the w-space. Such a measure was introduced in Section 3.3.4 in (3.20). We use the same measure here, namely

$$\delta_c(w) := \frac{\max(w)}{\min(w)}. \qquad (9.6)$$

Now we are ready to derive the desired result by adapting Theorem II.4 and its proof to the present case. With w fixed, for given $K \in \mathbb{R}$ the level set $\mathcal{L}_K$ of ϕ_w is defined by

$$\mathcal{L}_K = \left\{ (x, s) \ : \ x \in \mathcal{P}^+, \ s \in \mathcal{D}^+, \ \phi_w(x, s) \leq K \right\}.$$

Theorem III.1 *Let $w \in \mathbb{R}^n$ and $w > 0$. Then the following statements are equivalent:*
  (i) (P) and (D) satisfy the interior-point condition.
  (ii) There exists $K \geq 0$ such that the level set $\mathcal{L}_K$ is nonempty and compact.
  (iii) There exists a (unique) primal-dual pair (x, s) minimizing ϕ_w with x and s both positive.
  (iv) There exist (unique) $x, s \in \mathbb{R}^n$ and $y \in \mathbb{R}^m$ satisfying (9.2);
  (v) For each $K \geq 0$ the level set $\mathcal{L}_K$ is nonempty and compact.

Proof: $(i) \Rightarrow (ii)$: Assuming (i), there exists a positive $x^0 \in \mathcal{P}^+$ and a positive $s^0 \in \mathcal{D}^+$. With $K = \phi_w\left(x^0, s^0\right)$ the level set $\mathcal{L}_K$ contains the pair $\left(x^0, s^0\right)$. Thus, $\mathcal{L}_K$ is not empty, and we need to show that $\mathcal{L}_K$ is compact. Let $(x, s) \in \mathcal{L}_K$. Then, by the definition of $\mathcal{L}_K$,

$$\sum_{i=1}^{n} w_i \psi \left(\frac{x_i s_i}{w_i} - 1 \right) \leq K \max(w).$$

Since each term in the sum is nonnegative, this implies

$$\psi \left(\frac{x_i s_i}{w_i} - 1 \right) \leq \frac{K \max(w)}{\min(w)} = K \delta_c(w), \quad 1 \leq i \leq n.$$

Since ψ is strictly convex on its domain and goes to infinity at its boundaries, there exist unique positive numbers a and b, with $a < 1$, such that

$$\psi(-a) = \psi(b) = K \delta_c(w).$$

We conclude that

$$-a \leq \frac{x_i s_i}{w_i} - 1 \leq b, \quad 1 \leq i \leq n,$$

which gives

$$w_i(1 - a) \leq x_i s_i \leq w_i(1 + b), \quad 1 \leq i \leq n. \qquad (9.7)$$

From the right-hand side inequality we deduce that

$$x^T s \leq (1 + b) e^T w.$$

We proceed by showing that this and (i) imply that the coordinates of x and s can be bounded above. Since $A(x - x^0) = 0$, the vector $x - x^0$ belongs to the null space of

A. Similarly, $s - s^0 = A^T(y^0 - y)$ implies that $s - s^0$ lies in the row space of A. The row space and the null space of A are orthogonal and hence we have

$$(x - x^0)^T(s - s^0) = 0. \tag{9.8}$$

Writing this as

$$x^T s^0 + s^T x^0 = x^T s + (x^0)^T(s^0)$$

and using $x^T s \leq (1+b)e^T w$, we find

$$x^T s^0 + s^T x^0 \leq (1+b)e^T w + (x^0)^T(s^0). \tag{9.9}$$

Since $s^T x^0 \geq 0, x \geq 0$, and $s^0 > 0$, this implies for each index i that

$$x_i s_i^0 \leq x^T s^0 + s^T x^0 \leq (1+b)e^T w + (x^0)^T(s^0),$$

whence

$$x_i \leq \frac{(1+b)e^T w + (x^0)^T(s^0)}{s_i^0},$$

proving that the coordinates of the vector x are bounded above. The coordinates of the vector s are bounded above as well. This can be derived from (9.9) in exactly the same way as for the coordinates of x. Using $x^T s^0 \geq 0, s \geq 0$, and $x^0 > 0$, we obtain for each index i that

$$s_i \leq \frac{(1+b)e^T w + (x^0)^T(s^0)}{x_i^0}.$$

Thus we have shown that the level set $\mathcal{L}_K$ is bounded. We proceed by showing that $\mathcal{L}_K$ is compact. Each s_i being bounded above, the left inequality in (9.7) implies that x_i is bounded away from zero. In fact, we have

$$x_i \geq \frac{(1-a)w_i}{s_i} \geq \frac{(1-a)x_i^0 w_i}{(1+b)e^T w + (x^0)^T(s^0)}.$$

In the same way we derive that for each i,

$$s_i \geq \frac{(1-a)w_i}{x_i} \geq \frac{(1-a)s_i^0 w_i}{(1+b)e^T w + (x^0)^T(s^0)}.$$

We conclude that for each i there exist positive numbers α_i and β_i with $0 < \alpha_i \leq \beta_i$, such that

$$\alpha_i \leq x_i, s_i \leq \beta_i, \quad 1 \leq i \leq n.$$

Thus we have proved the inclusion

$$\mathcal{L}_K \subseteq \prod_{i=1}^n [\alpha_i, \beta_i] \times [\alpha_i, \beta_i].$$

The set on the right-hand side lies in the positive orthant of $\mathbb{R}^n \times \mathbb{R}^n$, and being the Cartesian product of closed intervals, it is compact. Since ϕ_w is continuous, and well defined on this set, it follows that $\mathcal{L}_K$ is compact. Thus we have shown that (ii) holds.

$(ii) \Rightarrow (iii)$: Suppose that (ii) holds. Then, for some nonnegative K the level set $\mathcal{L}_K$ is nonempty and compact. Since ϕ_w is continuous, it follows that ϕ_w has a minimizer (x, s) in $\mathcal{L}_K$. Moreover, since ϕ_w is strictly convex, this minimizer is unique. Finally, from the definition of ϕ_w, $\psi\left((x_i s_i / w_i) - 1\right)$ must be finite, and hence $x_i s_i > 0$ for each i. This implies that $x > 0$ and $s > 0$, proving (iii).

$(iii) \Rightarrow (iv)$: Suppose that (iii) holds. Then ϕ_w has a (unique) minimizer. Since the domain $\mathcal{P}^+ \times \mathcal{D}^+$ of ϕ_w is open, $(x, s) \in \mathcal{P}^+ \times \mathcal{D}^+$ is a minimizer of ϕ_w if and only if the gradient of ϕ_w is orthogonal to the linear space parallel to the smallest affine space containing $\mathcal{P}^+ \times \mathcal{D}^+$ (cf. Proposition A.1). This linear space is determined by the affine system

$$Ax = 0, \quad Hs = 0,$$

where H is a matrix such that its row space is the null space of A and vice versa. The gradient of ϕ_w with respect to the coordinates of x satisfies

$$\max(w)\nabla_x \phi_w(x, s) = s - \frac{w}{x},$$

and with respect to the coordinates of s we have

$$\max(w)\nabla_s \phi_w(x, s) = x - \frac{w}{s}.$$

Application of Proposition A.1 yields that $\nabla_x \phi_w(x, s)$ must lie in the row space of A and $\nabla_s \phi_w(x, s)$ must lie in the row space of H. These two spaces are orthogonal, and hence we obtain

$$\left(s - \frac{w}{x}\right)^T \left(x - \frac{w}{s}\right) = 0.$$

This can be rewritten as

$$\left(s - \frac{w}{x}\right)^T X S^{-1} \left(s - \frac{w}{x}\right) = 0.$$

Since XS^{-1} is a diagonal matrix with positive elements on the diagonal, this implies

$$\left\| s - \frac{w}{x} \right\| = 0.$$

Hence,

$$s - \frac{w}{x} = 0,$$

whence $xs = w$. This proves that (x, s) is a minimizer of ϕ_w if and only if (x, s) satisfies (9.2). Hence (iv) follows from (iii).

$(iv) \Rightarrow (i)$: Let (x, s) be a solution of (9.1). Since $w > 0$ and x and s are nonnegative, both x and s are positive. This proves that (P) and (D) satisfy the interior-point condition.

Thus it has been shown that statements (i) to (iv) in the theorem are equivalent. We finally prove that statement (v) is equivalent with each of these statements. Obviously (v) implies (ii). On the other hand, assuming that statements (i) to (iv) hold, let x and s solve (9.2). Then we have $x > 0$, $s > 0$ and $xs = w$. This implies that $\phi_w(x, s) = 0$, as easily follows by substitution. Now let K be any nonnegative number. Then the level set $\mathcal{L}_K$ contains the pair (x, s) and hence it is nonempty. Finally, from the above proof of the implication $(i) \Rightarrow (ii)$ it is clear that $\mathcal{L}_K$ is compact. This completes the proof of the theorem. □

If the interior-point condition is satisfied, then the target map Φ_{PD} provides a tool for representing any positive primal-dual pair (x, s) by the positive vector xs, which is the inverse image of the pair (x, s). The importance of this feature cannot be overestimated. It means that the interior of the nonnegative orthant in $\mathbb{R}^n$ can be used to represent all positive primal-dual pairs. As a consequence, the behavior of primal-dual methods that generate sequences of positive primal-dual pairs, can be described in the nonnegative orthant in $\mathbb{R}^n$. Obviously, the central paths of (P) and (D) are represented by the *bisector* $\{\mu e : \mu > 0\}$ of the w-space; in the sequel we refer to the bisector as the central path of the w-space. See Figure 9.1.

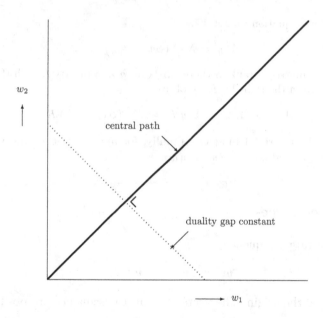

Figure 9.1 The central path in the w-space $(n = 2)$.

For central-path-following methods the target sequence is a sequence on this path converging to the origin. The iterates of these methods are positive primal-dual pairs 'close' to the target points on the central path, in the sense of some proximity measure. In the next sections we deal with target sequences that are not necessarily on the central path.

Remark III.2 We conclude this section with an interesting observation, namely that the target map of (P) and (D) contains so much information that we can reconstruct the data A, b and c from the target map.[4] This can be shown as follows. We take partial derivatives with

---

[4] This result was established by Crouzeix and Roos [57] in an unpublished note.

respect to the coordinates of w in the weighted KKT system (9.2). Denoting the Jacobians of x, y and s simply by x', y' and s' respectively, we have

$$x' = \frac{\partial x}{\partial w}, \quad y' = \frac{\partial y}{\partial w}, \quad s' = \frac{\partial s}{\partial w},$$

where the (i, j) entry of x' is the partial derivative $\partial x_i / \partial w_j$, etc. Note that x' and s' are $n \times n$ matrices and y' is an $m \times n$ matrix. Thus we obtain

$$
\begin{aligned}
Ax' &= 0, \\
A^T y' + s' &= 0, \\
X s' + S x' &= I_{nn},
\end{aligned}
\tag{9.10}
$$

where I denotes the identity matrix of size $n \times n$.[5] The third equation is equivalent to

$$s' = X^{-1} \left(I_{nn} - S x' \right).$$

Using also the second equation we get

$$A^T y' = X^{-1} \left(S x' - I_{nn} \right). \tag{9.11}$$

Since y' is an $m \times n$ matrix of rank m there exists an $n \times m$ matrix R such that $y' R = I_{mm}$. Multiplying (9.11) from the right by R we obtain

$$A^T = A^T I_{mm} = A^T y' R = X^{-1} \left(S x' - I_{nn} \right) R,$$

which determines the matrix A uniquely. Finally, for any positive w, the vectors b and c follow from $b = Ax(w)$ and $c = A^T y(w) + s(w)$.   •

9.3   Target sequences

Let us consider a target sequence

$$w^0, w^1, w^2, \ldots, w^k, \ldots \tag{9.12}$$

which converges to the origin. The vectors w^k in the sequence are positive and

$$\lim_{k \to \infty} w^k = 0.$$

As a consequence, for the duality gap $e^T w^k$ at w^k we have $\lim_{k \to \infty} e^T w^k = 0$; this implies that the accumulation points of the sequence

$$\Phi_{PD} \left(w^0 \right), \Phi_{PD} \left(w^1 \right), \Phi_{PD} \left(w^2 \right), \ldots, \Phi_{PD} \left(w^k \right), \ldots \tag{9.13}$$

are optimal primal-dual pairs.[6] In the sequel (x^*, s^*) denotes any such optimal primal-dual pair.

---

[5]   Since the matrix of system (9.10) is nonsingular, the implicit function theorem (cf. Proposition A.2 in Appendix A) implies the existence of all the relevant partial derivatives.

[6]   **Exercise 65** By definition, an accumulation point of the sequence (9.13) is a primal-dual pair that is the limiting point of some convergent subsequence of (9.13). Verify the existence of such a convergent subsequence.

We are especially interested in target sequences for which the accumulation pairs (x^*, s^*) are strictly complementary. We prove below that this happens if the target sequence lies in some *cone neighborhood* of the central path defined by

$$\delta_c(w) \le \tau,$$

where τ is fixed and $\tau \ge 1$. Recall that $\delta_c(w) \ge 1$, with equality if and only if w is on the central path. Also, $\delta_c(w)$ is homogeneous in w: for any positive scalar λ and for any positive vector w we have

$$\delta_c(\lambda w) = \delta_c(w).$$

As a consequence, the inequality $\delta_c(w) \le \tau$ determines a cone in the w-space.

In Theorem I.20 we showed for the self-dual model that the limiting pairs of any target sequence on the central path are strictly complementary optimal solutions. Our next result not only implies an analogous result for the standard format but it extends it to target sequences lying inside a cone around the central path in the w-space.

Theorem III.3 *Let $\tau \ge 1$ and let the target sequence (9.12) be such that $\delta_c(w^k) \le \tau$ for each k. Then every accumulation pair (x^*, s^*) of the sequence (9.13) is strictly complementary.*

Proof: For each $k = 1, 2, \ldots$, let $(x^k, s^k) := \Phi_{PD}(w^k)$. Then we have

$$x^k s^k = w^k, \quad k = 1, 2, \ldots.$$

Now let (x^*, s^*) be any accumulation point of the sequence (9.13). Then there exists a subsequence of the given sequence whose primal-dual pairs converge to (x^*, s^*). Without loss of generality we assume that the given sequence itself is such a subsequence. Since $x^k - x^*$ and $s^k - s^*$ belong respectively to the null space and the row space of A, these vectors are orthogonal. Hence,

$$\left(x^k - x^*\right)^T \left(s^k - s^*\right) = 0.$$

Expanding the product and rearranging terms, we get

$$(x^*)^T s^k + (s^*)^T x^k = \left(s^k\right)^T x^k + (s^*)^T x^*.$$

Using that $\left(s^k\right)^T x^k = e^T w^k$ and $(x^*)^T s^* = 0$, we arrive at

$$\sum_{j \in \sigma(x^*)} x_j^* s_j^k + \sum_{j \in \sigma(s^*)} s_j^* x_j^k = e^T w^k, \quad k = 1, 2, \ldots.$$

Here $\sigma(x^*)$ denotes the support of x^* and $\sigma(s^*)$ the support of s^*.[7] Using that $x^k s^k = w^k$, we can write the last equation as

$$\sum_{j \in \sigma(x^*)} \frac{w_j^k x_j^*}{x_j^k} + \sum_{j \in \sigma(s^*)} \frac{w_j^k s_j^*}{s_j^k} = e^T w^k, \quad k = 1, 2, \ldots.$$

_____

[7] The support of a vector is defined in Section 2.8, Definition I.19, page 36.

Now let ε be a (small) positive number such that

$$\frac{1+\varepsilon}{n\varepsilon} > \tau.$$

Then, since (x^*, s^*) is the limit of the sequence $(x^k, s^k)_{k=0}^{\infty}$, there exists a natural number K such that

$$\frac{x_j^*}{x_j^k} \le 1 + \varepsilon \quad \text{and} \quad \frac{s_j^*}{s_j^k} \le 1 + \varepsilon$$

for each j $(1 \le j \le n)$ and for all $k \ge K$. Hence, for $k \ge K$ we have

$$e^T w^k \le (1 + \varepsilon) \left(\sum_{j \in \sigma(x^*)} w_j^k + \sum_{j \in \sigma(s^*)} w_j^k \right).$$

If the pair (x^*, s^*) is not strictly complementary, there exists an index i that does not belong to the union $\sigma(x^*) \cup \sigma(s^*)$ of the supports of x^* and s^*. Then we have

$$\sum_{j \in \sigma(x^*)} w_j^k + \sum_{j \in \sigma(s^*)} w_j^k \le e^T w^k - w_i^k.$$

Substitution gives

$$e^T w^k \le (1 + \varepsilon) \left(e^T w^k - w_i^k \right).$$

This implies

$$(1 + \varepsilon) w_i^k \le \varepsilon e^T w^k.$$

The average value of the elements of w^k is $e^T w^k / n$. Since $\delta_c(w^k) \le \tau$, the average value does not exceed τw_i^k. Hence, $e^T w^k \le n\tau w_i^k$. Substituting this we obtain

$$(1 + \varepsilon) w_i^k \le n\varepsilon\tau w_i^k.$$

Now dividing both sides by w_i^k we arrive at the contradiction

$$1 + \varepsilon \le n\varepsilon\tau.$$

This proves that (x^*, s^*) is strictly complementary. $\qquad \square$

If a target sequence satisfies the condition in Theorem III.3 for some $\tau \ge 1$, it is clear that the ratios between the coordinates of the vectors w^k are bounded. In fact,

$$\frac{1}{\tau} \le \frac{w_i^k}{w_j^k} \le \tau$$

for all k and for all i and j. For target sequences on the central path these ratios are all equal to one, so the limits of the ratios exist if k goes to infinity. In general we are interested in target sequences for which the limits of these ratios exist when k goes to infinity. Since the ratios between the coordinates do not change if w^k is multiplied by a positive constant, this happens if and only if there exists a positive vector w^* such that

$$\lim_{k \to \infty} \frac{n w^k}{e^T w^k} = w^*, \tag{9.14}$$

and then the limiting values of the ratios are given by the ratios between the coordinates of w^*. Note that we have $e^T w^* = n$, because the sum of the coordinates of each vector $n w^k / e^T w^k$ is equal to n. Also note that if a target sequence satisfies (9.14), we may find a $\tau \geq 1$ such that $\delta_c(w^k) \leq \tau$ for each k. In fact, we may take

$$\tau = \max_{i,j,k} \frac{w_i^k}{w_j^k}.$$

Hence, by Theorem III.3, any accumulation pair (x^*, s^*) for such a sequence is strictly complementary.

Our next result shows that if (9.14) holds then the limiting pair (x^*, s^*) is unique and can be characterized as a *weighted-analytic* center of the optimal sets of (P) and (D). Let us first define this notion.

Definition III.4 (Weighted-analytic center) *Let the nonempty and bounded set T be the intersection of an affine space in $\mathbb{R}^p$ with the nonnegative orthant of $\mathbb{R}^p$. We define the support $\sigma(T)$ of T as the subset of the full index set $\{1, 2, \ldots, p\}$ given by*

$$\sigma(T) = \{i \; : \; \exists x \in T \; \text{such that} \; x_i > 0\}.$$

If w is any positive vector in $\mathbb{R}^p$ then the corresponding weighted-analytic center of T is defined as the zero vector if $\sigma(T)$ is empty, otherwise it is the vector in T that maximizes the product

$$\prod_{i \in \sigma(T)} x_i^{w_i}, \quad x \in T. \tag{9.15}$$

If the support of T is not empty then the convexity of T implies the existence of a vector $x \in T$ such that $x_{\sigma(T)} > 0$. Moreover, if $\sigma(T)$ is not empty then the maximum value of the product (9.15) exists since T is bounded. Since the product (9.15) is strictly concave, the maximum value is attained at a unique point of T. The above definition generalizes the notion of analytic center, as defined by Definition I.29 and it uniquely defines the weighted-analytic center (for any positive weighting vector w) for any bounded subset that is the intersection of an affine space in $\mathbb{R}^p$ with the nonnegative orthant of $\mathbb{R}^n$. [8]

Below we apply this notion to the optimal sets of (P) and (D). If a target sequence satisfies (9.14) then the next result states that the sequence of its primal-dual pairs converges to the pair of weighted-analytic centers of the optimal sets of (P) and (D).

Theorem III.5 *Let the target sequence (9.12) be such that (9.14) holds for some w^*, and let (x^*, s^*) be an accumulation point of the sequence (9.13). Then x^* is the weighted-analytic center of $\mathcal{P}^*$ with respect to w^*, and s^* is the weighted-analytic center of $\mathcal{D}^*$ with respect to w^*.*

Proof: We have already established that the limiting pair (x^*, s^*) is strictly complementary, from Theorem III.3. As a consequence, the support of the optimal set $\mathcal{P}^*$

---

[8] **Exercise 66** Let w be any positive vector in $\mathbb{R}^p$ and let the bounded set T be the intersection of an affine space in $\mathbb{R}^p$ with the nonnegative orthant of $\mathbb{R}^p$. Show that the weighted-analytic center (with w as weighting vector) of T coincides with the analytic center of T if and only if w is a scalar multiple of the all-one vector.

of (P) is equal to the support $\sigma(x^*)$ of x^*, and the support of the optimal set $\mathcal{D}^*$ of (D) is equal to the support $\sigma(s^*)$ of s^*.

Now let $\bar{x}$ be optimal for (P) and $\bar{s}$ for (D). Applying the orthogonality property to the pairs $(\bar{x}, \bar{s})$ and $(x^k, s^k) := \Phi_{PD}(w^k)$ we obtain

$$(x^k - \bar{x})^T (s^k - \bar{s}) = 0.$$

Expanding the product and rearranging terms, we get

$$(\bar{x})^T s^k + (\bar{s})^T x^k = \left(s^k\right)^T x^k + (\bar{s})^T \bar{x}.$$

Since $\left(s^k\right)^T x^k = e^T w^k$ and $(\bar{x})^T \bar{s} = 0$, we get

$$\sum_{j \in \sigma(x^*)} \bar{x}_j s_j^k + \sum_{j \in \sigma(s^*)} \bar{s}_j x_j^k = e^T w^k, \quad k = 1, 2, \ldots .$$

Here we have also used that $\sigma(\bar{x}) \subset \sigma(x^*)$ and $\sigma(\bar{s}) \subset \sigma(s^*)$. Using $x^k s^k = w^k$ we have

$$\sum_{j \in \sigma(x^*)} \frac{w_j^k \bar{x}_j}{x_j^k} + \sum_{j \in \sigma(s^*)} \frac{w_j^k \bar{s}_j}{s_j^k} = e^T w^k, \quad k = 1, 2, \ldots .$$

Multiplying both sides by $n/e^T w^k$ we get

$$\sum_{j \in \sigma(x^*)} \frac{n w_j^k}{e^T w^k} \frac{\bar{x}_j}{x_j^k} + \sum_{j \in \sigma(s^*)} \frac{n w_j^k}{e^T w^k} \frac{\bar{s}_j}{s_j^k} = n, \quad k = 1, 2, \ldots .$$

Letting $k \to \infty$, it follows that

$$\sum_{j \in \sigma(x^*)} \frac{w_j^* \bar{x}_j}{x_j^*} + \sum_{j \in \sigma(s^*)} \frac{w_j^* \bar{s}_j}{s_j^*} = n.$$

At this stage we apply the geometric inequality,[9] which states that for any two positive vectors α and β in $\mathbb{R}^n$,

$$\prod_{j=1}^n \left(\frac{\alpha_j}{\beta_j}\right)^{\beta_i} \leq \left(\frac{\sum_{j=1}^n \alpha_i}{\sum_{j=1}^n \beta_i}\right)^{\sum_{j=1}^n \beta_i}.$$

We apply this inequality with $\beta = w^*$ and

$$\alpha_j = \frac{w_j^* \bar{x}_j}{x_j^*} \quad (j \in \sigma(x^*)), \qquad \alpha_j = \frac{w_j^* \bar{s}_j}{s_j^*} \quad (j \in \sigma(s^*)).$$

Thus we obtain, using that the sum of the weights w_j^* equals n,

$$\prod_{j \in \sigma(x^*)} \left(\frac{\bar{x}_j}{x_j^*}\right)^{w_j^*} \prod_{j \in \sigma(s^*)} \left(\frac{\bar{s}_j}{s_j^*}\right)^{w_j^*} \leq \left(\frac{1}{n} \left(\sum_{j \in \sigma(x^*)} \frac{w_j^* \bar{x}_j}{x_j^*} + \sum_{j \in \sigma(s^*)} \frac{w_j^* \bar{s}_j}{s_j^*}\right)\right)^n = 1.$$

---

[9] When β is the all-one vector e, the geometric inequality reduces to the arithmetic-geometric-mean inequality. For a proof of the geometric inequality we refer to Hardy, Littlewood and Pólya [139].

Substituting $\bar{s} = s^*$ in the above inequality we get

$$\prod_{j \in \sigma(x^*)} \bar{x}_j^{w_j^*} \leq \prod_{j \in \sigma(x^*)} x_j^{* w_j^*},$$

and substituting $\bar{x} = x^*$ gives

$$\prod_{j \in \sigma(s^*)} \bar{s}_j^{w_j^*} \leq \prod_{j \in \sigma(s^*)} s_j^{* w_j^*}.$$

This shows that x^* maximizes the product

$$\prod_{i \in \sigma(x^*)} x_j^{w_j^*}$$

and s^* the product

$$\prod_{i \in \sigma(s^*)} s_j^{w_j^*}$$

over the optimal sets of (P) and (D) respectively. Hence the proof is complete. □

9.4 The target-following scheme

We are ready to describe more formally the main idea of the target-following approach. Assume we are given some positive primal-dual feasible pair (x^0, s^0). Put $w^0 := x^0 s^0$ and assume that we have a sequence

$$w^0, w^1, \ldots, w^k, \ldots, w^K \qquad (9.16)$$

of points w^k in the w-space with the following property:

Given the primal-dual pair for w^k, with $0 \leq k < K$, it is 'easy' to compute the primal-dual pair for w^{k+1}.

We call such a sequence a *traceable target sequence* of length K.

If a traceable sequence of length K is available, then we can solve the given problem pair (P) and (D), up to the precision level $e^T w^K$, in K iterations. The k-th iteration in this method would consist of the computation of the primal-dual target-pair corresponding to the target point w^k. Conceptually, the algorithm is described as follows (page 232).

Some remarks are in order. Firstly, in practice the primal-dual pair $(x(\bar{w}), s(\bar{w}))$ corresponding to an intermediate target $\bar{w}$ is not computed exactly. Instead we compute it approximately, but so that the approximating pair is close to $\bar{w}$ in the sense of a suitable proximity measure.

Secondly, the target sequence is not necessarily prescribed beforehand. It may be generated in the course of the algorithm. Both cases occurred in Chapter 7. For example, the primal-dual logarithmic barrier algorithm with full Newton steps in

---

Conceptual Target-following Algorithm

Input:
    A positive primal-dual pair (x^0, s^0);
    a final target vector $\tilde{w}$.
begin
    $w := x^0 s^0$;
    while w is not 'close' to $\tilde{w}$ **do**
    begin
        choose an 'intermediate' target $\bar{w}$;
        compute $x(\bar{w})$ and $s(\bar{w})$;
        $w := x(\bar{w})s(\bar{w})$;
    end
end

Section 7.5 uses intermediate targets of the form $w = \mu e$, and each subsequent target is given by $(1-\theta)w$, with θ fixed. The same is true for the primal-dual logarithmic barrier algorithm with large updates in Section 7.8. In contrast, the primal-dual logarithmic barrier algorithm with adaptive updates (cf. Section 7.6.1) defines its target points adaptively.

Thirdly, if we say that the primal-dual pair corresponding to a given target can be computed 'easily', we mean that we have an efficient numerical procedure for finding this primal-dual pair, at least approximately. The numerical method is always Newton's method, either for solving the KKT system defining the primal-dual pair, or for finding the minimizer of a suitable barrier function. When full Newton steps are taken, the target must be close to where we are, and one step must yield a sufficiently accurate approximation of the primal-dual pair for this target. In the literature, methods of this type are usually called *short-step methods* when the target sequence is prescribed, and *adaptive-step methods* if the target sequence is defined adaptively. We call them *full-step methods*. If subsequent targets are at a greater distance we are forced to use damped Newton steps. The number of Newton steps necessary to reach the next target (at least approximately) may then become larger than one. To achieve polynomiality we need to guarantee that this number can be bounded either by a constant or by some suitable function of n, e.g., $\mathcal{O}(\sqrt{n})$ or $\mathcal{O}(n)$. We refer to such methods as *multistep methods*. They appear in the literature as *medium-step methods* and *large-step methods*.

In general, a primal-dual target-following algorithm is based on some finite underlying target sequence $w^0, w^1, \ldots, w^k = \tilde{w}$. The final target $\tilde{w}$ is a vector with small duality gap $e^T\tilde{w}$ if we are optimizing, but other final targets are allowable as well; examples of both types of target sequence are given in Chapter 11 below. The general structure is as follows.

Generic (Primal-Dual) Target-following Algorithm

Input:
A positive primal-dual pair (x^0, s^0) such that $x^0 s^0 = w^0$;
a final target vector $\tilde{w}$.
begin
    $x = x^0$, $s = s^0$, $w := w^0$;
    while w is not 'close' to $\tilde{w}$ **do**
    begin
        replace w by the next target in the sequence;
        while xs is not 'close' to w **do**
        begin
            apply Newton steps at (x, s) with w as target
        end
    end
end

For each target in the sequence the next target can be prescribed (in advance), but it can also be defined adaptively. If it is close to the present target then a single (full) Newton step may suffice to reach the next target, otherwise we apply a multistep method, using damped Newton steps.

The target-following approach is more general than the standard central-path-following schemes that appear in the literature. The vast majority of the latter use target sequences on the central path.[10] We show below, in Chapter 11, that many classical results in the literature can be put in the target-following scheme and that this scheme often dramatically simplifies the analysis.

First, we derive the necessary numerical tools in the next chapter. This amounts to generalizing results obtained before in Part II for the case where the target is on the central path to the case where it is off the central path. We first analyze the full primal-dual Newton step method and the damped primal-dual Newton step method for computing the primal-dual pair corresponding to a given target vector. To this end we introduce a proximity measure, and we show that the full Newton step method is quadratically convergent. For the damped Newton method we show that a single step reduces the primal-dual barrier function by at least a constant, provided that the proximity measure is bounded below by a constant. We then have the basic ingredients

---

[10] There are so many papers on the subject that it is impossible to give an exhaustive list. We mention a few of them. Short-step methods along the central path can be found in Renegar [237], Gonzaga [118], Roos and Vial [245], Monteiro and Adler [218] and Kojima et al. [178]. We also refer the reader to the excellent survey of Gonzaga [124]. The concept of target-following methods was introduced by Jansen et al. [159]. Closely related methods, using so-called α-sequences, were considered before by Mizuno for the linear complementarity problem in [212] and [214]. The first results on multistep methods were those of Gonzaga [121, 122] and Roos and Vial [244]. We also mention den Hertog, Roos and Vial [146] and Mizuno, Todd and Ye [217]. The target-following scheme was applied first to multistep methods by Jansen et al. [158].

for the analysis of primal-dual target-following methods.

The results of the next chapter are used in Chapter 11 for the analysis of several interesting algorithms. There we restrict ourselves to full Newton step methods because they give the best complexity results. Later we show that the target-following concept is also useful when dealing with dual or primal methods. We also show that the primal-dual pair belonging to a target vector can be efficiently computed by such methods. This is the subject of Chapters 12 and 13.

10

The Primal-Dual Newton Method

10.1 Introduction

Suppose that a positive primal-dual feasible pair (x, s) is given as well as some *target vector* $w > 0$. Our aim is to find the primal-dual pair $(x(w), s(w))$. Recall that to the dual feasible slack vector s belongs a unique y such that $A^T y + s = c$. The vector in the y-space corresponding to $s(w)$ is denoted by $y(w)$. In this section we define search directions $\Delta x, \Delta y, \Delta s$ at the given pair (x, s) that are aimed to bring us closer to the *target pair* $(x(w), s(w))$ corresponding to w. The search directions in this section are obtained by applying Newton's method to the weighted KKT system (9.2), page 220. The approach closely resembles the treatment in Chapter 7. There the target was on the central path, but now the target may be any positive vector w. It will become clear that the results of Chapter 7 can be generalized almost straightforwardly to the present case. To avoid tiresome repetitions we to omit detailed arguments when they are similar to arguments used in Chapter 7.

10.2 Definition of the primal-dual Newton step

We want the iterates $x + \Delta x, y + \Delta y, s + \Delta s$ to satisfy the weighted KKT system (9.2) with respect to the target w. So we want $\Delta x, \Delta y$ and Δs to satisfy

$$
\begin{aligned}
A(x + \Delta x) &= b, & x + \Delta x \geq 0, \\
A^T(y + \Delta y) + s + \Delta s &= c, & s + \Delta s \geq 0, \\
(x + \Delta x)(s + \Delta s) &= w.
\end{aligned}
$$

Neglecting the inequality constraints, we can rewrite this as follows:

$$
\begin{aligned}
A\Delta x &= 0, \\
A^T \Delta y + \Delta s &= 0, \\
s\Delta x + x\Delta s + \Delta x \Delta s &= w - xs.
\end{aligned}
\tag{10.1}
$$

Newton's method amounts to linearizing this system by neglecting the second-order term $\Delta x \Delta s$ in the third equation. Thus we obtain the linear system

$$
\begin{aligned}
A\Delta x &= 0, \\
A^T \Delta y + \Delta s &= 0, \\
s\Delta x + x\Delta s &= w - xs.
\end{aligned}
\tag{10.2}
$$

Comparing this system with (7.2), page 150, in Chapter 7, we see that the only difference occurs in the third equation, where the target vector w replaces the target μe on the central path. In particular, both systems have the same matrix. Since this matrix is nonsingular (cf. Theorem II.42, page 150, and Exercise 46, page 151), system (10.2) determines the displacements $\Delta x, \Delta y$ and Δs uniquely. We call them the primal-dual Newton directions at (x, s) corresponding to the target w.[1,2,3] It may be worth pointing out that computation of the displacements $\Delta x, \Delta y$ and Δs amounts to solving a positive definite system with the matrix $AXS^{-1}A^T$, just like when the target is on the central path.

10.3 Feasibility of the primal-dual Newton step

In this section we investigate the feasibility of the (full) Newton step. As before, the result of the Newton step at (x, y, s) is denoted by (x^+, y^+, s^+), so

$$x^+ = x + \Delta x, \quad y^+ = y + \Delta y, \quad s^+ = s + \Delta s.$$

Since the new iterates satisfy the affine equations we only have to deal with the question of whether x^+ and s^+ are nonnegative or not. We have

$$x^+ s^+ = (x + \Delta x)(s + \Delta s) = xs + (s\Delta x + x\Delta s) + \Delta x \Delta s.$$

Since $s\Delta x + x\Delta s = w - xs$ this leads to

$$x^+ s^+ = w + \Delta x \Delta s. \tag{10.3}$$

Hence, x^+ and s^+ are feasible only if $w + \Delta x \Delta s$ is nonnegative. The converse is also true. This is the content of the next lemma.

Lemma III.6 *The primal-dual Newton step at (x, s) to the target w is feasible if and only if $w + \Delta x \Delta s \geq 0$.*

Proof: The proof uses exactly the same arguments as the proof of Lemma II.46; we simply need to replace the vector μe by w. We leave it to the reader to verify this. $\square$

Note that Newton's method is exact when the second-order term $\Delta x \Delta s$ vanishes. In that case we have $x^+ s^+ = w$. This means that the pair (x^+, s^+) is the image of w under the target map, whence $x^+ = x(w)$ and $s^+ = s(w)$.

In general $\Delta x \Delta s$ will not be zero and Newton's method will not be exact. However, the duality gap always assumes the correct value $e^T w$ after the Newton step.

---

[1] **Exercise 67** Prove that the system (10.2) has a unique solution, namely

$$\begin{aligned} \Delta y &= \left(AXS^{-1}A^T\right)^{-1}\left(b - AWs^{-1}\right) \\ \Delta s &= -A^T \Delta y \\ \Delta x &= ws^{-1} - x - xs^{-1}\Delta s. \end{aligned}$$

[2] **Exercise 68** When $w = 0$ in (10.2), the resulting directions coincide with the primal-dual affine-scaling directions introduced in Section 7.6.2. Verify this.

[3] **Exercise 69** When $w = \mu e$ and $\mu = x^T s/n$ in (10.2), the resulting directions coincide with the primal-dual centering directions introduced in Section 7.6.2. Verify this.

Lemma III.7 *If the primal-dual Newton step is feasible then* $(x^+)^T s^+ = e^T w$.

Proof: This is immediate from (10.3) because the vectors Δx and Δs are orthogonal. □

In the following sections we further analyze the primal-dual Newton method. This requires a quantity for measuring the progress of the Newton iterates on the way to the pair $\Phi_{PD}(w)$. As may be expected, two cases could occur. In the first case the present pair (x, s) is 'close' to $\Phi_{PD}(w)$ and full Newton steps are feasible. In that case the full Newton step method is (hopefully, and locally) quadratically convergent. In the second case the present pair (x, s) is 'far' from $\Phi_{PD}(w)$ and the full Newton step may not be feasible. Then we are forced to take damped Newton steps and we may expect no more than a linear convergence rate. In both cases we need a new quantity for measuring the proximity of the current iterate to the target vector w. The next section deals with the first case and the second case is considered in Section 10.5. It will be no surprise that we use the weighted primal-dual barrier function $\phi_w(x, s)$ in Section 10.5 to measure progress of the method.

10.4 Proximity and local quadratic convergence

Recall from (7.16), page 156, that in the analysis of the central-path-following primal-dual method we measured the distance of the pair (x, s) to the target μe by the quantity

$$\delta(x, s; \mu) = \frac{1}{2} \left\| \sqrt{\frac{\mu e}{xs}} - \sqrt{\frac{xs}{\mu e}} \right\|.$$

This can be rewritten as

$$\delta(x, s; \mu) = \frac{1}{2\sqrt{\mu}} \left\| \frac{\mu e - xs}{\sqrt{xs}} \right\|.$$

Note that the right-hand side measures, in some way, the distance in the w-space between the inverse image μe of the pair of μ-centers $(x(\mu e), s(\mu e))$ and the primal-dual pair (x, s).[4] For a general target vector w we adapt this measure to

$$\delta(xs, w) := \frac{1}{2\sqrt{\min(w)}} \left\| \frac{w - xs}{\sqrt{xs}} \right\|. \tag{10.4}$$

The quantity on the right measures the distance from the coordinatewise product xs to w. It is defined for (ordered) pairs of vectors in the w-space. Therefore, and because it will be more convenient in the future, we express this feature by using the notation

---

[4] This observation makes clear that the proximity measure $\delta(x, s; \mu)$ ignores the actual data of the problems (P) and (D), which is contained in A, b and c. Since the behavior of Newton's method does depend on these data, it follows that the effect of a (full) Newton step on the proximity measure depends on the data of the problem. This reveals the weakness of the analysis of the full-step method (cf. Chapter 6.7). It ignores the actual data of the problem and only provides a worst-case analysis. In contrast, with adaptive updates (cf. Chapter 6.8) the data of the problem are taken into account and, as a result, the performance of the method is improved.

238 _____ **III Target-following Approach**

$\delta(xs, w)$ instead of the alternative notation $\delta(x, s; w)$. We prove in this section that the Newton method is quadratically convergent in terms of this proximity measure.[5]

As before we use scaling vectors d and u. The definition of u needs to be adapted to the new situation:

$$d := \sqrt{\frac{x}{s}}, \quad u := \sqrt{\frac{xs}{w}}. \tag{10.5}$$

Note that $xs = w$ if and only if $u = e$. We also introduce a vector v according to

$$v = \sqrt{xs}.$$

With d we can rescale both x and s to the vector v:[6]

$$d^{-1}x = v, \quad ds = v.$$

Rescaling Δx and Δs similarly:

$$d^{-1}\Delta x =: d_x, \quad d\Delta s =: d_s, \tag{10.6}$$

we see that

$$\Delta x \Delta s = d_x d_s.$$

Consequently, the orthogonality of Δx and Δs implies that the scaled displacements d_x and d_s are orthogonal as well. Now we may reduce the left-hand side in the third equation of the KKT system as follows:

$$s\Delta x + x\Delta s = sdd^{-1}\Delta x + xd^{-1}d\Delta s = v(d_x + d_s),$$

so the third equation can be restated simply as

$$d_x + d_s = v^{-1}(w - xs).$$

---

[5] **Exercise 70** The definition (10.4) of the primal-dual proximity measure $\delta = \delta(xs, w)$ implies that

$$2\delta(xs, w) \geq \left\| \frac{w - xs}{\sqrt{w}\sqrt{xs}} \right\| = \left\| \sqrt{\frac{w}{xs}} - \sqrt{\frac{xs}{w}} \right\|.$$

Using this and Lemma II.62, prove

$$\frac{1}{\rho(\delta)} \leq \sqrt{\frac{x_i s_i}{w_i}} \leq \rho(\delta), \quad 1 \leq i \leq n.$$

[6] Here we deviate from the approach in Chapter 7. The natural generalization of the approach there would be to rescale x and s to u:

$$\frac{d^{-1}x}{\sqrt{w}} = u, \quad \frac{ds}{\sqrt{w}} =: u,$$

and then rescale Δx and Δs accordingly to

$$\frac{d^{-1}\Delta x}{\sqrt{w}} =: d_x, \quad \frac{d\Delta s}{\sqrt{w}} =: d_s.$$

But then we have

$$\Delta x \Delta s = wd_x d_s$$

and we lose the orthogonality of d_x and d_s with respect to the standard inner product. This could be resolved by changing the inner product in such a way that orthogonality is preserved. We leave it as an (interesting) exercise to the reader to work this out. Here the difficulty is circumvented by using a different scaling.

On the other hand, the first and second equations can be reformulated as $ADd_x = 0$ and $(AD)^T d_y + d_s = 0$, where $d_y = \Delta y$. We conclude that the scaled displacements d_x, d_y and d_s satisfy

$$
\begin{aligned}
ADd_x &= 0 \\
(AD)^T d_y + d_s &= 0 \\
d_x + d_s &= v^{-1}(w - xs).
\end{aligned}
\tag{10.7}
$$

Using the same arguments as in Chapter 7 we conclude that d_x and d_s form the components of $v^{-1}(w - xs)$ in the null space and the row space of AD, respectively. Note that $w - xs$ represents the move we want to make in the w-space. Therefore we denote it as Δw. It is also convenient to use a scaled version d_w of Δw, namely

$$
d_w := v^{-1}(w - xs) = v^{-1}\Delta w.
\tag{10.8}
$$

Then we have

$$
d_x + d_s = d_w
\tag{10.9}
$$

and, since d_x and d_s are orthogonal,

$$
\|d_x\|^2 + \|d_s\|^2 = \|d_w\|^2.
\tag{10.10}
$$

This makes clear that the scaled displacements d_x, d_s (and also d_y) are zero if and only if $d_w = 0$. In that case x, y and s coincide with their values at w. An immediate consequence of the definition (10.4) of the proximity $\delta(xs, w)$ is

$$
\delta(xs, w) = \frac{\|d_w\|}{2\sqrt{\min(w)}}.
\tag{10.11}
$$

The next lemma contains upper bounds for the 2-norm and the infinity norm of the second-order term $d_x d_s$.

Lemma III.8 _We have_ $\|d_x d_s\|_\infty \le \frac{1}{4}\|d_w\|^2$ _and_ $\|d_x d_s\| \le \frac{1}{2\sqrt{2}}\|d_w\|^2$.

Proof: The lemma follows immediately by applying the first uv-lemma (Lemma C.4) to the vectors d_x and d_s. □

Lemma III.9 _The Newton step is feasible if_ $\delta(xs, w) \le 1$.

Proof: Lemma III.6 guarantees feasibility of the Newton step if $w + \Delta x \Delta s \ge 0$. Since $\Delta x \Delta s = d_x d_s$ this certainly holds if the infinity norm of the quotient $d_x d_s/w$ does not exceed 1. Using Lemma III.8 and (10.11) we may write

$$
\left\|\frac{d_x d_s}{w}\right\|_\infty \le \frac{\|d_x d_s\|_\infty}{\min(w)} \le \frac{\|d_w\|^2}{4\min(w)} = \delta(xs, w)^2 \le 1.
$$

This implies the lemma. □

We are ready for the main result of this section, which is a perfect analogue of Theorem II.50, where the target is on the central path.

Theorem III.10 *If $\delta := \delta(xs; w) \leq 1$, then the primal-dual Newton step is feasible and $(x^+)^T s^+ = e^T w$. Moreover, if $\delta < 1$ then*

$$\delta(x^+ s^+, w) \leq \frac{\delta^2}{\sqrt{2(1 - \delta^2)}}.$$

Proof: The first part of the theorem is a restatement of Lemma III.9 and Lemma III.7. We proceed with the proof of the second statement. By definition,

$$\delta(x^+ s^+, w)^2 = \frac{1}{4 \min(w)} \left\| \frac{w - x^+ s^+}{\sqrt{x^+ s^+}} \right\|^2.$$

Recall from (10.3) that $x^+ s^+ = w + \Delta x \Delta s = w + d_x d_s$. Using also Lemma III.8 and (10.11), we write

$$\min(x^+ s^+) \geq \min(w) - \|d_x d_s\|_\infty \geq \min(w) - \frac{1}{4} \|d_w\|^2 = \min(w)\left(1 - \delta^2\right).$$

Thus we find, by substitution,

$$\delta(x^+ s^+, w)^2 \leq \frac{\|w - x^+ s^+\|^2}{4\left(1 - \delta^2\right)\min(w)^2} = \frac{\|d_x d_s\|^2}{4\left(1 - \delta^2\right)\min(w)^2}.$$

Finally, using the upper bound for $\|d_x d_s\|$ in Lemma III.8 and also using (10.11) once more, we obtain

$$\delta(x^+ s^+, w)^2 \leq \frac{\|d_w\|^4}{32\left(1 - \delta^2\right)\min(w)^2} = \frac{\delta^4}{2\left(1 - \delta^2\right)}.$$

This implies the theorem.[7]   □

It is clear that the above result has value only' if the given pair (x, s) is close enough to the target vector w. It guarantees quadratic convergence to the target if $\delta(xs, w) \leq 1/\sqrt{2}$. Convergence is guaranteed only if $\delta(xs, w) \leq \sqrt{2/3}$. For larger values of $\delta(xs, w)$ we need a different analysis. Then we measure progress of the iterates in terms of the barrier function $\phi_w(x, s)$ and we use damped Newton steps. This is the subject of the next section.

10.5 The damped primal-dual Newton method

As before, we are given a positive primal-dual pair (x, s) and a target vector $w > 0$. Let x^+ and s^+ result from a damped Newton step of size α at (x, s). In this section

---

[7] Recall from Lemma C.6 in Section 7.4.1 that we have the better estimate

$$\delta(x^+ s^+; w) \leq \frac{\delta^2}{\sqrt{2(1 - \delta^4)}}$$

if the target w is on the central path. We were not able to get the same result if w is off the central path. We leave this as a topic for further research.

we analyze the effect of a damped Newton step — at (x, s) and for the target w — on the value of the barrier function $\phi_w(x, s)$ (as defined on page 221). We have

$$x^+ = x + \alpha\Delta x, \quad s^+ = s + \alpha\Delta s,$$

where α denotes the step-size, and $0 \leq \alpha \leq 1$. Using the scaled displacements d_x and d_s as defined in (10.6), we may also write

$$x^+ = d\,(v + \alpha d_x), \quad s^+ = d^{-1}\,(v + \alpha d_s),$$

where $v = \sqrt{xs}$. As a consequence,

$$x^+ s^+ = (v + \alpha d_x)\,(v + \alpha d_s) = v^2 + \alpha v\,(d_x + d_s) + \alpha^2 d_x d_s.$$

Since

$$v\,(d_x + d_s) = w - xs = w - v^2,$$

we obtain

$$x^+ s^+ = v^2 + \alpha\,(w - v^2) + \alpha^2 d_x d_s. \tag{10.12}$$

Now, defining

$$v^+ := \sqrt{x^+ s^+},$$

we have

$$\left(v^+\right)^2 = (v + \alpha d_x)\,(v + \alpha d_s) \tag{10.13}$$

and

$$\left(v^+\right)^2 - v^2 = \alpha\,(w - v^2) + \alpha^2 d_x d_s. \tag{10.14}$$

The next theorem provides a lower bound for the decrease of the barrier function value during a damped Newton step. The bound coincides with the result of Lemma II.72 if w is on the central path and becomes worse if the 'distance' from w to the central path increases.

Theorem III.11 *Let* $\delta = \delta(xs, w)$ *and let* $\alpha = 1/\omega - 1/(\omega + 4\delta^2/\delta_c(w))$, *where*[8]

$$\omega := \sqrt{\left\|\frac{\Delta x}{x}\right\|^2 + \left\|\frac{\Delta s}{s}\right\|^2} = \sqrt{\left\|\frac{d_x}{v}\right\|^2 + \left\|\frac{d_s}{v}\right\|^2}.$$

Then the pair (x^+, s^+) *resulting from the damped Newton step of size* α *is feasible. Moreover,*

$$\phi_w(x, s) - \phi_w(x^+, s^+) \geq \psi\left(\frac{2\delta}{\delta_c(w)\rho(\delta)}\right).$$

Proof: It will be convenient to express $\max\,(w)\,\phi_w(x, s)$, given by (9.4), page 221, in terms of $v = \sqrt{xs}$. We obviously have

$$\max\,(w)\,\phi_w(x, s) = e^T v^2 - \sum_{j=1}^{n} w_j \log v_j^2 - e^T w + \sum_{j=1}^{n} w_j \log w_j.$$

---

[8] **Exercise 71** Verify that

$$\frac{\Delta x}{x} = \frac{d_x}{v}, \quad \frac{\Delta s}{s} = \frac{d_s}{v}.$$

Hence we have the following expression for $\max{(w)}\,\phi_w(x^+, s^+)$:

$$\max{(w)}\,\phi_w(x^+, s^+) = e^T\left(v^+\right)^2 - \sum_{j=1}^{n} w_j \log\left(v_j^+\right)^2 - e^T w + \sum_{j=1}^{n} w_j \log w_j.$$

With $\Delta := \phi_w(x, s) - \phi_w(x^+, s^+)$, subtracting both expressions yields

$$\max{(w)}\,\Delta = e^T\left(v^2 - \left(v^+\right)^2\right) + \sum_{j=1}^{n} w_j \log \frac{\left(v_j^+\right)^2}{v_j^2}.$$

Substitution of (10.13) and (10.14) gives

$$\max{(w)}\,\Delta = -\alpha e^T\left(w - v^2\right) + \sum_{j=1}^{n} w_j \log\left(1 + \alpha\frac{(d_x)_j}{v_j}\right) + \sum_{j=1}^{n} w_j \log\left(1 + \alpha\frac{(d_s)_j}{v_j}\right).$$

Here we took advantage of the orthogonality of d_x and d_s in omitting the term containing $e^T d_x d_s$. The definition of ψ implies

$$\log\left(1 + \alpha\frac{(d_x)_j}{v_j}\right) = \alpha\frac{(d_x)_j}{v_j} - \psi\left(\alpha\frac{(d_x)_j}{v_j}\right)$$

and a similar result for the terms containing entries of d_s. Substituting this we obtain

$$\max{(w)}\,\Delta \quad = \quad -\alpha e^T\left(w - v^2\right) + \alpha e^T\left(\frac{wd_x}{v}\right) + \alpha e^T\left(\frac{wd_s}{v}\right)$$

$$- \sum_{j=1}^{n} w_j\left(\psi\left(\alpha\frac{(d_x)_j}{v_j}\right) + \psi\left(\alpha\frac{(d_s)_j}{v_j}\right)\right).$$

The contribution of the terms on the left of the sum can be reduced to $\alpha\,\|d_w\|^2$. This follows because

$$-\left(w - v^2\right) + \frac{w\left(d_x + d_s\right)}{v} = -vd_w + \frac{wd_w}{v} = \frac{\left(w - v^2\right)d_w}{v} = \frac{vd_w^2}{v} = d_w^2.$$

It can easily be understood that the sum attains its maximal value if all the coordinates of the concatenation of the vectors $\alpha d_x/v$ and $\alpha d_x/v$ are zero except one, and the nonzero coordinate, for which w_j must be maximal, is equal to minus the norm of this concatenated vector. The norm of the concatenation of $\alpha d_x/v$ and $\alpha d_x/v$ being αw, we arrive at

$$\max{(w)}\,\Delta \quad \geq \quad \alpha\,\|d_w\|^2 - \max{(w)}\,\psi\left(-\alpha w\right)$$

$$= \quad 4\alpha\delta^2\min{(w)} - \max{(w)}\,\psi\left(-\alpha w\right).$$

This can be rewritten as

$$\Delta \geq \frac{4\alpha\delta^2}{\delta_c(w)} - \psi\left(-\alpha w\right) = \frac{4\alpha\delta^2}{\delta_c(w)} + \alpha w + \log\left(1 - \alpha w\right). \tag{10.15}$$

The derivative of the right-hand side expression with respect to α is

$$\frac{4\delta^2}{\delta_c(w)} + \omega - \frac{\omega}{1 - \alpha\omega},$$

and it vanishes only for the value of α specified in the lemma. As in the proof of Lemma II.72 (page 201) we conclude that the specified value of α maximizes the lower bound for Δ in (10.15), and, as a consequence, the damped Newton step of the specified size is feasible. Substitution in (10.15) yields, after some elementary reductions, the following bound for Δ:

$$\Delta \geq \frac{4\delta^2}{\omega\delta_c(w)} - \log\left(1 + \frac{4\delta^2}{\omega\delta_c(w)}\right) = \psi\left(\frac{4\delta^2}{\omega\delta_c(w)}\right).$$

In this bound we may replace ω by a larger value, since $\psi(t)$ is monotonically increasing for t nonnegative. An upper bound for ω can be obtained as follows:

$$\omega = \sqrt{\left\|\frac{d_x}{v}\right\|^2 + \left\|\frac{d_s}{v}\right\|^2} \leq \frac{\|d_w\|}{\min(v)} = \frac{2\delta\sqrt{\min(w)}}{\min(v)}.$$

Let the index k be such that $\min(v) = v_k$. Then we may write

$$\frac{2\delta\sqrt{\min(w)}}{\min(v)} = \frac{2\delta\sqrt{\min(w)}}{v_k} \leq \frac{2\delta\sqrt{w_k}}{v_k} = 2\delta\sqrt{\frac{w_k}{x_k s_k}} = 2\delta u_k^{-1},$$

where u denotes the vector defined in (10.5). The coordinates of u can be bounded nicely by using the function $\rho(\delta)$ defined in Lemma II.62 (page 182). This can be achieved by reducing $\delta = \delta(xs, w)$, as given in (10.4), in the following way:

$$\delta = \frac{1}{2\sqrt{\min(w)}}\left\|\frac{w - xs}{\sqrt{xs}}\right\| = \frac{1}{2}\left\|\sqrt{\frac{w}{\min(w)}}\left(\sqrt{\frac{w}{xs}} - \sqrt{\frac{xs}{w}}\right)\right\| \geq \frac{1}{2}\left\|u^{-1} - u\right\|.$$

Hence we have $\|u^{-1} - u\| \leq 2\delta$. Applying Lemma II.62 it follows that the coordinates of u and u^{-1} are bounded above by $\rho(\delta)$ (cf. Exercise 70, page 238). Hence we may conclude that

$$\omega \leq 2\delta\rho(\delta).$$

Substitution of this bound in the last lower bound for Δ yields

$$\Delta \geq \psi\left(\frac{2\delta}{\delta_c(w)\rho(\delta)}\right),$$

completing the proof. □

The damped Newton method will be used only if $\delta = \delta(xs, w) \geq 1/\sqrt{2}$, because for smaller values of δ full Newton steps give quadratic convergence to the target. For $\delta = \delta(xs, w) \geq 1/\sqrt{2}$ we have

$$\frac{2\delta}{\rho(\delta)} \geq \frac{\sqrt{2}}{\frac{1}{\sqrt{2}} + \sqrt{1 + \frac{1}{2}}} = \frac{2}{1 + \sqrt{3}} = \sqrt{3} - 1 = 0.73205,$$

so outside the region of quadratic convergence around the target w, a damped Newton step reduces the barrier function value by at least

$$\psi \left(\frac{0.73205}{\delta_c(w)} \right).$$

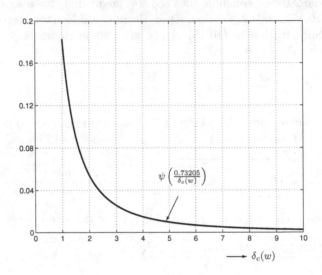

Figure 10.1 Lower bound for the decrease in ϕ_w during a damped Newton step.

The graph in Figure 10.1 depicts this function for $1 \le \delta_c(w) \le 10$.

Remark III.12 The above analysis is based on the barrier function $\phi_w(x, s)$ defined in (9.3). We showed in (9.4) and (9.5) that, up to a constant factor $\max(w)$, the variable part in this function is given by the weighted primal-dual logarithmic barrier function

$$x^T s - \sum_{j=1}^{n} w_j \log x_j s_j.$$

In this function the weights occur in the barrier term.

We want to point out that there exists an alternative way to analyze the damped Newton method by using a barrier function for which the weights occur in the objective term. Consider

$$\bar{\phi}_w(x, s) := e^T \left(\frac{xs}{w} - e \right) - \sum_{j=1}^{n} \log \frac{x_j s_j}{w_j} = \sum_{j=1}^{n} \psi \left(\frac{x_j s_j}{w_j} - 1 \right) = \Psi \left(\frac{xs}{w} - e \right). \qquad (10.16)$$

Clearly $\bar{\phi}_w(x, s)$ is defined for all positive primal-dual pairs (x, s). Moreover, $\bar{\phi}_w(x, s) \ge 0$ and the equality holds if and only if $xs = w$. Hence, the solution of the weighted KKT system (9.2) is characterized by the fact that it satisfies the equation $\bar{\phi}_w(x, s) = 0$. The variable part of $\bar{\phi}_w(x, s)$ is given by

$$e^T \frac{xs}{w} - \sum_{j=1}^{n} \log x_j s_j,$$

which has the weights in the objective term. It has recently been shown by de Klerk, Roos and Terlaky [172] that this function can equally well serve in the analysis of the damped Newton method. In fact, Theorem III.11 remains true if ϕ_w is replaced by $\bar{\phi}_w$. This might be surprising because, whereas ϕ_w is strictly convex on its domain, $\bar{\phi}_w$ is not convex unless w is on the central path.[9]

•

⸦

---

[9] **Exercise 72** Let (x, s) be any positive primal-dual pair. Show that

$$\bar{\phi}_w(x, s) \leq \phi_w(x, s).$$

11

Applications

11.1 Introduction

In this Chapter we present some examples of traceable target sequences. The examples are chosen to cover the most prominent primal-dual methods and results in the literature. We restrict ourselves to sequences that can be traced by full Newton steps.[1]

To keep the presentation simple, we make a further assumption, namely that Newton's method is exact in its region of quadratic convergence. In other words, we assume that the algorithm generates exact primal-dual pairs for the respective targets in the target sequence. In a practical algorithm the generated primal-dual pairs will never exactly match their respective targets. However, our assumption does not change the order of magnitude for the obtained iteration bounds. In fact, at the cost of a little more involved analysis we can obtain the same iteration bounds for a practical algorithm, except for a small constant factor. This can be understood from the following theorem, where we assume that we are given a 'good' approximation for the primal-dual pair $(x(w), s(w))$ corresponding to the target w and we consider the effect of an update of the target to $\bar{w}$. We make clear that $\delta(xs, \bar{w}) \approx \delta(w, \bar{w})$ if $\delta(xs, w)$ is small.

Thus, we assume that the proximity $\delta = \delta(xs, w)$ is small. Recall that the quadratic convergence property of Newton's method justifies this assumption. If $\delta \leq 1/\sqrt{2}$ then in no more than 6 full Newton steps we are sure that a primal-dual pair (x, s) is obtained for which $\delta(xs, w) \leq 10^{-10}$. Thus, if K denotes the length of the target sequence, $6K$ additional Newton steps are sufficient to work with 'exact' primal-dual pairs, at least from a computational point of view.

Theorem III.13 *Let the primal-dual pair (x, s) and the target w be such that $\delta = \delta(xs, w)$. Then, for any other target vector $\bar{w}$ we have*

$$\delta(xs, \bar{w}) \leq \sqrt{\frac{\min(w)}{\min(\bar{w})}}\, \delta + \rho(\delta)\, \delta(w, \bar{w}).$$

---

[1] The motivation for this choice is that full Newton steps give the best iteration bounds. The results in the previous chapter for the damped Newton step provide the ingredients for the analysis of target-following methods using the multistep strategy. Target sequences for multistep methods were treated extensively by Jansen in [151]. See also Jansen et al. [158].

Proof: We may write

$$\delta(xs, \bar{w}) = \frac{1}{2\sqrt{\min(\bar{w})}} \left\| \frac{xs - \bar{w}}{\sqrt{xs}} \right\| = \frac{1}{2\sqrt{\min(\bar{w})}} \left\| \frac{xs - w + w - \bar{w}}{\sqrt{xs}} \right\|.$$

Using the triangle inequality we get

$$\delta(xs, \bar{w}) \leq \frac{1}{2\sqrt{\min(\bar{w})}} \left\| \frac{xs - w}{\sqrt{xs}} \right\| + \frac{1}{2\sqrt{\min(\bar{w})}} \left\| \frac{w - \bar{w}}{\sqrt{xs}} \right\|.$$

This implies

$$\delta(xs, \bar{w}) \leq \sqrt{\frac{\min(w)}{\min(\bar{w})}} \delta(xs, w) + \frac{1}{2\sqrt{\min(\bar{w})}} \left\| \sqrt{\frac{w}{xs}} \frac{w - \bar{w}}{\sqrt{w}} \right\|.$$

From the result of Exercise 70 on page 238, this can be reduced to

$$\delta(xs, \bar{w}) \leq \sqrt{\frac{\min(w)}{\min(\bar{w})}} \delta + \rho(\delta)\delta(w, \bar{w}),$$

completing the proof. □

In the extreme case where $\delta(xs, w) = 0$, we have $xs = w$ and hence $\delta(xs, \bar{w}) = \delta(w, \bar{w})$. In that case the bound in the lemma is sharp, since $\delta = 0$ and $\rho(0) = 1$. If δ is small, then the first term in the bound for $\delta(xs, \bar{w})$ will be small compared to the second term. This follows by noting that the square root can be bounded by

$$\sqrt{\frac{\min(w)}{\min(\bar{w})}} \leq \sqrt{\frac{w_k}{\bar{w}_k}} \leq \rho(\delta(w, \bar{w})). \tag{11.1}$$

Here the index k is such that $\min(\bar{w}) = \bar{w}_k$.[2] Since $\rho(\delta) \approx 1$ if $\delta \approx 0$, we conclude that $\delta(xs, \bar{w}) \approx \delta(w, \bar{w})$ if δ is small.

11.2 Central-path-following method

Central-path-following methods were investigated extensively in Part II. The aim of this section is twofold. It provides a first (and easy) illustration of the use of the target-following approach, and it yields one of the main results of Part II in a relatively cheap way.

The target points have the form $w = \mu e, \mu > 0$. When at the target w, we let the next target point be given by

$$\bar{w} = (1 - \theta)w, \ 0 < \theta < 1.$$

---

[2] When combining the bounds in Theorem III.13 and (11.1) one gets the bound

$$\delta(xs, \bar{w}) \leq \rho(\delta(w, \bar{w})) \delta(xs, w) + \rho(\delta(xs, w)) \delta(w, \bar{w}),$$

which has a nice symmetry, but which is weaker than the bound of Theorem III.13.

Then some straightforward calculations yield $\delta(w, \bar{w})$:

$$\delta(w, \bar{w}) = \frac{1}{2\sqrt{\min(\bar{w})}} \left\| \frac{\bar{w} - w}{\sqrt{w}} \right\| = \frac{\theta \|\sqrt{\mu e}\|}{2\sqrt{(1 - \theta)\mu}} = \frac{\theta\sqrt{n}}{2\sqrt{1 - \theta}}.$$

Assuming that $n \geq 4$ we find that

$$\delta(w, \bar{w}) \leq \frac{1}{\sqrt{2}} \quad \text{if} \quad \theta = \frac{1}{\sqrt{n}}.$$

Hence, by Lemma I.36, a full Newton step method needs

$$\sqrt{n} \, \log \frac{n\mu^0}{\varepsilon}$$

iterations[3] to generate an ε-solution when starting at $w^0 = \mu^0 e$.

11.3 Weighted-path-following method

With a little extra effort, we can also analyze the case where the target sequence lies on the half line $w = \mu w^0$, $\mu > 0$, for some fixed positive vector w^0. This half line is a so-called *weighted path* in the w-space. The primal-dual pairs on it converge to weighted-analytic centers of the optimal sets of (P) and (D), due to Theorem III.5. Note that when using a target sequence of this type we can start the algorithm everywhere in the w-space. However, as we shall see, not using the central path diminishes the efficiency of the algorithm.

Letting the next target point be given by

$$\bar{w} = (1 - \theta)w, \, 0 < \theta < 1, \tag{11.2}$$

we obtain

$$\delta(w, \bar{w}) = \frac{1}{2\sqrt{\min(\bar{w})}} \left\| \frac{\bar{w} - w}{\sqrt{w}} \right\| = \frac{\|\theta\sqrt{w}\|}{2\sqrt{(1 - \theta)\min(w)}} = \frac{\theta}{2\sqrt{1 - \theta}} \left\| \sqrt{\frac{w}{\min(w)}} \right\|.$$

Using $\delta_c(w)$, as defined in (9.6), page 222, which measures the proximity of w to the central path, we may write

$$\left\| \sqrt{\frac{w}{\min(w)}} \right\| \leq \|e\| \sqrt{\frac{\max(w)}{\min(w)}} = \sqrt{n}\delta_c(w).$$

Thus we obtain

$$\delta(w, \bar{w}) \leq \frac{\theta\sqrt{n}\delta_c(w)}{2\sqrt{1 - \theta}}.$$

---

[3] Formally, we should round the iteration bound to the smallest integer exceeding it. For simplicity we omit the corresponding rounding operator in the iteration bounds in this chapter; this is common practice in the literature.

Assuming $n \geq 4$ again, we find that

$$\delta(w, \bar{w}) \leq \frac{1}{\sqrt{2}} \quad \text{if} \quad \theta = \frac{1}{\sqrt{n \delta_c(w)}}.$$

Hence, when starting at w^0, we are sure that the duality gap is smaller than ε after at most

$$\sqrt{n \delta_c(w^0)} \log \frac{e^T w^0}{\varepsilon} \tag{11.3}$$

iterations. Here we used the obvious identity $\delta_c(w^0) = \delta_c(w)$. Comparing this result with the iteration bound of the previous section we observe that we introduce a factor $\sqrt{\delta_c(w^0)} > 1$ into the iteration bound by not using the central path.[4]

The above result indicates that in some sense the central path is the best path to follow to the optimal set. When starting further from the central path the iteration bound becomes worse. This result gives us evidence of the very special status of the central path among all possible weighted paths to the optimal set.

11.4 Centering method

If we are given a primal-dual pair (x^0, s^0) such that $w^0 = x^0 s^0$ is not on the central path, then instead of following the weighted path through w^0 to the origin, we can use an alternative strategy. The idea is first to move to the central path and then follow the central path to the origin. We know already how to follow the central path. But the other problem, moving from some point w^0 in the w-space to the central path, is new. This problem has become known as the *centering problem*.[5,6]

The centering problem can be solved by using a target sequence starting at w^0 and ending on the central path. We shall propose a target sequence that converges in

$$\sqrt{n} \log \delta_c(w^0) \tag{11.4}$$

iterations.[7] The iteration bound (11.4) can be obtained as follows. Let $\bar{w}$ be obtained from some point w outside the central path by replacing each entry w_i such that

$$w_i < (1 + \theta) \min(w)$$

---

[4] Primal-dual weighted-path-following methods were first proposed and discussed by Megiddo [200]. Later they were also analyzed by Ding and Li [67]. A primal version was studied by Roos and den Hertog [241].

[5] The centering approach presented here was proposed independently by den Hertog [140] and Mizuno [212].

[6] **Exercise 73** The centering problem includes the problem of finding the analytic center of a polytope. Why?

[7] Note that the quantity $\delta_c(w^0)$ appears under a logarithm. This is very important from the viewpoint of complexity analysis. If the weights were initially determined from a primal-dual feasible pair (x^0, s^0), we can say that $\delta_c(w^0)$ has the same input length as the two points. It is reasonable to assume that this input length is at most equivalent to the input length of the data of the problem, but there is no real reason to state that it is strictly smaller. Since an algorithm is claimed to be polynomial only when the bound on the number of iterations is a function of the logarithm of the length of the input data, it is better to have the quantity $\delta_c(w^0)$ under the logarithm.

by $(1 + \theta) \min(w)$, where θ is some positive constant satisfying $1 + \theta \leq \delta_c(w)$. It then follows that

$$\delta_c(\bar{w}) = \frac{\delta_c(w)}{1 + \theta}.$$

Using that $0 \leq \bar{w}_i - w_i \leq \theta \min(w)$ we write

$$\delta(w, \bar{w}) = \frac{1}{2\sqrt{\min(\bar{w})}} \left\| \frac{\bar{w} - w}{\sqrt{w}} \right\| \leq \frac{1}{2\sqrt{(1+\theta)\min(w)}} \left\| \frac{\theta \min(w) e}{\sqrt{w}} \right\|.$$

This implies

$$\delta(w, \bar{w}) \leq \frac{\theta}{2\sqrt{(1+\theta)}} \left\| \frac{\sqrt{\min(w)} e}{\sqrt{w}} \right\| \leq \frac{\theta}{2\sqrt{(1+\theta)}} \|e\| = \frac{\theta\sqrt{n}}{2\sqrt{1+\theta}} \leq \frac{\theta\sqrt{n}}{2},$$

so we have

$$\delta(w, \bar{w}) \leq \frac{1}{\sqrt{2}} \quad \text{if} \quad \theta = \frac{\sqrt{2}}{\sqrt{n}}.$$

At each iteration, $\delta_c(w)$ decreases by the factor $1 + \theta$. Thus, when starting at w^0, we certainly have reached the central path if the iteration number k satisfies

$$(1 + \theta)^k \geq \delta_c(w^0).$$

Substituting the value of θ and then taking logarithms, we obtain

$$k \log \left(1 + \frac{\sqrt{2}}{\sqrt{n}} \right) \geq \log \delta_c(w^0).$$

If $n \geq 3$, this inequality is satisfied if[8]

$$\frac{k}{\sqrt{n}} \geq \log \delta_c(w^0).$$

Thus we find that no more than

$$\sqrt{n} \log \delta_c(w^0) \tag{11.5}$$

iterations bring the iterate onto the central path. This proves the iteration bound (11.4) for the centering problem.

The above-described target sequence ends at the point $\max(w) e$ on the central path. From there on we can follow the central path as described in Section 11.2 and we reach an ε-solution after a total of

$$\sqrt{n} \left(\log \delta_c(w^0) + \log \frac{n \max(w^0)}{\varepsilon} \right) \tag{11.6}$$

_____

[8] If $n \geq 3$ then we have

$$\log \left(1 + \frac{\sqrt{2}}{\sqrt{n}} \right) \geq \frac{1}{\sqrt{n}}.$$

iterations.

Note that this bound for a strategy that first centralizes and then optimizes is better than the one we obtained for the more direct strategy (11.2) of following a sequence along the weighted path. In fact the bound (11.6) is the best one known until now when the starting point lies away from the central path.

Remark III.14 The above centering strategy pushes the small coordinates of w^0 upward to max (w^0). We can also consider the more obvious strategy of moving the large coordinates of w^0 downward to min (w^0). Following a similar analysis we obtain

$$\delta(w, \bar{w}) \leq \frac{\theta\sqrt{n\delta_c(w^0)}}{2}.$$

Hence,

$$\delta(w, \bar{w}) \leq \frac{1}{\sqrt{2}} \quad \text{if} \quad \theta = \frac{\sqrt{2}}{\sqrt{n\delta_c(w^0)}}.$$

As a consequence, in the resulting iteration bound, which is proportional to $1/\theta$, the quantity $\delta_c(w^0)$ does not appear under the logarithm. This makes clear that we get a slightly worse result than (11.5) in this case.[9] •

11.5 Weighted-centering method

The converse of the centering problem consists in finding a primal-dual pair (x, s) such that the ratios between the coordinates of xs are prescribed, when a point on the central path is given. If w^1 is a positive vector whose coordinates have the prescribed weights, then we want to find feasible x and s such that $xs = \lambda w^1$ for some positive λ. In fact, the aim is not to solve this problem exactly; it is enough if we find a primal-dual pair such that $\delta(xs, \lambda w^1) \leq 1/\sqrt{2}$ for some positive λ. This problem is known as the *weighted-centering problem*.[10]

Let the primal-dual pair be given for the point $w^0 = \mu e$ on the central path, with $\mu > 0$. We first rescale the given vector w^1 by a positive scalar factor in such a way

---

[9] **Exercise 74** Another strategy for reaching the central path from a given vector w^0 can be defined as follows. When at w, we define $\bar{w}$ according to

$$\bar{w}_i = \begin{cases} (1+\theta)\min(w), & \text{if } w_i < (1+\theta)\min(w), \\ \max(w) - \theta\min(w), & \text{if } w_i > \max(w) - \theta\min(w), \\ w_i, & \text{otherwise.} \end{cases}$$

Analyze this strategy and show that the iteration bound is the same as (11.5), but when the central path is reached the duality gap is (in general) smaller, yielding a slight improvement of (11.6).

[10] The treatment of the weighted-centering problem presented here was first proposed by Mizuno [214]. It closely resembles our approach to the centering problem. See also Jansen et al. [159, 158] and Jansen [151]. A special case of the weighted-centering problem was considered by Atkinson and Vaidya [29] and later also by Freund [85] and Goffin and Vial [102]. Their objective was to find the weighted-analytic center of a polytope. Our approach generates the weighted-analytic center of the primal polytope $\mathcal{P}$ if we take $c = 0$, and the weighted-analytic center of the dual polytope $\mathcal{D}$ if we take $b = 0$. The approach of Atkinson and Vaidya was put into the target-following framework by Jansen et al. [158]. See also Jansen [151]. The last two references use two nested traceable target sequences. The result is a significantly simpler analysis as well as a better iteration bound than Atkinson and Vaidya's bound.

that
$$\max\left(w^1\right) = \mu,$$
and we construct a traceable target sequence from w^0 to w^1. When we put $w := w^0$, the coordinates of w corresponding to the largest coordinates of w^1 have their correct value. We gradually decrease the other coordinates of w to their correct value by using the same technique as in the previous section. Let $\bar{w}$ be obtained from w by redefining each entry w_i according to
$$w_i := \max\left(w_i^1, (1-\theta)w_i\right),$$
where θ is some positive constant smaller than one. Note that w_i can never become smaller than w_i^1 and if it has reached this value then it remains constant in subsequent target vectors. Hence, this process leaves the 'correct' coordinates of w — those have the larger values — invariant, and it decreases the other coordinates by a factor $1-\theta$, or less if undershooting should occur. Thus, we have
$$\min\left(\bar{w}\right) \geq (1-\theta)\min\left(w\right),$$
with equality, except possibly for the last point in the target sequence, and
$$0 \leq w_i - \bar{w}_i \leq \theta\min\left(w\right).$$
To make the sequence traceable, θ cannot be taken too large. Using the last two inequalities we write
$$\delta(w,\bar{w}) = \frac{1}{2\sqrt{\min\left(\bar{w}\right)}}\left\|\frac{\bar{w}-w}{\sqrt{w}}\right\| \leq \frac{1}{2\sqrt{(1-\theta)\min\left(w\right)}}\left\|\frac{\theta\min\left(w\right)e}{\sqrt{w}}\right\|.$$
This gives
$$\delta(w,\bar{w}) \leq \frac{\theta}{2\sqrt{(1-\theta)}}\left\|\frac{\sqrt{\min\left(w\right)}e}{\sqrt{w}}\right\| \leq \frac{\theta}{2\sqrt{(1-\theta)}}\|e\| = \frac{\theta\sqrt{n}}{2\sqrt{1-\theta}}.$$
As before, assuming $n \geq 4$ we get
$$\delta(w,\bar{w}) \leq \frac{1}{\sqrt{2}} \quad \text{if} \quad \theta = \frac{1}{\sqrt{n}}.$$
Before the final iteration, which puts all entries of w at their correct values, each iteration increases $\delta_c(w)$ by the factor $1/(1-\theta)$. We certainly have reached w^1 if the iteration number k satisfies
$$\frac{1}{(1-\theta)^k} \geq \delta_c(w^1).$$
Taking logarithms, this inequality becomes
$$-k\log\left(1-\theta\right) \geq \log\delta_c(w^1)$$
and this certainly holds if
$$k\theta \geq \log\delta_c(w^1),$$
since $\theta \leq -\log\left(1-\theta\right)$. Substitution of $\theta = 1/\sqrt{n}$ yields that no more than
$$\sqrt{n}\,\log\delta_c(w^1)$$
iterations bring the iterate to w^1.

11.6 Centering and optimizing together

In Section 11.4 we discussed a two-phase strategy for the case where the initial primal-dual feasible pair (x^0, s^0) is not on the central path. The first phase is devoted to centralizing and the second phase to optimizing. Although this strategy achieves the best possible iteration bound obtained so far, it is worth considering an alternative strategy that combines the two phases at the same time.

Let $w^0 := x^0 s^0$ and consider the function $f : \mathbb{R}_+ \to \mathbb{R}_+^n$ defined by

$$f(\theta) := \frac{w^0}{e + \theta w^0}, \; \theta \geq 0. \tag{11.7}$$

The image of f defines a path in the w-space starting at $f(0) = w^0$ and converging to the origin when θ goes to infinity. See Figure 11.1.

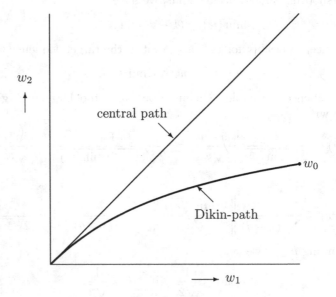

Figure 11.1   A Dikin-path in the w-space $(n = 2)$.

We refer to this path as the *Dikin-path* in the w-space starting at w^0.[11] It may easily be checked that if w^1 lies on the Dikin-path starting at w^0, then the Dikin-path

---

[11] Dikin, well known for his primal affine-scaling method for LO, did not consider primal-dual methods. Nevertheless, the discovery of this path in the w-space has been inspired by his work. Therefore, we gave his name to it. The relation with Dikin's work is as follows. The direction of the tangent to the Dikin-path is obtained by differentiating $f(\theta)$ with respect to θ. This yields

$$\frac{df(\theta)}{d\theta} = \frac{-(w^0)^2}{(e + \theta w^0)^2} = -f(\theta)^2.$$

This implies that the Dikin-path is a trajectory of the vector field $-w^2$ in the w-space. Without going further into it we refer the reader to Jansen, Roos and Terlaky [156] where this field was used to obtain the primal-dual analogue of the so-called primal affine-scaling direction of Dikin [63]. This is precisely the direction used in the Dikin Step Algorithm, in Appendix E.

starting at w^1 is just the continuation of the path starting at w^0.[12] Asymptotically, the Dikin-path becomes tangent to the central path, because for very large values of θ we have

$$f(\theta) \approx \frac{e}{\theta}.$$

We can easily establish that along the path the proximity to the central path is improving. This goes as follows. Let $w := f(\theta)$. Then, using that f preserves the ordering of the coordinates,[13] we may write

$$\delta_c(w) = \frac{\frac{\max(w^0)}{1+\theta\max(w^0)}}{\frac{\min(w^0)}{1+\theta\min(w^0)}} = \delta_c(w^0)\frac{1+\theta\min(w^0)}{1+\theta\max(w^0)} \leq \delta_c(w^0). \tag{11.8}$$

The last inequality is strict if $\delta_c(w^0) > 1$. Also, the duality gap is decreasing. This follows because

$$e^T w := e^T \frac{w^0}{e + \theta w^0} \leq \frac{e^T w^0}{1 + \theta\min(w^0)} < e^T w^0.$$

Consequently, the Dikin-path achieves the two goals that were assigned to it. It centralizes and optimizes at the same time.

Let us now try to devise a traceable target sequence along the Dikin-path. Suppose that w is a point on this path. Without loss of generality we may assume that $w = f(0) = w^0$. Let $\bar{w} := f(\theta)$ for some positive θ. Then we have

$$\delta(w, \bar{w}) = \frac{1}{2\sqrt{\min(\bar{w})}}\left\|\frac{\bar{w} - w}{\sqrt{w}}\right\| = \frac{1}{2\sqrt{\min(\bar{w})}}\left\|\frac{\frac{w}{e+\theta w} - w}{\sqrt{w}}\right\|,$$

which can be simplified to

$$\delta(w, \bar{w}) = \frac{1}{2\sqrt{\min(\bar{w})}}\left\|\frac{\theta w^{\frac{3}{2}}}{e + \theta w}\right\|.$$

Using that f preserves the ordering of the coordinates we further deduce

$$\delta(w, \bar{w}) = \frac{\sqrt{1 + \theta\min(w)}}{2\sqrt{\min(w)}}\left\|\frac{\theta w^{\frac{3}{2}}}{e + \theta w}\right\| \leq \frac{\sqrt{\max(w)}}{2\sqrt{\min(w)}}\left\|\frac{\theta w}{\sqrt{e + \theta w}}\right\|,$$

which gives

$$\delta(w, \bar{w}) \leq \frac{\sqrt{\delta_c(w)}}{2}\left\|\frac{\theta w}{\sqrt{e + \theta w}}\right\|.$$

Finally, since $e + \theta w > e$, we get

$$\delta(w, \bar{w}) \leq \frac{1}{2}\theta\sqrt{\delta_c(w)}\,\|w\|.$$

---

[12] **Exercise 75** Show that if w^1 lies on the Dikin-path starting at w^0, then the Dikin-path starting at w^1 is just the continuation of the path starting at w^0.

[13] **Exercise 76** Let $w_1^0 \leq w_2^0 \leq \ldots \leq w_n^0$ and $w := f(\theta)$, with $f(\theta)$ as defined in (11.7). Prove that for each positive θ we have $w_1 \leq w_2 \leq \ldots \leq w_n$.

So we have

$$\delta(w, \bar{w}) \le \frac{1}{\sqrt{2}} \quad \text{if} \quad \theta = \frac{\sqrt{2}}{\|w\| \sqrt{\delta_c(w)}}.$$

We established above that the duality gap is reduced by at least the factor $1 + \theta \min(w)$. Replacing θ by its value defined above, we have

$$1 + \theta \min(w) = 1 + \frac{\sqrt{2} \min(w)}{\|w\| \sqrt{\delta_c(w)}} \ge 1 + \frac{\sqrt{2} \min(w)}{\max(w) \sqrt{n \delta_c(w)}} = 1 + \frac{\sqrt{2}}{\delta_c(w) \sqrt{n \delta_c(w)}}.$$

Using $\delta_c(w) < \delta_c(w^0)$, we deduce in the usual way that after

$$\delta_c(w^0) \sqrt{n \delta_c(w^0)} \log \frac{e^T w^0}{\varepsilon} \tag{11.9}$$

iterations the duality gap is smaller than ε.

For large values of $\delta_c(w^0)$ this bound is significantly worse than the bounds obtained in the previous sections when starting off the central path. It is even worse — by a factor $\delta_c(w^0)$ — than the bound for the weighted-path-following method in Section 11.3. The reason for this weak result is that in the final step, just before (11.9), we replaced $\delta_c(w)$ by $\delta_c(w^0)$. Thus we did not fully explore the centralizing effect of the Dikin-path, which implies that in the final iterations $\delta_c(w)$ tends to 1.

To improve the bound we shall look at the process in a different way. Instead of directly estimating the number of target moves until a suitable duality gap is achieved, we shall concentrate on the number of steps that are required to get close to the central path, a state that can be measured for instance by $\delta_c(w) < 2$.

Using (11.8) and substituting the value of θ, we obtained

$$\delta_c(\bar{w}) = \delta_c(w) \frac{1 + \theta \min(w)}{1 + \theta \max(w)} = \delta_c(w) \frac{\|w\| \sqrt{\delta_c(w)} + \min(w) \sqrt{2}}{\|w\| \sqrt{\delta_c(w)} + \max(w) \sqrt{2}}.$$

This can be written as

$$\delta_c(\bar{w}) = \delta_c(w) \left(1 - \frac{\sqrt{2}(\max(w) - \min(w))}{\|w\| \sqrt{\delta_c(w)} + \max(w) \sqrt{2}} \right).$$

Using that $\|w\| \le \max(w) \sqrt{n}$ and $\max(w) = \delta_c(w) \min(w)$ we obtain

$$\delta_c(\bar{w}) \le \delta_c(w) \left(1 - \frac{\sqrt{2}(\delta_c(w) - 1)}{\delta_c(w) \left(\sqrt{n \delta_c(w)} + \sqrt{2} \right)} \right).$$

Now assuming $n \ge 6$ and $\delta_c(w) \ge 2$ we get

$$\frac{\sqrt{2}(\delta_c(w) - 1)}{\delta_c(w) \left(\sqrt{n \delta_c(w)} + \sqrt{2} \right)} \ge \frac{1}{2 \sqrt{n \delta_c(w)}}.$$

This can be verified by elementary means. As a consequence, under these assumptions,

$$\delta_c(\bar{w}) \le \delta_c(w) \left(1 - \frac{1}{2 \sqrt{n \delta_c(w)}} \right).$$

Hence, using that $\delta_c(w) \le \delta_c(w^0)$, after k iterations we have

$$\delta_c(\bar{w}) \le \left(1 - \frac{1}{2\sqrt{n\delta_c(w^0)}}\right)^k \delta_c(w^0).$$

By the usual arguments, it follows that $\delta_c(\bar{w}) \le 2$ after at most

$$2\sqrt{n\delta_c(w^0)} \log \frac{\delta_c(w^0)}{2}$$

iterations. The proximity to the central path is then at most 2. Now from (11.9) it follows that the number of iterations needed to reach an ε-solution does not exceed

$$2\sqrt{2n} \log \frac{e^T w^0}{\varepsilon}.$$

By adding the two numbers, we obtain the iteration bound

$$\sqrt{n} \left(2\sqrt{2} \log \frac{e^T w^0}{\varepsilon} + 2\sqrt{\delta_c(w^0)} \log \frac{\delta_c(w^0)}{2}\right).$$

Note that this bound is better than the previous bound (11.9) and also better than the bound (11.3) for following the weighted central path. But it is still worse than the bound (11.6) for the two-phase strategy.

11.7 Adaptive and large target-update methods

The complexity bounds derived in the previous sections are based on a worst-case analysis of full Newton step methods. Each target step is chosen to be short enough so that, in any possible instance, proximity will remain under control. Moreover, the target step is not at all influenced by the particular primal-dual feasible pair. As a consequence, for an implementation of a full-step target-following method the required running time may give rise to some disappointment.

It then becomes tempting to take larger target-updates. An obvious improvement would be to relate the target move to the primal-dual feasible pair and to make the move as large as possible while keeping proximity to the primal-dual feasible pair under control; in that case a full Newton step still yields a new primal-dual feasible pair closer to the target and the process may be repeated. This enhancement of the full-step strategy into the so-called *adaptive step* or *maximal step* strategy does not improve the overall theoretical complexity bound, but it has a dramatic effect on the efficiency, especially on the asymptotic convergence rate.[14]

Despite this nice asymptotic result, the steps in the adaptive-step method may in general be too short to produce a really efficient method. In practical applications it is often wise to work with larger target-updates. One obvious shortcoming of a large

---

[14] In a recent paper [125], Gonzaga showed that the maximal step method — with some additional safeguard steps — is asymptotically quadratically convergent; i.e., in the final iterations the duality gap converges to zero quadratically. Gonzaga also showed that the iterates converge to the analytic centers of the optimal sets of (P) and (D).

target-update is that the full Newton step may cause infeasibility. To overcome this difficulty one must use a damped Newton step. The progress is then measured by the primal-dual barrier logarithmic function $\phi_w(x, s)$ analyzed in Section 10.5. Using the results of that section, iteration bounds for the damped Newton method can be derived for large-update versions of the target sequences dealt with in this chapter. In accordance with the results in Chapter 7 for the logarithmic barrier central-path-following method, the iteration bounds are always a factor $\sqrt{n}$ worse than those for the full-step methods. We feel that it goes beyond the aim of this chapter to give a detailed report of the results obtained in this direction. We refer the reader to the references mentioned in the course of this chapter.[15]

---

[15] In this connection it may be useful to mention again the book of Jansen [151], which contains a thorough treatment of the target-following approach. Jansen also deals with methods using large target-updates. He provides some additional examples of traceable target sequences that can be used to simplify drastically the analysis of existing methods, such as the cone-affine-scaling method of Sturm and Zhang [260] and the shifted barrier method of Freund [84]. These results can also be found in Jansen et al. [158].

12

The Dual Newton Method

12.1 Introduction

The results in the previous sections have made clear that the image of a given target vector $w > 0$ under the target map $\Phi_{PD}(w)$ can be computed provided that we are given some positive primal-dual pair (x, s). If the given pair (x, s) is such that xs is close to w, Newton's method can be applied to the weighted KKT system (9.2). Starting at (x, s) this method generates a sequence of primal-dual pairs converging to $\Phi_{PD}(w)$. The distance from the pair (x, s) to w is measured by the proximity measure $\delta(xs, w)$ in (10.4):

$$\delta(xs, w) := \frac{1}{2\sqrt{\min(w)}} \left\| \frac{w - xs}{\sqrt{xs}} \right\|.$$

If $\delta(xs, w) \leq 1/\sqrt{2}$ then the primal-dual method converges quadratically to $\Phi_{PD}(w)$. For larger values of $\delta(xs, w)$ we could realize a linear convergence rate by using damped Newton steps of appropriate size. The sketched approach is called primal-dual because it uses search steps in both the x-space and the s-space at each iteration.

The aim of this chapter and the next is to show that the same goal can be realized by moving only in the primal space or the dual space. Assuming that we are given a positive primal feasible solution x, a *primal method* moves in the primal space until it reaches $x(w)$. Similarly, a *dual method* starts at some given dual feasible solution (y, s) with $s > 0$, and moves in the dual space until it reaches $(y(w), s(w))$. We deal with dual methods in the next sections, and consider primal methods in the next chapter. In both cases the search direction is obtained by applying Newton's method to a suitable weighted logarithmic barrier function. The general framework of a dual target-following algorithm is described on page 260. The underlying target sequence starts at w^0 and ends at $\tilde{w}$.

12.2 The weighted dual barrier function

The search direction in a dual method is obtained by applying Newton's method to the *weighted dual logarithmic barrier function* $\phi_w^d(y)$, given by

$$\phi_w^d(y) := \frac{-1}{\min(w)} \left(b^T y + \sum_{i=1}^n w_i \log s_i \right), \tag{12.1}$$

Generic Dual Target-following Algorithm

Input:
A dual feasible pair (y^0, s^0) such that $y^0 = y(w^0)$; $s^0 = s(w^0)$;
a final target vector $\tilde{w}$.

begin
$y := y^0; s = s^0; w := w^0$;
while w is not 'close' to $\tilde{w}$ **do**
begin
    replace w by the next target in the sequence;
    while (y, s) is not 'close' to $(y(w), s(w))$ **do**
    begin
        apply Newton steps at (y, s) to the target w
    end
end
end

---

with $s = c - A^T y$. In this section we prove that $\phi_w^d(y)$ attains its minimal value at $y(w)$. In the next section it turns out that $\phi_w^d(y)$ is strictly convex. The first property can easily be derived from the primal-dual logarithmic barrier function ϕ_w used in Section 10.5. With w fixed, we consider ϕ_w at the pair $(x(w), s)$. Starting from (9.4), page 221, and using $x(w)^T s = c^T x(w) - b^T y$ and $x(w)s(w) = w$ we write

$$
\begin{aligned}
\max{(w)}\, \phi_w(x(w), s) \;&=\; x(w)^T s - \sum_{j=1}^n w_j \log x_j(w) s_j - e^T w + \sum_{j=1}^n w_j \log w_j \\
&=\; x(w)^T s - \sum_{j=1}^n w_j \log s_j - e^T w + \sum_{j=1}^n w_j \log s_j(w) \\
&=\; c^T x(w) - b^T y - \sum_{j=1}^n w_j \log s_j - e^T w + \sum_{j=1}^n w_j \log s_j(w) \\
&=\; \min{(w)}\, \phi_w^d(y) + c^T x(w) - e^T w + \sum_{j=1}^n w_j \log s_j(w).
\end{aligned}
$$

Since w is fixed, this shows that $\min{(w)}\, \phi_w^d(y)$ and $\max{(w)}\phi_w(x(w), s)$ differ by a constant. Since $\phi_w(x(w), s)$ attains its minimal value at $s(w)$, it follows that $\phi_w^d(y)$ must attain its minimal value at $y(w)$.[1]

---

[1] **Exercise 77** For each positive primal-dual pair (x, s), prove that

$$
\phi_w(x, s) = \phi_w(x, s(w)) + \phi_w(x(w), s).
$$

12.3   Definition of the dual Newton step

Let y be dual feasible and $w > 0$. We denote the gradient of $\phi_w^d(y)$ at y by $g_w^d(y)$ and the Hessian by $H_w^d(y)$. These are

$$g_w^d(y) := \frac{-1}{\min(w)} \left(b - AWs^{-1} \right)$$

and

$$H_w^d(y) := \frac{1}{\min(w)} AWS^{-2}A^T,$$

as can be easily verified. Note that $H_w^d(y)$ is positive definite. It follows that $\phi_w^d(y)$ is a strictly convex function.

The Newton step at y is given by

$$\Delta y = -H_w^d(y)^{-1} g_w^d(y) = \left(AWS^{-2}A^T \right)^{-1} \left(b - AWs^{-1} \right). \qquad (12.2)$$

Since $y(w)$ is the minimizer of $\phi_w^d(y)$ we have $\Delta y = 0$ if and only if $y = y(w)$. We measure the proximity of y with respect to $y(w)$ by a suitable norm of Δy, namely the norm induced by the positive definite matrix $H_w^d(y)$:

$$\delta^d(y, w) := \|\Delta y\|_{H_w^d(y)}.$$

We call this the *Hessian norm* of Δy. We show below that it is an appropriate generalization of the proximity measure used in Section 6.5 (page 114) for the analysis of the dual logarithmic barrier approach. More precisely, we find that both measures coincide if w is on the central path.

Using the definition of the Hessian norm of $\Delta y = -H_w^d(y)^{-1} g_w^d(y)$ we may write

$$\delta^d(y, w) = \sqrt{\Delta y^T H_w^d(y) \Delta y} = \sqrt{g_w^d(y)^T H_w^d(y)^{-1} g_w^d(y)}. \qquad (12.3)$$

Remark III.15 The dual proximity measure $\delta^d(y, w)$ can be characterized in a different way as follows:

$$\delta^d(y, w) = \frac{1}{\sqrt{\min(w)}} \min_x \left\{ \left\| d^{-1} \left(x - \frac{w}{s} \right) \right\| : Ax = b \right\},$$

where

$$d := \frac{\sqrt{w}}{s}. \qquad (12.4)$$

We want to explain this here, because later on, for the primal method this characterization provides a natural way of defining a primal proximity measure.

Let x satisfy $Ax = b$. We do not require x to be nonnegative. Replacing b by Ax in the above expression (12.2) for Δy and using d from (12.4), we obtain

$$\Delta y = \left(AD^2A^T \right)^{-1} \left(Ax - AWs^{-1} \right).$$

This can be rewritten as

$$\Delta y = \left(AD^2A^T \right)^{-1} ADd^{-1} \left(x - ws^{-1} \right) = \left(AD^2A^T \right)^{-1} AD \frac{sx - w}{\sqrt{w}}.$$

The corresponding displacement in the slack space is given by $\Delta s = -A^T \Delta y$. This implies

$$d\Delta s = -(AD)^T \left(AD^2 A^T\right)^{-1} AD \frac{sx - w}{\sqrt{w}}.$$

This makes clear that $-d\Delta s$ is equal to the orthogonal projection of the vector $(sx - w)/\sqrt{w}$ into the row space of AD. Hence, we have

$$d\Delta s = -\frac{sx(s,w) - w}{\sqrt{w}},$$

where

$$x(s,w) = \text{argmin}_x \left\{ \left\| \frac{sx - w}{\sqrt{w}} \right\| : Ax = b \right\}.$$

Lemma III.16 below implies

$$\delta^d(y,w) = \frac{1}{\sqrt{\min(w)}} \|d\Delta s\|.$$

The claim follows.   •

12.4   Feasibility of the dual Newton step

Let y^+ result from the Newton step at y:

$$y^+ := y + \Delta y.$$

If we define

$$\Delta s := -A^T \Delta y,$$

the slack vector for y^+ is just $s + \Delta s$, as easily follows. The Newton step is feasible if and only if $s + \Delta s \geq 0$. It is convenient to introduce the vector v according to

$$v := \sqrt{\frac{w}{\min(w)}}. \tag{12.5}$$

Note that $v \geq e$ and $v = e$ if and only if w is on the central path. Now we can prove the next lemma. From this lemma it becomes clear that $\delta^d(y,w)$ coincides with the proximity measure $\delta(y,\mu)$, defined in (6.7), page 114, if $w = \mu e$.

Lemma III.16

$$\delta^d(y,w) = \left\| \frac{v\Delta s}{s} \right\| \geq \left\| \frac{\Delta s}{s} \right\| \geq \left\| \frac{\Delta s}{s} \right\|_\infty.$$

If $\delta^d(y,w) \leq 1$ then $y^* = y + \Delta y$ is dual feasible.

Proof: Using (12.3) and the above expression for $H_w^d(y)$, we write

$$\delta^d(y,w)^2 = \Delta y^T H_w^d(y) \Delta y = \frac{1}{\min(w)} \Delta y^T A W S^{-2} A^T \Delta y.$$

Replacing $A^T \Delta y$ by $-\Delta s$ and also using the definition (12.5) of v, we get

$$\delta^d(y,w)^2 = \Delta s^T V^2 S^{-2} \Delta s = \left\| \frac{v\Delta s}{s} \right\|^2.$$

Thus we obtain

$$\delta^d(y,w) = \left\| \frac{v\Delta s}{s} \right\| \geq \left\| \frac{\Delta s}{s} \right\| \geq \left\| \frac{\Delta s}{s} \right\|_\infty.$$

The first inequality follows because $v \geq e$, and the second inequality is trivial. This proves the first part of the lemma. For the second part, assume $\delta^d(y,w) \leq 1$. Then we derive from the last inequality in the first part of the lemma that $|\Delta s| \leq s$, which implies $s + \Delta s \geq 0$. The lemma is proved. $\qquad\square$

12.5 Quadratic convergence

The aim of this section is to generalize the quadratic convergence result of the dual Newton method in Theorem II.21, page 114, to the present case.[2]

Theorem III.17 $\delta^d(y^+, w) \leq \delta^d(y, w)^2.$

Proof: By definition

$$\delta^d(y^+, w)^2 = g_w^d(y^+)^T H_w^d(y^+)^{-1} g_w^d(y^+).$$

The main part of the proof consists of the calculation of $H_w^d(y^+)$ and $g_w^d(y^+)$.
It is convenient to work with the matrix

$$B := AV (S + \Delta S)^{-1}.$$

Using B we write

$$H_w^d(y^+) = AV^2 (S + \Delta S)^{-2} A^T = BB^T.$$

Note that BB^T is nonsingular because A has full row rank. For $g_w^d(y^+)$ we may write

$$
\begin{aligned}
g_w^d(y^+) &= \frac{-1}{\min(w)} \left(b - AW(s + \Delta s)^{-1} \right) \\
&= \frac{-1}{\min(w)} \left(b - AWs^{-1} + AW \left(\frac{e}{s} - \frac{e}{s + \Delta s} \right) \right).
\end{aligned}
$$

The first two terms form $g_w^d(y)$. Replacing W in the third term by $\min(w) V^2$, we obtain

$$g_w^d(y^+) = g_w^d(y) - AV^2 \frac{\Delta s}{s(s + \Delta s)}.$$

---

[2] An alternative proof of Theorem III.17 can be given by generalizing the proof of Theorem II.21; this approach is followed in Jansen et al. [157] and also in the next chapter, where we deal with the analogous result for primal target-following methods. The proof given here seems to be new, and is more straightforward.

Since

$$g_w^d(y) = -H_w^d(y)\Delta y = -AV^2 S^{-2} A^T \Delta y = AV^2 S^{-2}\Delta s$$

we get

$$g_w^d(y^+) = AV^2 \left(\frac{\Delta s}{s^2} - \frac{\Delta s}{s(s+\Delta s)} \right) = AV^2 \left(\frac{(\Delta s)^2}{s^2(s+\Delta s)} \right).$$

The definition of B enables us to rewrite this as

$$g_w^d(y^+) = BV \left(\frac{\Delta s}{s} \right)^2.$$

Substituting the derived expressions for $H_w^d(y^+)$ and $g_w^d(y^+)$ in the expression for $\delta^d(y^+, w)^2$ we find

$$\delta^d(y^+, w)^2 = \left(V \left(\frac{\Delta s}{s} \right)^2 \right)^T B^T (BB^T)^{-1} B\, V \left(\frac{\Delta s}{s} \right)^2.$$

Since $B^T (BB^T)^{-1} B$ is a projection matrix,[3] this implies

$$\delta^d(y^+, w)^2 \leq \left(V \left(\frac{\Delta s}{s} \right)^2 \right)^T V \left(\frac{\Delta s}{s} \right)^2 = \left\| V \left(\frac{\Delta s}{s} \right)^2 \right\|^2,$$

whence

$$\delta^d(y^+, w) \leq \left\| V \left(\frac{\Delta s}{s} \right)^2 \right\| \leq \left\| \frac{\Delta s}{s} \right\|_\infty \left\| \frac{V\Delta s}{s} \right\|.$$

Finally, using Lemma III.16, the theorem follows. $\qquad\square$

12.6 The damped dual Newton method

In this section we consider a damped Newton step to a target vector $w > 0$ at an arbitrary positive dual feasible y with positive slack vector $s = c - A^T y$. We use the damping factor α and move from y to $y^+ = y + \alpha \Delta y$. The resulting slack vector is $s^+ = c - A^T y^+$. Obviously $s^+ = s + \alpha \Delta s$, where $\Delta s = -A^T \Delta y$. We prove the following generalization of Lemma II.38.

Theorem III.18 *Let $\delta = \delta^d(y, w)$. If $\alpha = 1/(\delta_c(w) + \delta)$ then the damped Newton step of size α is feasible and*

$$\phi_w^d(y) - \phi_w^d(y^+) \geq \delta_c(w)\, \psi \left(\frac{\delta}{\delta_c(w)} \right).$$

---

[3] It may be worth mentioning here how the proof can be adapted to the case where A does not have full row rank. First, $\delta^d(y, w)$ can be redefined by replacing the inverse of the Hessian matrix $H_w^d(y)$ in (12.3) by its generalized inverse. Then, in the proof of Theorem III.17 we may use the generalized inverse of BB^T instead of its inverse. We then also have that

$$B^T (BB^T)^+ B$$

is a projection matrix and hence we can proceed in the same way.

Proof: Defining $\Delta := \phi_w^d(y) - \phi_w^d(y^+)$, we have

$$\Delta = \frac{-1}{\min{(w)}}\left(b^T y - b^T y^+ - \sum_{i=1}^n w_i \log \frac{s_i^+}{s_i}\right),$$

or equivalently,

$$\Delta = \frac{-1}{\min{(w)}}\left(-\alpha b^T \Delta y - \sum_{i=1}^n w_i \log \left(1 + \frac{\alpha \Delta s_i}{s_i}\right)\right).$$

Using the definition of the function ψ, we can write this as

$$\Delta = \frac{-1}{\min{(w)}}\left(-\alpha b^T \Delta y - \sum_{i=1}^n w_i \left(\frac{\alpha \Delta s_i}{s_i} - \psi\left(\frac{\alpha \Delta s_i}{s_i}\right)\right)\right).$$

Thus we obtain

$$\Delta = \frac{1}{\min{(w)}}\left(\alpha b^T \Delta y + \alpha w^T \frac{\Delta s}{s} - \sum_{i=1}^n w_i \psi\left(\frac{\alpha \Delta s_i}{s_i}\right)\right).$$

The first two terms between the outer brackets can be reduced to $\alpha \min{(w)} \delta^2$. To this end we write

$$b^T \Delta y + w^T \frac{\Delta s}{s} = (b - AWs^{-1})^T \Delta y = -\min{(w)} g_w^d(y)^T \Delta y.$$

Since $\Delta y = -H_w^d(y)^{-1} g_w^d(y)$, we get

$$b^T \Delta y + w^T \frac{\Delta s}{s} = \min{(w)}\delta^2,$$

proving the claim. Using the same argument as in the proof of Theorem III.11, it can easily be understood that the sum between the brackets attains its maximal value if all the coordinates of the vector $\alpha \Delta s/s$ are zero except one, and the nonzero coordinate, for which w_j must be maximal, is equal to minus the norm of this vector. Thus we obtain

$$\Delta \geq \frac{1}{\min{(w)}}\left(\alpha \min{(w)} \delta^2 - \max{(w)} \psi\left(-\alpha \left\|\frac{\Delta s}{s}\right\|\right)\right)$$

$$= \alpha \delta^2 - \delta_c(w) \psi\left(-\alpha \left\|\frac{\Delta s}{s}\right\|\right).$$

Now also using Lemma III.16 and the monotonicity of ψ we obtain

$$\Delta \geq \alpha \delta^2 - \delta_c(w) \psi(-\alpha \delta) = \alpha \delta^2 + \delta_c(w)(\alpha \delta + \log(1 - \alpha \delta)).$$

It is easily verified that the right-hand side expression is maximal if $\alpha = 1/(\delta_c(w) + \delta)$. Substitution of this value yields

$$\Delta \geq \delta + \delta_c(w) \log\left(1 - \frac{\delta}{\delta_c(w) + \delta}\right) = \delta - \delta_c(w) \log\left(1 + \frac{\delta}{\delta_c(w)}\right).$$

This can be written as

$$\Delta \geq \delta_c(w) \left(\frac{\delta}{\delta_c(w)} - \log \left(1 + \frac{\delta}{\delta_c(w)} \right) \right) = \delta_c(w) \, \psi \left(\frac{\delta}{\delta_c(w)} \right),$$

completing the proof. □

12.7   Dual target-updating

When analysing a dual target-following method we need to quantify the effect of an update of the target on the proximity measure. We derive the dual analogue of Theorem III.13 in this section. We assume that (y, s) is dual feasible and $\delta = \delta^d(y, w)$ for some target vector w, and letting w^* be any other target vector we derive an upper bound for $\delta^d(y, w^*)$. We have the following result, in which $\delta(w^*, w)$ measures the 'distance' from w^* to w according to the primal-dual proximity measure introduced in (10.4):

$$\delta(w^*, w) := \frac{1}{2\sqrt{\min(w)}} \left\| \frac{w - w^*}{\sqrt{w^*}} \right\|. \tag{12.6}$$

Theorem III.19

$$\delta^d(y, w^*) \leq \frac{\sqrt{\min(w)}}{\sqrt{\min(w^*)}} \left(\left\| \sqrt{\frac{w}{w^*}} \right\|_\infty \delta^d(y, w) + 2\delta\,(w^*, w) \right).$$

Proof: By definition $\delta^d(y, w^*)$ satisfies

$$\delta^d(y, w^*) = \left\| g_{w^*}^d(y) \right\|_{H_{w^*}^d(y)^{-1}} = \left\| \frac{-1}{\min(w^*)} \left(b - AW^* s^{-1} \right) \right\|_{H_{w^*}^d(y)^{-1}}.$$

This implies

$$\delta^d(y, w^*) = \frac{1}{\min(w^*)} \left\| b - AWs^{-1} - A\left(W^* - W \right) s^{-1} \right\|_{H_{w^*}^d(y)^{-1}}.$$

Using the triangle inequality we derive from this

$$\delta^d(y, w^*) \leq \frac{\min(w)}{\min(w^*)} \left\| g_w^d(y) \right\|_{H_{w^*}^d(y)^{-1}} + \frac{1}{\min(w^*)} \left\| A\left(W^* - W \right) s^{-1} \right\|_{H_{w^*}^d(y)^{-1}}.$$

We have[4]

$$
\begin{aligned}
H_{w^*}^d(y) \quad &= \quad \frac{1}{\min(w^*)} AW^* S^{-2} A^T = \frac{1}{\min(w^*)} A \frac{W^* W}{W} S^{-2} A^T \\
&\succeq \quad \frac{1}{\min(w^*)} \min\left(\frac{w^*}{w} \right) AW S^{-2} A^T \\
&= \quad \frac{\min(w)}{\min(w^*)} \min\left(\frac{w^*}{w} \right) H_w^d(y).
\end{aligned}
$$

---

[4] The meaning of the symbol '$\succeq$' below is as follows. For any two square matrices P and Q we write $P \succeq Q$ (or $P \preceq Q$) if the matrix $P - Q$ is positive semidefinite. If this holds and Q is nonsingular then P must also be nonsingular and $Q^{-1} \preceq P^{-1}$. This property is used here.

Hence

$$H_{w^*}^d(y)^{-1} \preceq \frac{\min(w^*)}{\min(w)} \left\| \frac{w}{w^*} \right\|_\infty H_w^d(y)^{-1}.$$

We use this inequality to estimate the first term in the above estimate for $\delta^d(y, w^*)$:

$$
\begin{aligned}
\frac{\min(w)}{\min(w^*)} \left\| g_w^d(y) \right\|_{H_{w^*}^d(y)^{-1}} &\leq \frac{\min(w)}{\min(w^*)} \sqrt{\frac{\min(w^*)}{\min(w)}} \left\| \frac{w}{w^*} \right\|_\infty \left\| g_w^d(y) \right\|_{H_w^d(y)^{-1}} \\
&= \sqrt{\frac{\min(w)}{\min(w^*)}} \left\| \frac{w}{w^*} \right\|_\infty \delta^d(y, w).
\end{aligned}
$$

For the second term it is convenient to use the positive vector v^* defined by

$$v^* = \sqrt{\frac{w^*}{\min(w^*)}},$$

and the matrix B defined by $B = AS^{-1}$. Then we have

$$H_{w^*}^d(y) = B(V^*)^2 B^T$$

and

$$A(W^* - W)s^{-1} = B(w^* - w),$$

so we may write

$$
\begin{aligned}
\left\| A(W^* - W)s^{-1} \right\|_{H_{w^*}^d(y)^{-1}}^2 &= (B(w - w^*))^T \left(B(V^*)^2 B^T \right)^{-1} B(w - w^*) \\
&= \left((V^*)^{-1}(w - w^*) \right)^T H (V^*)^{-1}(w - w^*),
\end{aligned}
$$

where

$$H = (BV^*)^T \left(B(V^*)^2 B^T \right)^{-1} BV^*.$$

Clearly, $H = H^2$. Thus, H is a projection matrix, whence $H \preceq I$. Therefore,

$$\left\| A(W^* - W)s^{-1} \right\|_{H_{w^*}^d(y)^{-1}}^2 \leq \left\| (V^*)^{-1}(w - w^*) \right\|^2 = \min(w^*) \left\| \frac{w - w^*}{\sqrt{w^*}} \right\|^2.$$

The last equality follows by using the definition of v^*. Thus we obtain

$$\frac{1}{\min(w^*)} \left\| A(W^* - W)s^{-1} \right\|_{H_{w^*}^d(y)^{-1}} \leq \frac{1}{\sqrt{\min(w^*)}} \left\| \frac{w - w^*}{\sqrt{w^*}} \right\|.$$

Substituting the obtained bounds we arrive at

$$\delta^d(y, w^*) \leq \sqrt{\frac{\min(w)}{\min(w^*)}} \left\| \frac{w}{w^*} \right\|_\infty \delta^d(y, w) + \frac{1}{\sqrt{\min(w^*)}} \left\| \frac{w - w^*}{\sqrt{w^*}} \right\|.$$

Finally, using the definition of the primal-dual proximity measure $\delta\left(w^*, w\right)$, according to (10.4), we may write

$$\frac{1}{\sqrt{\min\left(w^*\right)}}\left\|\frac{w - w^*}{\sqrt{w^*}}\right\| = \frac{2\delta\left(w^*, w\right)\sqrt{\min\left(w\right)}}{\sqrt{\min\left(w^*\right)}}, \tag{12.7}$$

and the theorem follows.   □

In the special case where $w^* = (1 - \theta)w$ the above result reduces to

$$\delta^d(y, w^*) \leq \frac{1}{\sqrt{1-\theta}}\left(\frac{\delta^d(y, w)}{\sqrt{1-\theta}} + \frac{\theta}{\sqrt{1-\theta}}\frac{\|w\|}{\min\left(w\right)}\right) = \frac{1}{1-\theta}\left(\delta^d(y, w) + \frac{\theta\|w\|}{\min\left(w\right)}\right).$$

Moreover, if $w = \mu e$, this gives

$$\delta^d(y, w^*) \leq \frac{1}{1-\theta}\left(\delta^d(y, w) + \theta\sqrt{n}\right).$$

13

The Primal Newton Method

13.1 Introduction

The aim of this chapter is to show that the idea of a target-following method can also be realized by moving only in the primal space. Starting at a given positive primal feasible solution x a *primal method* moves in the primal space until it reaches $x(w)$ where w denotes an intermediate (positive) target vector. The search direction follows by applying Newton's method to a weighted logarithmic barrier function. This function is introduced in the next section. Its minimizer is precisely $x(w)$. Hence, by taking (full or damped) Newton steps with respect to this function we can (approximately) compute $x(w)$. The general framework of a primal target-following algorithm is described below.

Generic Primal Target-following Algorithm

Input:
    A primal feasible vector x^0 such that $x^0 = x\left(w^0\right)$;
    a final target vector $\tilde{w}$.
begin
    $x := x^0; w := w^0$;
    while w is not 'close' to $\tilde{w}$ **do**
    begin
        Replace w by the next target in the sequence;
        while x is not 'close' to $x(w)$ **do**
        begin
            Apply Newton steps at x to the target w
        end
    end
end

The underlying target sequence starts at w^0 and ends — via some intermediate target vectors — at $\tilde{w}$.

13.2 The weighted primal barrier function

The search direction in a primal method is obtained by applying Newton's method to the *weighted primal barrier function* given by

$$\phi_w^p(x) := \frac{1}{\min(w)} \left(c^T x - \sum_{j=1}^n w_j \log x_j \right). \tag{13.1}$$

We first establish that $\phi_w^p(x)$ attains its minimal value at $x(w)$. This easily follows by using the barrier function ϕ_w in the same way as for the dual weighted barrier function. Starting from (9.4), on page 221, and using $x^T s(w) = c^T x - b^T y(w)$ and $x(w)s(w) = w$ we write

$$
\begin{aligned}
\max(w)\, \phi_w(x, s(w)) &= x^T s(w) - \sum_{j=1}^n w_j \log x_j s_j(w) - e^T w + \sum_{j=1}^n w_j \log w_j \\
&= x^T s(w) - \sum_{j=1}^n w_j \log x_j - e^T w + \sum_{j=1}^n w_j \log x_j(w) \\
&= c^T x - \sum_{j=1}^n w_j \log x_j - b^T y(w) - e^T w + \sum_{j=1}^n w_j \log x_j(w) \\
&= \min(w)\, \phi_w^p(x) - b^T y(w) - e^T w + \sum_{j=1}^n w_j \log x_j(w).
\end{aligned}
$$

This implies that $x(w)$ is a unique minimizer of $\phi_w^p(x)$.

13.3 Definition of the primal Newton step

Let x be primal feasible and let $w > 0$. We denote the gradient of $\phi_w^p(x)$ at x by $g_w^p(x)$ and the Hessian by $H_w^p(x)$. These are

$$g_w^p(x) := \frac{1}{\min(w)} \left(c - \frac{w}{x} \right)$$

and

$$H_w^p(x) := \frac{1}{\min(w)} W X^{-2} = V^2 X^{-2},$$

where $V = \operatorname{diag}(v)$, with v as defined in (12.5) in the previous chapter. Note that $H_w^p(x)$ is positive definite. It follows that $\phi_w^p(x)$ is a strictly convex function.

The calculation of the Newton step Δx is a little complicated by the fact that we want $x + \Delta x$ to stay in the affine space $Ax = b$. This means that Δx must satisfy $A\Delta x = 0$. The Newton step at x is then obtained by minimizing the second-order Taylor polynomial at x subject to this constraint. Thus, Δx is the solution of

$$\min_{\Delta x} \left\{ \Delta x^T g_w^p(x) + \frac{1}{2} \Delta x^T H_w^p(x) \Delta x \ : \ A\Delta x = 0 \right\}.$$

The optimality conditions for this minimization problem are

$$g_w^p(x) + H_w^p(x)\Delta x = A^T u$$
$$A\Delta x = 0,$$

where the coordinates of $u \in \mathbb{R}^m$ are Lagrange multipliers. We introduce the scaling vector d according to

$$d := \frac{x}{\sqrt{w}}.$$

Observe that $H_w^p(x) = D^{-2}/\min(w)$. The optimality conditions can be rewritten as

$$-d^{-1}\Delta x + \min(w)(AD)^T u = d\left(c - \frac{w}{x}\right)$$
$$AD(d^{-1}\Delta x) = 0,$$

which shows that $-d^{-1}\Delta x$ is the orthogonal projection of $d(c - w/x)$ into the null space of AD:

$$-d^{-1}\Delta x = P_{AD}\left(d\left(c - \frac{w}{x}\right)\right) \implies \Delta x = -DP_{AD}\left(\frac{xc - w}{\sqrt{w}}\right). \qquad (13.2)$$

Remark III.20 When $w = \mu e$ we have $d = x/\sqrt{\mu}$. Since AD and AX have the same null space, we have $P_{AD} = P_{AX}$. Therefore, in this case the Newton step is given by

$$\Delta x = -\frac{1}{\sqrt{\mu}}XP_{AX}\left(\frac{xc - \mu e}{\sqrt{\mu}}\right) = -XP_{AX}\left(\frac{Xc}{\mu} - e\right).$$

This search direction is used in the so-called *primal logarithmic barrier method*, which is obtained by applying the results of this chapter to the case where the targets are on the central path. It is the natural analogue of the dual logarithmic barrier method treated in Chapter 6. •

We introduce the following proximity measure to quantify the distance from x to $x(w)$:

$$\delta^p(x, w) = \frac{1}{\sqrt{\min(w)}} \min_{y,s}\left\{\left\|d\left(s - \frac{w}{x}\right)\right\| : A^T y + s = c\right\}. \qquad (13.3)$$

This measure is inspired by the measure (6.8) for the dual logarithmic barrier method, introduced in Section 6.5.[1] Let us denote by $s(x, w)$ the minimizing s in (13.3).

Lemma III.21 *We have*

$$\delta^p(x, w) = \left\|\frac{v\Delta x}{x}\right\| = \frac{1}{\sqrt{\min(w)}}\left\|\frac{x\,s(x, w) - w}{\sqrt{w}}\right\|.$$

Proof: For the proof of the first equality we eliminate s in (13.3) and write

$$\min_{y,s}\left\{\left\|d\left(s - \frac{w}{x}\right)\right\| : A^T y + s = c\right\} = \min_y\left\{\left\|d\left(c - \frac{w}{x}\right) - DA^T y\right\|\right\}.$$

---

[1] Similar proximity measures were used in Roos and Vial [245], and Hertog and Roos [142] for primal methods, and in Mizuno [212, 214] and Jansen et al. [159] for primal-dual methods.

Let $\bar{y}$ denote the solution of the last minimization problem. Then

$$d\left(c - \frac{w}{x}\right) = DA^T\bar{y} + P_{AD}\left(\left(d\left(c - \frac{w}{x}\right)\right)\right).$$

Thus we obtain

$$d\left(c - \frac{w}{x}\right) - DA^T\bar{y} = P_{AD}\left(d\left(c - \frac{w}{x}\right)\right).$$

From (13.2),

$$P_{AD}\left(d\left(c - \frac{w}{x}\right)\right) = -d^{-1}\Delta x.$$

Hence we get

$$\delta^p(x, w) = \frac{1}{\sqrt{\min(w)}}\|d^{-1}\Delta x\| = \frac{1}{\sqrt{\min(w)}}\left\|\frac{\sqrt{w}\Delta x}{x}\right\| = \left\|\frac{v\Delta x}{x}\right\|,$$

proving the first equality in the lemma. The second equality in the lemma follows from the definition of $s(x, w)$.[2] □

From the above proof and (13.2) we deduce that

$$d^{-1}\Delta x = -\frac{xs(x, w) - w}{\sqrt{w}}. \tag{13.4}$$

Also observe that the lemma implies that, just as in the dual case, the proximity measure is equal to the 'Hessian–norm' of the Newton step:

$$\delta^p(x, w) = \|\Delta x\|_{H^p_w(x)}.$$

13.4 Feasibility of the primal Newton step

Let x^+ result from the Newton step at x:

$$x^+ := x + \Delta x.$$

The Newton step is feasible if and only if $x + \Delta x \geq 0$. Now we can prove the next lemma.

Lemma III.22 *If $\delta^p(x, w) \leq 1$ then $x^* = x + \Delta x$ is primal feasible.*

Proof: From Lemma III.21 we derive

$$\delta^p(x, w) = \left\|\frac{v\Delta x}{x}\right\| \geq \left\|\frac{\Delta x}{x}\right\| \geq \left\|\frac{\Delta x}{x}\right\|_\infty.$$

Hence, if $\delta^p(x, w) < 1$, then $|\Delta x| \leq x$, which implies $x + \Delta x \geq 0$. The lemma follows. □

---

[2] **Exercise 78** If $\delta^p(x, w) \leq 1$ then $s(x, w)$ is dual feasible. Prove this.

13.5 Quadratic convergence

We proceed by showing that the primal Newton method is quadratically convergent.

Theorem III.23 $\delta^p(x^+, w) \leq \delta^p(x, w)^2$.

Proof: Using the definition of $\delta^p(x^+, w)$ we may write

$$
\begin{aligned}
\delta^p(x^+, w) \quad &= \quad \frac{1}{\sqrt{\min(w)}} \left\| \frac{x^+ s(x^+, w) - w}{\sqrt{w}} \right\| \\
&\leq \quad \frac{1}{\sqrt{\min(w)}} \left\| \frac{x^+ s(x, w) - w}{\sqrt{w}} \right\| \\
&\leq \quad \frac{1}{\left(\sqrt{\min(w)}\right)^2} \| x^+ s(x, w) - w \|.
\end{aligned}
$$

Denote $\bar{s} := s(x, w)$. From (13.4) we obtain

$$
\bar{s}\Delta x = \bar{s}dd^{-1}\Delta x = -d\bar{s}\frac{x\,\bar{s} - w}{\sqrt{w}} = -\frac{x\,\bar{s}(x\,\bar{s} - w)}{w}.
$$

This implies

$$
\| x^+ \bar{s} - w \| = \| (x + \Delta x)\bar{s} - w \| = \left\| x\bar{s} - w - \frac{x\bar{s}(x\bar{s} - w)}{w} \right\| = \left\| \frac{(x\bar{s} - w)^2}{w} \right\|.
$$

Combining the above relations, we get

$$
\delta^p(x^+, w) \leq \frac{1}{\left(\sqrt{\min(w)}\right)^2} \left\| \left(\frac{x\bar{s} - w}{\sqrt{w}} \right)^2 \right\| \leq \left(\frac{1}{\sqrt{\min(w)}} \left\| \frac{x\bar{s} - w}{\sqrt{w}} \right\| \right)^2 = \delta^p(x, w)^2.
$$

This completes the proof. $\square$

13.6 The damped primal Newton method

In this section we consider a damped primal Newton step to a target vector $w > 0$ at an arbitrary positive primal feasible x. The damping factor is again denoted by α and we move from x to $x^+ = x + \alpha \Delta x$. After Theorem III.18 it will be no surprise that we have the following result.

Theorem III.24 *Let* $\delta = \delta^p(x, w)$. *If* $\alpha = 1/(\delta_c(w) + \delta)$ *then the damped Newton step of size* α *is feasible and*

$$
\phi_w^p(x) - \phi_w^p(x^+) \geq \delta_c(w)\, \psi\left(\frac{\delta}{\delta_c(w)} \right).
$$

Proof: Defining $\Delta := \phi_w^p(x) - \phi_w^p(x^+)$, we have

$$\Delta = \frac{1}{\min(w)} \left(c^T x - c^T x^+ + \sum_{i=1}^{n} w_i \log \frac{x_i^+}{x_i} \right),$$

or equivalently,

$$\Delta = \frac{1}{\min(w)} \left(-\alpha c^T \Delta x + \sum_{i=1}^{n} w_i \log \left(1 + \frac{\alpha \Delta x_i}{x_i} \right) \right).$$

Using the definition of the function ψ, this can be rewritten as

$$\Delta = \frac{1}{\min(w)} \left(-\alpha c^T \Delta x + \sum_{i=1}^{n} w_i \left(\frac{\alpha \Delta x_i}{x_i} - \psi \left(\frac{\alpha \Delta x_i}{x_i} \right) \right) \right).$$

Thus we obtain

$$\Delta = \frac{1}{\min(w)} \left(-\alpha c^T \Delta x + \alpha w^T \frac{\Delta x}{x} - \sum_{i=1}^{n} w_i \psi \left(\frac{\alpha \Delta x_i}{x_i} \right) \right).$$

We reduce the first two terms between the outer brackets to $\alpha \min(w) \delta^2$:

$$-c^T \Delta x + w^T \frac{\Delta x}{x} = - \left(c - \frac{w}{x} \right)^T \Delta x,$$

and from (13.2),

$$-\left(c - \frac{w}{x} \right)^T \Delta x = d \left(c - \frac{w}{x} \right)^T P_{AD} \left(d \left(c - \frac{w}{x} \right) \right)$$

$$= \left\| P_{AD} \left(d \left(c - \frac{w}{x} \right) \right) \right\|^2 = \left\| d^{-1} \Delta x \right\|^2.$$

Since $d = x/\sqrt{w}$ this implies

$$-\left(c - \frac{w}{x} \right)^T \Delta x = \min(w) \delta^2,$$

proving the claim. The sum between the brackets can be estimated in the same way as for the dual method. Thus we obtain

$$\Delta \geq \frac{1}{\min(w)} \left(\alpha \min(w) \delta^2 - \max(w) \psi \left(-\alpha \left\| \frac{\Delta x}{x} \right\| \right) \right)$$

$$= \alpha \delta^2 - \delta_c(w) \psi \left(-\alpha \left\| \frac{\Delta x}{x} \right\| \right),$$

yielding exactly the same lower bound for Δ as in the dual case. Hence we can use the same arguments as we did there to complete the proof. $\square$

13.7 Primal target-updating

We derive the primal analogue of Theorem III.19 in this section. We assume that x is primal feasible and $\delta = \delta^p(x, w)$ for some target vector w. For any other target vector w^* we need to derive an upper bound for $\delta^p(x, w^*)$. The result is completely similar to Theorem III.19, but the proof must be adapted to the primal context.

Theorem III.25

$$\delta^p(x, w^*) \leq \frac{\sqrt{\min(w)}}{\sqrt{\min(w^*)}} \left(\left\| \sqrt{\frac{w}{w^*}} \right\|_\infty \delta^p(x, w) + 2\delta\left(w^*, w\right) \right).$$

Proof: By Lemma III.21,

$$\delta^p(x, w^*) = \frac{1}{\sqrt{\min(w^*)}} \left\| \frac{x\, s(x, w^*) - w^*}{\sqrt{w^*}} \right\|,$$

where $s(x, w^*)$ satisfies the affine dual constraint $A^T y + s = c$ and minimizes the above norm. Hence, since $s(x, w)$ satisfies the affine dual constraint, replacing $s(x, w^*)$ by $s(x, w)$ we obtain

$$
\begin{aligned}
\delta^p(x, w^*) &\leq \frac{1}{\sqrt{\min(w^*)}} \left\| \frac{x\, s(x, w) - w^*}{\sqrt{w^*}} \right\| \\
&= \frac{1}{\sqrt{\min(w^*)}} \left\| \frac{x\, s(x, w) - w + w - w^*}{\sqrt{w^*}} \right\|.
\end{aligned}
$$

Using the triangle inequality we derive from this

$$\delta^p(x, w^*) \leq \frac{1}{\sqrt{\min(w^*)}} \left\| \frac{x\, s(x, w) - w}{\sqrt{w^*}} \right\| + \frac{1}{\sqrt{\min(w^*)}} \left\| \frac{w - w^*}{\sqrt{w^*}} \right\|.$$

The second term can be reduced by using (12.7) and then the theorem follows if the first term on the right satisfies

$$\frac{1}{\sqrt{\min(w^*)}} \left\| \frac{x\, s(x, w) - w}{\sqrt{w^*}} \right\| \leq \sqrt{\frac{\min(w)}{\min(w^*)}} \left\| \frac{w}{w^*} \right\|_\infty \delta^p(x, w). \qquad (13.5)$$

This inequality can be obtained by writing

$$
\begin{aligned}
\frac{1}{\sqrt{\min(w^*)}} \left\| \frac{x\, s(x, w) - w}{\sqrt{w^*}} \right\| &= \frac{1}{\sqrt{\min(w^*)}} \left\| \frac{\sqrt{w}}{\sqrt{w^*}} \frac{x\, s(x, w) - w}{\sqrt{w}} \right\| \\
&\leq \frac{1}{\sqrt{\min(w^*)}} \sqrt{\left\| \frac{w}{w^*} \right\|_\infty} \left\| \frac{x\, s(x, w) - w}{\sqrt{w}} \right\| \\
&= \sqrt{\frac{\min(w)}{\min(w^*)}} \left\| \frac{w}{w^*} \right\|_\infty \delta^p(x, w).
\end{aligned}
$$

Hence the theorem follows.   □

14

Application to the Method of Centers

14.1 Introduction

Shortly after Karmarkar published his projective algorithm for linear optimization, some authors pointed out possible links with earlier literature. Gill et al. [97] noticed the close similarity between the search directions in Karmarkar's algorithm and in the logarithmic barrier approach extensively studied by Fiacco and McCormick [77]. At the same time, Renegar [237] proposed an algorithm with $\mathcal{O}(\sqrt{n}L)$ iterations, an improvement over Karmarkar's algorithm. Renegar's scheme was a clever implementation of Huard's *method of centers* [148]. Again, there were clear similarities, but equivalence was not established. For a while, the literature seemed to develop in three approximately independent directions. The first stream dealt with extensions of Karmarkar's algorithm and was identified with the notion of projective transformation and projective space.[1] This is the topic of the next chapter. The second stream of research was a revival and a new interpretation of the logarithmic approach. We amply elaborated on that approach in Part II of this book. The third stream prolonged Renegar's contribution. Not so much has been done in this framework.[2]

After a decade of active research, it has become apparent that the links between the three approaches are very tight. They only reflect different ways of looking at the same thing. From one point of view, the similarity between the method of centers and the logarithmic barrier approach is striking. In both cases, the progress towards optimality is triggered by a parameter that is gradually shifted to its optimal value. The iterations are performed in the primal, dual or primal-dual spaces; they are made of Newton steps or damped Newton steps that aim to catch up with the parameter variation. The parameter updates are either small enough to allow full Newton steps and the method is of a path-following type with an $\mathcal{O}(\sqrt{n}L)$ iteration bound; or, the updates are large and the method performs line searches along Newton's direction with the aim of reducing a certain potential. The parameter in the logarithmic barrier approach is the penalty coefficient attached to the logarithm; in the method of centers, the parameter is a bound on the optimal objective function value. In the logarithmic barrier approach, the parameter is gradually moved to zero. In the method of centers,

---

[1] For survey papers, we refer the reader to Anstreicher [17, 24], Goldfarb and Todd [109], Gonzaga [123, 124], den Hertog and Roos [142] and Todd [265].

[2] In this connection we cite den Hertog, Roos and Terlaky [143] and den Hertog [140].

the parameter is monotonically shifted to the optimal value of the LO problem.

A similar link exists between Renegar's method of centers and the variants of Karmarkar's method introduced by de Ghellinck and Vial [95] and Todd and Burrell [266]. Those variants use a parameter — a lower bound in case of a minimization problem — that is triggered to its optimal value. If this parameter is kept fixed, the projective algorithm computes an analytic center[3] that is the dual of the center used by Renegar. Consequently, there also exist path-following schemes for the projective algorithm, see Shaw and Goldfarb [254], and Goffin and Vial [103]; these are very close to Renegar's method.

In this chapter we concentrate on the method of centers. Our aim is to show that the method can be described and analyzed quite well in the target-following framework.[4]

14.2 Description of Renegar's method

The method of centers (or center method) can easily be described by considering the barrier function used by Renegar.[5] Assuming the knowledge of a strict lower bound z for the optimal value of the dual problem (D) he considers the function

$$\phi_R(y, z) := -q \log(b^T y - z) - \sum_{i=1}^{n} \log s_i,$$

where q is some positive number and $s = c - A^T y$. His method consists of finding (an approximation of) the minimizer $y(z)$ of this barrier function by using Newton's method. Then the lower bound z is enlarged to

$$\bar{z} = z + \theta(b^T y(z) - z) \tag{14.1}$$

---

[3] The computation of analytic centers can be performed via variants of the projective algorithm. In this connection, we cite Atkinson [29] and Goffin and Vial [102].

[4] The method of centers has an interest of its own. First, the approach formalizes Huard's scheme and supports Huard's intuition of an efficient interior-point algorithm. There are also close links with Karmarkar's method that are made explicit in Vial [285]. Second, the method of centers offers a natural framework for *cutting plane methods*. Cutting plane methods could be described in short as a way to solve an LO problem with so many (possibly infinite) inequality constraints that we cannot even enumerate them in a reasonable computational time. The only possibility is to generate them one at a time, as they seem needed to insure feasibility eventually. Generating cuts from a center, and in particular, from an analytic center, appears to be sound from both the theoretical and the practical point of views. The idea of using analytic centers in this context was alluded to by Sonnevend [257] and fully worked out by Goffin, Haurie and Vial [99]. See du Merle [209] and Gondzio et al. [115] for a detailed description of the method, and e.g., Bahn et al. [31] and Goffin et al. [98] for results on large scale programs. Let us mention that the complexity analysis of a conceptual method of analytic centers was given first by Atkinson and Vaidya [30] and Nesterov [225]. An implementable version of the method using approximate analytic centers is analyzed by Goffin, Luo and Ye [100], Luo [186], Ye [312], Goffin and Sharifi-Mokhtarian [101], Altman and Kiwiel [7], Kiwiel [168], and Goffin and Vial [104]. Besides, to highlight the similarity between the method of centers and the logarithmic barrier approach it is worth noting that logarithmic barrier methods also allow a natural cutting plane scheme based on adding and deleting constraints. We refer the reader to den Hertog [140], den Hertog, Roos and Terlaky [145], den Hertog et al. [141] and Kaliski et al. [164]. For a complexity analysis of a special variant of this method we refer the reader to Luo, Roos and Terlaky [187].

[5] The notation used here differs from the notation of Renegar. This is partly due to the fact that Renegar dealt with a solution method for the primal problem whereas we apply his approach to the dual problem.

for some positive θ such that $\bar{z}$ is again a strict lower bound for the optimal value and the process is repeated. Renegar showed that this scheme can be used to construct an ε-solution of (D) in at most

$$\mathcal{O}\left(\sqrt{n}\log\frac{b^T y^0 - z^0}{\varepsilon}\right)$$

iterations, where the superscript ^{0} refers to initial values, as usual. In this way he was the first to obtain this iteration bound.

The algorithm can be described as follows.

Renegar's Method of Centers

Input:
> A strict lower bound z^0 for the optimal value of (D);
> a dual feasible y^0 such that y^0 is 'close' to $y(z^0)$;
> a positive number $q \geq \sqrt{n}$;
> an update parameter θ, $0 < \theta < 1$.

begin
> $y := y^0; z := z^0;$
> **while** $b^T y - z \geq \varepsilon$ **do**
> **begin**
>> $z = z + \theta\left(b^T y - z\right);$
>> **while** y is not 'close' to $y(z)$ **do**
>> **begin**
>>> Apply Newton steps at y aiming at $y(z)$
>> **end**
> **end**

end

14.3 Targets in Renegar's method

Let us now look at how this approach fits into the target-following concept. First we observe that ϕ_R can be considered as the barrier term in a weighted barrier function for the dual problem when we add the constraint $b^T y \geq z$ to the dual constraints and give the extra constraint the weight q. Giving the extra constraint the index 0, and indexing the other constraints by 1 to n as usual, we have the vector of weights

$$w \doteq (q, 1, 1, \ldots, 1).$$

The second observation is that Renegar's barrier function is exactly the weighted dual barrier function ϕ_w^d (cf. (12.1) on page 259) for the problem

$$(DR) \qquad \max\left\{0^T y : A^T y + s = c, -b^T y + s^0 = -z, s \geq 0, s^0 \geq 0\right\}.$$

The feasible region of this problem is just the feasible region of (D) *cut by the objective constraint* $b^T y \geq z$. Since the objective function is trivial, each feasible point is optimal. As a consequence, the weighted central path of (DR) is a point and hence this point, which is the minimizer of ϕ_R, is just the weighted-analytic center (according to w) of the feasible region of (D) cut by the objective constraint (cf. Theorem III.5 on page 229).

The dual problem of (DR) is the following homogeneous problem:

$$(PR) \qquad \min \left\{ c^T \tilde{x} - \tilde{x}^0 z \; : \; A\tilde{x} - \tilde{x}^0 b = 0, \; \tilde{x} \geq 0, \; \tilde{x}^0 \geq 0 \right\}.$$

Applying Theorem III.1 (page 222), we see that the optimality conditions for $\phi_R(y, z) = \phi_w^d(y)$ are given by

$$
\begin{aligned}
A\tilde{x} - \tilde{x}^0 b &= 0, & \tilde{x}, x^0 &\geq 0, \\
A^T y + s &= c, & s &\geq 0, \\
b^T y - s^0 &= z, & s^0 &\geq 0, \\
\tilde{x} s &= e, & & \\
\tilde{x}^0 s^0 &= q. &
\end{aligned}
\tag{14.2}
$$

The third and fifth equations imply

$$\tilde{x}^0 = \frac{q}{s^0} = \frac{q}{b^T y - z}. \tag{14.3}$$

Hence, defining

$$x := \frac{\tilde{x}}{\tilde{x}^0} = \frac{b^T y - z}{q} \tilde{x}$$

we get

$$
\begin{aligned}
Ax &= b, & x &\geq 0, \\
A^T y + s &= c, & s &\geq 0, \\
xs &= \mu_z e,
\end{aligned}
\tag{14.4}
$$

where

$$\mu_z := \frac{b^T y(z) - z}{q}, \tag{14.5}$$

with $y(z)$ denoting the minimizer of Renegar's barrier function $\phi_R(y)$. We conclude that $y(z)$ can be characterized in two ways. First, it is the weighted-analytic center of the feasible region of (D) cut by the objective constraint $b^T y \geq z$ and, second, it is the point on the central path of (D) corresponding to the above barrier parameter value μ_z. Figure 14.1 depicts the situation.

In the course of the center method the lower bound z is gradually updated to the optimal value of (D) and after each update of the lower bound the corresponding minimizer $y(z)$ is (approximately) computed. Since $y(z)$ represents the dual part of the primal-dual pair belonging to the vector $\mu_z e$ in the w-space, we conclude that the center method can be considered as a central-path-following method.

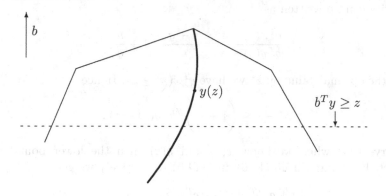

Figure 14.1 The center method according to Renegar.

14.4   Analysis of the center method

It will be clear that in the analysis of his method Renegar had to deal with the question of how far the value of the lower bound z can be enlarged — according to (14.1) — so that the minimizer $\bar{y}$ of $\phi_R(y, \bar{z})$ can be computed efficiently; hereby it may be assumed that the minimizer y of $\phi_R(y, z)$ is known.[6] The answer to this question determines the speed of convergence of the method. As we know, the answer depends on the proximity $\delta(\mu_z e, \mu_{\bar{z}} e)$ of the present target vector $\mu_z e$ to the new target vector $\mu_{\bar{z}} e$. Thus, we have to estimate the proximity $\delta(\mu_z e, \mu_{\bar{z}} e)$, where $\bar{z}$ is given by (14.1). Further analysis below is a little complicated by the fact that the new target vector $\mu_{\bar{z}} e$ is not known, since

$$\mu_{\bar{z}} = \frac{b^T y(\bar{z}) - \bar{z}}{q}$$

depends on the unknown minimizer $y(\bar{z})$ of $\phi_R(y, \bar{z})$. To cope with this complication we need some further estimates.

Let $(x(z), y(z), s(z))$ denote the solution of (14.4), so it is the point on the central path of (P) and (D) corresponding to the strict lower bound z for the optimal value. Then the duality gap at this point is given by

$$c^T x(z) - b^T y(z) = n\mu_z = \frac{n\left(b^T y(z) - z\right)}{q}.$$

---

[6] As far as the numerical procedure for the computation of the minimizer of Renegar's barrier function is concerned, it may be clear that there are a lot of possible choices. Renegar presented a dual method in [237]. His search direction is the Newton direction for minimizing ϕ_R. In our framework this amounts to applying the dual Newton method for the computation of the primal-dual pair corresponding to the target vector w for the problems (PR) and (DR); this method has been discussed in Section 12.2. Obviously, the same goal can be achieved by using any efficient computational — primal, dual or primal-dual — method for the computation of the primal-dual pair corresponding to the target vector $\mu_z e$ for (P) and (D).

This identity can be written as

$$\frac{c^T x(z) - z}{b^T y(z) - z} = \frac{n+q}{q} = 1 + \frac{n}{q}. \qquad (14.6)$$

Denoting the optimal value by z^* we have $c^T x(z) \geq z^*$. Hence

$$z^* - z \leq \left(1 + \frac{n}{q}\right)\left(b^T y(z) - z\right).$$

Also observe that when we know $x(z)$ and $y(z)$ then the lower bound z can be reconstructed: solving z from (14.6) and (14.5) respectively we get

$$z = \frac{(n+q)\, b^T y(z) - q\, c^T x(z)}{n} = b^T y(z) - q\mu_z.$$

For the updated lower bound $\bar{z}$ we thus find the expression

$$\bar{z} = b^T y(z) - q\mu_z + \theta\left(b^T y(z) - z\right) = b^T y(z) - q\mu_z + \theta q\mu_z = b^T y(z) - (1-\theta)\, q\mu_z.$$

Since $b^T y(z)$ is a lower bound for the optimal value, this relation makes clear that we are able to guarantee that $\bar{z}$ is a strict lower bound for the optimal value only if $\theta < 1$.

Lemma III.26 *The dual objective value $b^T y(z)$ is monotonically increasing, whereas the primal objective value $c^T x(z)$ and $b^T y(z) - z$ are monotonically decreasing if z increases.*[7]

Proof: We first prove the second part of the lemma. To this end we use the weighted primal barrier function for (PR),

$$\phi^p_{w,z}(\tilde{x}, \tilde{x}^0) = c^T \tilde{x} - \tilde{x}^0 z - q \log \tilde{x}^0 - \sum_{i=1}^n \log \tilde{x}_i.$$

The dependence of this function on the lower bound z is expressed by the corresponding subindex. Now let z and $\bar{z}$ be two strict lower bounds for the optimal value of (P) and (D) and $\bar{z} > z$. Since $(\tilde{x}(z), \tilde{x}^0(z))$ minimizes $\phi^p_{w,z}(\tilde{x}, \tilde{x}^0)$ and $(\tilde{x}(\bar{z}), \tilde{x}^0(\bar{z}))$ minimizes $\phi^p_{w,\bar{z}}(\tilde{x}, \tilde{x}^0)$ we have

$$\phi^p_{w,z}\left(\tilde{x}(z), \tilde{x}^0(z)\right) \leq \phi^p_{w,z}\left(\tilde{x}(\bar{z}), \tilde{x}^0(\bar{z})\right), \quad \phi^p_{w,\bar{z}}\left(\tilde{x}(\bar{z}), \tilde{x}^0(\bar{z})\right) \leq \phi^p_{w,\bar{z}}\left(\tilde{x}(z), \tilde{x}^0(z)\right).$$

Adding these inequalities, we get

$$\phi^p_{w,z}\left(\tilde{x}(z), \tilde{x}^0(z)\right) + \phi^p_{w,\bar{z}}\left(\tilde{x}(\bar{z}), \tilde{x}^0(\bar{z})\right) \leq \phi^p_{w,z}\left(\tilde{x}(\bar{z}), \tilde{x}^0(\bar{z})\right) + \phi^p_{w,\bar{z}}\left(\tilde{x}(z), \tilde{x}^0(z)\right).$$

Evaluating the expressions in these inequalities and omitting the common terms on both sides — the terms in which the parameters z and $\bar{z}$ do not occur — we find

$$-\tilde{x}^0(z)z - \tilde{x}^0(\bar{z})\bar{z} \leq -\tilde{x}^0(\bar{z})z - \tilde{x}^0(z)\bar{z},$$

---

[7] This lemma is taken from den Hertog [140]. The proof below is a slight variation on his proof. The proof technique is due to Fiacco and McCormick [77] and can be applied to obtain monotonicity of the objective value along the central path in a much wider class of convex problems. We refer the reader to den Hertog, Roos and Terlaky [144] and den Hertog [140].

or equivalently,
$$\left(\bar{z} - z\right)\left(\tilde{x}^0(\bar{z}) - \tilde{x}^0(z)\right) \geq 0.$$
This implies $\tilde{x}^0(\bar{z}) - \tilde{x}^0(z) \geq 0$, or
$$\tilde{x}^0(\bar{z}) \geq \tilde{x}^0(z).$$
By (14.3) this is equivalent to
$$b^T y(\bar{z}) - \bar{z} \leq b^T y(z) - z.$$
Thus we have shown that $b^T y(z) - z$ is monotonically decreasing if z increases. This implies that μ_z is also monotonically decreasing if z increases. The rest of the lemma follows because along the central path the dual objective value is increasing and the primal objective value is decreasing. The proof of this property of the central path can be found in Remark II.6 (page 95). □

Now let $\bar{z}$ be given by (14.1). Then we may write
$$\frac{c^T x(\bar{z}) - \bar{z}}{c^T x(z) - z} = \frac{c^T x(\bar{z}) - z - \theta\left(b^T y(z) - z\right)}{c^T x(z) - z}.$$
By the above lemma we have $c^T x(\bar{z}) \leq c^T x(z)$. Hence, using also (14.6) we get
$$\frac{c^T x(\bar{z}) - \bar{z}}{c^T x(z) - z} \leq 1 - \frac{\theta\left(b^T y(z) - z\right)}{c^T x(z) - z} = 1 - \frac{\theta q}{n + q}.$$
Using (14.6) once more we derive
$$\frac{b^T y(\bar{z}) - \bar{z}}{b^T y(z) - z} = \frac{c^T x(\bar{z}) - \bar{z}}{c^T x(z) - z},$$
and so
$$\frac{b^T y(\bar{z}) - \bar{z}}{b^T y(z) - z} \leq 1 - \frac{\theta q}{n + q}.$$
Therefore we obtain the following relation between $\mu_{\bar{z}}$ and μ_z:
$$\mu_{\bar{z}} \leq \left(1 - \frac{\theta q}{n + q}\right)\mu_z. \tag{14.7}$$

For the moment we deviate from Renegar's approach by taking as a new target the vector
$$\bar{w} := \left(1 - \frac{\theta q}{n + q}\right)w, \tag{14.8}$$
where $w = \mu_z e$. Instead of Renegar's target vector $\mu_{\bar{z}} e$ we use $\bar{w}$ as a target vector. Due to the inequality (14.7) this means that we slow down the progress to optimality compared with Renegar's approach. We show, however, that the modified strategy still yields an $\mathcal{O}(\sqrt{n}L)$ iteration bound, just as Renegar's approach. Assuming $n \geq 4$, the argument used in Section 11.2 implies that
$$\delta(\mu_z e, \bar{w}) \leq \frac{1}{\sqrt{2}} \quad \text{if} \quad \frac{\theta q}{n + q} = \frac{1}{\sqrt{n}}.$$

Hence, when

$$\theta = \frac{n+q}{q\sqrt{n}}, \tag{14.9}$$

the primal-dual pair belonging to the target $\bar{w}$ can be computed efficiently, to any desired accuracy.

Since the barrier parameter, and hence the duality gap, at the new target is reduced by the factor $1 - \theta q/(n+q)$ we obtain an ε-solution after at most

$$\frac{n+q}{\theta q} \log \frac{e^T w^0}{\varepsilon} = \sqrt{n} \log \frac{e^T w^0}{\varepsilon}$$

iterations. Here w^0 denotes the initial point in the w-space.

Note that the parameter q disappeared in the iteration bound. In fact, the above analysis, based on the updating scheme (14.8), works for every positive value of q and gives the same iteration bound for each value of q.

On the other hand, when using Renegar's scheme, the update goes via the strict lower bound z. As we established before, it is then necessary to keep $\theta < 1$. So Renegar's approach only works if q satisfies $n+q < q\sqrt{n}$. This amounts to the following condition on q:

$$q \geq \frac{n}{\sqrt{n}-1} > \sqrt{n}.$$

Renegar, in [237], recommended $q = n$ and $\theta = 1/(13\sqrt{q})$. Den Hertog [140], who simplified the analysis significantly, used $q \geq 2\sqrt{n}$ and $\theta = 1/(8\sqrt{q})$. In both cases the iteration bound is of the same order of magnitude as the bound derived above.[8]

14.5   Adaptive- and large-update variants of the center method

In the logarithmic barrier approach, we used a penalty parameter to trigger the algorithm. By letting the parameter go to zero in a controlled way, we could drive the pairs of dual solutions to optimality. The crux of the analysis was the updating scheme: small, adaptive or large updates, with results of variable complexity. Small or adaptive updates allow relatively small reductions of the duality gap — by a factor $1 - \mathcal{O}(1/\sqrt{n})$ — in $\mathcal{O}(1)$ Newton steps between two successive updates, and achieve global convergence in $\mathcal{O}(\sqrt{n}L)$ iterations. Large updates allow sharp decreases of the duality gap — by a factor $1 - \Theta(1)$ — but require more Newton steps (usually as many as $\mathcal{O}(n)$) between two successive updates and lead to global convergence in $\mathcal{O}(nL)$ iterations. A similar situation occurs for target-following methods, where the algorithm is triggered by the targets; the target sequence can be designed such that similar convergence results arise for small, adaptive and large updates respectively.

The method of this chapter, the (dual) center method of Renegar, has a different triggering mechanism: a lower bound on the optimal objective value. The idea is to

---

[8] For $q = n$ we obtain from (14.9) $\theta = 2/\sqrt{n}$ and for $q \geq 2\sqrt{n}$ we get $\theta \leq 1/2 + 1/\sqrt{n}$. These values for θ are larger than the respective values used by Renegar and Den Hertog. We should note however that this is, at least partly, due to the fact that the analysis of both Renegar and den Hertog is based on the use of approximate central solutions whereas we made the simplifying assumption that exact central solutions are computed for each value of μ_z.

move this bound up to the point where the objective is set near to its optimal value. For any such lower bound z the dual polytope $A^T y \leq c$ is cut by the objective constraint $b^T y \geq z$ and the (ideal) new iterate is a weighted-analytic center of the cut polytope. The weighting vector treats all the constraints in $A^T y \leq c$ equally but it gives extra emphasis to the objective constraint by the factor q. Enlarging q, pushes the new iterate in the direction of the optimal set. This opens the way to adaptive- and large-update versions of Renegar's method. Appropriate values for q can easily be found. To see this it suffices to recall from (14.7) that the duality gap between two successive updates of the lower bound reduces by at least the factor

$$1 - \frac{\theta q}{n + q}.$$

For example, $q = n$ and $\theta = 1/2$ give a reduction of the duality gap by at least $3/4$. It is clear that the reduction factor for the duality gap can be made arbitrarily small by choosing appropriate values for q and θ $(0 < \theta < 1)$. We then get out of the domain of quadratic convergence, but by using damped Newton steps we can reach the new weighted-analytic center in a controlled number of steps. From this it will be clear that the updates of the lower bound can be designed in such a way that adaptive- or large-update versions of the center method arise and that the complexity results will be similar to those for the logarithmic barrier method. These ideas can be worked out easily in the target-following framework. In fact, if Renegar's method is modified according to the updating scheme (14.8), the results immediately follow from the corresponding results for the logarithmic barrier approach.[9]

---

[9] Adaptive and large-update variants of the center method are analyzed by den Hertog [140].

Part IV

Miscellaneous Topics

15

Karmarkar's Projective Method

15.1 Introduction

It has been pointed out before that recent research in interior-point methods for LO has been motivated by the appearance of the seminal paper [165] of Karmarkar in 1984. Despite its extraordinary power of stimulation of the scientific community, Karmarkar's so-called *projective method* seemed to remain a very particular method, remotely related to the huge literature to which it gave rise. Significantly many papers appeared on the projective algorithm itself,[1] but the link with other methods, in particular Renegar's, has not drawn much attention up to recently.[2] The decaying interest for the primal projective method is also due to a poorer behavior on solving practical optimization problems.[3] In this chapter we provide a simplified description and analysis of the projective method and we also relate it to the other methods described in this book.

Karmarkar considered the very special problem

$$(PK) \qquad \min \left\{ c^T x \; : \; Ax = 0, \; e^T x = n, \; x \geq 0 \right\},$$

where, as before, A is an $m \times n$ matrix of rank m, and e denotes the all-one vector. Karmarkar made two seemingly restrictive assumptions, namely that the optimal value $c^T x^*$ of the problem is known and has value zero, and secondly, that the all-one vector e is feasible for (PK). Note that the problem (PK) is trivial if $c^T e = 0$. Then the all-one vector e is an optimal solution. So we assume throughout that this case is excluded. As a consequence we have

$$c^T e > 0. \tag{15.1}$$

---

[1] Papers in that stream were written by Anstreicher [14, 15, 16, 18, 19, 20, 21, 22, 23, 24], Freund [83, 85], de Ghellinck and Vial [95, 96], Goffin and Vial [102, 103], Goldfarb and Mehrotra [105, 106, 107], Goldfarb and Xiao [110], Goldfarb and Shaw [108], Shaw and Goldfarb [254], Gonzaga [117, 119], Roos [239], Vial [282, 283, 284], Xu, Yao and Chen [300], Yamashita [301], Ye [304, 305, 306, 307], Ye and Todd [315] and Todd and Burrell [266]. We also refer the reader to the survey papers Anstreicher [17, 24], Goldfarb and Todd [109], Gonzaga [123, 124], den Hertog and Roos [142] and Todd [265].

[2] See Vial [285, 286].

[3] In their comparison between the primal projective method and a primal-dual method, Fraley and Vial [80, 81] concluded to the superiority of the later for solving optimization problems. However, it is worth mentioning that the projective algorithm has been used with success in the computation of analytic centers in an interior-point cutting plane algorithm; in particular, Bahn et al. [31] and Goffin et al. [98] could solve very large decomposition problems with this approach.

Later on it is made clear that the model (PK) is general enough for our purpose. If it can be solved in polynomial time then the same is true for every LO problem.

15.2 The unit simplex Σ_n in $\mathbb{R}^n$

The feasible region of (PK) is contained in the unit simplex in $\mathbb{R}^n$. This simplex plays a crucial role in the projective method. We denote it by Σ_n:

$$\Sigma_n = \left\{ x \in \mathbb{R}^n \ : \ e^T x = n, \, x \geq 0 \right\}.$$

Obviously[4] the all-one vector e belongs to Σ_n and lies at the heart of it. The sphere in $\mathbb{R}^n$ centered at e and with radius ρ is denoted by $B(e, \rho)$. The analysis of the projective method requires knowledge of the smallest sphere $B(e, R)$ containing Σ_n as well as the largest sphere $B(e, r)$ whose intersection with the hyperplane $e^T x = n$ is contained in Σ_n.

It can easily be understood that R is equal to the Euclidean distance from the center e of Σ_n to the vertex $(n, 0, \ldots, 0)$. See Figure 15.1, which depicts Σ_3. We have

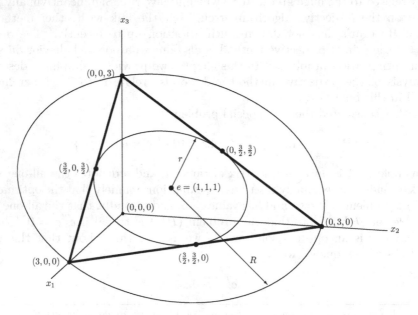

Figure 15.1   The simplex Σ_3.

$$R = \sqrt{(n-1)^2 + (n-1)1^2} = \sqrt{n(n-1)}.$$

Similarly, r is equal to the Euclidean distance from e to the center of one of the faces

---

4 It might be worthwhile to indicate that the dimension of the polytope Σ_n is $n-1$, since this is the dimension of the hyperplane $e^T x = n$, which is the smallest affine space containing Σ_n.

of Σ_n, such as $(0, \frac{n}{n-1}, \ldots, \frac{n}{n-1})$, and therefore

$$r = \sqrt{1 + (n-1)\left(\frac{n}{n-1} - 1\right)^2} = \sqrt{\frac{n}{n-1}}.$$

Assuming $n > 1$, we thus have

$$\frac{r}{R} = \frac{1}{n-1}.$$

15.3   The inner-outer sphere bound

As usual, let $\mathcal{P}$ denote the feasible region of the given problem (PK). Then we may write $\mathcal{P}$ as

$$\mathcal{P} = \Omega \cap \{x \in \mathbb{R}^n \; : \; x \geq 0\},$$

where Ω is the affine space determined by

$$\Omega = \{x \in \mathbb{R}^n \; : \; Ax = 0,\; e^T x = n\}.$$

Now consider the minimization problem

$$\min \{c^T x \; : \; x \in \Omega \cap B(e, r)\}.$$

This problem can be solved explicitly. Since Ω is an affine space containing the center e of the sphere $B(e, r)$, the intersection of the two sets is a sphere of radius r in a lower-dimensional space. Hence the minimum value of $c^T x$ over $\Omega \cap B(e, r)$ occurs uniquely at the point

$$z^1 := e - rp,$$

where p is the vector of unit length whose direction is obtained by projecting the vector c into the linear space parallel to Ω. Similarly, when x runs through $\Omega \cap B(e, R)$, the minimal value will be attained uniquely at the point

$$z^2 := e - Rp.$$

Since

$$\Omega \cap B(e, r) \subseteq \mathcal{P} \subseteq \Omega \cap B(e, R),$$

and the minimal value over $\mathcal{P}$ is given as zero, we must have

$$c^T z^2 \leq 0 \leq c^T z^1.$$

This can be rewritten as

$$c^T e - Rc^T p \leq 0 \leq c^T e - rc^T p.$$

The left inequality and (15.1) imply

$$c^T p \geq \frac{c^T e}{R} > 0.$$

Hence,

$$c^T x^1 = c^T e - r c^T p \le c^T e - \frac{r}{R} c^T e = \left(1 - \frac{1}{n-1}\right) c^T e.$$

Thus, starting at the feasible point e we may construct in this way the new feasible point z^1 whose objective value, compared with the value at e, is reduced by the factor $1 - 1/(n-1)$.

At this stage we note that we want the new point to be positive. The above procedure may end at the boundary of the simplex. This can be prevented by introducing a step-size $\alpha \in (0, 1)$ and using the point

$$z := e - \alpha r p$$

as the new iterate. Below $\alpha \approx 1/2$ will turn out to be a good choice. The objective value is then reduced by the factor

$$1 - \frac{\alpha}{n-1}.$$

It is clear that the above procedure can be used only once. The reduction factor for the objective value is $1 - r/R$, where r/R is the ratio between the radius of the largest inscribed sphere and the radius of the smallest circumscribed sphere for the feasible region. This ratio is maximal at the center e of the feasible region. If we approach the boundary of the region the ratio goes to zero and the reduction factor goes to 1 and we cannot make enough progress to get an efficient method.

Here Karmarkar made a brilliant contribution. His idea is to transform the problem to an equivalent problem by using a projective transformation that maps the new iterate back to the center e of the simplex Σ_n. We describe this transformation in the next section. After the transformation the procedure can be repeated and the objective value is reduced by the same factor. After sufficiently many iterations, a feasible point can be obtained with objective value as close to zero as we wish.

15.4   Projective transformations of Σ_n

Let $d > 0$ be any positive vector. With $\mathbb{R}^n_+$ denoting the set of nonnegative vectors in $\mathbb{R}^n$, the projective transformation $T_d : \mathbb{R}^n_+ \setminus \{0\} \to \Sigma_n$ is defined by

$$T_d : x \mapsto \frac{ndx}{d^T x} = \frac{ndx}{e^T (dx)}.$$

Note that T_d can be decomposed into two transformations: a coordinate-wise scaling $x \mapsto dx$ and a global scaling $x \mapsto nx/e^T x$. The first transformation is defined for each x, and is linear; the second transformation — which coincides with T_e — is only defined if $e^T x$ is nonzero, and is nonlinear. As a consequence, T_d is a nonlinear transformation.

It may easily be verified that T_d maps the simplex Σ_n into itself and that it is invertible on Σ_n; the inverse on Σ_n is simply

$$T_{d^{-1}} : x \mapsto \frac{nd^{-1}x}{e^T (d^{-1}x)}.$$

The projective transformation has some important properties.

Proposition IV.1 *For each $d > 0$ the projective transformation T_d is a one-to-one map of the simplex Σ_n onto itself. The intersection of Σ_n with the linear subspace $\{x : Ax = 0\}$ is mapped to the intersection of Σ_n with another subspace of the same dimension, namely $\{x : AD^{-1}x = 0\}$. Besides, the transformation is positively homogeneous of degree zero; that is, for any $\lambda > 0$,*

$$T_d(\lambda x) = T_d(x).$$

Proof: The first statement is immediate. To prove the second statement, let $x \in \Sigma_n$. Then $Ax = 0$ if and only if $Ad^{-1}dx = 0$, which is equivalent to $AD^{-1}T_d(x) = 0$. This implies the second statement. The last statement is immediate from the definition. $\square$

Now let z be a feasible and positive point. For any nonzero $x \in \mathcal{P}$ there exists a unique $\xi \in \Sigma_n$ such that $x = T_z(\xi)$. We have $Ax = 0$ if and only if $AZ\xi = 0$ and

$$c^T x = c^T T_z(\xi) = c^T \frac{nz\xi}{e^T(z\xi)} = \frac{n (Zc)^T \xi}{e^T(z\xi)}.$$

Hence the problem (PK) can be reformulated as

$$\min \left\{ \frac{n (Zc)^T \xi}{e^T(z\xi)} \ : \ AZ\xi = 0, \ e^T \xi = n, \ \xi \geq 0 \right\}.$$

Note that the objective of this problem is nonlinear. But we know that the optimal value is zero and this can happen only if $(Zc)^T \xi = 0$. So we may replace the nonlinear objective by the linear objective $(Zc)^T \xi$ and, changing the variable ξ back to x, we are left with the linear problem

$$(PKS) \qquad \min \left\{ (Zc)^T x \ : \ AZx = 0, \ e^T x = n, \ x \geq 0 \right\}.$$

Note that the feasibility of z implies $Az = 0$, whence $AZe = 0$, showing that e is feasible for the new problem. Thus we can use the procedure described in Section 15.3 to construct a new feasible point for the transformed problem so that the objective value is reduced by a factor $1 - \alpha/(n-1)$. The new point is obtained by minimizing the objective over the inscribed sphere with radius αr:

$$\min \left\{ (Zc)^T x \ : \ AZx = 0, \ e^T x = n, \ \|x - e\| \leq \alpha r \right\}.$$

15.5 The projective algorithm

We can now describe the algorithm as follows.

Projective Algorithm

Input:
An accuracy parameter $\varepsilon > 0$.

begin

$x := e$;

while $c^T x \geq \varepsilon$ **do**

begin

$z := \text{argmin}_\xi \left\{ (Xc)^T \xi \ : \ AX\xi = 0, \ e^T \xi = n, \ \|\xi - e\| \leq \alpha r \right\}$;

$x := T_x(z)$;

end

end

As long as the objective value at the current iterate x is larger than the threshold value ε, the problem is rescaled by the projective transformation $T_{x^{-1}}$. This makes the all-one vector feasible. Then the new iterate z for the transformed problem is obtained by minimizing the objective value over the inscribed sphere with radius αr. After this the inverse of the map $T_{x^{-1}}$ — that is T_x — is applied to z and we get a point that is feasible for the original problem (PK) again. This is repeated until the objective value is small enough. Figure 15.2 depicts one iteration of the algorithm.

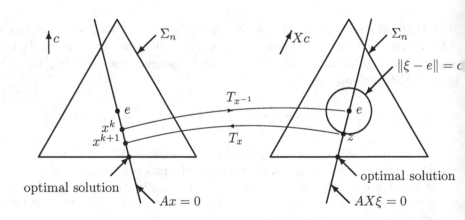

Figure 15.2 One iteration of the projective algorithm ($x = x^k$).

In the next section we derive an iteration bound for the algorithm. Unfortunately, the analysis of the algorithm cannot be based on the reduction of the objective value in each iteration. This is because the objective value is not preserved under the projective transformation. This is the price we pay for the linearization of the nonlinear problem after each projective transformation. Here, again, Karmarkar proposed an elegant solution. The progress of the method can be measured by a suitable potential function. We introduce this function in the next section.

15.6 The Karmarkar potential

Karmarkar used the following potential function in the analysis of his method.

$$\phi_K(x) = n \log c^T x - \sum_{i=1}^{n} \log x_i.$$

The usefulness of this function depends on two lemmas.

Lemma IV.2 *If $x \in \Sigma_n$ then*

$$c^T x \leq \exp\left(\frac{\phi_K(x)}{n}\right).$$

Proof: Since $e^T x = n$, using the geometric-arithmetic-mean inequality we may write

$$\sum_{i=1}^{n} \log x_i \leq n \log \frac{e^T x}{n} = n \log 1 = 0.$$

Therefore

$$\phi_K(x) = n \log c^T x - \sum_{i=1}^{n} \log x_i \geq n \log c^T x,$$

which implies the lemma.   □

Lemma IV.3 *Let x and z be positive vectors in Σ_n and $y = T_x(z)$. Then*

$$\phi_K(x) - \phi_K(y) = n \log \frac{(Xc)^T e}{(Xc)^T z} + \sum_{i=1}^{n} \log z_i.$$

Proof: First we observe that $\phi_K(x)$ is homogeneous of degree zero in x. In other words, for each positive λ we have

$$\phi_K(\lambda x) = \phi_K(x).$$

As a consequence we have

$$\phi_K(y) = \phi_K(T_x(z)) = \phi_K\left(\frac{nxz}{e^T(xz)}\right) = \phi_K(xz),$$

as follows by taking $\lambda = n/e^T(xz)$. Therefore,

$$\phi_K(x) - \phi_K(y) = \phi_K(x) - \phi_K(xz) = n \log \frac{c^T x}{c^T(xz)} - \sum_{i=1}^{n} \log \frac{x_i}{x_i z_i},$$

from which the lemma follows. □

Applying the above lemma with $z = e - \alpha r p$ we can prove that each iteration of the projective algorithm decreases the potential by at least 0.30685 when choosing α appropriately.

Lemma IV.4 *Taking $\alpha = 1/(1+r)$, each iteration of the projective algorithm decreases the potential function value by at least $1 - \log 2 = 0.30685$.*

Proof: By Lemma IV.3, at any iteration the potential function value decreases by the amount

$$\Delta = n \log \frac{(Xc)^T e}{(Xc)^T z} + \sum_{i=1}^{n} \log z_i.$$

Recall that Xc is the objective vector in the transformed problem. Since the objective value of the transformed problem is reduced by at least a factor $1 - \alpha r/R$ and $z = e - \alpha r p$, we obtain

$$\Delta \geq -n \log \left(1 - \frac{\alpha r}{R}\right) + \sum_{i=1}^{n} \log (1 - \alpha r p_i). \tag{15.2}$$

For the first term we write

$$-n \log \left(1 - \frac{\alpha r}{R}\right) = n \left(\frac{\alpha r}{R} + \psi \left(-\frac{\alpha r}{R}\right)\right) = \frac{\alpha n r}{R} + n \psi \left(-\frac{\alpha r}{R}\right) = \alpha r^2 + n \psi \left(-\frac{\alpha r}{R}\right).$$

Here, and below we use the function ψ as defined in (5.5), page 92. The second term in (15.2) can be written as

$$\sum_{i=1}^{n} \log (1 - \alpha r p_i) = -\alpha r e^T p - \sum_{i=1}^{n} \psi (-\alpha r p_i) = -\sum_{i=1}^{n} \psi (-\alpha r p_i).$$

Here we have used the fact that $e^T p = 0$. By the right-hand side inequality in (6.24), on page 134, the above sum can be bounded above by $\psi (-\alpha r \|p\|)$. Since $\|p\| = 1$ we obtain

$$\Delta \geq \alpha r^2 + n \psi \left(-\frac{\alpha r}{R}\right) - \psi (-\alpha r).$$

Omitting the second term, which is nonnegative, we arrive at

$$\Delta \geq \alpha r^2 - \psi (-\alpha r) = \alpha r^2 + \alpha r + \log (1 - \alpha r).$$

The right-hand side expression is maximal if $\alpha = 1/(1+r)$. Substitution of this value yields

$$\Delta \geq r + \log \left(1 - \frac{r}{1+r}\right) = r - \log (1 + r) = \psi (r).$$

Since $r = \sqrt{n/(n-1)} > 1$ we have $\psi (r) > \psi (1) = 1 - \log 2$, and the proof is complete.
□

15.7 Iteration bound for the projective algorithm

The convergence result is as follows.

Theorem IV.5 *After no more than*

$$\frac{n}{\psi(1)} \log \frac{c^T e}{\varepsilon}$$

iterations the algorithm stops with a feasible point x such that $c^T x \leq \varepsilon$.

Proof: After k iterations the iterate x satisfies

$$\phi_K(x) - \phi_K(e) < -k\psi(1).$$

Since $\phi_K(e) = n \log c^T e$,

$$\phi_K(x) < n \log c^T e - k\psi(1).$$

Using Lemma IV.2, we obtain

$$c^T x \leq \exp\left(\frac{\phi_K(x)}{n}\right) < \exp\left(\frac{n \log c^T e - k\psi(1)}{n}\right).$$

The stopping criterion is thus certainly met as soon as

$$\exp\left(\frac{n \log c^T e - k\psi(1)}{n}\right) \leq \varepsilon.$$

Taking logarithms of both sides we get

$$n \log c^T e - k\psi(1) \leq n \log \varepsilon,$$

or equivalently,

$$k \geq \frac{n}{\psi(1)} \log \frac{c^T e}{\varepsilon},$$

which yields the bound in the theorem. $\square$

15.8 Discussion of the special format

The problem (PK) solved by the Projective Method of Karmarkar has a special format that is called the *Karmarkar format*. Except for the so-called *normalizing constraint* $e^T x = n$, the constraints in (PK) are homogeneous. Furthermore, it is assumed that the optimal value is zero and that some positive feasible vector is given.[5] We may

---

[5] In fact, Karmarkar assumed that the all-one vector e is feasible, but it is sufficient if some given positive vector w is feasible. In that case we can use the projective transformation $T_{w^{-1}}$ as defined in Section 15.4, to transform the problem to another problem in the Karmarkar format and for which the all-one vector is feasible.

wonder how the Projective Method could be used to solve an arbitrary LO problem that is not given in the Karmarkar format.[6]

Clearly problem (PK) is in the standard format and, since its feasible region is contained in the unit simplex Σ_n in $\mathbb{R}^n$, the feasible region is bounded. Finally, since the all-one vector is feasible, (PK) satisfies the interior-point condition. In this section we first show that a problem (P) in standard format can easily be reduced to the Karmarkar format whenever the feasible region $\mathcal{P}$ of (P) is bounded and the interior-point condition is satisfied. Secondly, we discuss how a general LO problem can be put in the format of (PK).

Thus, let the feasible region $\mathcal{P}$ of the standard problem

$$(P) \qquad \min \left\{ c^T x \ : \ Ax = b, \ x \geq 0 \right\}$$

be bounded and let it contain a positive vector. Now let the pair $(\bar{y}, \bar{s})$ be optimal for the dual problem

$$(D) \qquad \max \left\{ b^T y \ : \ A^T y + s = c, \ s \geq 0 \right\}.$$

Then we have, for any primal feasible x,

$$\bar{s}^T x = c^T x - b^T \bar{y}.$$

So $\bar{s}^T x$ and $c^T x$ differ by the constant $b^T \bar{y}$ and hence the problem

$$(P') \qquad \min \left\{ \bar{s}^T x \ : \ Ax = b, \ x \geq 0 \right\}$$

has the same optimal set as (P). Since $\bar{s}$ is dual optimal, the optimal value of (P') is zero. Since the feasible region $\mathcal{P}$ is bounded, we deduce from Corollary II.14 that the row space of the constraint matrix A contains a positive vector. That is, there exists a $\lambda \in \mathbb{R}^m$ such that

$$v := A^T \lambda > 0.$$

Now, defining

$$\nu := b^T \lambda,$$

we have for any feasible x,

$$v^T x = \left(A^T \lambda \right)^T x = \lambda^T Ax = \lambda^T b = \nu.$$

---

[6] The first assumption on a known optimal value for a problem in the Karmarkar format was removed by Todd and Burrell [266]. They used a simple observation that for any ζ, the objective $c^T x - \zeta$ is equivalent to $(c - (\zeta/n) e)^T x$. If $\zeta = \zeta^*$, the optimal value of problem (PK), the assumption of a zero optimal value is verified for the problem with the new objective. If $\zeta < \zeta^*$, Todd and Burrell were able to show that the algorithm allows an update of the lower bound ζ by a simple linear ratio test after finitely many iterations; the overall procedure has the same complexity as the original algorithm of Karmarkar. The second assumption of a known interior feasible solution was removed by Ghellinck and Vial [95] by using a different projective embedding. They also used the same parametrization as Todd and Burrell and thus produced the first combined phase I – phase II interior-point algorithm, simultaneously resolving optimality and feasibility. They also pointed out that the projective algorithm was truly a Newton method. The update of the bound in their method is done by an awkward quadratic test. Fraley [79] was able to replace the quadratic test by a simpler linear ratio test. To remain consistent with Part I of the book, we shall not dwell upon those approaches, but rather use a homogeneous self-dual embedding, and analyze the behavior of Karmarkar algorithm on the embedding problem.

Since there exists a positive primal feasible x and v is positive, it follows that $\nu = v^T x > 0$. We may write

$$\nu A x = \nu b = \left(v^T x\right) b = b \left(v^T x\right) = \left(b v^T\right) x.$$

Hence,

$$\left(\nu A - b v^T\right) x = 0.$$

Defining

$$A' := \nu A - b v^T,$$

we conclude that

$$\mathcal{P} = \left\{ x \ : \ A'^T x = 0, \ v^T x = \nu \right\},$$

and hence (P') can be reformulated as

$$(P') \qquad \min \left\{ \bar{s}^T x \ : \ A'x = 0, \ v^T x = \nu, \ x \geq 0 \right\},$$

where $\nu > 0$. This problem can be rewritten as

$$(P'') \qquad \min \left\{ \left(\bar{s}v^{-1}\right)^T \bar{x} \ : \ \left(A'V^{-1}\right) \bar{x} = 0, \ e^T \bar{x} = n, \ \bar{x} \geq 0 \right\},$$

where the new variable $\bar{x}$ relates to the old variable x according to $\bar{x} = n v x / \nu$. Since (P) satisfies the interior-point condition, this condition is also satisfied by (P'). Hence, the problem (P'') is not only equivalent to the given standard problem (P), but it satisfies all the conditions of the Karmarkar format: except for the normalizing constraint the constraints are homogeneous, the optimal value is zero, and some positive feasible vector is given. Thus we have shown that any standard primal problem for which the feasible set is bounded has a representation in the Karmarkar format.[7]

Our second goal in this section is to point out that any given LO problem can be transformed to a problem in the Karmarkar format. Here we use some results from Chapter 2. First, the given problem can be put in the canonical format, where all constraints are inequality constraints and the variables are nonnegative (see Appendix D.1). Then we can embed the resulting canonical problem — and its dual problem — in a homogeneous self-dual problem, as described in Section 2.5 (cf. (2.15)). Thus we arrive at a problem of the form

$$\min \left\{ 0^T x \ : \ Mx \geq 0, \ x \geq 0 \right\},$$

where M is skew-symmetric $(M = -M^T)$ and we need to find a strictly complementary solution for this problem. We proceed by reducing this problem to the Karmarkar format.

First we use the procedure described in Section 2.5 to embed the above self-dual problem in a self-dual problem that satisfies the interior-point condition. As before, let the vector r be defined by

$$r := e - Me.$$

---

[7] It should be noted that this statement has only theoretical value; to reduce a given standard problem with bounded feasible region to the Karmarkar format we need a dual feasible pair $(\bar{y}, \bar{s})$ with $\bar{s} > 0$; in general such a pair will not be available beforehand.

Now consider the self-dual model in $\mathbb{R}^{n+1}$ given by

$$\min\left\{\begin{bmatrix} 0 \\ n+1 \end{bmatrix}^T \begin{bmatrix} x \\ \xi \end{bmatrix} : \begin{bmatrix} M & r \\ -r^T & 0 \end{bmatrix}\begin{bmatrix} x \\ \xi \end{bmatrix} + \begin{bmatrix} 0 \\ n+1 \end{bmatrix} \geq \begin{bmatrix} 0 \\ 0 \end{bmatrix}, \begin{bmatrix} x \\ \xi \end{bmatrix} \geq 0\right\}.$$

Taking

$$(x,\xi) = (e,1),$$

we get

$$\begin{bmatrix} M & r \\ -r^T & 0 \end{bmatrix}\begin{bmatrix} x \\ \xi \end{bmatrix} + \begin{bmatrix} 0 \\ n+1 \end{bmatrix} = \begin{bmatrix} Me+r \\ -r^T e + n + 1 \end{bmatrix} = \begin{bmatrix} e \\ 1 \end{bmatrix},$$

as can easily be verified. By introducing the surplus vector (s,η), we can write the inequality constraints as equality constraints and get the equivalent problem

$$\min\left\{\xi : \begin{bmatrix} M & r \\ -r^T & 0 \end{bmatrix}\begin{bmatrix} x \\ \xi \end{bmatrix} - \begin{bmatrix} s \\ \eta \end{bmatrix} = \begin{bmatrix} 0 \\ -n-1 \end{bmatrix}, \begin{bmatrix} x \\ \xi \end{bmatrix}, \begin{bmatrix} s \\ \eta \end{bmatrix} \geq 0\right\}. \quad (15.3)$$

We replaced the objective $(n+1)\xi$ by ξ; this is allowed since the optimal objective is 0. Note that the all-one vector $((x,\xi,s,\eta) = (e,1,e,1))$ is feasible for (15.3) and the optimal value is zero. When summing up all the constraints we obtain

$$e^T M x + e^T r \xi - e^T s - r^T x - \eta = -n - 1.$$

Since $r = e - Me$ and $e^T Me = 0$, this reduces to

$$e^T x + e^T s + \xi + \eta = (n+1)(1+\xi). \quad (15.4)$$

We can replace the last equality constraint in (15.3) by (15.4). Thus we arrive at the problem

$$\min\left\{\xi : \begin{bmatrix} M & r & -I & 0 \\ e^T & 1 & e^T & 1 \end{bmatrix}\begin{bmatrix} x \\ \xi \\ s \\ \eta \end{bmatrix} = \begin{bmatrix} 0 \\ (n+1)(1+\xi) \end{bmatrix}, \begin{bmatrix} x \\ \xi \\ s \\ \eta \end{bmatrix} \geq 0\right\}. \quad (15.5)$$

Instead of this problem we consider

$$\min\left\{\xi : \begin{bmatrix} M & r & -I & 0 \\ e^T & 1 & e^T & 1 \end{bmatrix}\begin{bmatrix} x \\ \xi \\ s \\ \eta \end{bmatrix} = \begin{bmatrix} 0 \\ 2(n+1) \end{bmatrix}, \begin{bmatrix} x \\ \xi \\ s \\ \eta \end{bmatrix} \geq 0\right\}. \quad (15.6)$$

We established above that the all-one vector is feasible for (15.5); obviously this implies that the all-one vector is also feasible for (15.6). It follows that the problem (15.6) is in the Karmarkar format and hence it can be solved by the projective method. Any optimal solution (x^*,ξ^*,s^*,η^*) of (15.6) has $\xi^* = 0$. It is easily verified that $(x^*,\xi^*,s^*,\eta^*)/2$ is feasible for (15.5) and also optimal.

Thus we have shown how any given LO problem can be embedded into a problem that has the Karmarkar format and for which the all-one vector is feasible. We should note however that solving the given problem by solving the embedding problem requires a strictly complementary solution of the embedding problem. Thus we are left with an important question, namely, does the Projective Method yield a strictly complementary solution? A positive answer to this question has been given by Muramatsu and Tsuchiya [223]. Their proof uses the fact that there is a close relation between Karmarkar's method and the primal affine-scaling method of Dikin[8] when applied to the homogeneous problem obtained by omitting the normalizing constraint in the Karmarkar format. The next two sections serve to highlight this relation. We first derive an explicit expression for the search direction in the Projective Method. The result is that this direction can be interpreted as a primal logarithmic barrier direction for the homogeneous problem. Then we show that the homogeneous problem has optimal value zero and that any strictly complementary solution of the homogeneous problem yields a solution of the Karmarkar format.

15.9 Explicit expression for the Karmarkar search direction

It may be surprising that in the discussion of Karmarkar's approach there is no mention of some issues that were crucial in the methods discussed in the rest of this book. The most striking example of this is the complete absence of the central path in Karmarkar's approach. Also, whereas the search direction in all the other methods is obtained by applying Newton's method — either to a logarithmic barrier function or to the centering conditions — the search direction in the Projective Method is obtained from a different perspective. The aim of this section is to derive an explicit expression for the search direction in the Projective Method. In this way we establish a surprising relation with the Newton direction in the primal logarithmic method for the homogeneous problem arising when the normalizing constraint in the Karmarkar format is neglected.

Let x be a positive vector that is feasible for (PK). Recall from Section 15.5 that the new iterate x^+ in the Projective Algorithm is obtained from $x^+ = T_x(z)$ where

$$z = \text{argmin}_\xi \left\{ (Xc)^T \xi \; : \; AX\xi = 0, \; e^T\xi = n, \; \|\xi - e\| \leq \alpha r \right\}.$$

Here r denotes the radius of the maximal inscribed sphere in the simplex Σ_n and α is the step-size. From this we can easily derive that[9]

$$z = e + \alpha \Delta z,$$

where

$$\Delta z = \text{argmin}_{\Delta\xi} \left\{ (Xc)^T \Delta\xi \; : \; AX\Delta\xi = 0, \; e^T\Delta\xi = 0, \; \|\Delta\xi\| = r \right\}.$$

---

[8] For a brief description of the primal affine-scaling method of Dikin we refer to the footnote on page 339.

[9] We assume throughout that $c^T x$ is not constant on the feasible region of (PK). With this assumption the vector z is uniquely defined.

By writing down the first-order optimality conditions for this minimization problem we obtain

$$
\begin{aligned}
AX\Delta z &= 0 \\
e^T \Delta z &= 0 \\
Xc &= XA^T y + \sigma e + \eta \Delta z \\
\|\Delta z\| &= r,
\end{aligned}
\tag{15.7}
$$

where $\sigma, \eta \in \mathbb{R}$ and $y \in \mathbb{R}^m$. Multiplying the third equation from the left by AX and using the first equation and

$$
AXe = Ax = 0,
\tag{15.8}
$$

we get

$$
AX^2 c = AX^2 A^T y,
$$

whence

$$
y = \left(AX^2 A^T\right)^{-1} AX^2 c.
$$

Substituting this in the third equation of (15.7) gives

$$
\begin{aligned}
\sigma e + \eta \Delta z &= Xc - XA^T \left(AX^2 A^T\right)^{-1} AX^2 c \\
&= \left(I - (AX)^T \left(AX^2 A^T\right)^{-1} AX\right) Xc \\
&= P_{AX}\left(Xc\right).
\end{aligned}
\tag{15.9}
$$

Taking the inner product with e on both sides, while using $e^T \Delta z = 0$ and $e^T e = n$, we get

$$
n\sigma = e^T P_{AX}\left(Xc\right).
$$

Since $AXe = 0$, according to (15.8), e belongs to the null space of AX and hence

$$
P_{AX}\left(e\right) = e.
\tag{15.10}
$$

Using this we write

$$
e^T P_{AX}\left(Xc\right) = (Xc)^T P_{AX}\left(e\right) = (Xc)^T e = c^T x.
\tag{15.11}
$$

Thus we obtain $n\sigma = c^T x$. Substituting this in (15.9) we get

$$
\eta \Delta z = P_{AX}\left(Xc\right) - \frac{c^T x}{n} e = P_{AX}\left(Xc - \frac{c^T x}{n} e\right).
$$

The second equality follows by using (15.10) once more. Up to its sign, the value of the factor η now follows from $\|\Delta z\| = r$. This implies

$$
\Delta z = \pm r \, \frac{P_{AX}\left(Xc - \frac{c^T x}{n} e\right)}{\left\|P_{AX}\left(Xc - \frac{c^T x}{n} e\right)\right\|}.
\tag{15.12}
$$

Here we assumed that the vector

$$
\chi := P_{AX}\left(Xc - \frac{c^T x}{n} e\right) = P_{AX}\left(Xc\right) - \frac{c^T x}{n} e
\tag{15.13}
$$

is nonzero. This is indeed true. We leave this fact as an exercise to the reader.[10] The sign in (15.12) follows by using that we are minimizing $(Xc)^T \Delta z$. So we must have $(Xc)^T \Delta z \leq 0$. In this respect the following observation is crucial. By using the Cauchy–Schwarz inequality we may write

$$c^T x = (Xc)^T e = (Xc)^T P_{AX}(e) = e^T P_{AX}(Xc) \leq \sqrt{n} \, \|P_{AX}(Xc)\|.$$

Note that this inequality holds with equality only if $P_{AX}(Xc)$ is a scalar multiple of e. This would imply that Δz is a scalar multiple of e. Since $e^T \Delta z = 0$ and $\|\Delta z\| = r > 0$ this case cannot occur. Thus we obtain

$$\|P_{AX}(Xc)\| > \frac{c^T x}{\sqrt{n}}.$$

As a consequence,

$$(Xc)^T P_{AX}\left(Xc - \frac{c^T x}{n} e\right) \quad = \quad (Xc)^T P_{AX}(Xc) - \frac{c^T x}{n}(Xc)^T P_{AX}(e)$$

$$= \quad \|P_{AX}(Xc)\|^2 - \frac{(c^T x)^2}{n} > 0.$$

We conclude from this that $(Xc)^T \Delta z \leq 0$ holds only for the minus sign in (15.12). Thus we find

$$\Delta z = -\frac{r\chi}{\|\chi\|}. \tag{15.14}$$

We proceed by deriving an expression for x^+. We have

$$x^+ = T_x(z) = \frac{nxz}{x^T z} = \frac{nx(e + \alpha \Delta z)}{x^T(e + \alpha \Delta z)} = x + \left(\frac{nx(e + \alpha \Delta z)}{x^T(e + \alpha \Delta z)} - x\right).$$

So the displacement in the x-space is given by the expression between the brackets. This expression can be reduced as follows. We have

$$\frac{nx(e + \alpha \Delta z)}{x^T(e + \alpha \Delta z)} - x = \frac{nx(e + \alpha \Delta z) - x^T(e + \alpha \Delta z)x}{x^T(e + \alpha \Delta z)} = \alpha \frac{nx(\Delta z) - x^T(\Delta z)x}{x^T(e + \alpha \Delta z)}.$$

Here we used that $e^T x = n$. Hence we may write

$$x^+ = x + \alpha \Delta x,$$

where

$$\Delta x = \frac{nx(\Delta z) - x^T(\Delta z)x}{x^T(e + \alpha \Delta z)}. \tag{15.15}$$

Using (15.14) the enumerator in the last expression can be reduced as follows:

$$nx(\Delta z) - x^T(\Delta z)x = -\frac{nrx\chi}{\|\chi\|} + \frac{r(x^T \chi)x}{\|\chi\|} = \frac{rx}{\|\chi\|}\left((x^T \chi)e - nx\right).$$

───────────────────────

[10] **Exercise 79** Show that the assumption (15.1) implies that $c^T x$ is positive on the (relative) interior of the feasible region of (PK). Derive from this that the vector χ is nonzero, for any feasible x with $x > 0$.

Using the definition (15.13) of χ and $e^T x = n$, we may write

$$
\begin{aligned}
\left(x^T \chi\right) e - n\chi &= \left(x^T P_{AX}(Xc) - c^T x\right) e - n P_{AX}(Xc) + \left(c^T x\right) e \\
&= \left(x^T P_{AX}(Xc)\right) e - n P_{AX}(Xc) \\
&= n\mu\, P_{AX}\left(e - \frac{Xc}{\mu}\right)
\end{aligned}
$$

where

$$
\mu = \frac{x^T P_{AX}(Xc)}{n}. \tag{15.16}
$$

So we have

$$
nx\left(\Delta z\right) - x^T\left(\Delta z\right) x = \frac{rn\mu}{\|\chi\|} X P_{AX}\left(e - \frac{Xc}{\mu}\right).
$$

Substituting this relation in the above expression (15.15) for Δx gives

$$
\Delta x = \frac{rn\mu}{\|\chi\|\, x^T\left(e + \alpha \Delta z\right)} X P_{AX}\left(e - \frac{Xc}{\mu}\right) \tag{15.17}
$$

Thus we have found an explicit expression for the search direction Δx used in the Projective Method of Karmarkar.[11] Note that this direction is a scalar multiple of

$$
-X P_{AX}\left(\frac{Xc}{\mu} - e\right)
$$

and that this is precisely the primal logarithmic barrier direction[12] at x for the barrier parameter value μ, given by (15.16), for the homogeneous problem

$$
(PKH) \qquad \min\left\{c^T x \ : \ Ax = 0,\ x \geq 0\right\}.
$$

Note also that problem (PKH) arises when the normalizing constraint in (PK) is neglected. We consider the problem (PKH) in more detail in the next section.

15.10   The homogeneous Karmarkar format

In this section we want to point out a relation between the primal logarithmic barrier method when applied to the homogeneous problem (PKH) and the Projective Method of Karmarkar. It is assumed throughout that (PK) satisfies the assumptions of the Karmarkar format. Recall that (PKH) is given by

$$
(PKH) \qquad \min\left\{c^T x \ : \ Ax = 0,\ x \geq 0\right\}.
$$

We first show that the optimal value of (PKH) is zero. Otherwise there exists a nonnegative vector x satisfying $Ax = 0$ such that $c^T x < 0$. But then

$$
T_e(x) = \frac{nx}{e^T x}
$$

---

[11] Show that $c^T \Delta x$, with Δx given by (15.17), is negative if and only if

$$
\left(c^T x\right) x^T P_{AX}(Xc) > n \left\| P_{AX}(Xc) \right\|^2.
$$

[12] See Remark III.20 on page 271.

is feasible for (PK) and satisfies $c^T T_e(x) < 0$, contradicting the fact that the optimal value of (PK) is zero. The claim follows.[13]

It is clear that any optimal solution of (PK) is nonzero and optimal for (PKH). So (PK) will have a nonzero optimal solution x. Now, if x is optimal then λx is optimal as well for any nonnegative λ. Therefore, since (PKH) has a nonzero optimal solution, the optimal set of (PKH) is unbounded. This implies, by Corollary II.12 (page 102), that the dual problem (DKH) of (PKH), given by

$$(DKH) \qquad \max \left\{ 0^T y \; : \; A^T y + s = c, \, s \geq 0 \right\},$$

does not contain a strictly feasible solution. Thus, (DKH) cannot satisfy the interior-point condition. As a consequence, the central paths of (PKH) and (DKH) do not exist.

Note that any nonzero feasible solution x of (PKH) can be rescaled to $T_e(x)$ so that it becomes feasible for (PK). All scalar multiples λx, with $\lambda \geq 0$, are feasible for (PKH), so we have a one-to-one correspondence between feasible solutions of (PK) and feasible rays in (PKH). Therefore, we can neglect the normalizing constraint in (PK) and just look for a nonzero optimal solution of (PKH). The behavior of the affine-scaling direction on (PKH) has been carefully analyzed by Tsuchiya and Muramatsu [273]. The results of this paper form the basis of the paper [223] by the same authors in which they prove that the Projective Method yields a strictly complementary solution of (PK).[14]

---

[13] A different proof of the claim can be obtained as follows. The dual problem of (PK) is

$$(DK) \qquad \max \left\{ 0^T y + n\zeta \; : \; A^T y + \zeta e + s = c, \, s \geq 0 \right\}.$$

This problem has an optimal solution and, due to Karmarkar's assumption, its optimal value is zero. Thus it follows that (y, ζ) is optimal for (DK) if and only if $\zeta = 0$ and y is an optimal solution of

$$(DKH) \qquad \max \left\{ 0^T y \; : \; A^T y + s = c, \, s \geq 0 \right\}.$$

By dualizing (DKH) we regain the problem (PKH), and hence, by the duality theorem the optimal value of (PKH) must be zero.

[14] **Exercise 80** Let x be feasible for (PKH) and positive and let $\mu > 0$. Then, defining the number $\delta(x, \mu)$ by

$$\delta(x, \mu) := \min_{y,s} \left\{ \left\| \frac{xs}{\mu} - e \right\| \; : \; A^T y + s = c \right\},$$

we have $\delta(x, \mu) \geq 1$. Prove this.

16

More Properties of the Central Path

16.1 Introduction

In this chapter we reconsider the self-dual problem

$$(SP) \qquad \min \left\{ q^T x \ : \ Mx \geq -q, \ x \geq 0 \right\},$$

where the matrix M is of size $n \times n$ and skew-symmetric, and the vector q is nonnegative.

We assume that the central path of (SP) exists, and our aim is to further investigate the behavior of the central path, especially as μ tends to 0. As usual, we denote the μ-center by $x(\mu)$ and its surplus vector by $s(\mu) = s(x(\mu))$. From Theorem I.30 (on page 45) we know that the central path converges to the analytic center of the optimal set $\mathcal{SP}^*$ of (SP). The limit point x^* and $s^* := s(x^*)$ form a strictly complementary optimal solution pair, and hence determine the optimal partition of (SP), which is denoted by $\pi = (B, N)$.

We first deal with the derivatives of $x(\mu)$ and $s(\mu)$ with respect to μ. In the next section we prove their existence. In Section 16.2.2 we show that the derivatives are bounded, and we also investigate the limits of the derivatives when μ approaches zero.

In a final section we show that there exist two homothetic ellipsoids that are centered at the μ-center and which respectively contain, and are contained in, an appropriate level set of the objective value $q^T x$.

16.2 Derivatives along the central path

16.2.1 Existence of the derivatives

A fundamental result in the theory of interior point methods is the existence and uniqueness of the solution of the system

$$F(w, x, s) = \begin{bmatrix} Mx + s - q \\ xs - w \end{bmatrix} = 0$$

for all positive w.[1] The solution is denoted by $x(w)$ and $s(w)$.

Remark IV.6 It is possible to give an elementary proof of the fact that the equation $F(w, x, s) = 0$ cannot have more than one solution. This goes as follows. Let x^1, s^1 and x^2, s^2 denote two solutions of the equation. Define $\Delta x := x^2 - x^1$, and $\Delta s := s^2 - s^1$. Then it follows from $Mx^1 + s^1 = Mx^2 + s^2 = q$ that

$$M\Delta x + \Delta s = 0.$$

Since M is skew symmetric it follows that $\Delta x^T \Delta s = -\Delta x^T M \Delta x = 0$, so

$$e^T (\Delta x \Delta s) = 0. \tag{16.1}$$

From $x^1 s^1 = x^2 s^2 = w$ we derive that if $x_j^1 = x_j^2$ holds for some j then also $s_j^1 = s_j^2$, and vice versa. In other words,

$$(\Delta x)_j = 0 \Leftrightarrow (\Delta s)_j = 0, \quad j = 1, \cdots, n. \tag{16.2}$$

Also, if $x_j^1 \leq x_j^2$ for some j, then $s_j^1 \geq s_j^2$, and if $x_j^1 \geq x_j^2$ then $s_j^1 \leq s_j^2$. Thus we have

$$(\Delta x)_j (\Delta s)_j \leq 0, \quad j = 1, \cdots, n.$$

Using (16.1) we obtain

$$(\Delta x)_j (\Delta s)_j = 0, \quad j = 1, \cdots, n.$$

This, together with (16.2) yields that $(\Delta x)_j = 0$ and $(\Delta s)_j = 0$, for each j. Hence we conclude that $\Delta x = \Delta s = 0$. Thus we have shown $x^2 = x^1$ and $s^2 = s^1$, proving the claim.[2] ●

With $z = (x, s)$, the gradient matrix (or Jacobian) of $F(w, x, s)$ with respect to z is

$$\nabla_z F(w, x, s) = \begin{bmatrix} M & I \\ S & X \end{bmatrix},$$

where S and X are the diagonal matrices corresponding to x and s, respectively. This matrix is independent of w and depends continuously on x, s and is nonsingular. Hence we may apply the implicit function theorem.[3] Since $F(w, x, s)$ is infinitely many times differentiable the same is true for $x(w)$ and $s(w)$, and we have

$$\begin{bmatrix} \frac{\partial x}{\partial w} \\ \frac{\partial s}{\partial w} \end{bmatrix} = \begin{bmatrix} M & I \\ S & X \end{bmatrix}^{-1} \begin{bmatrix} 0 \\ I \end{bmatrix}.$$

On the central path we have $w = \mu e$, with $\mu \in (0, \infty)$. Let us consider the more general situation where w is a function of a parameter t, such that $w(t) > 0$ for all $t \in T$ with T an open interval $T \subseteq \mathbb{R}^n$. Moreover, we assume that w is in the class C^∞ of infinitely differentiable functions. Then the first-order derivatives of $x(t)$ and $s(t)$ with respect to t are given by

$$\begin{bmatrix} x'(t) \\ s'(t) \end{bmatrix} = \begin{bmatrix} M & I \\ S(t) & X(t) \end{bmatrix}^{-1} \begin{bmatrix} 0 \\ w'(t) \end{bmatrix}. \tag{16.3}$$

---

[1] This result follows from Theorem II.4 if $w = \mu e$. For arbitrary $w > 0$ a proof similar to that of Theorem III.1 can be given.

[2] A more general result, for the case where M is a so-called P_0-matrix, is proven in Kojima et al. [175].

[3] Cf. Proposition A.2.

Changing notation, and denoting $x'(t)$ by $x^{(1)}$, and similar for s and w, using induction we can easily obtain the higher-order derivatives. Actually, we have

$$
\begin{bmatrix} x^{(k)} \\ s^{(k)} \end{bmatrix} = \begin{bmatrix} M & I \\ S(t) & X(t) \end{bmatrix}^{-1} \begin{bmatrix} 0 \\ \tilde{w} \end{bmatrix}
$$

where

$$
\tilde{w} = w^{(k)} - \sum_{i=1}^{k-1} \binom{k}{i} x^{(k-i)} s^{(i)}.
$$

If w is analytic in t then so are x and s.[4]

When applying the above results to the case where $x = x(\mu)$ and $s = s(\mu)$, with $\mu \in (0, \infty)$, it follows that all derivatives with respect to μ exist and that x and s depend analytically on μ.

16.2.2 *Boundedness of the derivatives*

Recall that the point $x(\mu)$ and its surplus vector $s(\mu)$ are uniquely determined by the system of equations

$$
\begin{aligned}
Mx + q &= s, \quad x \geq 0, \, s \geq 0, \\
xs &= \mu e.
\end{aligned}
\tag{16.4}
$$

Taking derivatives with respect to μ in (16.4) we find, as a special case of (16.3),

$$
\begin{aligned}
M\dot{x} &= \dot{s} \\
x\dot{s} + s\dot{x} &= e.
\end{aligned}
\tag{16.5}
$$

The derivatives of $x(\mu)$ and $s(\mu)$ with respect to μ are now denoted by $\dot{x}$ and $\dot{s}$ respectively. In this section we derive bounds for the derivatives.[5] These bounds are used in the next section to study the asymptotic behavior of the derivatives when μ approaches zero. Since we are interested only in the asymptotic behavior, we assume in this section that μ is bounded above by some fixed positive number $\bar{\mu}$.

Table 16.1. (page 310) summarizes some facts concerning the order of magnitude of the components of various vectors of interest. We are interested in the dependence on μ. All other problem dependent data (like the condition number σ_{SP}, the dimension n of the problem, etc.) are considered as constants in the analysis below.

From Table 16.1. we read that, e.g., $x_B(\mu) = \Theta(1)$ and $\dot{x}_N(\mu) = \mathcal{O}(1)$. For the meaning of the symbols Θ and $\mathcal{O}$ we refer to Section 1.7.4. See also page 190. It is important to stress that the constants hidden in the order symbols are independent

---

[4] This follows from an extension of the implicit function theorem. We refer the reader to, e.g., Fiacco [76], Theorem 2.4.2, page 36. See also Halická [137], Wechs [290] and Zhao and Zhu [321].

[5] We restrict ourselves to first-order derivatives. The asymptotic behavior of the derivatives has been considered by Adler and Monteiro [3], Witzgall, Boggs and Domich [294] and Ye et al. [313]. We also mention Güler [131], who also considers the higher-order derivatives and their asymptotic behavior, both when μ goes to zero and when μ goes to infinity. A very interesting result in his paper is that all the higher-order derivatives vanish if μ approaches infinity, which indicates that the central path is asymptotically linear at infinity.

	Vector	B	N
1	$x(\mu)$	$\Theta(1)$	$\Theta(\mu)$
2	$s(\mu)$	$\Theta(\mu)$	$\Theta(1)$
3	$d(\mu)$	$\Theta(\frac{1}{\sqrt{\mu}})$	$\Theta(\sqrt{\mu})$
4	$\dot{x}(\mu)$	$\mathcal{O}(1)$	$\mathcal{O}(1)$
5	$\dot{s}(\mu)$	$\mathcal{O}(1)$	$\mathcal{O}(1)$

Table 16.1. Asymptotic orders of magnitude of some relevant vectors.

of the vectors x, s and of the value μ of the barrier parameter. They depend only on the problem data M and q and the upper bound $\bar{\mu}$ for μ.

The statements in the first two lines of the table almost immediately follow from Lemma I.43 on page 57. For example, for $i \in B$ the lemma states $n x_i(\mu) \geq \sigma_{SP}$, where σ_{SP} is the condition number of (SP). This means that $x_i(\mu)$ is bounded below by a constant. But, since $x_i(\mu)$ is bounded on the finite section $0 < \mu \leq \bar{\mu}$ of the central path, as a consequence of Lemma I.9 (page 24), $x_i(\mu)$ is also bounded above by a constant. This justifies the statement $x_i(\mu) = \Theta(1)$. Since, $x_i(\mu) s_i(\mu) = \mu$, this also implies $s_i(\mu) = \Theta(\mu)$. This explains the first two lines for the B-part. The estimates for the N-parts of $x_i(\mu)$ and $s_i(\mu)$ are derived in the same way.

The third line shows order estimates for the vector $d(\mu)$, given by

$$d(\mu) = \sqrt{\frac{x(\mu)}{s(\mu)}}.$$

These estimates immediately follow from the first two lines of the table. It remains to deal with the last two lines in the table, which concern the derivatives.

In the rest of this section we omit the argument μ and write simply x instead of $x(\mu)$. This gives no rise to confusion. We start by writing the second equation in (16.5) in a different way. Dividing both sides by $\sqrt{xs}$, and using that $xs = \mu e$ we get

$$d\dot{s} + d^{-1}\dot{x} = \frac{e}{\sqrt{\mu}}. \tag{16.6}$$

Note that the orthogonality of $\dot{x}$ and $\dot{s}$ — which is immediate from the first equation in (16.5) since M is skew-symmetric — implies that the vectors $d\dot{s}$ and $d^{-1}\dot{x}$ are orthogonal as well. Hence we have

$$\left\| d\dot{s} \right\|^2 + \left\| d^{-1}\dot{x} \right\|^2 = \left\| \frac{e}{\sqrt{\mu}} \right\|^2 = \frac{n}{\mu}.$$

Consequently

$$\left\| d\dot{s} \right\| \leq \frac{\sqrt{n}}{\sqrt{\mu}}, \quad \left\| d^{-1}\dot{x} \right\| \leq \frac{\sqrt{n}}{\sqrt{\mu}}.$$

This implies

$$\|d_B \dot{s}_B\| \leq \frac{\sqrt{n}}{\sqrt{\mu}}, \quad \|d_N^{-1} \dot{x}_N\| \leq \frac{\sqrt{n}}{\sqrt{\mu}}.$$

The third line in the table gives $d_B = \Theta(1/\sqrt{\mu})$. This together with the left-hand side inequality implies $\dot{s}_B = \mathcal{O}(1)$. Similarly, the right-hand side inequality implies that $\dot{x}_N = \mathcal{O}(1)$. Thus we have settled the derivatives of the small coordinates.

It remains to deal with the estimates for the derivatives of the large coordinates, $\dot{x}_B$ and $\dot{s}_N$. This is the harder part. We need to characterize the scaled derivatives $d\dot{s}$ and $d^{-1}\dot{x}$ in a different way.

Lemma IV.7 *Let $\tilde{x}$ be any vector in $\mathbb{R}^n$ and $\tilde{s} = s(\tilde{x})$. Then*

$$\begin{bmatrix} d^{-1}\dot{x} \\ d\dot{s} \end{bmatrix} = \frac{1}{\mu} P_{(MD \, -D^{-1})} \begin{bmatrix} d\tilde{s} \\ d^{-1}\tilde{x} \end{bmatrix}.$$

Here $P_{(MD \, -D^{-1})}$ denotes the orthogonal projection onto the null space of the matrix $(MD - D^{-1})$, where D is the diagonal matrix of d.[6]

Proof: Letting I denote the identity matrix of size $n \times n$, we multiply both sides in (16.6) by $DMD - I$. This gives

$$(DMD - I)(d\dot{s} + d^{-1}\dot{x}) = (DMD - I)\frac{e}{\sqrt{\mu}}.$$

By expanding the products we get

$$DM\dot{x} - d\dot{s} + DMD^2\dot{s} - d^{-1}\dot{x} = DMD\frac{e}{\sqrt{\mu}} - \frac{e}{\sqrt{\mu}}.$$

With $M\dot{x} = \dot{s}$ this simplifies to

$$DMD^2\dot{s} - d^{-1}\dot{x} = DMD\frac{e}{\sqrt{\mu}} - \frac{e}{\sqrt{\mu}},$$

and this can be rewritten as

$$\begin{aligned} \frac{e}{\sqrt{\mu}} - d^{-1}\dot{x} &= DMD\frac{e}{\sqrt{\mu}} - DMD^2\dot{s} = DMD\left(\frac{e}{\sqrt{\mu}} - d\dot{s}\right) \\ &= -DM^T D\left(\frac{e}{\sqrt{\mu}} - d\dot{s}\right). \end{aligned}$$

---

[6] **Exercise 81** Using the notation of Lemma IV.7, let $\tilde{x}$ run through all vectors in $\mathbb{R}^n$. Then the vector

$$\begin{bmatrix} d\tilde{s} \\ d^{-1}\tilde{x} \end{bmatrix}$$

runs through an affine space parallel to the row space of the matrix $\begin{bmatrix} MD & -D^{-1} \end{bmatrix}$. This space intersects the null space of $\begin{bmatrix} MD & -D^{-1} \end{bmatrix}$ in a unique point. Using this, prove that there exists a vector $\tilde{x}$ in $\mathbb{R}^n$ such that

$$\begin{aligned} \mu \dot{x} s(\mu) &= x(\mu)\tilde{s} \\ \mu \dot{s} x(\mu) &= s(\mu)\tilde{x}, \end{aligned}$$

where $\tilde{s} = s(\tilde{x})$.

Using this we write

$$\begin{bmatrix} \frac{e}{\sqrt{\mu}} - d^{-1}\dot{x} \\ \frac{e}{\sqrt{\mu}} - d\dot{s} \end{bmatrix} = \begin{bmatrix} -DM^T \\ D^{-1} \end{bmatrix} D\left(\frac{e}{\sqrt{\mu}} - d\dot{s}\right)$$

$$= -\begin{bmatrix} MD & -D^{-1} \end{bmatrix}^T D\left(\frac{e}{\sqrt{\mu}} - d\dot{s}\right).$$

This shows that the vector on the left belongs to the row space of the matrix $(MD - D^{-1})$. Observing that, on the other hand,

$$\begin{bmatrix} MD & -D^{-1} \end{bmatrix} \begin{bmatrix} d^{-1}\dot{x} \\ d\dot{s} \end{bmatrix} = 0,$$

which means that the vector of scaled derivatives

$$\begin{bmatrix} d^{-1}\dot{x} \\ d\dot{s} \end{bmatrix} \tag{16.7}$$

belongs to the null space of the matrix $(MD - D^{-1})$, we conclude that the vector (16.7) can be characterized as the orthogonal projection of the vector

$$\begin{bmatrix} \frac{e}{\sqrt{\mu}} \\ \frac{e}{\sqrt{\mu}} \end{bmatrix}$$

into the null space of $(MD - D^{-1})$. In other words,

$$\begin{bmatrix} d^{-1}\dot{x} \\ d\dot{s} \end{bmatrix} = P_{(MD - D^{-1})} \begin{bmatrix} \frac{e}{\sqrt{\mu}} \\ \frac{e}{\sqrt{\mu}} \end{bmatrix}.$$

Since $xs = \mu e$, we may replace the vector e by $\sqrt{xs/\mu}$. Now using that $\sqrt{xs} = ds = d^{-1}x$, we get

$$\begin{bmatrix} d^{-1}\dot{x} \\ d\dot{s} \end{bmatrix} = \frac{1}{\mu} P_{(MD - D^{-1})} \begin{bmatrix} ds \\ d^{-1}x \end{bmatrix}.$$

Finally, let $\tilde{x}$ be any vector in $\mathbb{R}^n$ and $\tilde{s} = s(\tilde{x})$. Then we may write

$$\begin{bmatrix} d\tilde{s} \\ d^{-1}\tilde{x} \end{bmatrix} - \begin{bmatrix} ds \\ d^{-1}x \end{bmatrix} = \begin{bmatrix} d(\tilde{s} - s) \\ d^{-1}(\tilde{x} - x) \end{bmatrix} = \begin{bmatrix} DM(\tilde{x} - x) \\ d^{-1}(\tilde{x} - x) \end{bmatrix}$$

$$= \begin{bmatrix} -DM^T(\tilde{x} - x) \\ d^{-1}(\tilde{x} - x) \end{bmatrix} = -\begin{bmatrix} MD & -D^{-1} \end{bmatrix}^T (\tilde{x} - x).$$

The last vector is in the row space of $\begin{bmatrix} MD & -D^{-1} \end{bmatrix}$, and hence we have

$$P_{(MD - D^{-1})} \begin{bmatrix} ds \\ d^{-1}x \end{bmatrix} = P_{(MD - D^{-1})} \begin{bmatrix} d\tilde{s} \\ d^{-1}\tilde{x} \end{bmatrix},$$

proving the lemma. □

Using Lemma IV.7 with $\tilde{x} = x^*$ and $\tilde{s} = s^*$ we have

$$\mu \begin{bmatrix} d^{-1}\dot{x} \\ d\dot{s} \end{bmatrix} = \operatorname{argmin}_{h,k \in \mathbb{R}^n} \left\{ \left\| \begin{bmatrix} ds^* - h \\ d^{-1}x^* - k \end{bmatrix} \right\|^2 : \begin{bmatrix} MD & -D^{-1} \end{bmatrix} \begin{bmatrix} h \\ k \end{bmatrix} = 0 \right\}.$$

Hence, the unique solution of the above least squares problem is given by

$$h = \mu d^{-1}\dot{x}, \quad k = \mu d\dot{s}.$$

The left-hand side of the constraint in the problem can be split as follows:

$$\begin{bmatrix} M_B D_B & -D_N^{-1} \end{bmatrix} \begin{bmatrix} h_B \\ k_N \end{bmatrix} + \begin{bmatrix} M_N D_N & -D_B^{-1} \end{bmatrix} \begin{bmatrix} h_N \\ k_B \end{bmatrix} = 0.$$

Substituting the optimal values for h_N and k_B we find that h_B and k_N need to satisfy

$$\begin{bmatrix} M_B D_B & -D_N^{-1} \end{bmatrix} \begin{bmatrix} h_B \\ k_N \end{bmatrix} = -\mu \begin{bmatrix} M_N D_N & -D_B^{-1} \end{bmatrix} \begin{bmatrix} d_N^{-1}\dot{x}_N \\ d_B\dot{s}_B \end{bmatrix}$$

$$= -\mu \begin{bmatrix} M_N & -I_B \end{bmatrix} \begin{bmatrix} \dot{x}_N \\ \dot{s}_B \end{bmatrix}.$$

Since $x_N^* = 0$ and $s_B^* = 0$ we obtain the following characterization of the derivatives for the large coordinates:

$$\mu \begin{bmatrix} d_B^{-1}\dot{x}_B \\ d_N\dot{s}_N \end{bmatrix} =$$

$$\operatorname{argmin}_{h_B,k_N} \left\{ \left\| \begin{bmatrix} h_B \\ k_N \end{bmatrix} \right\|^2 : \begin{bmatrix} M_B D_B & -D_N^{-1} \end{bmatrix} \begin{bmatrix} h_B \\ k_N \end{bmatrix} = -\mu \begin{bmatrix} M_N & -I_B \end{bmatrix} \begin{bmatrix} \dot{x}_N \\ \dot{s}_B \end{bmatrix} \right\}.$$

$$(16.8)$$

Now let $z = (z_B, z_N)$ be the least norm solution of the equation

$$\begin{bmatrix} M_B & -I_N \end{bmatrix} \begin{bmatrix} z_B \\ z_N \end{bmatrix} = -\mu \begin{bmatrix} M_N & -I_B \end{bmatrix} \begin{bmatrix} \dot{x}_N \\ \dot{s}_B \end{bmatrix}.$$

Then we have

$$\begin{bmatrix} z_B \\ z_N \end{bmatrix} = -\mu \begin{bmatrix} M_B & -I_N \end{bmatrix}^+ \begin{bmatrix} M_N & -I_B \end{bmatrix} \begin{bmatrix} \dot{x}_N \\ \dot{s}_B \end{bmatrix} \qquad (16.9)$$

where $\begin{bmatrix} M_B & -I_N \end{bmatrix}^+$ denotes the pseudo-inverse[7] of the matrix $\begin{bmatrix} M_B & -I_N \end{bmatrix}$. It is obvious that

$$h_B = d_B^{-1} z_B, \quad k_N = d_N z_N$$

---

[7] See Appendix B.

is feasible for (16.8). It follows that

$$\left\| \begin{bmatrix} \mu d_B^{-1} \dot{x}_B \\ \mu d_N \dot{s}_N \end{bmatrix} \right\| \le \begin{bmatrix} d_B^{-1} z_B \\ d_N z_N \end{bmatrix}.$$

From Table 16.1. we know that $d_B^{-1} = \Theta(\sqrt{\mu})$ and $d_N = \Theta(\sqrt{\mu})$, so it follows that

$$\left\| \begin{bmatrix} \mu d_B^{-1} \dot{x}_B \\ \mu d_N \dot{s}_N \end{bmatrix} \right\| \le \Theta(\sqrt{\mu}) \left\| \begin{bmatrix} z_B \\ z_N \end{bmatrix} \right\|.$$

Moreover, we have already established that

$$\dot{x}_N = \mathcal{O}(1), \quad \dot{s}_B = \mathcal{O}(1).$$

Hence, using also (16.9),

$$z = \mu \, \mathcal{O}(1),$$

where the constant in the order symbol now also contains the norm of the matrix $(M_B - I_N)^+ (M_N - I_B)$. Note that this matrix, and hence also its norm, depends only on the data of the problem. Substitution yields, after dividing both sides by μ,

$$\begin{bmatrix} d_B^{-1} \dot{x}_B \\ d_N \dot{s}_N \end{bmatrix} = \Theta(\sqrt{\mu}) \, \mathcal{O}(1).$$

Using once more $d_B^{-1} = \Theta(\sqrt{\mu})$ and $d_N = \Theta(\sqrt{\mu})$, we finally obtain

$$\begin{bmatrix} \dot{x}_B \\ \dot{s}_N \end{bmatrix} = \mathcal{O}(1),$$

completing the proof of the estimates in the table.

16.2.3 Convergence of the derivatives

Consider the second equation in (16.5):

$$x\dot{s} + s\dot{x} = e.$$

Recall that $\dot{x}$ and $\dot{s}$ are orthogonal. Since $xs = \mu e$, the vectors $x\dot{s}$ and $s\dot{x}$ are orthogonal as well, so this equation represents an orthogonal decomposition of the all-one vector e. It is interesting to consider this decomposition as μ goes to zero. This is done in the next theorem. Its proof uses the results of the previous section, which are summarized in Table 16.1..

Theorem IV.8 *If μ approaches zero, then $x\dot{s}$ and $s\dot{x}$ converge to complementary $\{0,1\}$-vectors. The supports of their limits are B and N, respectively.*

Proof: For each index i we have

$$x_i \dot{s}_i + s_i \dot{x}_i = 1.$$

Now let $i \in B$ and let μ approach zero. Then $s_i \to 0$. Since $\dot{x}_i$ is bounded, it follows that $s_i \dot{x}_i \to 0$. Therefore, $x_i \dot{s}_i \to 1$. Similarly, if $i \in N$, then $x_i \to 0$. Since $\dot{s}_i$ is bounded, $x_i \dot{s}_i \to 0$ and hence $s_i \dot{x}_i \to 1$. This implies the theorem. □

The next theorem is an immediate consequence of Theorem IV.8 and requires no further proof. It establishes that the derivatives of the small variables converge if μ approaches zero.[8]

Theorem IV.9 *We have* $\lim_{\mu \downarrow 0} \dot{x}_N = (s_N^*)^{-1}$ *and* $\lim_{\mu \downarrow 0} \dot{s}_B = (x_B^*)^{-1}$. □

16.3 Ellipsoidal approximations of level sets

In this section we discuss another property of μ-centers. Namely, that there exist two homothetic ellipsoids that are centered at the μ-center and which respectively contain, and are contained in, an appropriate level set of the objective function $q^T x$. In this section we keep $\mu > 0$ fixed.

For any $K \geq 0$ we define the level set

$$\mathcal{M}_K := \left\{ x \ : \ x \geq 0, \ s(x) = Mx + q \geq 0, \ q^T x \leq K \right\}. \tag{16.10}$$

Since $q^T x(\mu) = n\mu$, we have $x(\mu) \in \mathcal{M}_K$ if and only if $n\mu \leq K$. Note that $\mathcal{M}_0$ represents the set of optimal solutions of (SP), since $q^T x \leq 0$ if and only if $q^T x = 0$. Hence $\mathcal{M}_0 = \mathcal{SP}^*$.

For any number $r \geq 0$ we also define the ellipsoid

$$\mathcal{E}(\mu, r) := \left\{ x \ : \ \left\| \frac{x}{x(\mu)} - e \right\|^2 + \left\| \frac{s(x)}{s(\mu)} - e \right\|^2 \leq r^2 \right\}.$$

Note that the norms in the defining inequality of this ellipsoid vanish if $x = x(\mu)$, so the analytic center $x(\mu)$ is the center of the ellipsoid $\mathcal{E}(\mu, r)$.

Theorem IV.10 $\mathcal{E}(\mu, 1) \subseteq \mathcal{M}_{\mu(n+\sqrt{n})}$ *and* $\mathcal{M}_0 \subseteq \mathcal{E}(\mu, n)$.

Proof: Assume $x \in \mathcal{E}(\mu, 1)$. We denote $s(x)$ simply by s. To prove the first inclusion we need to show that $x \geq 0$, $s = s(x) \geq 0$ and $q^T x \leq \mu(n + \sqrt{n})$.

To simplify the notation we make use once more of the vectors h_x and h_s introduced in Section 6.9.2, namely

$$h_x = \frac{x}{x(\mu)} - e, \quad h_s = \frac{s}{s(\mu)} - e, \tag{16.11}$$

or equivalently,

$$h_x = \frac{x - x(\mu)}{x(\mu)}, \quad h_s = \frac{s - s(\mu)}{s(\mu)}.$$

---

[8] Theorem IV.9 gives only the limiting values of the derivatives of the small variables and says nothing about convergence of the derivatives for the large coordinates. For this we refer to Güler [131], who shows that all derivatives converge when μ approaches zero along a weighted path. In fact, he extends this result to all higher-order derivatives and he gets similar results for the case where μ approaches infinity.

Obviously, h_x and h_s are orthogonal. Hence, defining

$$h := h_x + h_s,$$

we find

$$\|h\|^2 = \|h_x\|^2 + \|h_s\|^2 \leq 1.$$

Hence $\|h_x\| \leq 1$. We easily see that this implies $x \geq 0$. Similarly, $\|h_s\| \leq 1$ implies $s \geq 0$. Thus it remains to show that $q^T x \leq \mu(n + \sqrt{n})$. Since

$$(h_x + e)(h_s + e) = \frac{xs}{x(\mu)s(\mu)} = \frac{xs}{\mu},$$

and on the other hand

$$(h_x + e)(h_s + e) = h_x h_s + h_x + h_s + e = h_x h_s + h + e,$$

we get

$$h = \frac{xs}{\mu} - e - h_x h_s. \tag{16.12}$$

Taking the inner product of both sides with the all-one vector, while using once more that h_x and h_s are orthogonal, we arrive at

$$e^T h = \frac{x^T s}{\mu} - e^T e = \frac{q^T x}{\mu} - n. \tag{16.13}$$

This gives

$$q^T x = \mu(n + e^T h).$$

Finally, applying the Cauchy–Schwarz inequality to $e^T h$, while using $\|h\| \leq 1$, we get

$$q^T x \leq \mu(n + \|e\|) = \mu(n + \sqrt{n}),$$

proving the first inclusion in the theorem.

To prove the second inclusion, let x be optimal for (SP). Then $q^T x = 0$, and hence, from (16.13), $e^T h = -n$. Since $x \geq 0$ and $s \geq 0$, (16.11) gives $h_x \geq -e$ and $h_s \geq -e$. Thus we find $h \geq -2e$. Now consider the maximization problem

$$\max_{h} \left\{ \|h\|^2 \ : \ e^T h = -n, \ h \geq -2e \right\}, \tag{16.14}$$

and let $\bar{h}$ be a solution of this problem. Then, for arbitrary i and j, with $1 \leq i < j \leq n$, $\bar{h}_i$ and $\bar{h}_j$ solve the problem

$$\max_{h_i, h_j} \left\{ h_i^2 + h_j^2 \ : \ h_i + h_j = \bar{h}_i + \bar{h}_j, \ h_i \geq -2, \ h_j \geq -2 \right\},$$

We easily understand that this implies that either $h_i = -2$ or $h_j = -2$. Thus, $\bar{h}$ must have $n - 1$ coordinates equal to -2 and the remaining coordinate equal to $-n - (n - 1)(-2) = n - 2$, and hence,

$$\|\bar{h}\|^2 = (n - 1)4 + (n - 2)^2 = n^2.$$

Therefore, $\|h\| \leq n$. This means that $x \in \mathcal{E}(\mu, n)$, and hence the theorem follows.[9] $\quad\square$

---

[9] **Exercise 82** Using the notation of this section, prove that

$$\mathcal{M}_{n\mu} \subseteq \mathcal{E}(\mu, 2n).$$

17

Partial Updating

17.1 Introduction

In this chapter we deal with a technique that can be applied to almost every interior-point algorithm to enhance the theoretical efficiency by a factor $\sqrt{n}$. The technique is called *partial updating*, and was introduced by Karmarkar in [165]. His projective algorithm, as presented in Chapter 15, needs $\mathcal{O}(nL)$ iterations and $\mathcal{O}(n^3)$ arithmetic operations per iteration. Thus, in total the projective algorithm requires $\mathcal{O}(n^4 L)$ arithmetic operations. Karmarkar showed that this complexity bound can be reduced to $\mathcal{O}(n^{3.5} L)$ arithmetic operations by using partial updating. It has since become apparent that the same technique can be applied to many other interior-point algorithms with the same effect: a reduction of the complexity bound by a factor $\sqrt{n}$.[1]

The partial updating technique can be described as follows. In an interior-point method for solving the problems (P) and (D) — in the standard format of Part II — each search direction is obtained by solving a linear system involving a matrix of the form $AD^2 A^T$, where the *scaling matrix* $D = \mathrm{diag}(d)$ is a positive diagonal matrix depending on the method. In a primal-dual method we have $D^2 = XS^{-1}$, in a primal method $D = X$, and in a dual method $D = S^{-1}$. The matrix D varies from iteration to iteration, due to the variations in x and/or s. We assume that A is $m \times n$ with rank m. Without partial updating the computation of the search directions requires at each iteration $\mathcal{O}(n^3)$ arithmetic operations for factorization of the matrix $AD^2 A^T$ and only $\mathcal{O}(n^2)$ operations for all the other required arithmetic operations.

Although the matrix $AD^2 A^T$ varies from iteration to iteration, it seems reasonable to expect that the variations are not too large, and that the matrix at the next iteration is related in some sense to the current matrix. In other words, the calculation of the search direction in the next iteration might benefit from earlier calculations. In some way, that goal is achieved by the use of partial updating.

To simplify the discussion we assume for the moment that at some iteration the scaling matrix is the identity matrix I and at the next iteration D. Then, if a_i denotes the i-th column of A, we may write

$$AD^2 A^T = \sum_{i=1}^{n} d_i^2 a_i a_i^T = \sum_{i=1}^{n} a_i a_i^T + \sum_{i=1}^{n} \left(d_i^2 - 1 \right) a_i a_i^T.$$

---

[1] See for example Anstreicher [20], Anstreicher and Bosch [25, 26], Bosch [46], Bosch and Anstreicher [47], den Hertog, Roos and Vial [146], Gonzaga [118], Kojima, Mizuno and Yoshise [177], Mehrotra [204], Mizuno [213], Monteiro and Adler [219], Roos [240], Vaidya [276] and Ye [306].

Hence

$$AD^2A^T = AA^T + \sum_{i=1}^{n} \left(d_i^2 - 1\right) a_i a_i^T,$$

showing that AD^2A^T arises by adding the n rank-one matrices $\left(d_i^2 - 1\right) a_i a_i^T$ to AA^T. Now consider the hypothetical situation that $d_i = 1$ for every i, except for $i = 1$. Then we have

$$AD^2A^T = AA^T + \left(d_1^2 - 1\right) a_1 a_1^T$$

and AD^2A^T is a so-called *rank-one modification* of AA^T. By the well known Sherman-Morrison formula[2] we then have

$$\left(AD^2A^T\right)^{-1} = \left(AA^T\right)^{-1} - \left(d_1^2 - 1\right) \frac{\left(AA^T\right)^{-1} a_1 a_1^T \left(AA^T\right)^{-1}}{1 + \left(d_1^2 - 1\right) a_1^T \left(AA^T\right)^{-1} a_1}.$$

This expression makes clear that the inverse of AD^2A^T is equal to the inverse of AA^T plus a scalar multiple of the rank-one matrix vv^T, where

$$v = \left(AA^T\right)^{-1} a_1.$$

We say that $\left(AD^2A^T\right)^{-1}$ is a *rank-one update* of $\left(AA^T\right)^{-1}$. The computation of a rank-one update requires $\mathcal{O}(n^2)$ arithmetic operations, as may easily be verified.

In the general situation, when all the entries of d differ from 1, the inverse of the matrix AD^2A^T can be obtained by applying n rank-one updates to the inverse of AA^T. This still requires $\mathcal{O}(n^3)$ arithmetic operations.

The underlying idea for the partial updating technique is to perform only those rank-one updates that correspond to coordinates i of d for which $\left|d_i^2 - 1\right|$ exceeds some threshold value. A partial updating algorithm maintains an approximation $\tilde{d}$ of d and uses $A\tilde{D}^2A^T$ instead of AD^2A^T; the value of $\tilde{d}_i$ is updated to its correct value if it deviates too much from d_i. Each update of an entry in $\tilde{d}$ necessitates modification of the inverse (or factorization) of $A\tilde{D}^2A^T$. But each such modification can be accomplished by a rank-one update, and this requires only $\mathcal{O}(n^2)$ arithmetic operations.[3] The success of the partial updating technique comes from the fact that it can reduce the total number of rank-one updates in the course of an algorithm by a factor $\sqrt{n}$.

The analysis of an interior-point algorithm with partial updating consists of two parts. First we need to show that the modified search directions, obtained by using the scaling matrix $\tilde{d}$ instead of d, are sufficiently accurate to maintain the polynomial complexity of the original algorithm; this amounts to showing that the modified algorithm has a worst-case iteration count of the same order of magnitude as the

---

[2] **Exercise 83** Let Q, R, S, T be matrices such that the matrices Q and $Q + RS^T$ are nonsingular and R and S are $n \times k$ matrices of rank $k \leq n$. Prove that

$$(Q + RS^T)^{-1} = Q^{-1} - Q^{-1}R(I + S^TQ^{-1}R)^{-1}S^TQ^{-1}.$$

The Sherman-Morrison formula arises by taking $R = S = a$, where a is a nonzero vector [136].

[3] We refer the reader to Shanno [251] for more details of rank-one updates of a Cholesky factorization of a matrix of the form AD^2A^T.

original algorithm. Then, secondly, we have to count the total number of rank-one updates in the modified algorithm.

As indicated above, the partial updating technique can be applied to a wide class of interior-point algorithms. Below we demonstrate its use only for the dual logarithmic barrier method with full Newton steps, which was analyzed in Chapter 6.

17.2 Modified search direction

Recall from Exercise 35 (page 111) that the search direction in the dual logarithmic barrier method is given by

$$\Delta y = \left(AS^{-2}A^T\right)^{-1}\left(\frac{b}{\mu} - AS^{-1}e\right).$$

More precisely, this is the search direction at y, with $s = c - A^Ty > 0$, and for the barrier parameter value μ. In the sequel we use instead

$$\Delta y = \left(A\tilde{S}^{-2}A^T\right)^{-1}\left(\frac{b}{\mu} - AS^{-1}e\right),$$

where $\tilde{s}$ is such that $\tilde{s} = \lambda s$ with

$$\lambda_i \in \left(\frac{1}{\sigma}, \sigma\right), \quad 1 \leq i \leq n, \tag{17.1}$$

for some fixed real constant $\sigma > 1$. The corresponding displacement in the s-space is given by

$$\Delta s = -A^T\Delta y = -A^T\left(A\tilde{S}^{-2}A^T\right)^{-1}\left(\frac{b}{\mu} - AS^{-1}e\right). \tag{17.2}$$

Letting $\bar{x}$ be such that $A\bar{x} = b$ we may write

$$\tilde{s}^{-1}\Delta s = -\left(A\tilde{S}^{-1}\right)^T\left(A\tilde{S}^{-2}A^T\right)^{-1}A\tilde{S}^{-1}\Lambda\left(\frac{s\bar{x}}{\mu} - e\right),$$

showing that $-\tilde{s}^{-1}\Delta s$ equals the orthogonal projection of the vector

$$\Lambda\left(\frac{s\bar{x}}{\mu} - e\right)$$

into the row space of the matrix $A\tilde{S}^{-1}$. Since the row space of the matrix $A\tilde{S}^{-1}$ is equal to the null space of $H\tilde{S}$, where H is the same matrix as used in Chapter 6 — and defined in Section 5.8, page 111 — we have

$$\tilde{s}^{-1}\Delta s = -P_{H\tilde{S}}\left(\Lambda\left(\frac{s\bar{x}}{\mu} - e\right)\right). \tag{17.3}$$

Note that if $\lambda = e$ then the above expression coincides with the expression for the dual Newton step in (6.1). Defining

$$\tilde{x}(s,\mu) = \text{argmin}_x\left\{\left\|\Lambda\left(\frac{sx}{\mu} - e\right)\right\| : Ax = b\right\}, \tag{17.4}$$

and using the same arguments as in Section 6.5 we can easily verify that

$$P_{H\widetilde{S}}\left(\Lambda\left(\frac{s\bar{x}}{\mu}-e\right)\right)=\Lambda\left(\frac{s\tilde{x}(s,\mu)}{\mu}-e\right)=\frac{\tilde{s}\tilde{x}(s,\mu)}{\mu}-\lambda,$$

yielding the following expression for the modified Newton step:

$$\tilde{s}^{-1}\Delta s=\Lambda\left(e-\frac{s\tilde{x}(s,\mu)}{\mu}\right). \tag{17.5}$$

17.3 Modified proximity measure

The proximity of s to $s(\mu)$ is measured by the quantity

$$\tilde{\delta}(s,\mu):=\left\|\frac{\Delta s}{\tilde{s}}\right\|. \tag{17.6}$$

From (17.5) it follows that the modified Newton step Δs vanishes if and only if $s\tilde{x}(s,\mu)=\mu e$, which holds if and only if $\tilde{x}(s,\mu)=x(\mu)$ and $s=s(\mu)$. As a consequence we have

$$\tilde{\delta}(s,\mu)=0 \Longleftrightarrow s=s(\mu).$$

An immediate consequence of (17.4) and (17.5) is

$$\tilde{\delta}(s,\mu)=\left\|\Lambda\left(e-\frac{s\tilde{x}(s,\mu)}{\mu}\right)\right\|=\min_{x}\left\{\left\|\Lambda\left(\frac{sx}{\mu}-e\right)\right\| : Ax=b\right\}. \tag{17.7}$$

The next lemma shows that the modified proximity $\tilde{\delta}(s,\mu)$ has a simple relation with the standard proximity measure $\delta(s,\mu)$.

Lemma IV.11

$$\frac{\delta(s,\mu)}{\sigma}\leq\tilde{\delta}(s,\mu)\leq\sigma\delta(s,\mu).$$

Proof: Using (17.7) and $\max(\lambda)\leq\sigma$ we may write

$$
\begin{aligned}
\tilde{\delta}(s,\mu) \quad &= \quad \left\|\Lambda\left(e-\frac{s\tilde{x}(s,\mu)}{\mu}\right)\right\|\leq\left\|\Lambda\left(e-\frac{sx(s,\mu)}{\mu}\right)\right\| \\
&\leq \quad \|d\|_{\infty}\left\|e-\frac{sx(s,\mu)}{\mu}\right\|\leq\sigma\delta(s,\mu).
\end{aligned}
$$

On the other hand we have

$$
\begin{aligned}
\delta(s,\mu) \quad &= \quad \left\|e-\frac{sx(s,\mu)}{\mu}\right\|\leq\left\|\Lambda^{-1}\Lambda\left(e-\frac{s\tilde{x}(s,\mu)}{\mu}\right)\right\| \\
&\leq \quad \|d^{-1}\|_{\infty}\left\|\Lambda\left(e-\frac{s\tilde{x}(s,\mu)}{\mu}\right)\right\|\leq\sigma\tilde{\delta}(s,\mu).
\end{aligned}
$$

This implies the lemma. □

The next lemma generalizes Lemma II.26 in Section 6.7.

Lemma IV.12 *Assuming $\tilde{\delta}(s,\mu) \le 1$, let s^+ be obtained from s by moving along the modified Newton step Δs at s for the barrier parameter value μ, and let $\mu^+ = (1-\theta)\mu$. Assuming that s^+ is feasible, we have*

$$\tilde{\delta}(s^+,\mu^+) \le \sigma \sqrt{\tilde{\delta}(s,\mu)^4 + \frac{\theta^2 n}{(1-\theta)^2} + \frac{(\sigma^2 - 1)\,\tilde{\delta}(s,\mu)}{1-\theta}}.$$

Proof: By definition,

$$\delta(s^+,\mu^+) = \min_x \left\{ \left\| e - \frac{s^+ x}{\mu^+} \right\| \; : \; Ax = b \right\}.$$

Substituting for x the vector $\tilde{x}(s,\mu)$ and replacing μ^+ by $\mu(1-\theta)$ we obtain the following inequality:

$$\delta(s^+,\mu^+) \le \left\| e - \frac{s^+ \tilde{x}(s,\mu)}{\mu(1-\theta)} \right\|. \tag{17.8}$$

Simplifying the notation by using

$$h := \frac{\Delta s}{\tilde{s}} = \frac{\Delta s}{\lambda s}, \tag{17.9}$$

we may rewrite (17.5) as

$$s\tilde{x}(s,\mu) = \mu\left(e - \lambda^{-1}h\right). \tag{17.10}$$

Using this and (17.9) we get

$$
\begin{aligned}
s^+ \tilde{x}(s,\mu) &= (s + \Delta s)\,\tilde{x}(s,\mu) = (s + \lambda sh)\,\tilde{x}(s,\mu) \\
&= (e + \lambda h)\, s\tilde{x}(s,\mu) = \mu\,(e + \lambda h)\left(e - \lambda^{-1}h\right).
\end{aligned}
$$

Substituting this into (17.8) we obtain

$$\delta(s^+,\mu^+) \le \left\| e - \frac{(e + \lambda h)\left(e - \lambda^{-1}h\right)}{1-\theta} \right\| = \left\| e - \frac{e - h^2 + (\lambda - \lambda^{-1})\,h}{1-\theta} \right\|.$$

This can be rewritten as

$$\delta(s^+,\mu^+) \le \left\| h^2 - \frac{\theta\left(e - h^2\right)}{1-\theta} - \frac{(\lambda - \lambda^{-1})\,h}{1-\theta} \right\|.$$

The triangle inequality now yields

$$\delta(s^+,\mu^+) \le \left\| h^2 - \frac{\theta\left(e - h^2\right)}{1-\theta} \right\| + \left\| \frac{(\lambda - \lambda^{-1})\,h}{1-\theta} \right\|. \tag{17.11}$$

The first norm resembles (6.12) and, since $\|h\| \le 1$, can be estimated in the same way. This gives

$$\left\| h^2 - \frac{\theta\left(e - h^2\right)}{1-\theta} \right\|^2 \le \|h\|^4 + \frac{\theta^2 n}{(1-\theta)^2}.$$

For the second norm in (17.11) we write

$$\left\| \frac{(\lambda - \lambda^{-1})\, h}{1 - \theta} \right\| \le \| \lambda - \lambda^{-1} \|_\infty \frac{\|h\|}{1 - \theta} \le \frac{(\sigma - \sigma^{-1})\, \|h\|}{1 - \theta}.$$

Substituting the last two bounds in (17.11), while using $\|h\| = \tilde{\delta}(s, \mu)$, we find

$$\delta(s^+, \mu^+) \le \sqrt{\tilde{\delta}(s, \mu)^4 + \frac{\theta^2 n}{(1 - \theta)^2}} + \frac{(\sigma - \sigma^{-1})\, \tilde{\delta}(s, \mu)}{1 - \theta}.$$

Finally, Lemma IV.11 gives $\tilde{\delta}(s^+, \mu^+) \le \sigma \delta(s^+, \mu^+)$ and the bound in the lemma follows. $\qquad\square$

Lemma IV.13 *Let $n \ge 3$. Using the notation of Lemma IV.12 and taking $\sigma = 9/8$ and $\theta = 1/(6\sqrt{n})$, we have*

$$\tilde{\delta}(s, \mu) \le \frac{1}{2} \Rightarrow \tilde{\delta}(s^+, \mu^+) \le \frac{1}{2},$$

and the new iterate s^+ is feasible.

Proof: The implication in the lemma follows by substituting the given values in the bound for $\tilde{\delta}(s^+, \mu^+)$ in Lemma IV.12. If $n \ge 3$ this gives

$$\tilde{\delta}(s^+, \mu^+) \le 0.49644 < 0.5,$$

yielding the desired result. By Lemma IV.11 this implies $\delta(s^+, \mu^+) \le \sigma/2 = 9/16$. From this the feasibility of s^+ follows. $\qquad\square$

The above lemma shows that for the specified values of the parameters σ and θ the modified Newton steps keep the iterates close to the central path. The value of the barrier update parameter θ in Lemma IV.13 is a factor of two smaller than in the algorithm of Section 6.7. Hence we must expect that the iteration bound for an algorithm based on these parameter values will be a factor of two worse. This is the price we pay for using the modified Newton direction. On the other hand, in terms of the number of arithmetic operations required to reach an ε-solution, the gain is much larger. This will become clear in the next section.

The modified algorithm is described on page 323.

Note that in this algorithm the vector λ may be arbitrarily at each iteration, subject to (17.1). The next theorem specifies values for the parameters τ, θ and σ for which the algorithm is well defined and has a polynomial iteration bound.

Theorem IV.14 *If $\tau = 1/2$, $\theta = 1/(6\sqrt{n})$ and $\sigma = 9/8$, then the Dual Logarithmic Barrier Algorithm with Modified Full Newton Steps requires at most*

$$6\sqrt{n} \log \frac{n\mu^0}{\varepsilon}$$

iterations. The output is a primal-dual pair (x, s) such that $x^T s \le 2\varepsilon$.

Dual Log. Barrier Algorithm with Modified Full Newton Steps

Input:
   A proximity parameter τ, $0 \leq \tau < 1$;
   an accuracy parameter $\varepsilon > 0$;
   $(y^0, s^0) \in \mathcal{D}$ and $\mu^0 > 0$ such that $\delta(s^0, \mu^0) \leq \tau$;
   a barrier update parameter θ, $0 < \theta < 1$;
   a threshold value σ, $\sigma > 1$.
begin
   $s := s^0$; $\mu := \mu^0$;
   while $n\mu \geq (1 - \theta)\varepsilon$ **do**
   begin
      Choose any λ satisfying (17.1);
      $s := s + \Delta s$, Δs from (17.2);
      $\mu := (1 - \theta)\mu$;
   end
end

Proof: According to Lemma IV.13 the algorithm is well defined. The iteration bound is an immediate consequence of Lemma I.36. Finally, the duality gap of the final iterate can be estimated as follows. For the final iterate s we have $\tilde{\delta}(s, \mu) \leq 1/2$, with $n\mu \leq \varepsilon$. Taking $x = \tilde{x}(s, \mu)$ it follows from (17.10) that

$$s^T \tilde{x}(s, \mu) = n\mu - \mu h^T \lambda^{-1}.$$

Since

$$\left| h^T \lambda^{-1} \right| \leq \| \lambda^{-1} \| \, \| h \| \leq \sigma \tilde{\delta}(s, \mu) \sqrt{n} \leq 9n/16 \leq n,$$

we obtain

$$s^T \tilde{x}(s, \mu) \leq 2n\mu \leq \varepsilon.$$

The proof is complete. □

17.4   Algorithm with rank-one updates

We now present a variant of the algorithm in the previous section in which the vector λ used in the computation of the modified Newton step is prescribed. See page 324.

Note that at each iteration the vector $\tilde{s}$ is updated in such a way that the vector λ used in the computation of the modified Newton step satisfies (17.1). As a consequence, the iteration bound for the algorithm is given by Theorem IV.14. Hence, the algorithm yields an exact solution of (D) in $\mathcal{O}\left(\sqrt{n}L \right)$ iterations. Without using partial updates — which corresponds to giving the threshold parameter σ the value 1 — the bound for

---

Full Step Dual Log. Barrier Algorithm with Rank-One Updates

Input:
  A proximity parameter τ, $\tau = 1/2$;
  an accuracy parameter $\varepsilon > 0$;
  $(y^0, s^0) \in \mathcal{D}$ and $\mu^0 > 0$ such that $\delta(s^0, \mu^0) \leq \tau$;
  a barrier update parameter θ, $\theta = 1/(6\sqrt{n})$;
  a threshold value σ, $\sigma = 9/8$.
begin
  $s := s^0$; $\mu := \mu^0$; $\tilde{s} = s$;
  while $n\mu \geq (1 - \theta)\varepsilon$ **do**
  begin
    $\lambda := \tilde{s}s^{-1}$;
    $s := s + \Delta s$, Δs from (17.2);
    for $i := 1$ to n **do**
    begin
      if $\frac{s_i}{\tilde{s}_i} \notin \left(\frac{1}{\sigma}, \sigma\right)$ **then** $\tilde{s}_i := s_i$
    end
      $\mu := (1 - \theta)\mu$;
  end
end

---

the total number of arithmetic operations becomes $\mathcal{O}\left(n^{3.5}L\right)$. Recall that the extra factor n^3 can be interpreted as being due to n rank-one updates per iteration, with $\mathcal{O}\left(n^2\right)$ arithmetic operations per rank-one update.

The total number of rank-one updates in the above algorithm is equal to the number of times that a coordinate of the vector $\tilde{s}$ is updated. We estimate this number in the next section, and we show that on the average it is not more than $\mathcal{O}\left(\sqrt{n}\right)$ per iteration, instead of n. Thus the overall bound for the total number of arithmetical operations becomes $\mathcal{O}\left(n^3 L\right)$.

17.5 Count of the rank-one updates

We need to count (or estimate) the number of times that a coordinate of the vector $\tilde{s}$ changes. Let s^k and $\tilde{s}^k$ denote the values assigned to s and to $\tilde{s}$, respectively, at iteration k of the algorithm. We use also the superscript ^{k} to refer to values assigned to other relevant entities during the k-th iteration. For example, the value assigned to λ at iteration k is denoted by λ^k and satisfies

$$\lambda^k = \frac{\tilde{s}^{k-1}}{s^{k-1}}, \quad k \geq 1.$$

Moreover, denoting the modified Newton step on iteration k by Δs^k, we have

$$\Delta s^k = s^k - s^{k-1} = \tilde{s}^{k-1} h^k = \lambda^k s^{k-1} h^k, \quad k \geq 1. \tag{17.12}$$

Note that the algorithm is initialized so that $s^0 = \tilde{s}^0$ and these are the values of s and $\tilde{s}$ just before the first iteration.

Now consider the i-th coordinate of $\tilde{s}$. Suppose that $\tilde{s}_i$ is updated at iteration $k_1 \geq 0$ and next updated at iteration $k_2 > k_1$. Then the updating rule implies that the sequence

$$\frac{s_i^{k_1+1}}{\tilde{s}_i^{k_1}}, \frac{s_i^{k_1+2}}{\tilde{s}_i^{k_1+1}}, \ldots, \frac{s_i^{k_2-1}}{\tilde{s}_i^{k_2-2}}, \frac{s_i^{k_2}}{\tilde{s}_i^{k_2-1}}$$

has the property that the last entry lies outside the interval $(1/\sigma, \sigma)$ whereas all the other entries lie inside this interval. Since

$$s_i^{k_1} = \tilde{s}_i^{k_1} = \tilde{s}_i^{k_1+1} = \ldots = \tilde{s}_i^{k_2-1}$$

we can rewrite the above sequence as

$$\frac{s_i^{k_1+1}}{s_i^{k_1}}, \frac{s_i^{k_1+2}}{s_i^{k_1}}, \ldots, \frac{s_i^{k_2-1}}{s_i^{k_1}}, \frac{s_i^{k_2}}{s_i^{k_1}}. \tag{17.13}$$

Hence, with

$$p^j := s_i^{k_1+j}, \quad 0 \leq j \leq K := k_2 - k_1, \tag{17.14}$$

the sequence

$$p_0, p_1, \ldots, p_K \tag{17.15}$$

has the property

$$\frac{p_j}{p_0} \in \left(\frac{1}{\sigma}, \sigma\right), \quad 1 \leq j < K \tag{17.16}$$

and

$$\frac{p_K}{p_0} \notin \left(\frac{1}{\sigma}, \sigma\right). \tag{17.17}$$

Our estimate of the number of rank-one updates in the algorithm depends on a technical lemma on such sequences. The proof of this lemma (Lemma IV.15 below) requires another technical lemma that can be found in Appendix C (Lemma C.3).

Lemma IV.15 *Let $\sigma \geq 1$ and let $p_0, p_1, \ldots, p_K$ be a finite sequence of positive numbers satisfying (17.16) and (17.17). Then*

$$\sum_{j=0}^{K-1} \left| \frac{p_{j+1} - p_j}{p_j} \right| \geq 1 - \frac{1}{\sigma}. \tag{17.18}$$

Proof: We start with $K = 1$. Then we need to show

$$\left| \frac{p_1 - p_0}{p_0} \right| = \left| \frac{p_1}{p_0} - 1 \right| \geq 1 - \frac{1}{\sigma}.$$

If $p_1/p_0 \leq 1/\sigma$ then

$$\left| \frac{p_1}{p_0} - 1 \right| = 1 - \frac{p_1}{p_0} \geq 1 - \frac{1}{\sigma}$$

and if $p_1/p_0 \geq \sigma$ then

$$\left| \frac{p_1}{p_0} - 1 \right| = \frac{p_1}{p_0} - 1 \geq \sigma - 1 \geq \frac{\sigma - 1}{\sigma} = 1 - \frac{1}{\sigma}.$$

We proceed with $K \geq 2$. It is convenient to denote the left-hand side expression on (17.18) by $g(p_0, p_1, \ldots, p_K)$. We start with an easy observation: if $p_{j+1} = p_j$ for some j ($0 \leq j < K$) then $g(p_0, p_1, \ldots, p_K)$ does not change if we remove p_{j+1} from the sequence. So without loss of generality we may assume that no two subsequent elements in the given sequence $p_0, p_1, \ldots, p_K$ are equal.

Now let the given sequence $p_0, p_1, \ldots, p_K$ be such that $g(p_0, p_1, \ldots, p_K)$ is minimal. For $0 < j < K$ we consider the two terms in $g(p_0, p_1, \ldots, p_K)$ that contain p_j. The contribution of these two terms is given by

$$\left| \frac{p_j - p_{j-1}}{p_{j-1}} \right| + \left| \frac{p_{j+1} - p_j}{p_j} \right| = \left| 1 - \frac{p_j}{p_{j-1}} \right| + \left| 1 - \frac{\frac{p_{j+1}}{p_{j-1}}}{\frac{p_j}{p_{j-1}}} \right|. \tag{17.19}$$

Since $p_0, p_1, \ldots, p_K$ minimizes $g(p_0, p_1, \ldots, p_K)$, when fixing p_{j-1} and p_{j+1}, p_j must minimize (17.19). If $p_{j+1} \leq p_{j-1}$ then Lemma C.3 (page 437) implies that

$$\frac{p_j}{p_{j-1}} = 1 \text{ or } \frac{p_j}{p_{j-1}} = \frac{p_{j+1}}{p_{j-1}}.$$

This means that

$$p_j = p_{j-1} \text{ or } p_j = p_{j+1}.$$

Hence, in this case the sequence has two subsequent elements that are equal, which has been excluded above. We conclude that $p_{j+1} > p_{j-1}$. Applying Lemma C.3 once more, we obtain

$$\frac{p_j}{p_{j-1}} = \sqrt{\frac{p_{j+1}}{p_{j-1}}}.$$

Thus it follows that

$$p_{j-1} < p_j = \sqrt{p_{j-1} p_{j+1}} < p_{j+1}$$

for each j, $0 < j < K$, showing that the sequence $p_0, p_1, \ldots, p_K$ is strictly increasing and each entry p_j in the sequence, with $0 < j < K$, is the geometric mean of the surrounding entries. This implies that the sequence $p_j/p_0, 1 \leq j \leq K$, is geometric and we have

$$\frac{p_j}{p_0} = \alpha^j, \quad 1 \leq j \leq K,$$

where

$$\alpha = \sqrt{\frac{p_2}{p_0}} > 1.$$

In that case we must have $p_K \geq \sigma$ and hence α satisfies $\alpha^K \geq \sigma$. Since

$$g(p_0, p_1, \ldots, p_K) = \sum_{j=0}^{K-1} \left| \frac{p_{j+1}}{p_j} - 1 \right| = \sum_{j=0}^{K-1} (\alpha - 1) = K(\alpha - 1),$$

the inequality in the lemma follows if

$$K(\alpha - 1) \geq 1 - \frac{1}{\alpha^K}.$$

This inequality holds for each natural number K and for each real number $\alpha \geq 1$. This can be seen by reducing the right-hand side as follows:

$$
\begin{aligned}
1 - \frac{1}{\alpha^K} &= \frac{\alpha^K - 1}{\alpha^K} = \frac{(\alpha - 1) \cdot (\alpha^{K-1} + \ldots + \alpha + 1)}{\alpha^K} \\
&= (\alpha - 1)\left(\alpha^{-1} + \alpha^{-2} \ldots + \alpha^{-K}\right) < K(\alpha - 1).
\end{aligned}
$$

This completes the proof. □

Now the next lemma follows easily.

Lemma IV.16 *Suppose that the component $\tilde{s}_i$ of $\tilde{s}$ is updated at iteration k_1 and next updated at iteration $k_2 > k_1$. Then*

$$\sum_{k=k_1}^{k_2-1} \left| \frac{\Delta s_i^{k+1}}{s_i^k} \right| \geq 1 - \frac{1}{\sigma},$$

where Δs_i^{k+1} denotes the i-th coordinate of the modified Newton step at iteration $k+1$.

Proof: Applying Lemma IV.15 to the sequence $p_0, p_1, \ldots, p_K$ defined by (17.14) we get

$$\sum_{k=k_1}^{k_2-1} \left| \frac{s_i^{k+1} - s_i^k}{s_i^k} \right| \geq 1 - \frac{1}{\sigma}.$$

Since $s^{k+1} - s^k = \Delta s^{k+1}$ by definition, the lemma follows. □

Theorem IV.17 *Let N denote the total number of iterations of the algorithm and n_i the total number of updates of $\tilde{s}_i$. Then*

$$\sum_{i=1}^{n} n_i \leq 6N\sqrt{n}.$$

Proof: Recall from (17.12) that

$$\Delta s^{k+1} = \lambda^{k+1} s^k h^{k+1}.$$

Hence, for $1 \leq i \leq n$,

$$\frac{\Delta s_i^{k+1}}{s_i^k} = \lambda_i^{k+1} h_i^{k+1}.$$

Now Lemma IV.16 implies

$$\sum_{k=1}^{N} |\lambda_i^k h_i^k| = \sum_{k=0}^{N-1} |\lambda_i^{k+1} h_i^{k+1}| \geq n_i \left(1 - \frac{1}{\sigma}\right).$$

Taking the sum over i we obtain

$$\sum_{i=1}^{n} n_i \leq \frac{\sigma}{\sigma - 1} \sum_{k=1}^{N} \sum_{i=1}^{n} \left| \lambda_i^k h_i^k \right|.$$

The inner sum can be bounded above by

$$\sum_{i=1}^{n} \left| \lambda_i^k h_i^k \right| \leq \sigma \sum_{i=1}^{n} \left| h_i^k \right| = \sigma \left\| h^k \right\|_1 \leq \sigma \left\| h^k \right\| \sqrt{n}.$$

Since $\left\| h^k \right\| = \tilde{\delta}(s^k, \mu^k) \leq \tau$ we obtain

$$\sum_{i=1}^{n} n_i \leq \frac{\sigma}{\sigma - 1} \sum_{k=1}^{N} \sigma \tau \sqrt{n} = \frac{N \sigma^2 \tau \sqrt{n}}{\sigma - 1}.$$

Substituting the values of σ and τ specified in the algorithm proves the theorem. $\square$

Finally, using the iteration bound of Theorem IV.14 and that each rank-one update requires $\mathcal{O}(n^2)$ arithmetic operations, we may state our final result without further proof.

Theorem IV.18 *The Full Step Dual Logarithmic Barrier Algorithm with Rank-One Updates requires at most*

$$36n^3 \log \frac{n\mu^0}{\varepsilon}$$

arithmetic operations. The output is a primal-dual pair (x, s) such that $x^T s \leq 2\varepsilon$.

18
Higher-Order Methods

18.1 Introduction

In a target-following method the Newton directions Δx and Δs to a given target point w in the w-space,[1] and at a given positive primal-dual pair (x, s), are obtained by solving the system (10.2):

$$
\begin{aligned}
A\Delta x &= 0, \\
A^T \Delta y + \Delta s &= 0, \\
s\Delta x + x\Delta s &= \Delta w,
\end{aligned}
\tag{18.1}
$$

where $\Delta w = w - xs$. Recall that this system was obtained by neglecting the second-order term $\Delta x \Delta s$ in the third equation of the nonlinear system (10.1), given by

$$
\begin{aligned}
A\Delta x &= 0, \\
A^T \Delta y + \Delta s &= 0, \\
s\Delta x + x\Delta s + \Delta x \Delta s &= \Delta w.
\end{aligned}
\tag{18.2}
$$

An exact solution — $(\Delta^e x, \Delta^e y, \Delta^e s)$ say — of (18.2) would yield the primal-dual pair corresponding to the target w, because

$$
(x + \Delta^e x)(s + \Delta^e s) = w,
$$

as can easily be verified. Unfortunately, finding an exact solution of the nonlinear system (18.2) is hard from a computational point of view. Therefore, following a classical approach in mathematics when dealing with nonlinearity, we linearize the system, and use the solutions of the linearized system (18.1). Denoting its solution simply by $(\Delta x, \Delta y, \Delta s)$, the primal-dual pair $(x + \Delta x, s + \Delta s)$ satisfies

$$
(x + \Delta x)(s + \Delta s) = w - \Delta x \Delta s,
$$

and hence, the 'error' after the step is given by $\Delta x \Delta s$. Thus, this error represents the price we have to pay for using a solution of the linearized system (18.1). We refer to it henceforth as the *second-order effect*.

---

[1] We defined the w-space in Section 9.1, page 220; it is simply the interior of the nonnegative orthant in $\mathbb{R}^n$.

Clearly, the second-order effect strongly depends on the actual data of the problem under consideration.[2]

It would be very significant if we could eliminate the above described second-order effect, or at least minimize it in some way or another. One way to do this is to use so-called *higher-order methods*.[3] The Newton method used so far is considered to be a *first-order method*. In the next section the search directions for higher-order methods are introduced. Then we devote a separate section (Section 18.3) to the estimate of the (higher-order) error term $E^r(\alpha)$, where $r \geq 1$ denotes the order of the search direction and α the step-size. The results of Section 18.3 are applied in two subsequent sections. In Section 18.4 we first discuss and extend the definition of the primal-dual Dikin direction, as introduced in Appendix E for the self-dual problem, to a primal-dual Dikin direction for the problems (P) and (D) in standard format. Then we consider a higher-order version of this direction, and we show that the iteration bound can be reduced by the factor $\sqrt{n}$ without increasing the complexity per iteration. Then, in Section 18.5 we apply the results of Section 18.3 to the primal-dual logarithmic barrier method, as considered in Chapter 7 of Part II. This section is based on a paper of Zhao [320]. Here the use of higher-order search directions does not improve the iteration bound when compared with the (first-order) full Newton step method. Recall that in the full Newton step method the iterates stay very close to the central path. This can be expressed by saying this method keeps the iterates in a 'narrow cone' around the central path. We shall see that a higher-order method allows the iterates to stay further away from the central path, which makes such a method a 'wide cone' method.

18.2 Higher-order search directions

Suppose that we are given a positive primal-dual pair (x, s) and we want to find the primal-dual pair corresponding to $\bar{w} := xs + \Delta w$ for some displacement Δw in the w-space. Our aim is to generate suitable search directions Δx and Δs at (x, s). One way to derive such directions is to consider the linear line segment in the w-space connecting xs with $\bar{w}$. A parametric representation of this segment is given by

$$xs + \alpha \Delta w, \quad 0 \leq \alpha \leq 1.$$

---

[2] In the w-space the ideal situation is that the curve $(x + \alpha \Delta x)(s + \alpha \Delta s)$, $0 \leq \alpha \leq 1$, moves from xs in a straight line to the target w. As a matter of fact, the second-order effect 'blows' the curve away from this straight line segment. Considering α as a time parameter, we can think of the iterate $(x + \alpha \Delta x)(s + \alpha \Delta s)$ as the trajectory of a vessel sailing from xs to w and of the second-order effect as a wind blowing it away from its trajectory. To reach the target w the bargeman can put over the helm now and then, which in our context is accomplished by updating the search direction. In practice, a bargeman will anticipate the fact that the wind is (locally) constant and he can put the helm in a fixed position that prevents the vessel being driven from its correct course. It may be interesting to mention a computer game called *Schiet Op^{TM}*, designed by Brinkhuis and Draisma, that is based on this phenomenon [51]. It requires the player to find an optimal path in the w-space to the origin.

[3] The idea of using higher-order search directions as presented in this chapter is due to Monteiro, Adler and Resende [220], and was later investigated by Zhang and Zhang [318], Hung and Ye [150], Jansen et al. [160] and Zhao [320]. The idea has been applied also in the context of a predictor-corrector method by Mehrotra [202, 205].

To any point of this segment belongs a primal-dual pair and we denote this pair by $(x(\alpha), s(\alpha))$.[4] Since $x(\alpha)$ and $s(\alpha)$ depend analytically[5] on α there exist $x^{(i)}$ and $s^{(i)}$, with $i = 0, 1, \ldots$, such that

$$x(\alpha) = \sum_{i=0}^{\infty} x^{(i)} \alpha^i, \quad s(\alpha) = \sum_{i=0}^{\infty} s^{(i)} \alpha^i, \quad 0 \le \alpha \le 1. \tag{18.3}$$

Obviously, $x(0) = x = x^{(0)}$ and $s(0) = s = s^{(0)}$. From $Ax(\alpha) = b$, for each $\alpha \in [0, 1]$, we derive

$$Ax^{(0)} = b, \quad Ax^{(i)} = 0, \quad i \ge 1. \tag{18.4}$$

Similarly, there exist unique $y^{(i)}$ and $s^{(i)}$, $i = 0, 1, \ldots$, such that

$$A^T y^{(0)} + s^{(0)} = c, \quad A^T y^{(i)} + s^{(i)} = 0, \quad i \ge 1. \tag{18.5}$$

Furthermore, from

$$x(\alpha) s(\alpha) = xs + \alpha \Delta w,$$

by expanding $x(\alpha)$ and $s(\alpha)$ and then equating terms with equal powers of α, we get the following relations:

$$x^{(0)} s^{(0)} = xs \tag{18.6}$$

$$x^{(0)} s^{(1)} + s^{(0)} x^{(1)} = \Delta w \tag{18.7}$$

$$\sum_{i=0}^{k} x^{(i)} s^{(k-i)} = 0, \quad k = 2, 3, \ldots . \tag{18.8}$$

The first relation implies once more that $x^{(0)} = x$ and $s^{(0)} = s$. Using this and (18.4), (18.5) and (18.7) we obtain

$$
\begin{aligned}
Ax^{(1)} &= 0 \\
A^T y^{(1)} + s^{(1)} &= 0 \\
sx^{(1)} + xs^{(1)} &= \Delta w.
\end{aligned}
\tag{18.9}
$$

---

[4] In other chapters of this book $x(\alpha)$ denotes the α-center on the primal central path. To avoid any misunderstanding it might be appropriate to emphasize that in this chapter $x(\alpha)$ — as well as $s(\alpha)$ — has a different meaning, as indicated.

[5] Note that $x(\alpha)$ and $s(\alpha)$ are uniquely determined by the relations

$$
\begin{aligned}
Ax(\alpha) &= b, & x &> 0, \\
A^T y(\alpha) + s(\alpha) &= c, & s &> 0, \\
x(\alpha) s(\alpha) &= xs + \alpha \Delta w.
\end{aligned}
$$

Obviously, the right-hand sides in these relations depend linearly (and hence analytically) on α. Since the Jacobian matrix with respect to α of the left-hand sides is nonsingular, the implicit function theorem (cf. Proposition A.2 in Appendix A) implies that $x(\alpha), y(\alpha)$ and $s(\alpha)$ depend analytically on α. See also Section 16.2.1.

This shows that $x^{(1)}$ and $s^{(1)}$ are just the primal-dual Newton directions at (x,s) for the target $\bar{w} = xs + \Delta w$.[6] Using (18.4), (18.5) and (18.8) we find that the higher-order coefficients $x^{(k)}, y^{(k)}$ and $s^{(k)}$, with $k \geq 2$, satisfy the linear system

$$
\begin{aligned}
Ax^{(k)} &= 0 \\
A^T y^{(k)} + s^{(k)} &= 0 \\
sx^{(k)} + xs^{(k)} &= -\sum_{i=1}^{k-1} x^{(i)} s^{(k-i)}, \quad k = 2, 3, \ldots,
\end{aligned}
\tag{18.10}
$$

thus finding a recursive expression for the higher-order coefficients. The remarkable thing here is that the coefficient matrix in (18.10) is the same as in (18.9). This has the important consequence that as soon as the standard (first-order) Newton directions $x^{(1)}$ and $s^{(1)}$ have been calculated from the linear system (18.9), the second-order terms $x^{(2)}$ and $s^{(2)}$ can be computed from a linear system with the same coefficient matrix. Having $x^{(2)}$ and $s^{(2)}$, we can compute $x^{(3)}$ and $s^{(3)}$, and so on. Hence, from a computational point of view the higher-order terms $x^{(k)}$ and $s^{(k)}$, with $k \geq 2$, can be obtained relatively cheaply.

Assuming that the computation of the Newton directions requires $\mathcal{O}(n^3)$ arithmetic operations, the computation of each subsequent pair $\left(x^{(k)}, s^{(k)}\right)$ of higher-order coefficients requires $\mathcal{O}(n^2)$ arithmetic operations. For example, if we compute the pairs $\left(x^{(k)}, s^{(k)}\right)$ for $k = 1, 2, \ldots, n$, this doubles the computational cost per iteration. There is some reason to expect, however, that we will obtain a more accurate search direction; this may result in a speedup that justifies the extra computational burden in the computation.

By truncating the expansion (18.3), we define the *primal-dual Newton directions of order r at (x,s) with step-size α* by

$$
\Delta^{r,\alpha} x := \sum_{i=1}^{r} x^{(i)} \alpha^i, \quad \Delta^{r,\alpha} s := \sum_{i=1}^{r} s^{(i)} \alpha^i.
\tag{18.11}
$$

Moving along these directions we arrive at

$$
x^r(\alpha) := x + \Delta^{r,\alpha} x, \quad s^r(\alpha) := s + \Delta^{r,\alpha} s.
$$

Recall that we started this section by taking $\bar{w} = xs + \Delta w$ as the target point in the w-space. Now that we have introduced the step-size α it is more natural to consider

$$
\bar{w}(\alpha) := xs + \alpha \Delta w
$$

as the target. In the following lemma we calculate $x^r(\alpha)\, s^r(\alpha)$, which is the next iterate in the w-space, and hence obtain an expression for the deviation from the target $\bar{w}(\alpha)$ after the step.

---

[6] **Exercise 84** Verify that $y^{(1)}$ can be solved from (18.9) by the formula

$$
y^{(1)} = -\left(AXS^{-1}A^T\right)^{-1} AS^{-1}\Delta w.
$$

This generalizes the expression for the logarithmic barrier direction in Exercise 35, page 111. Given $y^{(1)}$, $s^{(1)}$ and $x^{(1)}$ follow from

$$
s^{(1)} = -A^T y^{(1)}, \quad x^{(1)} = S^{-1}\left(\Delta w - xs^{(1)}\right).
$$

Lemma IV.19

$$x^r(\alpha)\, s^r(\alpha) = xs + \alpha \Delta w + \sum_{k=r+1}^{2r} \alpha^k \left(\sum_{i=k-r}^{r} x^{(i)} s^{(k-i)} \right).$$

Proof: We may write

$$x^r(\alpha) := x + \Delta^{r,\alpha} x = \sum_{i=0}^{r} x^{(i)} \alpha^i,$$

and we have a similar expression for $s^r(\alpha)$. Therefore,

$$x^r(\alpha)\, s^r(\alpha) = \left(\sum_{i=0}^{r} x^{(i)} \alpha^i \right) \left(\sum_{i=0}^{r} s^{(i)} \alpha^i \right). \tag{18.12}$$

The right-hand side can be considered as a polynomial in α of degree $2r$. We consider the coefficient of α^k for $0 \le k \le 2r$. If $0 \le k \le r$ then the coefficient of α^k is given by

$$\sum_{i=0}^{k} x^{(i)} s^{(k-i)},$$

By (18.8), this expression vanishes if $k \ge 2$. Furthermore, if $k = 1$ the expression is equal to Δw, by (18.7) and if $k = 0$ it is equal to xs, by (18.6). So it remains to consider the coefficient of α^k on the right-hand side of (18.12) for $r + 1 \le k \le 2r$. For these values of k the corresponding coefficient in (18.12) is given by

$$\sum_{i=k-r}^{r} x^{(i)} s^{(k-i)}.$$

Hence, collecting the above results, we get

$$x^r(\alpha)\, s^r(\alpha) = xs + \alpha \Delta w + \sum_{k=r+1}^{2r} \alpha^k \left(\sum_{i=k-r}^{r} x^{(i)} s^{(k-i)} \right). \tag{18.13}$$

This completes the proof. □

In the next section we further analyze the error term

$$E^r(\alpha) := \sum_{k=r+1}^{2r} \alpha^k \left(\sum_{i=k-r}^{r} x^{(i)} s^{(k-i)} \right). \tag{18.14}$$

We conclude this section with two observations. First, taking $r = 1$ we get

$$E^1(\alpha) = \alpha^2 x^{(1)} s^{(1)} = \alpha^2 \Delta x \Delta s,$$

where Δx and Δs are the standard primal-dual Newton directions at (x, s). This is in accordance with earlier results (see, e.g., (10.12)). If we use a first-order Newton step then the error is of order two in α. In the general case, of a step of order r, the error term $E^r(\alpha)$ is of order $r+1$ in α.

The second observation concerns the orthogonality of the search directions in the x- and s-spaces. It is immediate from the first two equations in (18.9) and (18.10) that

$$\left(x^{(i)}\right)^T s^{(j)} = 0, \quad \forall i \geq 1, \forall j \geq 1.$$

As a consequence,

$$\left(\Delta^{r,\alpha} x\right)^T \Delta^{r,\alpha} s = 0,$$

and also, from Lemma IV.19,

$$\left(x^r(\alpha)\right)^T s^r(\alpha) = e^T \left(xs + \alpha \Delta w\right) = e^T \bar{w}(\alpha).$$

Thus, after the step with size α, the duality gap is equal to the gap at the target $\bar{w}(\alpha)$.

Figure 18.1 illustrates the use of higher-order search directions.

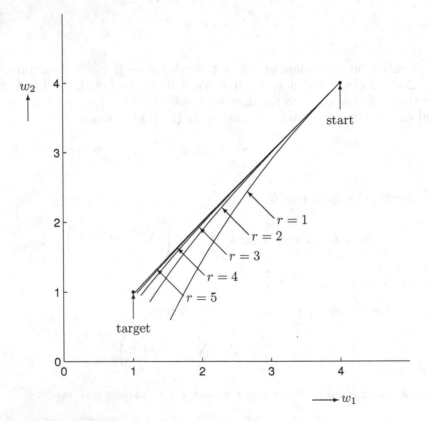

Figure 18.1 Trajectories in the w-space for higher-order steps with $r = 1, 2, 3, 4, 5$.

18.3   Analysis of the error term

The main task in the analysis of the higher-order method is to estimate the error term $E^r(\alpha)$, given by (18.14). Our first estimation is very loose. We write

$$\|E^r(\alpha)\| \le \sum_{k=r+1}^{2r} \alpha^k \left\| \sum_{i=k-r}^{r} x^{(i)} s^{(k-i)} \right\| \le \sum_{k=r+1}^{2r} \alpha^k \sum_{i=k-r}^{r} \left\| x^{(i)} s^{(k-i)} \right\|, \qquad (18.15)$$

and we concentrate on estimating the norms in the last sum. We use the vectors d and v introduced in Section 10.4:

$$d = \sqrt{\frac{x}{s}}, \quad v = \sqrt{xs}. \qquad (18.16)$$

Then the third equation in (18.9) can be rewritten as

$$d^{-1}x^{(1)} + ds^{(1)} = \frac{\Delta w}{v} \qquad (18.17)$$

and the third equation in (18.10) as

$$d^{-1}x^{(k)} + ds^{(k)} = -v^{-1} \sum_{i=1}^{k-1} x^{(i)} s^{(k-i)}, \quad k = 2, 3, \ldots. \qquad (18.18)$$

Since $x^{(k)}$ and $s^{(k)}$ are orthogonal for $k \ge 1$, the vectors $d^{-1}x^{(k)}$ and $ds^{(k)}$ are orthogonal as well. Therefore,

$$\left\| d^{-1}x^{(k)} \right\|^2 + \left\| ds^{(k)} \right\|^2 = \left\| d^{-1}x^{(k)} + ds^{(k)} \right\|^2, \quad k \ge 1.$$

Hence, defining

$$q^{(k)} := d^{-1}x^{(k)} + ds^{(k)}, \quad k \ge 1, \qquad (18.19)$$

we have for each $k \ge 1$,

$$\left\| d^{-1}x^{(k)} \right\| \le \left\| q^{(k)} \right\|, \quad \left\| ds^{(k)} \right\| \le \left\| q^{(k)} \right\|,$$

and as a consequence, for $1 \le i \le k-1$ we may write

$$\left\| x^{(i)} s^{(k-i)} \right\| = \left\| d^{-1}x^{(i)} ds^{(k-i)} \right\| \le \left\| d^{-1}x^{(i)} \right\| \left\| ds^{(k-i)} \right\| \le \left\| q^{(i)} \right\| \left\| q^{(k-i)} \right\|. \qquad (18.20)$$

Substitution of these inequalities in the bound (18.15) for $E^r(\alpha)$ yields

$$\|E^r(\alpha)\| \le \sum_{k=r+1}^{2r} \alpha^k \sum_{i=k-r}^{r} \left\| q^{(i)} \right\| \left\| q^{(k-i)} \right\|. \qquad (18.21)$$

We proceed by deriving upper bounds for $\left\| q^{(k)} \right\|$, $k \ge 1$.

Lemma IV.20 *For each* $k \geq 1$,

$$\left\|q^{(k)}\right\| \leq \varphi_k \left\|v^{-1}\right\|_\infty^{k-1} \left\|q^{(1)}\right\|^k, \tag{18.22}$$

where the integer sequence $\varphi_1, \varphi_2, \ldots$ *is defined recursively by* $\varphi_1 = 1$ *and*

$$\varphi_k = \sum_{i=1}^{k-1} \varphi_i \varphi_{k-i}. \tag{18.23}$$

Proof: The proof is by induction on k. Note that (18.22) holds trivially if $k = 1$. Assume that (18.22) holds for $\left\|q^{(\ell)}\right\|$ if $1 \leq \ell < k$. We complete the proof by deducing from this assumption that the lemma is also true for $\left\|q^{(k)}\right\|$. For $k \geq 2$ we obtain from the definition (18.19) of $q^{(k)}$ and (18.18) that

$$q^{(k)} = -v^{-1} \sum_{i=1}^{k-1} x^{(i)} s^{(k-i)}.$$

Hence, using (18.20),

$$\left\|q^{(k)}\right\| \leq \left\|v^{-1}\right\|_\infty \sum_{i=1}^{k-1} \left\|q^{(i)}\right\| \left\|q^{(k-i)}\right\|.$$

At this stage we apply the induction hypothesis to the last two norms, yielding

$$\left\|q^{(k)}\right\| \leq \left\|v^{-1}\right\|_\infty \sum_{i=1}^{k-1} \varphi_i \left\|v^{-1}\right\|_\infty^{i-1} \left\|q^{(1)}\right\|^i \varphi_{k-i} \left\|v^{-1}\right\|_\infty^{k-i-1} \left\|q^{(1)}\right\|^{k-i},$$

which can be simplified to

$$\left\|q^{(k)}\right\| \leq \left\|v^{-1}\right\|_\infty^{k-1} \left\|q^{(1)}\right\|^k \sum_{i=1}^{k-1} \varphi_i \varphi_{k-i}.$$

Finally, using (18.23) the lemma follows. $\qquad\square$

The solution of the recursion (18.23) with $\varphi_1 = 1$ is given by[7]

$$\varphi_k = \frac{1}{k} \binom{2k-2}{k-1}. \tag{18.24}$$

This enables us to prove our next result.

Lemma IV.21 *For each* $k = r+1, \ldots, 2r$,

$$\sum_{i=k-r}^{r} \left\|q^{(i)}\right\| \left\|q^{(k-i)}\right\| \leq \frac{2^{2k-3}}{k} \left\|v^{-1}\right\|_\infty^{k-2} \left\|q^{(1)}\right\|^k.$$

---

[7] **Exercise 85** Prove that (18.24) is the solution of the recursion in (18.23) satisfying $\varphi_1 = 1$ (cf., e.g., Liu [184]).

Proof: Using Lemma IV.20 we may write

$$\sum_{i=k-r}^{r} \left\| q^{(i)} \right\| \left\| q^{(k-i)} \right\| \leq \sum_{i=k-r}^{r} \varphi_i \left\| v^{-1} \right\|_{\infty}^{i-1} \left\| q^{(1)} \right\|^{i} \varphi_{k-i} \left\| v^{-1} \right\|_{\infty}^{k-i-1} \left\| q^{(1)} \right\|^{k-i},$$

which is equivalent to

$$\sum_{i=k-r}^{r} \left\| q^{(i)} \right\| \left\| q^{(k-i)} \right\| \leq \left\| v^{-1} \right\|_{\infty}^{k-2} \left\| q^{(1)} \right\|^{k} \sum_{i=k-r}^{r} \varphi_i \varphi_{k-i}.$$

For the last sum we use again a loose bound:

$$\sum_{i=k-r}^{r} \varphi_i \varphi_{k-i} \leq \sum_{i=1}^{k-1} \varphi_i \varphi_{k-i} = \varphi_k.$$

Using (18.24) and $k \geq 2$ we can easily derive that

$$\varphi_k \leq \frac{2^{2k-3}}{k}, \quad k \geq 2.$$

Substituting this we obtain

$$\sum_{i=k-r}^{r} \left\| q^{(i)} \right\| \left\| q^{(k-i)} \right\| \leq \frac{2^{2k-3}}{k} \left\| v^{-1} \right\|_{\infty}^{k-2} \left\| q^{(1)} \right\|^{k},$$

proving the lemma. □

Now we are ready for the main result of this section.

Theorem IV.22 *We have*

$$\| E^r(\alpha) \| \leq \frac{1}{r+1} \sum_{k=r+1}^{2r} \alpha^k 2^{2k-3} \left\| v^{-1} \right\|_{\infty}^{k-2} \left\| q^{(1)} \right\|^{k}.$$

Proof: From (18.21) we recall that

$$\| E^r(\alpha) \| \leq \sum_{k=r+1}^{2r} \alpha^k \sum_{i=k-r}^{r} \left\| q^{(i)} \right\| \left\| q^{(k-i)} \right\|.$$

Replacing the second sum by the upper bound in Lemma IV.21 and using that $k \geq r+1$ in the first sum, we obtain the result. □

18.4   Application to the primal-dual Dikin direction

18.4.1   Introduction

The Dikin direction, described in Appendix E, is one of the directions that can be used for solving the self-dual problem. In the next section we show that its definition can easily be adapted to problems (P) and (D) in standard format. It will become clear that the analysis of the self-dual model also applies to the standard model and vice versa. Although we don't work it out here, we mention that use of the (first-order) Dikin direction leads to an algorithm for solving the standard model that requires at most

$$\tau n \, \log \frac{\left(x^0\right)^T s^0}{\varepsilon}$$

iterations, where (x^0, s^0) denotes the initial primal-dual pair, ε is an upper bound for the duality gap upon termination of the algorithm and τ an upper bound for the distance of the iterates to the central path.[8] The complexity per iteration is $\mathcal{O}(n^3)$, as usual. This is in accordance with the bounds in Appendix E for the self-dual model. By using higher-order versions of the Dikin direction the complexity can be improved by a factor $(\tau n)^{\frac{r-1}{2r}}$. Note that this factor goes to $\sqrt{\tau n}$ if r goes to infinity. The complexity per iteration is $\mathcal{O}(n^3 + rn^2)$. Hence, when taking $r = n$, the complexity per iteration remains $\mathcal{O}(n^3)$. In that case we show that the iteration bound is improved by the factor $\sqrt{\tau n}$. When $\tau = \mathcal{O}(1)$, which can be assumed without loss of generality, we obtain the iteration bound

$$\mathcal{O}\left(\sqrt{n} \, \log \frac{\left(x^0\right)^T s^0}{\varepsilon}\right),$$

which is the best iteration bound for interior point methods known until now.

18.4.2   The (first-order) primal-dual Dikin direction

Let (x, s) be a positive primal-dual pair for (P) and (D) and let Δx and Δs denote displacements in the x-space and the s-space. Moving along Δx and Δs we arrive at

$$x^+ := x + \Delta x, \quad s^+ := s + \Delta s.$$

The new iterates will be feasible if

$$A\Delta x = 0, \quad A^T \Delta y + \Delta s = 0,$$

where Δy represents the displacement in the y-space corresponding to Δs, and both x^+ and s^+ are nonnegative. Since Δx and Δs are orthogonal, the new duality gap is given by

$$\left(x^+\right)^T s^+ = x^T s + x^T \Delta s + s^T \Delta x.$$

---

[8] Originally, the Dikin direction was introduced for the standard format. See Jansen, Roos and Terlaky [156].

Replicating Dikin's idea, just as in Section E.2, we replace the nonnegativity conditions by the condition[9]

$$\left\| \frac{\Delta x}{x} + \frac{\Delta s}{s} \right\| \le 1.$$

This can be rewritten as

$$\left\| \frac{x^+ - x}{x} + \frac{s^+ - s}{s} \right\| \le 1,$$

showing that the new iterates are sought within an ellipsoid, called the *Dikin ellipsoid* at the given pair (x, s). Since our aim is to minimize the duality gap, we consider the optimization problem

$$\min \left\{ e^T (s\Delta x + x\Delta s) \; : \; A\Delta x = 0, \; A^T \Delta y + \Delta s = 0, \; \left\| \frac{\Delta x}{x} + \frac{\Delta s}{s} \right\| \le 1 \right\}. \quad (18.25)$$

The crucial observation is that (18.25) determines the displacements Δx and Δs uniquely. The arguments are almost the same as in Section E.2. Using the vectors d and v in (18.16), x and s can be rescaled to the same vector v:

$$d^{-1}x = v, \quad ds = v.$$

As usual, we rescale Δx and Δs accordingly to

$$d_x := d^{-1}\Delta x, \quad d_s := d\Delta s. \quad (18.26)$$

Then

$$\frac{\Delta x}{x} = \frac{d_x}{v}, \quad \frac{\Delta s}{s} = \frac{d_s}{v},$$

and moreover,

$$\Delta x \Delta s = d_x d_s.$$

---

[9]

  Dikin introduced the so-called *primal affine-scaling* direction at a primal feasible x $(x > 0)$ by minimizing the primal objective $c^T (x + \Delta x)$ over the ellipsoid

$$\left\| \frac{\Delta x}{x} \right\| \le 1,$$

  subject to $A\Delta x = 0$. So the primal affine-scaling direction is determined as the unique solution of

$$\min \left\{ c^T \Delta x \; : \; A\Delta x = 0, \; \left\| \frac{\Delta x}{x} \right\| \le 1 \right\}.$$

  Dikin showed convergence of the primal affine-scaling method ([63, 64, 65]) under some non-degeneracy assumptions. Later, without nondegeneracy assumptions, Tsuchiya [268, 270] proved convergence of the method with damped steps. Dikin and Roos [66] proved convergence of a full-step method for the special case that the given problem is homogeneous. Despite many attempts, until now it has not been possible to show that the method is polynomial. For a recent survey paper we refer the reader to Tsuchiya [272]. The approach in this section seems to be the natural generalization to the primal-dual framework.

Also, the scaled displacements d_x and d_s are orthogonal. Now the vector occurring in the ellipsoidal constraint in (18.25) can be reduced to

$$\frac{\Delta x}{x} + \frac{\Delta s}{s} = \frac{d_x + d_s}{v}.$$

Moreover, the variable vector in the objective of problem (18.25) can be written as

$$s\Delta x + x\Delta s = xs\left(\frac{\Delta x}{x} + \frac{\Delta s}{s}\right) = v\left(d_x + d_s\right).$$

With

$$d_w := d_x + d_s, \tag{18.27}$$

the vectors d_x and d_s are uniquely determined as the orthogonal components of d_w in the null space and row space of AD, so we have

$$
\begin{aligned}
d_x &= P_{AD}\left(d_w\right) & (18.28)\\
d_s &= d_w - d_x. & (18.29)
\end{aligned}
$$

Thus we can solve the problem (18.25) by solving the much simpler problem

$$\min\left\{v^T d_w \ : \ \left\|\frac{d_w}{v}\right\| \leq 1\right\}. \tag{18.30}$$

The solution of (18.30) is given by

$$d_w = -\frac{v^3}{\|v^2\|} = -\frac{(xs)^{\frac{3}{2}}}{\|xs\|}.$$

It follows that d_x and d_s are uniquely determined by the system

$$
\begin{aligned}
ADd_x &= 0\\
(AD)^T d_y + d_s &= 0\\
d_x + d_s &= d_w.
\end{aligned}
$$

In terms of the unscaled displacements this can be rewritten as

$$
\begin{aligned}
A\Delta x &= 0\\
A^T \Delta y + \Delta s &= 0 \qquad (18.31)\\
s\Delta x + x\Delta s &= \Delta w,
\end{aligned}
$$

where

$$\Delta w = v d_w = -\frac{v^4}{\|v^2\|} = -\frac{(xs)^2}{\|xs\|}. \tag{18.32}$$

We conclude that the solution of the minimization problem (18.25) is uniquely determined by the linear system (18.31). Hence the (first-order) Dikin directions Δx and Δs are the Newton directions at (x, s) corresponding to the displacement Δw in the w-space, as given by (18.32). We therefore call Δw the Dikin direction in the w-space.

In the next section we consider an algorithm using higher-order Dikin directions. Using the estimates of the error term $E^r(\alpha)$ in the previous section we analyze this algorithm in subsequent sections.

18.4.3 Algorithm using higher-order Dikin directions

For the rest of this section, Δw denotes the Dikin direction in the w-space as given by (18.32). For $r = 1, 2, \ldots$ and for some fixed step-size α that is specified later, the corresponding higher-order Newton steps of order r at (x, s) are given by (18.11). The iterates after the step depend on the step-size α. To express this dependence we denote them as $x(\alpha)$ and $s(\alpha)$ as in Section 18.2. We consider the following algorithm.

Higher-Order Dikin Step Algorithm for the Standard Model

Input:
    An accuracy parameter $\varepsilon > 0$;
    a step-size parameter α, $0 < \alpha \leq 1$;
    a positive primal-dual pair (x^0, s^0).

begin
    $x := x^0$; $s := s^0$;
    while $x^T s \geq \varepsilon$ **do**
    begin
        $x := x(\alpha) = x + \Delta^{r,\alpha} x$;
        $s := s(\alpha) = s + \Delta^{r,\alpha} s$
    end
end

Below we analyze this algorithm. Our aim is to keep the iterates (x, s) within some cone

$$\delta_c(xs) = \frac{\max{(xs)}}{\min{(xs)}} \leq \tau$$

around the central path, for some fixed $\tau > 1$; τ is chosen such that

$$\delta_c(x^0 s^0) \leq \tau.$$

18.4.4 Feasibility and duality gap reduction

As before, we use the superscript $^+$ to refer to entities after the higher-order Dikin step of size α at (x, s). Thus,

$$
\begin{aligned}
x^+ &:= & x(\alpha) = x + \Delta^{r,\alpha} x, \\
s^+ &:= & s(\alpha) = s + \Delta^{r,\alpha} s,
\end{aligned}
$$

and from Lemma IV.19,

$$x^+ s^+ = x(\alpha)s(\alpha) = xs + \alpha \Delta w + E^r(\alpha), \tag{18.33}$$

where the higher-order error term $E^r(\alpha)$ is given by (18.14).

The step-size α is feasible if the new iterates are positive. Using the same (simple continuity) argument as in the proof of Lemma E.2, page 455, we get the following result.

Lemma IV.23 *If $\bar{\alpha}$ is such that $x(\alpha)s(\alpha) > 0$ for all α satisfying $0 \leq \alpha \leq \bar{\alpha}$, then the step-size $\bar{\alpha}$ is feasible.*

Lemma IV.23 implies that the step-size $\bar{\alpha}$ is feasible if

$$xs + \alpha\Delta w + E^r(\alpha) \geq 0, \quad 0 \leq \alpha \leq \bar{\alpha}.$$

At the end of Section 18.2 we established that after the step the duality gap attains the value $e^T(xs + \alpha\Delta w)$. This leads to the following lemma.

Lemma IV.24 *If the step-size α is feasible then*

$$\left(x^+\right)^T s^+ \leq \left(1 - \frac{\alpha}{\sqrt{n}}\right) x^T s.$$

Proof: We have

$$\left(x^+\right)^T s^+ = e^T\left(xs - \alpha\frac{(xs)^2}{\|xs\|}\right) = x^T s - \alpha\|xs\|.$$

The Cauchy–Schwarz inequality implies

$$\|xs\| = \frac{1}{\sqrt{n}}\|e\|\,\|xs\| \geq \frac{e^T(xs)}{\sqrt{n}} = \frac{x^T s}{\sqrt{n}},$$

and the lemma follows. $\qquad\qquad\qquad\qquad\qquad\qquad\qquad\qquad\qquad\qquad\qquad\qquad\square$

18.4.5 Estimate of the error term

By Theorem IV.22 the error term $E^r(\alpha)$ satisfies

$$\|E^r(\alpha)\| \leq \frac{1}{8(r+1)} \sum_{k=r+1}^{2r} \alpha^k 2^{2k} \left\|v^{-1}\right\|_\infty^{k-2} \left\|q^{(1)}\right\|^k.$$

In the present case we have, from (18.19), (18.17) and (18.32),

$$q^{(1)} = \frac{\Delta w}{v} = -\frac{v^3}{\|v^2\|}.$$

Hence

$$\left\|q^{(1)}\right\| = \left\|\frac{v^3}{\|v^2\|}\right\| \leq \|v\|_\infty \left\|\frac{v^2}{\|v^2\|}\right\| = \|v\|_\infty = \max(v).$$

Therefore,

$$\left\|v^{-1}\right\|_\infty \left\|q^{(1)}\right\| \leq \frac{\max(v)}{\min(v)} = \sqrt{\frac{\max(xs)}{\min(xs)}} = \sqrt{\delta_c(xs)} \leq \sqrt{\tau}.$$

Substituting this we get

$$\|E^r(\alpha)\| \leq \frac{1}{8(r+1)} \left\|v^{-1}\right\|_\infty^{-2} \sum_{k=r+1}^{2r} \alpha^k 2^{2k} \left(\sqrt{\tau}\right)^k = \frac{\min(xs)}{8(r+1)} \sum_{k=r+1}^{2r} \left(4\alpha\sqrt{\tau}\right)^k.$$

$$(18.34)$$

18.4.6 Step size

Assuming $\delta_c(x, s) \leq \tau$, with $\tau > 1$, we establish a bound for the step-size α such that this property is maintained after a higher-order Dikin step. The analysis follows the same lines as the analysis in Section E.4 of the algorithm for the self-dual model with first-order Dikin steps. As there, we derive from $\delta_c(x, s) \leq \tau$ the existence of positive numbers τ_1 and τ_2 such that

$$\tau_1 e \leq xs \leq \tau_2 e, \qquad \text{with } \tau_2 = \tau \tau_1. \tag{18.35}$$

Without loss of generality we take

$$\tau_1 = \min (xs).$$

The following lemma generalizes Lemma E.4.

Lemma IV.25 *Let* $\tau > 1$. *Suppose that* $\delta_c(xs) \leq \tau$ *and let* τ_1 *and* τ_2 *be such that (18.35) holds. If the step-size* α *satisfies*

$$\alpha \leq \min \left\{ \frac{\|xs\|}{2\tau_2}, \ \frac{1}{4\sqrt{\tau}}, \ \frac{1}{4\sqrt{\tau}} \sqrt[r]{\frac{2\tau_1\sqrt{\tau}}{\|xs\|}} \right\},$$

then we have $\delta_c(x^+ s^+) \leq \tau$.

Proof: Using (18.33) and the definition of Δw we obtain

$$x^+ s^+ = x(\alpha)s(\alpha) = xs + \alpha \Delta w + E^r(\alpha) = xs - \frac{\alpha (xs)^2}{\|xs\|} + E^r(\alpha).$$

Using the first bound on α in the lemma, we can easily verify that the map

$$t \mapsto t - \frac{\alpha t^2}{\|xs\|}$$

is an increasing function for $t \in [0, \tau_2]$. Application of this map to each component of the vector xs gives

$$\left(\tau_1 - \frac{\alpha \tau_1^2}{\|xs\|} \right) e \leq xs - \frac{\alpha (xs)^2}{\|xs\|} \leq \left(\tau_2 - \frac{\alpha \tau_2^2}{\|xs\|} \right) e.$$

It follows that

$$\left(\tau_1 - \frac{\alpha \tau_1^2}{\|xs\|} \right) e + E^r(\alpha) \leq x^+ s^+ \leq \left(\tau_2 - \frac{\alpha \tau_2^2}{\|xs\|} \right) e + E^r(\alpha).$$

Hence, assuming for the moment that the Dikin step of size α is feasible, we certainly have $\delta(x^+ s^+) \leq \tau$ if

$$\tau \left(\left(\tau_1 - \frac{\alpha \tau_1^2}{\|xs\|} \right) e + E^r(\alpha) \right) \geq \left(\tau_2 - \frac{\alpha \tau_2^2}{\|xs\|} \right) e + E^r(\alpha).$$

Since $\tau_2 = \tau\tau_1$, this reduces to

$$\alpha\left(\frac{\tau_2^2 - \tau\tau_1^2}{\|xs\|}\right)e + (\tau - 1)E^r(\alpha) \geq 0.$$

Since $\tau_2^2 - \tau\tau_1^2 = (\tau - 1)\tau_1\tau_2$ we can divide by $\tau - 1$, thus obtaining

$$\frac{\alpha\tau_1\tau_2}{\|xs\|}e + E^r(\alpha) \geq 0.$$

This inequality is certainly satisfied if

$$\|xs\|\,\|E^r(\alpha)\| \leq \alpha\tau_1\tau_2.$$

Using the upper bound (18.34) for $E^r(\alpha)$ it follows that we have $\delta(x^+s^+) \leq \tau$ if α is such that

$$\frac{\|xs\|\min{(xs)}}{8(r+1)}\sum_{k=r+1}^{2r}\left(4\alpha\sqrt{\tau}\right)^k \leq \alpha\tau_1\tau_2.$$

Since $\min{(xs)} = \tau_1$, this inequality simplifies to

$$\frac{\|xs\|}{8(r+1)}\sum_{k=r+1}^{2r}\left(4\alpha\sqrt{\tau}\right)^k \leq \alpha\tau_2.$$

The second bound in the lemma implies that $4\alpha\sqrt{\tau} < 1$. Therefore, the last sum is bounded above by

$$\sum_{k=r+1}^{2r}\left(4\alpha\sqrt{\tau}\right)^k \leq r\left(4\alpha\sqrt{\tau}\right)^{r+1}.$$

Substituting this we arrive at the inequality

$$\frac{r\,\|xs\|\left(4\alpha\sqrt{\tau}\right)^{r+1}}{8(r+1)} \leq \alpha\tau_2.$$

Omitting the factor $r/(r+1)$, we can easily check that this inequality certainly holds if

$$\alpha \leq \frac{1}{4\sqrt{\tau}}\sqrt[r]{\frac{2\tau_1\sqrt{\tau}}{\|xs\|}},$$

which is the third bound on α in the lemma. Thus we have shown that for each step-size α satisfying the bounds in the lemma, we have $\delta(x^+s^+) \leq \tau$. But this implies that the coordinates of x^+s^+ do not vanish for any of these step-sizes. By Lemma IV.23 this also implies that the given step-size α is feasible. Hence the lemma follows. $\quad\square$

18.4.7 Convergence analysis

With the result of the previous section we can now derive an upper bound for the number of iterations needed by the algorithm.

Lemma IV.26 *Let $4/n \leq \tau \leq 4n$. Then, with the step-size*

$$\alpha = \frac{1}{4\sqrt{\tau}} \sqrt[r]{\frac{2}{\sqrt{\tau n}}},$$

the Higher-Order Dikin Step Algorithm for the Standard Model requires at most

$$4\sqrt{\tau n} \sqrt[r]{\frac{\sqrt{\tau n}}{2}} \log \frac{\left(x^0\right)^T s^0}{\varepsilon}$$

iterations.[10] *The output is a feasible primal-dual pair (x, s) such that $\delta_c(xs) \leq \tau$ and $x^T s \leq \varepsilon$.*

Proof: Initially we are given a feasible primal-dual pair (x^0, s^0) such that $\delta_c(x^0 s^0) \leq \tau$. The given step-size α guarantees that these properties are maintained after each iteration. This can be deduced from Lemma IV.25, as we now show. It suffices to show that the specified value of α meets the bound in Lemma IV.25. Since $\tau n \geq 4$ we have

$$\alpha = \frac{1}{4\sqrt{\tau}} \sqrt[r]{\frac{2}{\sqrt{\tau n}}} \leq \frac{1}{4\sqrt{\tau}},$$

showing that α meets the second bound. Since $\|xs\| \leq \tau_2 \sqrt{n}$ we have

$$\frac{2\tau_1 \sqrt{\tau}}{\|xs\|} \geq \frac{2\tau_1 \sqrt{\tau}}{\tau_2 \sqrt{n}} = \frac{2\sqrt{\tau}}{\tau \sqrt{n}} = \frac{2}{\sqrt{\tau n}},$$

which implies that α also meets the third bound in Lemma IV.25. Finally, for the first bound in the lemma, we may write

$$\frac{\|xs\|}{2\tau_2} \geq \frac{\tau_1 \sqrt{n}}{2\tau_2} = \frac{\sqrt{n}}{2\tau} \geq \frac{1}{4\sqrt{\tau}}.$$

The last inequality follows because $\tau \leq 4n$. Thus we have shown that α meets the bounds in Lemma IV.25. As a consequence, the property $\delta_c(xs) \leq \tau$ is maintained during the course of the algorithm. This also implies that the algorithm is well defined and, hence, the only remaining task is to derive the iteration bound in the lemma. By Lemma IV.24, each iteration reduces the duality gap by a factor $1 - \theta$, where

$$\theta = \frac{\alpha}{\sqrt{n}} = \frac{1}{4\sqrt{\tau n}} \sqrt[r]{\frac{2}{\sqrt{\tau n}}}.$$

---

[10] When $r = 1$ the step-size becomes

$$\alpha = \frac{1}{2\tau \sqrt{n}},$$

which is a factor of 2 smaller than the step-size in Section E.5. As a consequence the iteration bound is a factor of 2 worse than in Section E.5. This is due to a weaker estimate of the error term.

Hence, by Lemma I.36, the duality gap satisfies $x^T s \leq \varepsilon$ after at most

$$\frac{1}{\theta} \log \frac{n\mu^0}{\varepsilon} = 4\sqrt{\tau n} \sqrt[r]{\frac{\sqrt{\tau n}}{2}} \log \frac{\left(x^0\right)^T s^0}{\varepsilon}$$

iterations. This completes the proof. □

Recall that each iteration requires $\mathcal{O}\left(n^3 + rn^2\right)$ arithmetic operations. In the rest of this section we take the order r of the search direction equal to $r = n$. Then the complexity per iteration is still $\mathcal{O}\left(n^3\right)$ just as in the case of a first-order method. The iteration bound of Lemma IV.26 then becomes

$$4\sqrt{\tau n} \sqrt[n]{\frac{\sqrt{\tau n}}{2}} \log \frac{\left(x^0\right)^T s^0}{\varepsilon}.$$

Now, assuming $\tau \leq 4n$, we have

$$\sqrt[n]{\frac{\sqrt{\tau n}}{2}} \leq \sqrt[n]{n}.$$

The last expression is maximal for $n = 3$ and is then equal to 1.44225. Thus we may state without further proof the following theorem.

Theorem IV.27 *Let* $4/n \leq \tau \leq 4n$ *and* $r = n$. *Then the Higher-Order Dikin Step Algorithm for the Standard Model stops after at most*

$$6\sqrt{\tau n} \log \frac{\left(x^0\right)^T s^0}{\varepsilon}$$

iterations. Each iteration requires $\mathcal{O}(n^3)$ *arithmetic operations.*

For $\tau = 2$, which can be taken without loss of generality, the iteration bound of Theorem IV.27 becomes

$$\mathcal{O}\left(\sqrt{n} \log \frac{\left(x^0\right)^T s^0}{\varepsilon}\right),$$

which is the best obtainable bound.

18.5 Application to the primal-dual logarithmic barrier method

18.5.1 Introduction

In this section we apply the higher-order approach to the (primal-dual) logarithmic barrier method. If the target value of the barrier parameter is μ, then the search direction in the w-space at a given primal-dual pair (x, s) is given by

$$\Delta w = \mu e - xs.$$

We measure the proximity from (x, s) to the target μe by the usual measure

$$\delta(xs, \mu) = \frac{1}{2} \left\| \sqrt{\frac{xs}{\mu e}} - \sqrt{\frac{\mu e}{xs}} \right\| = \frac{1}{2\sqrt{\mu}} \left\| \frac{\mu e - xs}{\sqrt{xs}} \right\|. \tag{18.36}$$

In this chapter we also use an infinity-norm based proximity of the central path, namely

$$\delta_\infty(xs, \mu) := \left\| \sqrt{\frac{\mu e}{xs}} \right\|_\infty = \max_i \left| \sqrt{\frac{\mu e}{xs}} \right| = \max_i \left| \frac{\sqrt{\mu}}{v_i} \right|. \tag{18.37}$$

Recall from Lemma II.62 that we always have

$$\delta_\infty(xs, \mu) \le \rho\left(\delta(xs, \mu)\right). \tag{18.38}$$

Just as in the previous section, where we used the Dikin direction, our aim is to consider a higher-order logarithmic barrier method that keeps the iterates within some cone around the central path. The cone is obtained by requiring that the primal-dual pairs (x, s) generated by the method are such that there exists a $\mu > 0$ such that

$$\delta(xs, \mu) \le \tau, \quad \text{and} \quad \delta_\infty(xs, \mu) \le \zeta \tag{18.39}$$

where τ and ζ denote some fixed positive numbers that specify the 'width' of the cone around the central path in which the iterates are allowed to move.

When $\zeta = \rho(\tau)$ it follows from (18.38) that

$$\delta(xs, \mu) \le \tau \quad \Rightarrow \quad \delta_\infty(xs, \mu) \le \zeta.$$

Hence, the logarithmic barrier methods considered in Part II fall within the present framework with $\zeta = \rho(\tau)$. The full Newton step method considered in Part II uses $\tau = 1/\sqrt{2}$. In the large-update methods of Part II the updates of the barrier parameter μ reduce μ by a factor $1 - \theta$, where $\theta = \mathcal{O}(1)$. As a consequence, after a barrier update we have $\delta(xs, \mu) = \mathcal{O}(\sqrt{n})$. Hence, we may say that the full Newton step methods in Part II keep the iterates in a cone with $\tau = \mathcal{O}(1)$, and the large-update methods in a wider cone with $\tau = \mathcal{O}(\sqrt{n})$. Recall that the method using the wider cone — the large-update methods — are multistep methods. Each single step is a damped (first-order) Newton step and the progress is measured by the decrease of the (primal-dual) logarithmic barrier function.

In this section we consider a method that works within a 'wide' cone, with $\tau = \mathcal{O}(\sqrt{n})$ and $\zeta = \mathcal{O}(1)$, but we use higher-order Newton steps instead of damped first-order steps. The surprising feature of the method is that progress can be controlled by using the proximity measures $\delta(xs, \mu)$ and $\delta_\infty(xs, \mu)$. We show that after an update of the barrier parameter a higher-order step reduces the proximity $\delta(xs, \mu)$ by a factor smaller than one and keeps the proximity $\delta_\infty(xs, \mu)$ under a fixed threshold value $\zeta \ge 2$. Then the barrier parameter value can be decreased to a smaller value while respecting the cone condition (18.39). In this way we obtain a 'wide-cone method' whose iteration bound is $\mathcal{O}(\sqrt{n} \log \log (x^0)^T s^0 / \varepsilon)$. Each iteration consists of a single higher-order Newton step.

Below we need to analyze the effect of a higher-order Newton step on the proximity measures. For that purpose the error term must be estimated.

18.5.2 Estimate of the error term

Recall from Lemma IV.19 that the error term $E^r(\alpha)$ is given by

$$\|E^r(\alpha)\| \le \frac{1}{8(r+1)} \sum_{k=r+1}^{2r} \alpha^k 2^{2k} \|v^{-1}\|_\infty^{k-2} \|q^{(1)}\|^k, \tag{18.40}$$

where $v = \sqrt{xs}$. In the present case, (18.19) and (18.17) give

$$q^{(1)} = \frac{\Delta w}{v} = \frac{\mu e - xs}{\sqrt{xs}}.$$

Hence, using (18.36) and denoting $\delta(xs, \mu)$ by δ, we find

$$\left\| q^{(1)} \right\| = 2\sqrt{\mu}\,\delta \le 2\sqrt{\mu}\,\tau.$$

Furthermore by using (18.37) and putting $\delta_\infty := \delta_\infty(xs, \mu)$ we have

$$\left\| v^{-1} \right\|_\infty = \frac{1}{\sqrt{\mu}} \left\| \sqrt{\frac{\mu}{xs}} \right\|_\infty = \frac{\delta_\infty}{\sqrt{\mu}} \le \frac{\zeta}{\sqrt{\mu}}.$$

Substituting these in (18.40) we get[11]

$$\| E^r(\alpha) \| \le \frac{\mu}{8(r+1)\delta_\infty^2} \sum_{k=r+1}^{2r} (8\alpha\delta\delta_\infty)^k \le \frac{\mu}{8(r+1)\zeta^2} \sum_{k=r+1}^{2r} (8\alpha\tau\zeta)^k. \qquad (18.41)$$

Below we always make the natural assumption that $\alpha \le 1$. Moreover, δ and δ_∞ always denote $\delta(xs, \mu)$ and $\delta_\infty(xs, \mu)$ respectively.

Lemma IV.28 *Let the step-size be such that $\alpha \le 1/(8\delta\delta_\infty)$. Then*

$$\| E^r(\alpha) \| \le \frac{r\mu}{r+1} \frac{(8\alpha\delta\delta_\infty)^{r+1}}{8\delta_\infty^2}.$$

Proof: Since $8\alpha\delta\delta_\infty \le 1$, we have

$$\sum_{k=r+1}^{2r} (8\alpha\delta\delta_\infty)^k \le r (8\alpha\delta\delta_\infty)^{r+1}.$$

Substitution in (18.41) gives the lemma. □

Corollary IV.29 *Let $\delta \le \tau$, $\delta_\infty \le \zeta$ and $\alpha \le 1/(8\tau\zeta)$. Then*

$$\| E^r(\alpha) \| \le \frac{r\mu}{r+1} \frac{(8\alpha\tau\zeta)^{r+1}}{8\zeta^2}.$$

---

[11] For $r = 1$ the derived bound for the error term gives $\left\| E^1(1) \right\| \le 4\mu\delta^2$, as follows easily. It is interesting to compare this bound with the error bound in Section 7.4 (cf. Lemma II.49), which amounts to $\left\| E^1(1) \right\| \le \mu\delta^2\sqrt{2}$. Although the present bound is weaker by a factor of $2\sqrt{2}$ for $r = 1$, it is sharp enough for our present purpose. It is also sharp enough to derive an $\mathcal{O}(\sqrt{n})$ complexity bound for $r = 1$ with some worse constant than before. Our main interest here is the case where $r > 1$.

18.5.3 Reduction of the proximity after a higher-order step

Recall from (18.13) that after a higher-order step of size α we have

$$x^r(\alpha)s^r(\alpha) = xs + \alpha\Delta w + E^r(\alpha) = xs + \alpha(\mu e - xs) + E^r(\alpha).$$

We consider

$$\bar{w}(\alpha) := xs + \alpha(\mu e - xs)$$

as the (intermediate) target during the step. The new iterate in the w-space is denoted by $w(\alpha)$, so

$$w(\alpha) = x^r(\alpha)s^r(\alpha).$$

As a consequence,

$$w(\alpha) = \bar{w}(\alpha) + E^r(\alpha). \qquad (18.42)$$

The proximities of the new iterate with respect to the μ-center are given by

$$\delta(w(\alpha), \mu) = \frac{1}{2}\left\|\sqrt{\frac{w(\alpha)}{\mu}} - \sqrt{\frac{\mu}{w(\alpha)}}\right\|$$

and

$$\delta_\infty(w(\alpha), \mu) = \left\|\sqrt{\frac{\mu}{w(\alpha)}}\right\|_\infty.$$

Ideally the proximities after the step would be $\delta(\bar{w}(\alpha), \mu)$ and $\delta_\infty(\bar{w}(\alpha), \mu)$. We first derive an upper bound for $\delta(\bar{w}(\alpha), \mu)$ and $\delta_\infty(\bar{w}(\alpha), \mu)$ respectively in terms of τ, ζ and the step-size α.

Lemma IV.30 *We have*

  (i) $\delta(\bar{w}(\alpha), \mu) \leq \sqrt{1-\alpha}\,\delta$,

  (ii) $\delta_\infty(\bar{w}(\alpha), \mu) \leq \sqrt{\alpha + (1-\alpha)\delta_\infty^2}$.

Proof: It is easily verified that for any positive vector w, by their definitions (18.36) and (18.37), both $\delta(w, \mu)^2$ and $\delta_\infty(w, \mu)^2$ are convex functions of w. Since

$$\bar{w}(\alpha) = xs + \alpha(\mu e - xs) = \alpha(\mu e) + (1 - \alpha)xs, \quad 0 \leq \alpha \leq 1,$$

$\bar{w}(\alpha)$ is a convex combination of μe and xs. Hence, by the convexity of $\delta(w, \mu)^2$,

$$\delta(\bar{w}(\alpha), \mu)^2 \leq \alpha\,\delta(\mu e, \mu)^2 + (1 - \alpha)\,\delta(xs, \mu)^2.$$

Since $\delta(\mu e, \mu) = 0$, the first statement of the lemma follows.

The proof of the second claim is analogous. The convexity of $\delta_\infty(xs, \mu)^2$ gives

$$\delta_\infty(\bar{w}(\alpha), \mu)^2 \leq \alpha\,\delta_\infty(\mu e, \mu)^2 + (1 - \alpha)\,\delta_\infty(xs, \mu)^2.$$

Since $\delta_\infty(\mu e, \mu) = 1$, the lemma follows. $\qquad\square$

It is very important for our purpose that when the pair (x, s) satisfies the cone condition (18.39) for $\mu > 0$, then after a higher-order step at (x, s) to the μ-center, the new iterates also satisfy the cone condition. The next corollary of Lemma IV.30 is a first step in this direction. It shows that $\bar{w}(\alpha)$ satisfies the cone condition. Recall that $\bar{w}(\alpha) = w(\alpha)$ if the higher-order step is exact. Later we deal with the case where the higher-order step is not exact (cf. Theorem IV.35 below). This requires careful estimation of the error term $E^r(\alpha)$.

Corollary IV.31 *Let $\delta \leq \tau$ and $\delta_\infty \leq \zeta$, with $\zeta \geq 2$. Then we have*

(i) $\delta(\bar{w}(\alpha), \mu) \leq \sqrt{1-\alpha}\, \tau \leq (1 - \frac{\alpha}{2})\tau;$

(ii) $\delta_\infty(\bar{w}(\alpha), \mu) \leq \sqrt{\alpha + (1-\alpha)\zeta^2} \leq (1 - \frac{3\alpha}{8})\zeta \leq \zeta.$

Proof: The first claim is immediate from the first part of Lemma IV.30, since $\delta \leq \tau$ and $\sqrt{1-\alpha} \leq 1 - \alpha/2$. For the proof of the second statement we write, using the second part of Lemma IV.30 and $\zeta \geq 2$,

$$\sqrt{\alpha + (1-\alpha)\delta_\infty^2} \quad \leq \quad \sqrt{\alpha + (1-\alpha)\zeta^2} \leq \sqrt{\alpha\frac{\zeta^2}{4} + (1-\alpha)\zeta^2}$$

$$= \quad \sqrt{\left(1 - \frac{3\alpha}{4}\right)\zeta^2} \leq \left(1 - \frac{3\alpha}{8}\right)\zeta \leq \zeta.$$

Thus the corollary has been proved. $\qquad\qquad\qquad\qquad\qquad\qquad\qquad\qquad\qquad\qquad\qquad\square$

The next lemma provides an expression for the 'error' in the proximities after the step. We use the following relation, which is an obvious consequence of (18.42):

$$\frac{w(\alpha)}{\mu} = \frac{\bar{w}(\alpha)}{\mu}\left(e + \frac{E^r(\alpha)}{\bar{w}(\alpha)}\right). \qquad\qquad (18.43)$$

Lemma IV.32 *Let α be such that*

$$\left\|\frac{E^r(\alpha)}{\bar{w}(\alpha)}\right\|_\infty \leq \frac{\sqrt{5}-1}{2}.$$

Then we have

(i) $\delta(w(\alpha), \mu) \leq \delta(\bar{w}(\alpha), \mu) + \sqrt{1 + \delta(\bar{w}(\alpha), \mu)^2}\, \left\|\frac{E^r(\alpha)}{\bar{w}(\alpha)}\right\|;$

(ii) $\delta_\infty(w(\alpha), \mu) \leq \delta_\infty(\bar{w}(\alpha), \mu)\left(1 + \left\|\frac{E^r(\alpha)}{\bar{w}(\alpha)}\right\|_\infty\right).$

Proof: Using (18.43) we may write

$$\delta(w(\alpha), \mu) = \frac{1}{2}\left\|\sqrt{\frac{\bar{w}(\alpha)}{\mu}\left(e + \frac{E^r(\alpha)}{\bar{w}(\alpha)}\right)} - \sqrt{\frac{\mu}{\bar{w}(\alpha)}\left(e + \frac{E^r(\alpha)}{\bar{w}(\alpha)}\right)^{-1}}\right\|.$$

To simplify the notation we omit the argument α in the rest of the proof and we introduce the notation

$$\lambda := \frac{E^r(\alpha)}{\bar{w}(\alpha)},$$

so that
$$\delta(w(\alpha), \mu) = \frac{1}{2} \left\| \sqrt{\frac{\bar{w}}{\mu}} (e + \lambda) - \sqrt{\frac{\mu}{\bar{w}}} (e + \lambda)^{-1} \right\|.$$

Since
$$\sqrt{\frac{\bar{w}}{\mu}} (e + \lambda) - \sqrt{\frac{\mu}{\bar{w}}} (e + \lambda)^{-1} =$$

$$\sqrt{\frac{\bar{w}}{\mu}} - \sqrt{\frac{\mu}{\bar{w}}} + \sqrt{\frac{\bar{w}}{\mu}} \left((e + \lambda)^{\frac{1}{2}} - e \right) - \sqrt{\frac{\mu}{\bar{w}}} \left((e + \lambda)^{-\frac{1}{2}} - e \right),$$

application of the triangle inequality gives

$$\delta(w(\alpha), \mu) \le \delta(\bar{w}(\alpha), \mu) + \frac{1}{2} \left\| \sqrt{\frac{\bar{w}}{\mu}} \left((e + \lambda)^{\frac{1}{2}} - e \right) - \sqrt{\frac{\mu}{\bar{w}}} \left((e + \lambda)^{-\frac{1}{2}} - e \right) \right\|. \quad (18.44)$$

Denoting the i-th coordinate of the vector under the last norm by z_i, we have

$$z_i = \sqrt{\frac{\bar{w}_i}{\mu}} \left((1 + \lambda_i)^{\frac{1}{2}} - 1 \right) - \sqrt{\frac{\mu}{\bar{w}_i}} \left((1 + \lambda_i)^{-\frac{1}{2}} - 1 \right).$$

This implies

$$|z_i| \le \sqrt{\frac{\bar{w}_i}{\mu}} \left| (1 + \lambda_i)^{\frac{1}{2}} - 1 \right| + \sqrt{\frac{\mu}{\bar{w}_i}} \left| (1 + \lambda_i)^{-\frac{1}{2}} - 1 \right|.$$

The hypothesis of the lemma implies $|\lambda_i| \le (\sqrt{5} - 1)/2$. Now using some elementary inequalities,[12] we get

$$|z_i| \le \sqrt{\frac{\bar{w}_i}{\mu}} |\lambda_i| + \sqrt{\frac{\mu}{\bar{w}_i}} |\lambda_i| = \left(\sqrt{\frac{\bar{w}_i}{\mu}} + \sqrt{\frac{\mu}{\bar{w}_i}} \right) |\lambda_i|.$$

Since
$$\left(\sqrt{\frac{\bar{w}_i}{\mu}} + \sqrt{\frac{\mu}{\bar{w}_i}} \right)^2 = 4 + \left(\sqrt{\frac{\bar{w}_i}{\mu}} - \sqrt{\frac{\mu}{\bar{w}_i}} \right)^2 \le 4 + \left\| \sqrt{\frac{\bar{w}}{\mu}} - \sqrt{\frac{\mu}{\bar{w}}} \right\|_{\infty}^2$$

and
$$\left\| \sqrt{\frac{\bar{w}}{\mu}} - \sqrt{\frac{\mu}{\bar{w}}} \right\|_{\infty}^2 \le \left\| \sqrt{\frac{\bar{w}}{\mu}} - \sqrt{\frac{\mu}{\bar{w}}} \right\|^2 = 4\delta(\bar{w}(\alpha), \mu)^2,$$

we conclude that
$$|z_i| \le 2\sqrt{1 + \delta(\bar{w}(\alpha), \mu)^2} |\lambda_i|, \quad 1 \le i \le n.$$

Hence
$$\|z\| \le 2\sqrt{1 + \delta(\bar{w}(\alpha), \mu)^2} \|\lambda\|.$$

---

[12] **Exercise 86** Prove the following inequalities:

$$\left| (1 + \lambda)^{\frac{1}{2}} - 1 \right| \quad \le \quad |\lambda|, \quad -1 \le \lambda \le 1,$$

$$\left| (1 + \lambda)^{-\frac{1}{2}} - 1 \right| \quad \le \quad |\lambda|, \quad \frac{1 - \sqrt{5}}{2} \le \lambda \le 1.$$

Substituting this in (18.44) proves the first statement of the lemma.

The proof of the second statement in the lemma is analogous. We write

$$
\begin{aligned}
\delta_\infty(w(\alpha), \mu) \;&=\; \left\| \sqrt{\frac{\mu}{w(\alpha)}} \right\|_\infty = \left\| \sqrt{\frac{\mu}{\bar{w}(\alpha)}} \left(e + \frac{E^r(\alpha)}{\bar{w}(\alpha)} \right)^{-1} \right\|_\infty \\[2mm]
&=\; \left\| \sqrt{\frac{\mu}{\bar{w}(\alpha)}} \, (e + \lambda)^{-1} \right\|_\infty \\[2mm]
&=\; \left\| \sqrt{\frac{\mu}{\bar{w}(\alpha)}} + \sqrt{\frac{\mu}{\bar{w}(\alpha)}} \left((e+\lambda)^{-\frac{1}{2}} - e \right) \right\|_\infty \\[2mm]
&\leq\; \left\| \sqrt{\frac{\mu}{\bar{w}(\alpha)}} \right\|_\infty + \left\| \sqrt{\frac{\mu}{\bar{w}(\alpha)}} \right\|_\infty \left\| (e+\lambda)^{-\frac{1}{2}} - e \right\|_\infty .
\end{aligned}
$$

Using again the results of Exercise 86 we can simplify this to

$$
\begin{aligned}
\delta_\infty(w(\alpha), \mu) \;&\leq\; \delta_\infty(\bar{w}(\alpha), \mu) + \delta_\infty(\bar{w}(\alpha), \mu) \, \|\lambda\|_\infty \\[2mm]
&=\; \delta_\infty(\bar{w}(\alpha), \mu) \left(1 + \left\| \frac{E^r(\alpha)}{\bar{w}(\alpha)} \right\|_\infty \right),
\end{aligned}
$$

proving the lemma. □

The following corollary easily follows from Lemma IV.32 and Corollary IV.31.

Corollary IV.33 *Let* $\delta \leq \tau$ *and* $\delta_\infty \leq \zeta$, *with* $\zeta \geq 2$. *If* α *is such that*

$$
\left\| \frac{E^r(\alpha)}{\bar{w}(\alpha)} \right\|_\infty \leq \frac{\sqrt{5} - 1}{2},
$$

then we have

(i) $\delta(w(\alpha), \mu) \leq (1 - \frac{\alpha}{2})\tau + \sqrt{1 + \tau^2} \left\| \frac{E^r(\alpha)}{\bar{w}(\alpha)} \right\|$;

(ii) $\delta_\infty(w(\alpha), \mu) \leq (1 - \frac{3}{8}\alpha) \left(1 + \left\| \frac{E^r(\alpha)}{\bar{w}(\alpha)} \right\|_\infty \right) \zeta$.

We proceed by finding a step-size α that satisfies the hypothesis of Lemma IV.32 and Corollary IV.33.

Lemma IV.34 *With* δ *and* ζ *as in Corollary IV.33, let the step-size* α *be such that* $\alpha \leq 1/(8\tau\zeta)$. *Then*

$$
\left\| \frac{E^r(\alpha)}{\bar{w}(\alpha)} \right\| \leq \frac{r}{8(r+1)} (8\alpha\tau\zeta)^{r+1}
$$

and α *satisfies the hypothesis of Lemma IV.32 and Corollary IV.33.*

Proof: We may write

$$
\left\| \frac{E^r(\alpha)}{\bar{w}(\alpha)} \right\| \leq \left\| \frac{e}{\bar{w}(\alpha)} \right\|_\infty \|E^r(\alpha)\| = \frac{1}{\mu} \left\| \sqrt{\frac{\mu e}{\bar{w}(\alpha)}} \right\|_\infty^2 \|E^r(\alpha)\| \leq \frac{\zeta^2}{\mu} \|E^r(\alpha)\|,
$$

where the last inequality follows from Corollary IV.31. Now using Corollary IV.29 we have

$$\left\| \frac{E^r(\alpha)}{\bar{w}(\alpha)} \right\| \leq \frac{r}{8(r+1)} (8\alpha\tau\zeta)^{r+1},$$

proving the first part of the lemma. The second part follows from the first part by using $8\alpha\tau\zeta \leq 1$:

$$\left\| \frac{E^r(\alpha)}{\bar{w}(\alpha)} \right\|_\infty \leq \left\| \frac{E^r(\alpha)}{\bar{w}(\alpha)} \right\| \leq \frac{r}{8(r+1)} < \frac{\sqrt{5}-1}{2},$$

completing the proof. □

Equipped with the above results we can prove the next theorem.

Theorem IV.35 *Let* $\delta \leq \tau$, $\delta_\infty \leq \zeta$, *with* $\zeta \geq 2$, *and* $\alpha \leq 1/(8\tau\zeta)$. *Then*

(i) $\delta(w(\alpha), \mu) \leq \left(1 - \frac{\alpha}{2}\right)\tau + \frac{r}{8(r+1)}\sqrt{1+\tau^2}\,(8\alpha\tau\zeta)^{r+1}$;

(ii) $\delta_\infty(w(\alpha), \mu) \leq \left(1 - \frac{3}{8}\alpha\right)\left(1 + \frac{r}{8(r+1)}(8\alpha\tau\zeta)^{r+1}\right)\zeta$.

Proof: For the given step-size the hypothesis of Corollary IV.33 is satisfied, by Lemma IV.34. From Lemma IV.34 we also deduce the second inequality in

$$\left\| \frac{E^r(\alpha)}{\bar{w}(\alpha)} \right\|_\infty \leq \left\| \frac{E^r(\alpha)}{\bar{w}(\alpha)} \right\| \leq \frac{r}{8(r+1)}(8\alpha\tau\zeta)^{r+1}.$$

Substituting these inequalities in Corollary IV.33 yields the theorem. □

18.5.4 The step-size

In the sequel the step-size α is given the value

$$\alpha = \frac{1}{8\tau\zeta\sqrt[r]{(r+1)\zeta}\sqrt{1+\tau^2}}, \tag{18.45}$$

where $\delta = \delta(xs, \mu) \leq \tau$ and $\delta_\infty = \delta_\infty(xs, \mu) \leq \zeta$. It is assumed that $\zeta \geq 2$. The next theorem makes clear that after a higher-order step with the given step-size α the proximity δ is below a fixed fraction of τ and the proximity δ_∞ below a fixed fraction of ζ.

Theorem IV.36 *If the step-size is given by (18.45) then*

$$\delta(w(\alpha), \mu) \leq \left(1 - \frac{\alpha(r^2+1)}{2(r+1)^2}\right)\tau.$$

Moreover,

$$\delta_\infty(w(\alpha), \mu) \leq \left(1 - \frac{\alpha}{8}\right)\zeta.$$

Proof: The proof uses Theorem IV.35. This theorem applies because for the given value of α we have

$$\alpha = \frac{1}{8\tau\zeta \sqrt[r]{(r+1)\zeta\sqrt{1+\tau^2}}} \leq \frac{1}{8\tau\zeta},$$

whence $8\alpha\tau\zeta \leq 1$. Hence, by the first statement in Theorem IV.35,

$$\delta(w(\alpha),\mu) \leq \left(1 - \frac{\alpha}{2}\right)\tau + \frac{r}{8(r+1)}\sqrt{1+\tau^2}\,(8\alpha\tau\zeta)^{r+1}. \tag{18.46}$$

The second term on the right can be reduced by using the definition of α:

$$
\begin{aligned}
\delta(w(\alpha),\mu) \quad &\leq \quad \left(1 - \frac{\alpha}{2}\right)\tau + \frac{r}{8(r+1)}\sqrt{1+\tau^2}\,\frac{8\alpha\tau\zeta}{(r+1)\zeta\sqrt{1+\tau^2}} \\[2mm]
&= \quad \left(1 - \frac{\alpha}{2} + \frac{r\alpha}{(r+1)^2}\right)\tau \\[2mm]
&= \quad \left(1 - \frac{r^2+1}{2(r+1)^2}\alpha\right)\tau.
\end{aligned}
$$

This proves the first statement. The second claim follows in a similar way from the second statement in Theorem IV.35:

$$
\begin{aligned}
\delta_\infty(w(\alpha),\mu) \quad &\leq \quad \left(1 - \frac{3}{8}\alpha\right)\left(1 + \frac{r}{8(r+1)}(8\alpha\tau\zeta)^{r+1}\right)\zeta \\[2mm]
&= \quad \left(1 - \frac{3}{8}\alpha\right)\left(1 + \frac{r}{8(r+1)}\frac{8\alpha\tau\zeta}{(r+1)\zeta\sqrt{1+\tau^2}}\right)\zeta \\[2mm]
&= \quad \left(1 - \frac{3}{8}\alpha\right)\left(1 + \frac{r}{(r+1)^2}\frac{\alpha\tau}{\sqrt{1+\tau^2}}\right)\zeta \\[2mm]
&< \quad \left(1 - \frac{3}{8}\alpha\right)\left(1 + \frac{r\alpha}{(r+1)^2}\right)\zeta \\[2mm]
&\leq \quad \left(1 - \frac{3}{8}\alpha\right)\left(1 + \frac{1}{4}\alpha\right)\zeta \\[2mm]
&\leq \quad \left(1 - \frac{\alpha}{8}\right)\zeta.
\end{aligned}
$$

In the last but one inequality we used that $r/(r+1)^2$ is monotonically decreasing if r increases (for $r \geq 1$). $\qquad\square$

18.5.5 Reduction of the barrier parameter

In this section we assume that $\delta = \delta(xs,\mu) \leq \tau$, where τ is any positive number. After a higher-order step with step-size α, given by (18.45), we have by Theorem IV.36,

$$\delta(w(\alpha),\mu) \leq (1 - \beta)\,\delta,$$

where

$$\beta = \frac{\alpha(r^2+1)}{2(r+1)^2}. \tag{18.47}$$

Below we investigate how far μ can be decreased after the step while keeping the proximity δ less than or equal to τ. Before doing this we observe that

$$\delta_\infty(xs, \mu) = \left\| \sqrt{\frac{\mu e}{xs}} \right\|_\infty$$

is monotonically decreasing as μ decreases. Hence, we do not have to worry about δ_∞ when μ is reduced. Defining

$$\mu^+ := (1 - \theta)\mu,$$

we first deal with a lemma that later gives an upper bound for $\delta(w(\alpha), \mu^+)$.[13]

Lemma IV.37 *Let (x, s) be a positive primal-dual pair and suppose $\mu > 0$. If $\delta := \delta(xs, \mu)$ and $\mu^+ = (1 - \theta)\mu$ then*

$$\delta(xs, \mu^+) \leq \frac{2\delta + \theta\sqrt{n}}{2\sqrt{1 - \theta}}.$$

Proof: By the definition of $\delta(xs, \mu^+)$,

$$\delta(xs, \mu^+) = \frac{1}{2} \left\| \sqrt{\frac{xs}{\mu^+ e}} - \sqrt{\frac{\mu^+ e}{xs}} \right\|.$$

To simplify the notation in the proof we use $u = \sqrt{xs/\mu}$. Then we may write

$$\delta(xs, \mu^+) = \frac{1}{2} \left\| \frac{u}{\sqrt{1 - \theta}} - \sqrt{1 - \theta}\, u^{-1} \right\| = \frac{1}{2} \left\| \sqrt{1 - \theta}\, (u - u^{-1}) + \frac{\theta u}{\sqrt{1 - \theta}} \right\|.$$

Using the triangle inequality and[14] also

$$\|u\| \leq \|u - u^{-1}\| + \sqrt{n} = 2\delta + \sqrt{n},$$

we get

$$\delta(xs, \mu^+) \leq \delta\sqrt{1 - \theta} + \frac{\theta \|u\|}{2\sqrt{1 - \theta}} \leq \delta\sqrt{1 - \theta} + \frac{\theta(2\delta + \sqrt{n})}{2\sqrt{1 - \theta}} = \frac{2\delta + \theta\sqrt{n}}{2\sqrt{1 - \theta}},$$

proving the lemma. □

⎯⎯

[13] A similar result was derived in Lemma II.54, but under the assumption that $x^T s = n\mu$. This assumption will in general not be satisfied in the present context, and hence we have a weaker bound.

[14] **Exercise 87** For each positive number ξ we have

$$|\xi| \leq \left| \xi - \frac{1}{\xi} \right| + 1.$$

Prove this and derive that for each positive vector u the following inequality holds:

$$\|u\| \leq \|u - u^{-1}\| + \sqrt{n}.$$

Theorem IV.38 *Let* $\delta = \delta(xs, \mu) \leq \tau$, $\delta_\infty = \delta_\infty(xs, \mu) \leq \zeta$, *with* $\zeta \geq 2$. *Taking first a higher-order step at* (x, s), *with* α *according to (18.45), and then updating the barrier parameter to* $\mu^+ = (1 - \theta)\mu$, *where*

$$\theta = \frac{2\beta\tau}{2\tau + \sqrt{n}} = \frac{\alpha\tau(r^2 + 1)}{(r+1)^2(2\tau + \sqrt{n})}, \qquad (18.48)$$

we have $\delta(w(\alpha), \mu^+) \leq \tau$ *and* $\delta_\infty(w(\alpha), \mu^+) \leq \zeta$.

Proof: The second part of Theorem IV.36 implies that after a step of the given size $\delta_\infty(w(\alpha), \mu) \leq \zeta$. We established earlier that δ_∞ monotonically decreases when μ decreases. As a result we have $\delta_\infty(w(\alpha), \mu^+) \leq \zeta$. Now let us estimate $\delta(w(\alpha), \mu^+)$. After a higher-order step with step-size α as given by (18.45), we have by the first part of Theorem IV.36,

$$\delta(w(\alpha), \mu) \leq \left(1 - \frac{\alpha(r^2 + 1)}{2(r+1)^2}\right)\delta(xs, \mu) = (1 - \beta)\,\delta,$$

with β as defined in (18.47). Also using Lemma IV.37 we obtain

$$\delta(w(\alpha), \mu^+) \leq \frac{2\delta(w(\alpha), \mu) + \theta\sqrt{n}}{2\sqrt{1-\theta}} \leq \frac{2(1 - \beta)\,\delta + \theta\sqrt{n}}{2\sqrt{1-\theta}}.$$

Since $\delta \leq \tau$, we certainly have $\delta(w(\alpha), \mu^+) \leq \tau$ if

$$\frac{2(1 - \beta)\tau + \theta\sqrt{n}}{2\sqrt{1-\theta}} \leq \tau.$$

This inequality can be rewritten as

$$2(1 - \beta)\tau + \theta\sqrt{n} \leq 2\tau\sqrt{1-\theta}.$$

Using $\sqrt{1-\theta} \geq 1 - \theta$ the above inequality certainly holds if

$$2(1 - \beta)\tau + \theta\sqrt{n} \leq 2\tau(1 - \theta).$$

It is easily verified that the value of θ in (18.48) satisfies this inequality with equality. Thus the proof is complete. □

18.5.6 *A higher-order logarithmic barrier algorithm*

Formally the logarithmic barrier algorithm using higher-order Newton steps can be described as below.

Higher-Order Logarithmic Barrier Algorithm

Input:
   A natural number r, the order of the search directions;
   a positive number τ, specifying the cone;
   a primal-dual pair (x^0, s^0) and $\mu^0 > 0$ such that $\delta(x^0 s^0, \mu^0) \leq \tau$.
   $\zeta := \max\left(2, \delta_\infty(x^0 s^0, \mu^0)\right);$
   a step-size parameter α, from (18.45);
   an update parameter θ, from (18.48);
   an accuracy parameter $\varepsilon > 0$;
begin
   $x := x^0;\ s := s^0;\ \mu := \mu^0;$
   while $x^T s \geq \varepsilon$ **do**
   begin
      $x := x(\alpha) = x + \Delta^{r,\alpha} x;$
      $s := s(\alpha) = s + \Delta^{r,\alpha} s;$
      $\mu := (1 - \theta)\mu$
   end
end

A direct consequence of the specified values of the step-size α and update parameter θ is that the properties $\delta(xs, \mu) \leq \tau$ and $\delta_\infty(xs, \mu) \leq \zeta$ are maintained in the course of the algorithm. This follows from Theorem IV.38 and makes the algorithm well-defined.

18.5.7 Iteration bound

In the further analysis of the algorithm we choose

$$\tau = \sqrt{n} \quad \text{and} \quad r = n.$$

At the end of each iteration of the algorithm we have

$$\delta(xs, \mu) \leq \tau = \sqrt{n}.$$

As a consequence (cf. Exercise 62),

$$x^T s \leq \left(1 + \frac{2\tau\rho(\tau)}{\sqrt{n}}\right) n\mu = \left(1 + 2\rho\left(\sqrt{n}\right)\right) n\mu \leq 4\left(1 + \sqrt{n}\right) n\mu.$$

Hence $x^T s \leq \varepsilon$ holds if

$$4\left(1 + \sqrt{n}\right) n\mu \leq \varepsilon,$$

or

$$\mu \leq \frac{\varepsilon}{4\left(1 + \sqrt{n}\right)}.$$

Recall that at each iteration the barrier parameter is reduced by a factor $1 - \theta$, with

$$\theta = \frac{\alpha\tau(r^2 + 1)}{(r + 1)^2 (2\tau + \sqrt{n})} = \frac{\alpha(n^2 + 1)}{3(n + 1)^2} \geq \frac{\alpha}{6}. \tag{18.49}$$

The last inequality holds for all $n \geq 1$. Using Lemma I.36 we find that the number of iterations does not exceed

$$\frac{6}{\alpha} \log \frac{4(1 + \sqrt{n}) n\mu^0}{\varepsilon}.$$

Substituting α in (18.45) and $\tau = \sqrt{n}$, we get

$$\frac{6}{\alpha} = 48\zeta\sqrt{n} \sqrt[n]{(n + 1)\zeta\sqrt{1 + n}}.$$

For $n \geq 1$ we have

$$\sqrt[n]{(n + 1)\sqrt{1 + n}} \leq 2\sqrt{2} = 2.8284,$$

with equality only if $n = 1$. Thus we find

$$\frac{6}{\alpha} \leq 136\,\zeta^{\frac{n+1}{n}} \sqrt{n}.$$

Thus we may state the next theorem without further proof.

Theorem IV.39 *The Higher-Order Logarithmic Barrier Algorithm needs at most*

$$136\,\zeta^{\frac{n+1}{n}} \sqrt{n}\, \log \frac{4(1 + \sqrt{n}) n\mu^0}{\varepsilon}$$

iterations. Each iteration requires $\mathcal{O}(n^3)$ arithmetic operations. The output is a primal-dual pair (x, s) such that $x^T s \leq \varepsilon$.

When starting the algorithm on the central path, with $\mu^0 = (x^0)^T s^0/n$, we have $\zeta = 2$. In that case $\delta_\infty(xs, \mu) \leq 2$ at each iteration and the iteration bound of Theorem IV.39 becomes

$$544\sqrt{n}\, \log \frac{4(1 + \sqrt{n}) n\mu^0}{\varepsilon} = \mathcal{O}\left(\sqrt{n}\, \log \frac{\sqrt{n}n\mu^0}{\varepsilon}\right). \tag{18.50}$$

In fact, as long as $\zeta = \mathcal{O}(1)$ the iteration bound is given by the right-hand expression in (18.50). Note that this bound has the same order of magnitude as the best known iteration complexity bound.

When (x^0, s^0) is far from the central path, the value of ζ may be so large that the iteration bound of Theorem IV.39 becomes very poor. Note that ζ can be as large as $\rho(\tau)$, which would give an extra factor $\mathcal{O}\left(n^{\frac{n+1}{2n}}\right)$ in (18.50). However, a more careful analysis yields a much better bound, as we show in the next section.

18.5.8 *Improved iteration bound*

In this section we consider the situation where the algorithm starts with a high value of ζ. Recall from the previous section that if $\tau = \sqrt{n}$ then ζ is always bounded by

$\zeta \leq \rho(\sqrt{n}) = \mathcal{O}(\sqrt{n})$. Now the second part of Theorem IV.36 implies that after a higher-order step at (x, s) to the μ-center we have

$$\delta_\infty(w(\alpha), \mu) \leq \left(1 - \frac{\alpha}{8}\right)\zeta.$$

Reducing μ to $\mu^+ = (1 - \theta)\mu$ we get

$$\delta_\infty(w(\alpha), \mu^+) \leq (1 - \theta)\left(1 - \frac{\alpha}{8}\right)\zeta.$$

Now using the lower bound (18.49) for θ it follows that

$$\delta_\infty(w(\alpha), \mu^+) \leq \left(1 - \frac{\alpha}{6}\right)\left(1 - \frac{\alpha}{8}\right)\zeta.$$

Since $0 \leq \alpha \leq 1$ we have $\left(1 - \frac{\alpha}{6}\right)\left(1 - \frac{\alpha}{8}\right) \leq \left(1 - \frac{\alpha}{4}\right)$. Hence

$$\delta_\infty(w(\alpha), \mu^+) \leq \left(1 - \frac{\alpha}{4}\right)\zeta.$$

Substituting the value of α, while using

$$8\sqrt[n]{(n+1)\zeta\sqrt{1+n}} \leq 8\sqrt[n]{(n+1)\rho(\sqrt{n})\sqrt{1+n}} \leq 55,$$

we obtain

$$\delta_\infty(w(\alpha), \mu^+) \leq \zeta - \frac{\zeta}{220\tau\zeta} = \zeta - \frac{1}{220\sqrt{n}},$$

showing that $\delta_\infty(xs, \mu)$ decreases by at least $1/(220\sqrt{n})$ in one iteration. Obviously, we can redefine ζ according to

$$\zeta := \max\left(2, \delta_\infty(w(\alpha), \mu^+)\right) \leq \max\left(2, \zeta - \frac{1}{220\sqrt{n}}\right)$$

in the next iteration and continue the algorithm with this new value. In this way ζ reaches the value 2 in no more than

$$220\sqrt{n}\left(\zeta^0 - 2\right) = \mathcal{O}\left(\zeta^0\sqrt{n}\right)$$

iterations, where $\zeta^0 = \delta_\infty(x^0 s^0, \mu^0)$. From then on ζ keeps the value 2, and the number of additional iterations is bounded by (18.50). Hence we may state the following improvement of Theorem IV.39 without further proof.

Theorem IV.40 *The Higher-Order Logarithmic Barrier Algorithm needs at most*

$$\mathcal{O}\left(\zeta^0\sqrt{n} + \sqrt{n}\,\log\frac{4\sqrt{n}n\mu^1}{\varepsilon}\right).$$

iterations. Each iteration requires $\mathcal{O}(n^3)$ arithmetic operations. The output is a primal-dual pair (x, s) such that $x^T s \leq \varepsilon$.

In this theorem μ^1 denotes the value of the barrier parameter attained at the first iteration for which $\zeta = 2$. Obviously, $\mu^1 \leq \mu^0$.

19

Parametric and Sensitivity Analysis

19.1 Introduction

Many commercial optimization packages for solving LO problems not only solve the problem at hand, but also provide additional information on the solution. This added information concerns the sensitivity of the solution produced by the package to perturbations in the data of the problem. In this chapter we deal with a problem (P) in standard format:

$$(P) \qquad \min\left\{c^T x \; : \; Ax = b, \, x \geq 0\right\}.$$

The dual problem (D) is written as

$$(D) \qquad \max\left\{b^T y \; : \; A^T y + s = c, \, s \geq 0\right\}.$$

The input data for both problems consists of the matrix A, which is of size $m \times n$, and the vectors $b \in \mathbb{R}^m$ and $c \in \mathbb{R}^n$. The optimal value of (P) and (D) is denoted by $z_A(b, c)$, with $z_A(b, c) = -\infty$ if (P) is unbounded and (D) infeasible, and $z_A(b, c) = \infty$ if (D) is unbounded and (P) infeasible. If (P) and (D) are both infeasible then $z_A(b, c)$ is undefined. We call z_A the *optimal-value function* for the matrix A.

The extra information provided by solution packages concerns only changes in the vectors b and c. We also restrict ourselves to such changes. It will follow from the results below that $z_A(b, c)$ depends continuously on the vectors b and c. In contrast, the effect of changes in the matrix A is not necessarily continuous. The next example provides a simple illustration of this phenomenon.[1]

Example IV.41 Consider the problem

$$\min\left\{x_2 \; : \; \alpha x_1 + x_2 = 1, \, x_1 \geq 0, \, x_2 \geq 0\right\},$$

where $\alpha \in \mathbb{R}$. In this example we have $A = (\alpha \; 1)$, $b = (1)$ and $c = (0 \; 1)^T$. We can easily verify that $z_A(b, c) = 0$ if $\alpha > 0$ and $z_A(b, c) = 1$ if $\alpha \leq 0$. Thus, if $z_A(b, c)$ is considered a function of α, a discontinuity occurs at $\alpha = 0$. $\diamond$

Thus. the dependence of $z_A(b, c)$ on the entries in b and c is more simple than the dependence of $z_A(b, c)$ on the entries in A.

---

[1] For some results on the effect of changes in A we refer the reader to Mills [210] and Gal [89].

We develop some theory in this chapter for the analysis of one-dimensional parametric perturbations of the vectors b and c. Given a pair of optimal solutions for (P) and (D), we present an algorithm in Section 19.4.5 for the computation of the optimal-value function under such a perturbation. Then, in Section 19.5 we consider the special case of *sensitivity analysis*, also called *postoptimal analysis*. This classical topic is treated in almost all (text-)books on LO and implemented in almost all commercial optimization packages for LO. We show in Section 19.5.1 that the so-called *ranges* and *shadow prices* of the coefficients in b and c can be obtained by solving auxiliary LO problems. In Section 19.5.3 we briefly discuss the classical approach to sensitivity analysis, which is based on the use of an *optimal basic solution* and the corresponding *optimal basis*. Although the classical approach is much cheaper from a computational point of view, it yields less information and can easily be misinterpreted. This is demonstrated in Section 19.5.4, where we provide a striking example of the inherent weaknesses of the classical approach.

19.2 Preliminaries

The feasible regions of (P) and (D) are denoted by

$$
\begin{aligned}
\mathcal{P} &:= \{x \ : \ Ax = b, \ x \geq 0\}, \\
\mathcal{D} &:= \{(y, s) \ : \ A^T y + s = c, \ s \geq 0\}.
\end{aligned}
$$

Assuming that (P) and (D) are both feasible, the optimal sets of (P) and (D) are denoted by $\mathcal{P}^*$ and $\mathcal{D}^*$. We define the index sets B and N by

$$ B := \{i \ : \ x_i > 0 \text{ for some } x \in \mathcal{P}^*\}, $$

$$ N := \{i \ : \ s_i > 0 \text{ for some } (y, s) \in \mathcal{D}^*\}. $$

The Duality Theorem (Theorem II.2) implies that $B \cap N = \emptyset$, and the Goldman–Tucker Theorem (Theorem II.3) that

$$ B \cup N = \{1, 2, \ldots, n\}. $$

Thus, B and N form a partition of the full index set. This (ordered) partition, denoted by $\pi = (B, N)$, is the *optimal partition* of problems (P) and (D). It is obvious that the optimal partition depends on b and c.

19.3 Optimal sets and optimal partition

In the rest of this chapter we assume that b and c are such that (P) and (D) have optimal solutions, and $\pi = (B, N)$ denotes the optimal partition of both problems. By definition, the optimal partition is determined by the sets of optimal solutions for (P) and (D). In this section it is made clear that, conversely, the optimal partition provides essential information on the optimal solution sets $\mathcal{P}^*$ and $\mathcal{D}^*$. The next lemma follows immediately from the Duality Theorem and is stated without proof.

Lemma IV.42 *Let $x^* \in \mathcal{P}^*$ and $(y^*, s^*) \in \mathcal{D}^*$. Then*

$$\mathcal{P}^* = \{x : x \in \mathcal{P}, x^T s^* = 0\},$$
$$\mathcal{D}^* = \{(y, s) : (y, s) \in \mathcal{D}, s^T x^* = 0\}.$$

As before, we use the notation x_B and x_N to refer to the restriction of a vector $x \in \mathbb{R}^n$ to the coordinate sets B and N respectively. Similarly, A_B denotes the restriction of A to the columns in B and A_N the restriction of A to the columns in N. Now the sets $\mathcal{P}^*$ and $\mathcal{D}^*$ can be described in terms of the optimal partition.

Lemma IV.43 *Given the optimal partition (B, N) of (P) and (D), the optimal sets of both problems are given by*

$$\mathcal{P}^* = \{x : x \in \mathcal{P}, x_N = 0\},$$
$$\mathcal{D}^* = \{(y, s) : (y, s) \in \mathcal{D}, s_B = 0\}.$$

Proof: Let x^*, s^* be any strictly complementary pair of solutions of (P) and (D), and (x, s) an arbitrary pair of feasible solutions. Then, from Lemma IV.42, x is optimal for (P) if and only if $x^T s^* = 0$. Since $s_B^* = 0$ and $s_N^* > 0$, we have $x^T s^* = 0$ if and only if $x_N = 0$, thus proving that $\mathcal{P}^*$ consists of all primal feasible x for which $x_N = 0$. Similarly, if $(y, s) \in \mathcal{D}$ then this pair is optimal if and only if $s^T x^* = 0$. Since $x_B^* > 0$ and $x_N^* = 0$, this occurs if and only if $s_B = 0$, thus proving that $\mathcal{D}^*$ consists of all dual feasible s for which $s_B = 0$. $\qquad\square$

To illustrate the meaning of Lemma IV.43 we give an example.

Example IV.44 Figure 19.1 shows a network with given arc lengths, and we ask for a shortest path from node s to node t.

Denoting the set of nodes in this network by V and the set of arcs by E, any path from s to t can be represented by a 0-1 vector x of length $|E|$, whose coordinates are indexed by the arcs, such that $x_e = 1$ if and only if arc e belongs to the path. The length of the path is then given by

$$\sum_{e \in E} c_e x_e, \tag{19.1}$$

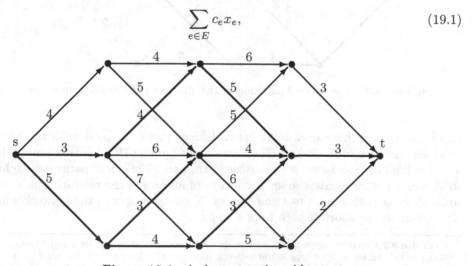

Figure 19.1   A shortest path problem.

where c_e denote the length of arc e, for all $e \in E$. Furthermore, denoting $e = (v, w)$ if arc e points from node v to node w (with $v \in V$ and $w \in V$), and denoting x_e by x_{vw}, x will satisfy the following *balance equations*:

$$
\begin{array}{rcl}
\displaystyle\sum_{v \in V} x_{sv} &=& 1 \\[2mm]
\displaystyle\sum_{v \in V} x_{vu} &=& \displaystyle\sum_{v \in V} x_{uv}, \quad u \in V \setminus \{s, t\} \\[2mm]
\displaystyle\sum_{v \in V} x_{vt} &=& 1
\end{array}
\tag{19.2}
$$

Now consider the LO problem consisting of minimizing the linear function (19.1) subject to the linear equality constraints in (19.2), with all variables x_e, $e \in E$, nonnegative. This problem has the standard format: it is a minimization problem with equality constraints and nonnegative variables. Solving this problem with an interior-point method we find a strictly complementary solution, and hence the optimal partition of the problem. In this way we have computed the optimal partition (B, N) of the problem. Since in this example there is a 1-to-1 correspondence between the arcs and the variables, we may think of B and N as a partition of the arcs in the network.

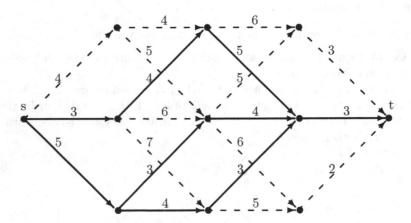

Figure 19.2 The optimal partition of the shortest path problem in Figure 19.1.

In Figure 19.2 we have drawn the network once more, but now with the arcs in B solid and the arcs in N dashed. The meaning of Lemma IV.43 is that any path from s to t using only solid arcs is a shortest path, and all shortest paths use exclusively solid arcs. In other words, the set B consists of all arcs in the network which occur in some shortest path from s to t and the set N contains arcs in the network which do not belong to any shortest path from s to t.[2]   ◇

---

[2] **Exercise 88** Consider any network with node set V and arc set E and let s and t be two distinct nodes in this network. If all arcs in the network have positive length, then the set B, consisting of all arcs in the network which occur in at least one shortest path from s to t, does not contain a (directed) circuit. Prove this.

The next result deals with the dimensions of the optimal sets $\mathcal{P}^*$ and $\mathcal{D}^*$. Here, as usual the (affine) dimension of a subset of $\mathbb{R}^k$ is the dimension of the smallest affine subspace in $\mathbb{R}^k$ containing the subset.

Lemma IV.45 *We have*

$$\begin{aligned} \dim \mathcal{P}^* &= |B| - \operatorname{rank}(A_B) \\ \dim \mathcal{D}^* &= m - \operatorname{rank}(A_B). \end{aligned}$$

Proof: The optimal set of (P) is given by

$$\mathcal{P}^* = \{x \ : \ Ax = b, \ x_B \geq 0, \ x_N = 0\},$$

and hence the smallest affine subspace of $\mathbb{R}^n$ containing $\mathcal{P}^*$ is given by

$$\{x \ : \ A_B x_B = b, \ x_N = 0\}.$$

The dimension of this affine space is equal to the dimension of the null space of A_B. Since this dimension is given by $|B| - \operatorname{rank}(A_B)$, the first statement follows.

For the proof of the second statement we use that the dual optimal set can be described by

$$\mathcal{D}^* = \{(y, s) \ : \ A^T y + s = c, \ s_B = 0, \ s_N \geq 0\}.$$

This is equivalent to

$$\mathcal{D}^* = \{(y, s) \ : \ A_B^T y = c_B, \ A_N^T y + s_N = c_N, \ s_B = 0, \ s_N \geq 0\}.$$

The smallest affine subspace containing this set is

$$\{(y, s) \ : \ A_B^T y = c_B, \ A_N^T y + s_N = c_N, \ s_B = 0\}.$$

Obviously s_N is uniquely determined by y, and any y satisfying $A_B^T y = c_B$ yields a point in this affine space. Hence the dimension of the affine space is equal to the dimension of the null space of A_B^T. Since m is the number of columns of A_B^T, the dimension of the null space of A_B^T equals $m - \operatorname{rank}(A_B)$. This completes the proof. $\square$

Lemma IV.45 immediately implies that (P) has a unique solution if and only if $\operatorname{rank}(A_B) = |B|$. Clearly this happens if and only if the columns in A_B are linearly independent. Also, (D) has a unique solution if and only if $\operatorname{rank}(A_B) = m$, which happens if and only if the rows in A_B are linearly independent.[3]

---

[3] It has become common practice in the literature to call the problem (P) *degenerate* if (P) or (D) have multiple optimal solutions. Degeneracy is an important topic in LO. In the context of the Simplex Method it is well known as a source of difficulties. This is especially true when dealing with sensitivity analysis. See, e.g., Gal [90] and Greenberg [128]. But also in the context of interior-point methods the occurrence of degeneracy may influence the behavior of the method. We mention some references: Gonzaga [120], Güler et al. [132], Todd [263], Tsuchiya [269, 271], Hall and Vanderbei [138].

19.4   Parametric analysis

In this section we start to investigate the effect of changes in b and c on the optimal-value function $z_A(b,c)$. We consider one-dimensional parametric perturbations of b and c. So we want to study

$$z_A(b + \beta\Delta b, c + \gamma\Delta c)$$

as a function of the parameters β and γ, where Δb and Δc are given *perturbation vectors*. From now on the vectors b and c are fixed, and the variations come from the parameters β and γ. In fact, we restrict ourselves to the cases that the variations occur only in one of the two vectors b and c. In other words, taking $\gamma = 0$ we consider variations in β and taking $\beta = 0$ we consider variations in γ.

If $\gamma = 0$, then (P_β) will denote the perturbed primal problem and (D_β) its dual. The feasible regions of these problems are denoted by $\mathcal{P}_\beta$ and $\mathcal{D}_\beta$. Similarly, if $\beta = 0$, then (D_γ) will denote the perturbed dual problem and (P_γ) its dual and the feasible regions of these problems are $\mathcal{D}_\gamma$ and $\mathcal{P}_\gamma$. Observe that the feasible region of (D_β) is simply $\mathcal{D}$ and the feasible region of (P_γ) is simply $\mathcal{P}$. We use the superscript $*$ to refer to the optimal set of each of these problems.

We assume that b and c are such that (P) and (D) are both feasible. Then $z_A(b,c)$ is well defined and finite. It is convenient to introduce the following notations:

$$b(\beta) := b + \beta\Delta b, \quad c(\gamma) := c + \gamma\Delta c,$$

$$f(\beta) := z_A(b(\beta), c), \quad g(\gamma) := z_A(b, c(\gamma)).$$

Here the domain of the parameters β and γ is taken as large as possible. Let us consider the domain of f. This function is defined as long as $z_A(b(\beta), c)$ is well defined. Since the feasible region of (D_β) is constant when β varies, and since we assume that (D_β) is feasible for $\beta = 0$, it follows that (D_β) is feasible for all values of β. Therefore, $f(\beta)$ is well defined if the dual problem (D_β) has an optimal solution and $f(\beta)$ is not defined (or infinity) if the dual problem (D_β) is unbounded. By the Duality Theorem it follows that $f(\beta)$ is well defined if and only if the primal problem (P_β) is feasible. In exactly the same way it can be understood that the domain of g consists of all γ for which (D_γ) is feasible (and (P_γ) bounded).

Lemma IV.46 *The domains of f and g are convex.*

Proof: We give the proof for f. The proof for g is similar and therefore omitted. Let $\beta_1, \beta_2 \in \text{dom}(f)$ and $\beta_1 < \beta < \beta_2$. Then $f(\beta_1)$ and $f(\beta_2)$ are finite, which means that both $\mathcal{P}_{\beta_1}$ and $\mathcal{P}_{\beta_2}$ are nonempty. Let $x^1 \in \mathcal{P}_{\beta_1}$ and $x^2 \in \mathcal{P}_{\beta_2}$. Then x^1 and x^2 are nonnegative and

$$Ax^1 = b + \beta_1\Delta b, \quad Ax^2 = b + \beta_2\Delta b.$$

Now consider

$$x := x^1 + \frac{\beta - \beta_1}{\beta_2 - \beta_1}\left(x^2 - x^1\right) = \frac{(\beta_2 - \beta)\,x^1 + (\beta - \beta_1)\,x^2}{\beta_2 - \beta_1}.$$

Note that x is a convex combination of x^1 and x^2 and hence x is nonnegative. We proceed by showing that $x \in \mathcal{P}_\beta$. Using that $A\left(x^2 - x^1\right) = (\beta_2 - \beta_1)\Delta b$ this goes as follows:

$$
\begin{aligned}
Ax &= Ax^1 + \frac{\beta - \beta_1}{\beta_2 - \beta_1}A\left(x^2 - x^1\right) \\
&= b + \beta_1 \Delta b + \frac{\beta - \beta_1}{\beta_2 - \beta_1}(\beta_2 - \beta_1)\Delta b \\
&= b + \beta_1 \Delta b + (\beta - \beta_1)\Delta b \\
&= b + \beta \Delta b.
\end{aligned}
$$

This proves that (P_β) is feasible and hence $\beta \in \mathrm{dom}\,(f)$, completing the proof. $\quad\square$

The domains of f and g are in fact closed intervals on the real line. This follows from the above lemma, and the fact that the complements of the domains of f and g are open subsets of the real line. The last statement is the content of the next lemma.

Lemma IV.47 *The complements of the domains of f and g are open subsets of the real line.*

Proof: As in the proof of the previous lemma we omit the proof for g because it is similar to the proof for f. We need to show that the complement of $\mathrm{dom}\,(f)$ is open. Let $\beta \notin \mathrm{dom}\,(f)$. This means that (D_β) is unbounded. This is equivalent to the existence of a vector z such that

$$
A^T z \leq 0, \qquad (b + \beta\Delta b)^T z > 0.
$$

Fixing z and considering β as a variable, the set of all β satisfying the strict inequality $(b + \beta\Delta b)^T z > 0$ is an open interval. For all β in this interval (D_β) is unbounded. Hence the domain of f is open. This proves the lemma. $\quad\square$

A consequence of the last two lemmas is the next theorem, which requires no further proof.

Theorem IV.48 *The domains of f and g are closed intervals on the real line.*[4] $\quad\square$

Example IV.49 Let (D) be the problem

$$
\max_{y=(y_1,y_2)} \{y_2 \ : \ y_2 \leq 1\}.
$$

In this case $b = (0,1)$ and $c = (1)$. Note that (D) is feasible and bounded. The set of all optimal solutions consists of all $(y_1, 1)$ with $y_1 \in \mathbb{R}$. Now let $\Delta b = (1,0)$, and consider the effect of replacing b by $b + \beta\Delta b$, and let $f(\beta)$ be as defined above. Then

$$
f(\beta) = \max_{y=(y_1,y_2)} \{y_2 + \beta y_1 \ : \ y_2 \leq 1\}.
$$

---

[4] To avoid misunderstanding we point out that a singleton $\{a\}$ $(a \in \mathbb{R})$ is also considered as a closed interval.

We can easily verify that the perturbed problem is unbounded for all nonzero β. Hence the domain of f is the singleton $\{0\}$.[5] $\diamond$

19.4.1 The optimal-value function is piecewise linear

In this section we show that the functions $f(\beta)$ and $g(\gamma)$ are piecewise linear on their domains. We start with $g(\gamma)$.

Theorem IV.50 $g(\gamma)$ is continuous, concave and piecewise linear.

Proof: By definition,

$$g(\gamma) = \min\left\{c(\gamma)^T x \ : \ x \in \mathcal{P}\right\}.$$

For each γ the minimum value is attained at the central solution of the perturbed problem (P_γ). This solution is uniquely determined by the optimal partition of (P_γ). Since the number of partitions of the full index set $\{1, 2, \ldots, n\}$ is finite, we may write

$$g(\gamma) = \min\left\{c(\gamma)^T x \ : \ x \in \mathcal{T}\right\},$$

where $\mathcal{T}$ is a finite subset of $\mathcal{P}$. For each $x \in \mathcal{T}$ we have

$$c(\gamma)^T x = c^T x + \gamma \Delta c^T x,$$

which is a linear function of γ. Thus, $g(\gamma)$ is the minimum of a finite set of linear functions.[6] This implies that $g(\gamma)$ is continuous, concave and piecewise linear, proving the theorem.   □

Theorem IV.51 $f(\beta)$ is continuous, convex and piecewise linear.

Proof: The proof goes in the same way as for Theorem IV.50. By definition,

$$f(\beta) = \max\left\{b(\beta)^T y \ : \ y \in \mathcal{D}\right\}.$$

For each β the maximum value is attained at a central solution (y^*, s^*) of (D). Now s^* is uniquely determined by the optimal partition of (D) and $b(\beta)^T y^*$ is constant for all optimal y^*. Associating one particular y^* with any possible slack s^* arising in this way, we obtain that

$$f(\beta) = \max\left\{b(\beta)^T y \ : \ y \in \mathcal{S}\right\},$$

where S is a finite subset of $\mathcal{D}$. For each $y \in \mathcal{S}$, we have

$$b(\beta)^T y = b^T y + \beta \Delta b^T y,$$

---

[5] **Exercise 89** With (D) and $f(\beta)$ as defined in Example IV.49 we consider the effect on the domain of f when some constraints are added. When the constraint $y_1 \geq 0$ is added to (D), the domain of f becomes $(-\infty, 0]$. When the constraint $y_1 \leq 0$ is added to (D), the domain of f becomes $[0, \infty)$ and when both constraints are added the domain of f becomes $(-\infty, \infty)$. Prove this.

[6] **Exercise 90** Prove that the minimum of a finite family of linear functions, each defined on the same closed interval, is continuous, concave and piecewise linear.

which is a linear function of β. This makes clear that $f(\beta)$ is the maximum of a finite set of linear functions. Therefore, $f(\beta)$ is continuous, convex and piecewise linear, as required. $\square$

The values of β where the slope of the optimal-value function $f(\beta)$ changes are called *break points* of f, and any interval between two successive break points of f is called a *linearity interval* of f. In a similar way we define break points and linearity intervals for g.

Example IV.52 For any $\gamma \in \mathbb{R}$ consider the problem (P_γ) defined by

$$(P_\gamma) \qquad \min \quad x_1 + (3 + \gamma)x_2 + (1 - \gamma)x_3$$
$$\text{s.t.} \quad x_1 + x_2 + x_3 = 4, \quad x_1, x_2, x_3 \geq 0.$$

In this case b is constant and the perturbation vector for $c = (1, 3, 1)$ is

$$\Delta c = (0, 1, -1).$$

The dual problem is

$$(D_\gamma) \qquad \max\{4y : y \leq 1, \ y \leq 3 + \gamma, \ y \leq 1 - \gamma\}.$$

From this it is obvious that the optimal value is given by

$$g(\gamma) = 4 \min(1, 3 + \gamma, 1 - \gamma).$$

The graph of the optimal-value function $g(\gamma)$ is depicted in Figure 19.3. Note that

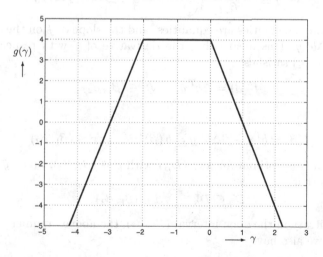

Figure 19.3 The optimal-value function $g(\gamma)$.

$g(\gamma)$ is piecewise linear and concave. The break points of g occur for $\gamma = -2$ and $\gamma = 0$. $\diamond$

19.4.2   Optimal sets on a linearity interval

For any β in the domain of f we denote the optimal set of (P_β) by $\mathcal{P}_\beta^*$ and the optimal set of (D_β) by $\mathcal{D}_\beta^*$.

Theorem IV.53 *If $f(\beta)$ is linear on the interval $[\beta_1, \beta_2]$, where $\beta_1 < \beta_2$, then the dual optimal set $\mathcal{D}_\beta^*$ is constant (i.e. invariant) for $\beta \in (\beta_1, \beta_2)$.*

Proof: Let $\bar{\beta} \in (\beta_1, \beta_2)$ be arbitrary and let $\bar{y} \in \mathcal{D}_{\bar{\beta}}^*$ be arbitrary as well. Since $\bar{y}$ is optimal for $(D_{\bar{\beta}})$ we have

$$f(\bar{\beta}) = b(\bar{\beta})^T \bar{y} = b^T \bar{y} + \bar{\beta} \Delta b^T \bar{y},$$

and, since $\bar{y}$ is dual feasible for all β,

$$b\left(\beta_1\right)^T \bar{y} = b^T \bar{y} + \beta_1 \Delta b^T \bar{y} \leq f(\beta_1), \quad b\left(\beta_2\right)^T \bar{y} = b^T \bar{y} + \beta_2 \Delta b^T \bar{y} \leq f(\beta_2).$$

Hence we find

$$f(\beta_1) - f(\bar{\beta}) \geq \left(\beta_1 - \bar{\beta}\right) \Delta b^T \bar{y}, \quad f(\beta_2) - f(\bar{\beta}) \geq \left(\beta_2 - \bar{\beta}\right) \Delta b^T \bar{y}.$$

The linearity of f on $[\beta_1, \beta_2]$ implies

$$\frac{f(\bar{\beta}) - f(\beta_1)}{\bar{\beta} - \beta_1} = \frac{f(\beta_2) - f(\bar{\beta})}{\beta_2 - \bar{\beta}}.$$

Now using that $\beta_2 - \bar{\beta} > 0$ and $\beta_1 - \bar{\beta} < 0$ we obtain

$$\Delta b^T \bar{y} \leq \frac{f(\beta_2) - f(\bar{\beta})}{\beta_2 - \bar{\beta}} = \frac{f(\bar{\beta}) - f(\beta_1)}{\bar{\beta} - \beta_1} \leq \Delta b^T \bar{y}.$$

Hence, the last two inequalities are equalities, and the slope of f on the closed interval $[\beta_1, \beta_2]$ is just $\Delta b^T \bar{y}$. This means that the derivative of f with respect to β on the open interval (β_1, β_2) satisfies

$$f'(\bar{\beta}) = \Delta b^T \bar{y}, \quad \forall \beta \in (\beta_1, \beta_2),$$

or equivalently,

$$f(\beta) = b^T \bar{y} + \beta \Delta b^T \bar{y} = b\left(\beta\right)^T \bar{y}, \quad \forall \beta \in (\beta_1, \beta_2).$$

We conclude that $\bar{y}$ is optimal for any (D_β) with $\beta \in (\beta_1, \beta_2)$. Since $\bar{y}$ was arbitrary in $\mathcal{D}_{\bar{\beta}}^*$, it follows that

$$\mathcal{D}_{\bar{\beta}}^* \subseteq \mathcal{D}_\beta^*, \quad \forall \beta \in (\beta_1, \beta_2).$$

Since $\bar{\beta}$ was arbitrary in the open interval (β_1, β_2), the above argument applies to any $\tilde{\beta} \in (\beta_1, \beta_2)$; so we also have

$$\mathcal{D}_{\tilde{\beta}}^* \subseteq \mathcal{D}_\beta^*, \quad \forall \beta \in (\beta_1, \beta_2).$$

We may conclude that $\mathcal{D}_{\bar{\beta}}^* \subseteq \mathcal{D}_{\tilde{\beta}}^*$ and $\mathcal{D}_{\tilde{\beta}}^* \subseteq \mathcal{D}_{\bar{\beta}}^*$, which gives $\mathcal{D}_{\bar{\beta}}^* = \mathcal{D}_{\tilde{\beta}}^*$. The theorem follows. $\square$

The above proof reveals that $\Delta b^T y$ must have the same value for all $y \in \mathcal{D}_\beta^*$ and for all $\beta \in (\beta_1, \beta_2)$. So we may state the following.

Corollary IV.54 *Under the hypothesis of Theorem IV.53,*

$$f'(\beta) = \Delta b^T y, \quad \forall \beta \in (\beta_1, \beta_2), \forall y \in \mathcal{D}_\beta^*.$$

By continuity we may write

$$f(\beta) = b^T \bar{y} + \beta \Delta b^T \bar{y} = b(\beta)^T \bar{y}, \quad \forall \beta \in [\beta_1, \beta_2].$$

This immediately implies another consequence.

Corollary IV.55 *Under the hypothesis of Theorem IV.53 let* $\mathcal{D}_{(\beta_1,\beta_2)}^* := \mathcal{D}_\beta^*$ *for arbitrary* $\beta \in (\beta_1, \beta_2)$. *Then*

$$\mathcal{D}_{(\beta_1,\beta_2)}^* \subseteq \mathcal{D}_{\beta_1}^*, \quad \mathcal{D}_{(\beta_1,\beta_2)}^* \subseteq \mathcal{D}_{\beta_2}^*.$$

In the next result we deal with the converse of the implication in Theorem IV.53.

Theorem IV.56 *Let* β_1 *and* $\beta_2 > \beta_1$ *be such that* $\mathcal{D}_{\beta_1}^* = \mathcal{D}_{\beta_2}^*$. *Then* $\mathcal{D}_\beta^*$ *is constant for all* $\beta \in [\beta_1, \beta_2]$ *and* $f(\beta)$ *is linear on the interval* $[\beta_1, \beta_2]$.

Proof: Let $\bar{y} \in \mathcal{D}_{\beta_1}^* = \mathcal{D}_{\beta_2}^*$. Then

$$f(\beta_1) = b(\beta_1)^T \bar{y}, \quad f(\beta_2) = b(\beta_2)^T \bar{y}.$$

Consider the linear function h:

$$h(\beta) = b(\beta)^T \bar{y} = (b + \beta \Delta b)^T \bar{y}, \quad \forall \beta \in [\beta_1, \beta_2].$$

Then h coincides with f at β_1 and β_2. Since f is convex this implies

$$f(\beta) \leq h(\beta), \quad \forall \beta \in [\beta_1, \beta_2].$$

Now $\bar{y}$ is feasible for all $\beta \in [\beta_1, \beta_2]$. Since $f(\beta)$ is the optimal value of (D_β), it follows that

$$f(\beta) \geq b(\beta)^T \bar{y} = (b + \beta \Delta b)^T \bar{y} = h(\beta).$$

Therefore, f coincides with h on $[\beta_1, \beta_2]$. As a consequence, f is linear on $[\beta_1, \beta_2]$ and $\bar{y}$ is optimal for (D_β) whenever $\beta \in [\beta_1, \beta_2]$. Since $\bar{y}$ is arbitrary in $\mathcal{D}_{\beta_1}^* = \mathcal{D}_{\beta_2}^*$ this implies that $\mathcal{D}_{\beta_1}^* = \mathcal{D}_{\beta_2}^*$ is a subset of $\mathcal{D}_\beta^*$ for any $\beta \in (\beta_1, \beta_2)$. By Theorem IV.53, and Corollary IV.55 we also have the converse inclusion. The dual optimal set on (β_1, β_2) is therefore constant, and the proof is complete. $\qquad \square$

Each of the above results about $f(\beta)$ has its analogue for $g(\gamma)$. We state these results without further proof.[7] The omitted proofs are straightforward modifications of the above proofs.

Theorem IV.57 *If* $g(\gamma)$ *is linear on the interval* $[\gamma_1, \gamma_2]$, *where* $\gamma_1 < \gamma_2$, *then the primal optimal set* $\mathcal{P}_\gamma^*$ *is constant for* $\gamma \in (\gamma_1, \gamma_2)$.

---

[7] **Exercise 91** Prove Theorem IV.57, Corollary IV.58, Corollary IV.59 and Theorem IV.60.

Corollary IV.58 *Under the hypothesis of Theorem IV.57,*

$$g'(\gamma) = \Delta c^T x, \quad \forall \gamma \in (\gamma_1, \gamma_2), \forall x \in \mathcal{P}_\gamma^*.$$

Corollary IV.59 *Under the hypothesis of Theorem IV.57 let* $\mathcal{P}_{(\gamma_1,\gamma_2)}^* := \mathcal{P}_\gamma^*$ *for arbitrary* $\gamma \in (\gamma_1, \gamma_2)$. *Then*

$$\mathcal{P}_{(\gamma_1,\gamma_2)}^* \subseteq \mathcal{P}_{\gamma_1}^*, \quad \mathcal{P}_{(\gamma_1,\gamma_2)}^* \subseteq \mathcal{P}_{\gamma_2}^*.$$

Theorem IV.60 *Let* γ_1 *and* $\gamma_2 > \gamma_1$ *be such that* $\mathcal{P}_{\gamma_1}^* = \mathcal{P}_{\gamma_2}^*$. *Then* $\mathcal{P}_\gamma^*$ *is constant for all* $\gamma \in [\gamma_1, \gamma_2]$ *and* $g(\gamma)$ *is linear on the interval* $[\gamma_1, \gamma_2]$.

19.4.3 Optimal sets in a break point

Returning to the function f, we established in the previous section that if $\beta \in \text{dom}(f)$ is not a break point of f then the quantity $\Delta b^T y$ is constant for all $y \in \mathcal{D}_\beta^*$. In this section we will see that this property is characteristic for 'nonbreak' points.

If the domain of f has a right extreme point then we may consider the right derivative at this point to be ∞, and if the domain of f has a left extreme point the left derivative at this point may be taken as $-\infty$. Then β is a break point of f if and only if the left and the right derivatives of f at β are different. This follows from the definition of a break point. Denoting the left and the right derivatives by $f'_-(\beta)$ and $f'_+(\beta)$ respectively, the convexity of f implies that at a break point β we have

$$f'_-(\beta) < f'_+(\beta).$$

If $\text{dom}(f)$ has a right extreme point, it is convenient to consider the open interval at the right of this point as a linearity interval where both f and its derivative are ∞. Similarly, if $\text{dom}(f)$ has a left extreme point, we may consider the open interval at the left of this point as a linearity interval where f is ∞ and its derivative $-\infty$. Obviously, these extreme linearity intervals are characterized by the fact that on the intervals the primal problem is infeasible and the dual problem unbounded. The dual problem is unbounded if and only if the set $\mathcal{D}_\beta^*$ of optimal solutions is empty.

Lemma IV.61 [8] *Let* β, β^- *and* β^+ *belong to the interior of* $\text{dom}(f)$ *such that* β^+ *belongs to the open linearity interval just to the right of* β *and* β^- *to the open linearity interval just to the left of* β. *Moreover, let* $y^+ \in \mathcal{D}_{\beta^+}^*$ *and* $y^- \in \mathcal{D}_{\beta^-}^*$. *Then*

$$f'_-(\beta) = \min_y \left\{ \Delta b^T y : y \in \mathcal{D}_\beta^* \right\} = \Delta b^T y^-$$

$$f'_+(\beta) = \max_y \left\{ \Delta b^T y : y \in \mathcal{D}_\beta^* \right\} = \Delta b^T y^+.$$

Proof: We give the proof for $f'_+(\beta)$. The proof for $f'_-(\beta)$ goes in the same way and is omitted. Since y^+ is optimal for $\mathcal{D}_{\beta^+}^*$ we have

$$(b + \beta^+ \Delta b)^T y^+ = f(\beta^+) \geq (b + \beta^+ \Delta b)^T y, \forall y \in \mathcal{D}_\beta^*.$$

---

[8] This lemma can also be obtained as a special case of a result of Mills [210]. His more general result gives the directional derivatives of the optimal-value function with respect to any 'admissible' perturbation of A, b and c; when only b is perturbed it gives the same result as the lemma.

We also have $y^+ \in \mathcal{D}_\beta^*$, from Theorem IV.53 and Corollary IV.55. Therefore,

$$(b + \beta \Delta b)^T y^+ = (b + \beta \Delta b)^T y, \; \forall y \in \mathcal{D}_\beta^*.$$

Subtracting both sides of this equality from the corresponding sides in the last inequality gives

$$(\beta^+ - \beta) \, \Delta b^T y^+ \geq (\beta^+ - \beta) \, \Delta b^T y, \; \forall y \in \mathcal{D}_\beta^*.$$

Dividing both sides by the positive number $\beta^+ - \beta$ we get

$$\Delta b^T y^+ \geq \Delta b^T y, \; \forall y \in \mathcal{D}_\beta^*,$$

thus proving that

$$\max_y \left(\Delta b^T y \; : \; y \in \mathcal{D}_\beta^* \right) = \Delta b^T y^+.$$

Since $f'_+(\beta) = \Delta b^T y^+$, from Corollary IV.54, the lemma follows. $\qquad \square$

The above lemma admits a nice generalization that is also valid if β is an extreme point of the domain of f.

Theorem IV.62 *Let $\beta \in \mathrm{dom}\,(f)$ and let x^* be any optimal solution of (P_β). Then the derivatives at β satisfy*

$$f'_-(\beta) = \min_{y,s} \left\{ \Delta b^T y \; : \; A^T y + s = c, \, s \geq 0, \, s^T x^* = 0 \right\}$$

$$f'_+(\beta) = \max_{y,s} \left\{ \Delta b^T y \; : \; A^T y + s = c, \, s \geq 0, \, s^T x^* = 0 \right\}.$$

Proof: As in the previous lemma, we give the proof for $f'_+(\beta)$ and omit the proof for $f'_-(\beta)$. Consider the optimization problem

$$\max_{y,s} \left\{ \Delta b^T y \; : \; A^T y + s = c, \, s \geq 0, \, s^T x^* = 0 \right\}. \tag{19.3}$$

First we establish that if β belongs to the interior of $\mathrm{dom}\,(f)$ then this is exactly the same problem as the maximization problem in Lemma IV.61. This follows because if $A^T y + s = c$, $s \geq 0$, then (y, s) is optimal for (D_β) if and only if $s^T x^* = 0$, since x^* is an optimal solution of the dual problem (P_β) of (D_β). If β belongs to the interior of $\mathrm{dom}\,(f)$ then the theorem follows from Lemma IV.61. Hence it remains to deal with the case where β is an extreme point of $\mathrm{dom}\,(f)$. It is easily verified that if β is the left extreme point of $\mathrm{dom}\,(f)$ then we can repeat the arguments in the proof of Lemma IV.61. Thus it remains to prove the theorem if β is the right extreme point of $\mathrm{dom}\,(f)$. Since $f'_+(\beta) = \infty$ in that case, we need to show that the above maximization problem (19.3) is unbounded.

Let β be the right extreme point of $\mathrm{dom}\,(f)$ and suppose that the problem (19.3) is not unbounded. Let us point out first that (19.3) is feasible. Its feasible region is just the optimal set of the dual (D_β) of (P_β). Since (P_β) has as an optimal solution, (D_β) has an optimal solution as well. This implies that (D_β) is feasible. Therefore, (19.3) is feasible as well. Hence, if (19.3) is not unbounded, the problem itself and its dual have optimal solutions. The dual problem is given by

$$\min_{\xi,\lambda} \left\{ c^T \xi \; : \; A\xi = \Delta b, \, \xi + \lambda x^* \geq 0 \right\}.$$

We conclude that there exists a vector $\xi \in \mathbb{R}^n$ and a scalar λ such that $A\xi = \Delta b$, $\xi + \lambda x^* \geq 0$. This implies that we cannot have $\xi_i < 0$ and $x_i^* = 0$. In other words,

$$x_i^* = 0 \Rightarrow \xi_i \geq 0.$$

Hence, there exists a positive ε such that $\bar{x} := x^* + \varepsilon\xi \geq 0$. Now we have

$$A\bar{x} = A\left(x^* + \varepsilon\xi\right) = Ax^* + \varepsilon A\xi = b + (\beta + \varepsilon)\,\Delta b.$$

Thus we find that $(P_{\beta+\varepsilon})$ admits $\bar{x}$ as a feasible point. This contradicts the assumption that β is the right extreme point of dom (f). We conclude that (19.3) is unbounded, proving the theorem. $\qquad \square$

The picture becomes more complete now. Note that Theorem IV.62 is valid for any value of β in the domain of f. The theorem reestablishes that at a 'nonbreak' point, where the left and right derivative of f are equal, the value of $\Delta b^T y$ is constant when y runs through the dual optimal set $\mathcal{D}_\beta^*$. But it also makes it clear that at a break point, where the two derivatives are different, $\Delta b^T y$ is not constant when y runs through the dual optimal set $\mathcal{D}_\beta^*$. Then the extreme values of $\Delta b^T y$ yield the left and the right derivatives of f at β; the left derivative is the minimum and the right derivative the maximal value of $\Delta b^T y$ when y runs through the dual optimal set $\mathcal{D}_\beta^*$.

It is worth pointing out another consequence of Lemma IV.61 and Theorem IV.62. Using the notation of the lemma we have the inclusions

$$\mathcal{D}_{\beta-}^* \subseteq \mathcal{D}_\beta^*, \quad \mathcal{D}_{\beta+}^* \subseteq \mathcal{D}_\beta^*,$$

which follow from Corollary IV.55 if β is not an extreme point of dom (f). If β is the right extreme point then $\mathcal{D}_{\beta+}^*$ is empty, and if it is the left extreme point then $\mathcal{D}_{\beta-}^*$ is empty as well; hence the above inclusions hold everywhere. Now suppose that β is a nonextreme break point of f. Then letting y run through the set $\mathcal{D}_{\beta-}^*$ we know that $\Delta b^T y$ is constant and equal to the left derivative of f at β, and if y runs through $\mathcal{D}_{\beta+}^*$ then $\Delta b^T y$ is constant and equal to the right derivative of f at β and, finally, if y runs through $\mathcal{D}_\beta^*$ then $\Delta b^T y$ is not constant. Thus the three sets must be mutually different. As a consequence, the above inclusions must be strict. Moreover, since the left and the right derivatives at β are different, the sets $\mathcal{D}_{\beta-}^*$ and $\mathcal{D}_{\beta+}^*$ are disjoint. Thus we may state the following.

Corollary IV.63 *Let β be a nonextreme break point of f and let β^+ and β^- be as defined in Lemma IV.61. Then we have*

$$\mathcal{D}_{\beta-}^* \subset \mathcal{D}_\beta^*, \quad \mathcal{D}_{\beta+}^* \subset \mathcal{D}_\beta^*, \quad \mathcal{D}_{\beta-}^* \cap \mathcal{D}_{\beta+}^* = \emptyset,$$

where the inclusions are strict.[9]

---

[9] **Exercise 92** Using the notation of Lemma IV.61 and Corollary IV.63, we have

$$\mathcal{D}_{\beta-}^* \cup \mathcal{D}_{\beta+}^* \subseteq \mathcal{D}_\beta^*.$$

Show that the inclusion is always strict. (Hint: use the central solution of (D_β).)

Two other almost obvious consequences of the above results are the following corollaries.[10]

Corollary IV.64 *Let β be a nonextreme break point of f and let β^+ and β^- be as defined in Lemma IV.61. Then*

$$\mathcal{D}_{\beta^-}^* = \left\{ y \in \mathcal{D}_\beta^* \; : \; \Delta b^T y = \Delta b^T y^- \right\}, \quad \mathcal{D}_{\beta^+}^* = \left\{ y \in \mathcal{D}_\beta^* \; : \; \Delta b^T y = \Delta b^T y^+ \right\}.$$

Corollary IV.65 *Let β be a nonextreme break point of f and let β^+ and β^- be as defined in Lemma IV.61. Then*

$$\dim \mathcal{D}_{\beta^-}^* < \dim \mathcal{D}_\beta^*, \quad \dim \mathcal{D}_{\beta^+}^* < \dim \mathcal{D}_\beta^*.$$

Remark IV.66 It is interesting to consider the dual optimal set $\mathcal{D}_\beta^*$ when β runs from $-\infty$ to ∞. To the left of the smallest break point (the break point for which β is minimal) the set $\mathcal{D}_\beta^*$ is constant. It may happen that $\mathcal{D}_\beta^*$ is empty there, due to the absence of optimal solutions for these small values of β. This occurs if (D_β) is unbounded (which means that (P_β) is infeasible) for the values of β on the farthest left open linearity interval. Then, at the first break point, the set $\mathcal{D}_\beta^*$ increases to a larger set, and as we pass to the next open linearity interval the set $\mathcal{D}_\beta^*$ becomes equal to a proper subset of this enlarged set. This process repeats itself at every new break point: at a break point of f the dual optimal set expands itself, and as we pass to the next open linearity interval it shrinks to a proper subset of the enlarged set. Since the derivative of f is monotonically increasing when β runs from $-\infty$ to ∞, every new dual optimal set arising in this way differs from all previous ones. In other words, every break point of f and every linearity interval of f has its own dual optimal set.[11]   •

We state the dual analogues of Lemma IV.61 and Theorem IV.62 and their corollaries without further proof.[12]

Lemma IV.67 *Let γ, γ^- and γ^+ belong to the interior of $\mathrm{dom}\,(g)$, γ^+ to the open linearity interval just to the right of γ, and γ^- to the open linearity interval just to the left of γ. Moreover, let $x^+ \in \mathcal{P}_{\gamma^+}^*$ and $x^- \in \mathcal{P}_{\gamma^-}^*$. Then*

$$
\begin{aligned}
g_-'(\gamma) &= \max_x \left\{ \Delta c^T x \; : \; x \in \mathcal{P}_\gamma^* \right\} = \Delta c^T x^- \\
g_+'(\gamma) &= \min_x \left\{ \Delta c^T x \; : \; x \in \mathcal{P}_\gamma^* \right\} = \Delta c^T x^+.
\end{aligned}
$$

Theorem IV.68 *Let $\gamma \in \mathrm{dom}\,(g)$ and let (y^*, s^*) be any optimal solution of (D_γ). Then the derivatives at γ satisfy*

$$
\begin{aligned}
g_-'(\gamma) &= \max_x \left\{ \Delta c^T x \; : \; Ax = b, \, x \geq 0, \, x^T s^* = 0 \right\} \\
g_+'(\gamma) &= \min_x \left\{ \Delta c^T x \; : \; Ax = b, \, x \geq 0, \, x^T s^* = 0 \right\}.
\end{aligned}
$$

---

[10] **Exercise 93** Prove Corollary IV.64 and Corollary IV.65.

[11] **Exercise 94** The dual optimal sets belonging to two different open linearity intervals of f are disjoint. Prove this. (Hint: use that the derivatives of f on the two intervals are different.)

[12] **Exercise 95** Prove Lemma IV.67, Theorem IV.68, Corollary IV.69, Corollary IV.70 and Corollary IV.71.

Corollary IV.69 *Let γ be a nonextreme break point of g and let γ^+ and γ^- be as defined in Lemma IV.67. Then*

$$\mathcal{P}^*_{\gamma^-} \subset \mathcal{P}^*_{\gamma}, \quad \mathcal{P}^*_{\gamma^+} \subset \mathcal{P}^*_{\gamma}, \quad \mathcal{P}^*_{\gamma^-} \cap \mathcal{P}^*_{\gamma^+} = \emptyset,$$

where the inclusions are strict.[13]

Corollary IV.70 *Let γ be a nonextreme break point of g and let γ^+ and γ^- be as defined in Lemma IV.67. Then*

$$\mathcal{P}^*_{\gamma^-} = \left\{ x \in \mathcal{P}^*_{\gamma} \; : \; \Delta c^T x = \Delta c^T x^- \right\}, \quad \mathcal{P}^*_{\gamma^+} = \left\{ x \in \mathcal{P}^*_{\gamma} \; : \; \Delta c^T x = \Delta c^T x^+ \right\}.$$

Corollary IV.71 *Let γ be a nonextreme break point of g and let γ^+ and γ^- be as defined in Lemma IV.67. Then*

$$\dim \mathcal{P}^*_{\gamma^-} < \dim \mathcal{P}^*_{\gamma}, \quad \dim \mathcal{P}^*_{\gamma^+} < \dim \mathcal{P}^*_{\gamma}.$$

The next example illustrates the results of this section.

Example IV.72 We use the same problem as in Example IV.52. For any $\gamma \in \mathbb{R}$ the problem (P_γ) is defined by

$$(P_\gamma) \qquad \min \quad x_1 + (3 + \gamma)x_2 + (1 - \gamma)x_3$$

$$\text{s.t.} \quad x_1 + x_2 + x_3 = 4, \qquad x_1, x_2, x_3 \geq 0,$$

and the dual problem is

$$(D_\gamma) \qquad \max \left\{ 4y \; : \; y \leq 1, \, y \leq 3 + \gamma, \, y \leq 1 - \gamma \right\}.$$

The perturbation vector for $c = (1, 3, 1)$ is

$$\Delta c = (0, 1, -1).$$

The graph of g is depicted in Figure 19.3 (page 369). The break points of g occur at $\gamma = -2$ and $\gamma = 0$.

For $\gamma < -2$ the optimal solution of (P_γ) is $x = (0, 4, 0)$, and then $\Delta c^T x = 4$. At the break point $\gamma = -2$ the primal optimal solution set is given by

$$\left\{ x = (x_1, x_2, 0) \; : \; x_1 + x_2 = 4, \, x_1 \geq 0, \, x_2 \geq 0 \right\}.$$

The extreme values of $\Delta c^T x$ on this set are 4 and 0. The maximal value occurs for $x = (0, 4, 0)$ and the minimal value for $x = (4, 0, 0)$. Hence, the left and right derivatives of g at $\gamma = -2$ are given by these values. If $-2 < \gamma < 0$ then the optimal solution of the primal problem is given by $x = (4, 0, 0)$ and $\Delta c^T x = 0$, so the derivative of g is 0 in this region. At the break point $\gamma = 0$ the primal optimal solution set is given by

$$\left\{ x = (x_1, 0, x_3) \; : \; x_1 + x_3 = 4, \, x_1 \geq 0, \, x_3 \geq 0 \right\}.$$

The extreme values of $\Delta c^T x$ on this set are 0 and -4. The left and right derivatives of g at $\gamma = 0$ are given by these values. The maximal value occurs for $x = (4, 0, 0)$ and the minimal value for $x = (0, 0, 4)$. Observe that in this example the primal optimal solution set at every break point has dimension 1, whereas in the open linearity intervals the optimal solution is always unique. $\diamond$

---

[13] **Exercise 96** Find an example where $\mathcal{P}^*_{\gamma^-} = \emptyset$ and $\mathcal{P}^*_{\gamma} \neq \emptyset$.

19.4.4 Extreme points of a linearity interval

In this section we assume that $\bar{\beta}$ belongs to the interior of a linearity interval $[\beta_1, \beta_2]$. Given an optimal solution of $(D_{\bar{\beta}})$ we show how the extreme points β_1 and β_2 of the linearity interval containing $\bar{\beta}$ can be found by solving two auxiliary LO problems.

Theorem IV.73 *Let $\bar{\beta}$ be arbitrary and let (y^*, s^*) be any optimal solution of $(D_{\bar{\beta}})$. Then the extreme points of the linearity interval $[\beta_1, \beta_2]$ containing $\bar{\beta}$ follow from*

$$\beta_1 \;\;=\;\; \min_{\beta,x}\left\{\beta \;:\; Ax = b + \beta\Delta b, \; x \geq 0, \; x^T s^* = 0\right\}$$

$$\beta_2 \;\;=\;\; \max_{\beta,x}\left\{\beta \;:\; Ax = b + \beta\Delta b, \; x \geq 0, \; x^T s^* = 0\right\}.$$

Proof: We only give the proof for β_1.[14] Consider the minimization problem

$$\min_{\beta,x}\left\{\beta \;:\; Ax = b + \beta\Delta b, \; x \geq 0, \; x^T s^* = 0\right\}. \tag{19.4}$$

We first show that this problem is feasible. Since $(D_{\bar{\beta}})$ has an optimal solution, its dual problem $(P_{\bar{\beta}})$ has an optimal solution as well. Letting $\bar{x}$ be optimal for $(P_{\bar{\beta}})$, we can easily verify that $\beta = \bar{\beta}$ and $x = \bar{x}$ are feasible for (19.4).

We proceed by considering the case where (19.4) is unbounded. For any $\beta \leq \bar{\beta}$ there exists a vector x that satisfies $Ax = b + \beta\Delta b$, $x \geq 0$, $x^T s^* = 0$. Now (y^*, s^*) is feasible for (D_β) and x is feasible for (P_β). Since $x^T s^* = 0$, x is optimal for (P_β) and (y^*, s^*) is optimal for (D_β). The optimal value of both problems is given by $b(\beta)^T y^* = b^T y^* + \beta\Delta b^T y^*$. This means that β belongs to the linearity interval containing $\bar{\beta}$. Since this holds for any $\beta \leq \bar{\beta}$, the left boundary of this linearity interval is $-\infty$, as it should be.

It remains to deal with the case where (19.4) has an optimal solution, say (β^*, x^*). We then have $Ax^* = b + \beta^*\Delta b = b(\beta^*)$, so x^* is feasible for (P_{β^*}). Since (y^*, s^*) is feasible for (D_{β^*}) and $x^{*T} s^* = 0$ it follows that x^* is optimal for (P_{β^*}) and (y^*, s^*) is optimal for (D_{β^*}). The optimal value of both problems is given by $b(\beta^*)^T y^* = b^T y^* + \beta^*\Delta b^T y^*$. This means that β^* belongs to the linearity interval containing $\bar{\beta}$, and it follows that $\beta^* \geq \beta_1$.

On the other hand, from Corollary IV.55 the pair (y^*, s^*) is optimal for (D_{β_1}). Now let $\bar{x}$ be optimal for (P_{β_1}). Then we have

$$A\bar{x} = b(\beta_1) = b + \beta_1\Delta b, \; x \geq 0, \quad \bar{x}^T s^* = 0,$$

which shows that the pair $(\beta_1, \bar{x})$ is feasible for the above minimization problem. This implies that $\beta^* \leq \beta_1$. Hence we obtain that $\beta^* = \beta_1$. This completes the proof. □

If $\bar{\beta}$ is not a break point then there is only one linearity interval containing $\bar{\beta}$, and hence this must be the linearity interval $[\beta_1, \beta_2]$, as given by Theorem IV.73.

It is worth pointing out that if $\bar{\beta}$ is a break point there are three linearity intervals containing $\bar{\beta}$, namely the singleton interval $[\bar{\beta}, \bar{\beta}]$ and the two surrounding linearity intervals. In the singleton case, the linearity interval $[\beta_1, \beta_2]$ given by Theorem IV.73 may be any of these three intervals, and which one it is depends on the given optimal

_____
[14] **Exercise 97** Prove the second part (on β_2) of Theorem IV.73.

solution (y^*, s^*) of $(D_{\bar{\beta}})$. It can easily be understood that the linearity interval at the right of $\bar{\beta}$ will be found if (y^*, s^*) happens to be optimal on the right linearity interval. This occurs when $\Delta b^T y^* = f'_+(\bar{\beta})$, due to Corollary IV.64. Similarly, the linearity interval at the left of $\bar{\beta}$ will be found if (y^*, s^*) is optimal on the left linearity interval and this occurs when $\Delta b^T y^* = f'_-(\bar{\beta})$, also due to Corollary IV.64. Finally, if

$$f'_-(\bar{\beta}) < \Delta b^T y^* < f'_+(\bar{\beta}), \tag{19.5}$$

then we have $\beta_1 = \beta_2 = \bar{\beta}$ in Theorem IV.73. The last situation seems to be most informative. It clearly indicates that $\bar{\beta}$ is a break point of f, which is not apparent in the other two situations. Knowing that $\bar{\beta}$ is a break point of f we can find the two one-sided derivatives of f at $\bar{\beta}$ as well as optimal solutions for the two intervals surrounding $\bar{\beta}$ from Theorem IV.62. In the light of this discussion the following result is of interest. It shows that the above ambiguity can be avoided by the use of strictly complementary optimal solutions.

Theorem IV.74 *Let $\bar{\beta}$ be a break point and let (y^*, s^*) be a strictly complementary optimal solution of $(D_{\bar{\beta}})$. Then the numbers β_1 and β_2 given by Theorem IV.73 satisfy $\beta_1 = \beta_2 = \bar{\beta}$.*

Proof: If (y^*, s^*) is a strictly complementary optimal solution of $(D_{\bar{\beta}})$ then it uniquely determines the optimal partition of $(D_{\bar{\beta}})$ and this partition differs from the optimal partitions corresponding to the optimal sets of the linearity intervals surrounding $\bar{\beta}$. Hence (y^*, s^*) does not belong to the optimal sets of the linearity intervals surrounding $\bar{\beta}$. From Corollary IV.64 it follows that $\Delta b^T y^*$ satisfies (19.5), and the theorem follows. □

It is not difficult to state the corresponding results for g. We do this below, omitting the proofs, and then provide an example of their use.[15]

Theorem IV.75 *Let $\bar{\gamma}$ be arbitrary and let x^* be any optimal solution of $(P_{\bar{\gamma}})$. Then the extreme points of the linearity interval $[\gamma_1, \gamma_2]$ containing $\bar{\gamma}$ follow from*

$$\gamma_1 = \min_{\gamma, y, s} \{ \gamma : A^T y + s = c + \gamma \Delta c, \ s \geq 0, \ s^T x^* = 0 \}$$

$$\gamma_2 = \max_{\gamma, y, s} \{ \gamma : A^T y + s = c + \gamma \Delta c, \ s \geq 0, \ s^T x^* = 0 \}.$$

Theorem IV.76 *Let $\bar{\gamma}$ be a break point and let x^* be a strictly complementary optimal solution of $(P_{\bar{\gamma}})$. Then the numbers γ_1 and γ_2 given by Theorem IV.75 satisfy $\gamma_1 = \gamma_2 = \bar{\gamma}$.*

Example IV.77 We use the same problem as in Example IV.72. Using the notation of Theorem IV.75 we first determine the linearity interval for $\bar{\gamma} = -1$. We can easily verify that $x = (4, 0, 0)$ is optimal for (P_{-1}). Hence the extreme points γ_1 and γ_2 of the linearity interval containing $\bar{\gamma}$ follow by minimizing and maximizing γ over the region

$$\{\gamma : y \leq 1, \ y \leq 3 + \gamma, \ y \leq 1 - \gamma, \ 4(1 - y) = 0\}.$$

---

[15] **Exercise 98** Prove Theorem IV.75 and Theorem IV.76.

The last constraint implies $y = 1$, so the other constraints reduce to $1 \leq 3 + \gamma$ and $1 \leq 1 - \gamma$, which gives $-2 \leq \gamma \leq 0$. Hence the linearity interval containing $\bar{\gamma} = -1$ is $[-2, 0]$.

When $\bar{\gamma} = 1$, $x = (0, 0, 4)$ is optimal for (P_1), and the linearity interval containing $\bar{\gamma}$ follows by minimizing and maximizing γ over the region

$$\{\gamma \; : \; y \leq 1, \, y \leq 3 + \gamma, \, y \leq 1 - \gamma, \, 4(1 - \gamma - y) = 0\} \, .$$

The last constraint implies $y = 1 - \gamma$. Now the other constraints reduce to $1 - \gamma \leq 1$ and $1 - \gamma \leq 3 + \gamma$, which is equivalent to $\gamma \geq 0$. So the linearity interval containing $\bar{\gamma} = 1$ is $[0, \infty)$.

When $\bar{\gamma} = -3$, $x = (0, 4, 0)$ is optimal for (P_{-3}), and the linearity interval containing $\bar{\gamma}$ follows by minimizing and maximizing γ over the region

$$\{\gamma \; : \; y \leq 1, \, y \leq 3 + \gamma, \, y \leq 1 - \gamma, \, 4(3 + \gamma - y) = 0\} \, .$$

The last constraint implies $y = 3 + \gamma$, and the other constraints reduce to $3 + \gamma \leq 1$ and $3 + \gamma \leq 1 - \gamma$, which is equivalent to $\gamma \leq -2$. Thus, the linearity interval containing $\bar{\gamma} = -3$ is $(-\infty, -2]$.

Observe that the linearity intervals just calculated agree with Figure 19.3.

Finally we demonstrate the use of Theorem IV.76 at a break point. Taking $\bar{\gamma} = 0$, we see that $x = (4, 0, 0)$ is optimal for (P_0), and we need to minimize and maximize γ over the region

$$\{\gamma \; : \; y \leq 1, \, y \leq 3 + \gamma, \, y \leq 1 - \gamma, \, 4(1 - y) = 0\} \, .$$

This gives $-2 \leq \gamma \leq 0$ and we find the linearity interval $[-2, 0]$ left from 0. This is because $x = (4, 0, 0)$ is also optimal on this interval. Recall from Example IV.72 that the optimal set at $\gamma = 0$ is given by

$$\{x = (x_1, 0, x_3) \; : \; x_1 + x_3 = 4, \, x_1 \geq 0, \, x_3 \geq 0\} \, .$$

Thus, instead of the optimal solution $x = (4, 0, 0)$ we may equally well use the strictly complementary solution $x = (2, 0, 2)$. Then we need to minimize and maximize γ over the region

$$\{\gamma \; : \; y \leq 1, \, y \leq 3 + \gamma, \, y \leq 1 - \gamma, \, 2(1 - y) + 2(1 - \gamma - y) = 0\} \, .$$

The last constraint amounts to $\gamma = 2 - 2y$. Substitution in the third constraint yields $y \leq -1 + 2y$ or $y \geq 1$. Because of the first constraint we get $y = 1$, from which it follows that $\gamma = 0$. Thus, $\gamma_1 = \gamma_2 = 0$ in accordance with Theorem IV.76. $\qquad \diamond$

19.4.5 *Running through all break points and linearity intervals*

Using the results of the previous sections, we present in this section an algorithm that yields the optimal-value function for a one-dimensional perturbation of the vector b or the vector c. We first deal with a one-dimensional perturbation of b by a scalar multiple of the vector Δb; we state the algorithm for the calculation of the optimal-value function and then prove that the algorithm finds all break points and linearity intervals. It will then be clear how to treat a one-dimensional perturbation of c; we state the corresponding algorithm and its convergence result without further proof. We provide examples for both cases.

Assume that we are given optimal solutions x^* of (P) and (y^*, s^*) of (D). In the notation of the previous sections, the problem (P_β) and its dual (D_β) arise by replacing the vector b by $b(\beta) = b + \beta\Delta b$; the optimal value of these problems is denoted by $f(\beta)$. So we have $f(0) = c^T x^* = b^T y^*$. The domain of the optimal-value function is $(-\infty, \infty)$ and $f(\beta) = \infty$ if and only if (D_β) is unbounded. Recall from Theorem IV.51 that $f(\beta)$ is convex and piecewise linear. Below we present an algorithm that determines f on the nonnegative part of the real line. We leave it to the reader to find some straightforward modifications of the algorithm, yielding an algorithm that generates f on the other part of the real line.[16] The algorithm is as follows.[17]

The Optimal Value Function $f(\beta)$, $\beta \geq 0$

Input:
    An optimal solution (y^*, s^*) of (D);
    a perturbation vector Δb.
begin
    $k := 1; y^0 := y^*; s^0 = s^*$; ready:=false;
    while not ready **do**
    begin
        Solve $\max_{\beta,x} \{\beta \ : \ Ax = b + \beta\Delta b, \ x \geq 0, \ x^T s^{k-1} = 0\}$;
        if this problem is unbounded: ready:=true
        else let (β_k, x^k) be an optimal solution;
        begin
            Solve $\max_{y,s} \{\Delta b^T y \ : \ A^T y + s = c, \ s \geq 0, \ s^T x^k = 0\}$;
            if this problem is unbounded: ready := true
            else let (y^k, s^k) be an optimal solution;
            $k := k + 1$;
        end
    end
end

The next theorem states that the above algorithm finds the successive break points of f on the nonnegative part of the real line, as well as the slopes of f on the successive linearity intervals.

Theorem IV.78 *The algorithm terminates after a finite number of iterations. If K is the number of iterations upon termination then $\beta_1, \beta_2, \ldots, \beta_K$ are the successive*

---

[16] **Exercise 99** When the two maximization problems in the algorithm are changed into minimization problems, the algorithm yields the break points and linearity intervals for negative values of β. Prove this.

[17] After the completion of this section the same algorithm appeared in a recent paper of Monteiro and Mehrotra [221] and the authors became aware of the fact that these authors already published the algorithm in 1992 [207].

break points of f on the nonnegative real line. The optimal value at β_k $(1 \leq k \leq K)$ is given by $c^T x^k$ and the slope of f on the interval (β_k, β_{k+1}) $(1 \leq k < K)$ by $\Delta b^T y^k$.

Proof: In the first iteration the algorithm starts by solving

$$\max_{\beta, x} \left\{ \beta \ : \ Ax = b + \beta \Delta b, \ x \geq 0, \ x^T s^0 = 0 \right\},$$

where s^0 is the slack vector in the given optimal solution $(y^0, s^0) = (y^*, s^*)$ of $(D) = (D_0)$. This problem is feasible, because (P) has an optimal solution x^* and $(\beta, x) = (0, x^*)$ satisfies the constraints. Hence the first auxiliary problem is either unbounded or it has an optimal solution (β_1, x^1). By Theorem IV.73 β_1 is equal to the extreme point at the right of the linearity interval containing 0. If the problem is unbounded (when $\beta_1 = \infty$) then f is linear on $(0, \infty)$ and the algorithm stops; otherwise β_1 is the first break point to the right of 0. (Note that it may happen that $\beta_1 = 0$. This certainly occurs if 0 is a break point of f and the starting solution (y^*, s^*) is strictly complementary.) Clearly x^1 is primal feasible at $\beta = \beta_1$. Since (y^1, s^1) is dual feasible at $\beta = \beta_1$ and $(x^1)^T s^1 = 0$ we see that x^1 is optimal for (P_{β_1}). Hence $f(\beta_1) = c^T x^1$. Also observe that (y^1, s^1) is dual optimal at β_1. (This also follows from Corollary IV.55.)

Assuming that the second half of the algorithm occurs, when the above problem has an optimal solution, the algorithm proceeds by solving a second auxiliary problem, namely

$$\max_{y, s} \left\{ \Delta b^T y \ : \ A^T y + s = c, \ s \geq 0, \ s^T x^1 = 0 \right\}.$$

By Theorem IV.62 the maximal value is equal to the right derivative of f at β_1. If the problem is unbounded then β_1 is the largest break point of f on $(0, \infty)$ and $f(\beta) = \infty$ for $\beta > \beta_1$. In that case we are done and the algorithm stops. Otherwise, when the problem is bounded, the optimal solution (y^1, s^1) is such that $\Delta b^T y^1$ is equal to the slope on the linearity interval to the right of β_1, by Lemma IV.61. Moreover, from Corollary IV.64, (y^1, s^1) is dual optimal on the open linearity interval to the right of β_1. Hence, at the start of the second iteration (y^1, s^1) is an optimal solution at the open interval to the right of the first break point on $[0, \infty)$. Thus we can start the second iteration and proceed as in the first iteration. Since each iteration produces a linearity interval, and f has only finitely many such intervals, the algorithm terminates after a finite number of iterations. □

Example IV.79 Consider the primal problem

$$(P) \qquad \min \left\{ x_1 + x_2 + x_3 \ : \ x_1 - x_2 = 0, \ x_3 = 1, \ x = (x_1, x_2, x_3) \geq 0 \right\}$$

and its dual

$$(D) \qquad \max \left\{ y_2 \ : \ -1 \leq y_1 \leq 1, \ y_2 \leq 1 \right\}.$$

Hence, in this case we have

$$A = \begin{bmatrix} 1 & -1 & 0 \\ 0 & 0 & 1 \end{bmatrix}, \quad c = \begin{bmatrix} 1 \\ 1 \\ 1 \end{bmatrix}, \quad b = \begin{bmatrix} 0 \\ 1 \end{bmatrix}.$$

We perturb the vector b by a scalar multiple of

$$\Delta b = \begin{bmatrix} 1 \\ -1 \end{bmatrix}$$

to

$$b(\beta) = b + \beta \Delta b = \begin{bmatrix} 0 \\ 1 \end{bmatrix} + \beta \begin{bmatrix} 1 \\ -1 \end{bmatrix} = \begin{bmatrix} \beta \\ 1 - \beta \end{bmatrix},$$

and use the algorithm to find the break points and linearity intervals of $f(\beta) = z(b(\beta), c)$.

Optimal solutions of (P) and (D) are given by

$$x^* = (0, 0, 1), \quad y^* = (0, 1), \quad s^* = (1, 1, 0).$$

Thus, entering the first iteration of the algorithm we consider

$$\max_{\beta, x} \{\beta \; : \; x_1 - x_2 = \beta, \, x_3 = 1 - \beta, \, x \geq 0, \, x_1 + x_2 = 0\}.$$

From $x \geq 0$, $x_1 + x_2 = 0$ we deduce that $x_1 = x_2 = 0$ and hence $\beta = 0$. Thus we find the first break point and the optimal value at this break point:

$$\beta_1 = 0, \quad x^1 = (0, 0, 1), \quad f(\beta_1) = c^T x^1 = 1.$$

We proceed with the second auxiliary problem:

$$\max_{y} \{y_1 - y_2 \; : \; -1 \leq y_1 \leq 1, \, y_2 \leq 1, \, 1 - y_2 = 0\}.$$

It follows that $y_2 = 1$ and $y_1 - y_2 = y_1 - 1$ is maximal if $y_1 = 1$. Thus we find an optimal solution (y^1, s^1) for the linearity interval just to the right of β_1 and the slope of f on this interval:

$$y^1 = (1, 1), \quad s^1 = (0, 2, 0), \quad f'_+(\beta_1) = \Delta b^T y_1 = 0.$$

In the second iteration the first auxiliary problem is

$$\max_{\beta, x} \{\beta \; : \; x_1 - x_2 = \beta, \, x_3 = 1 - \beta, \, x \geq 0, \, 2x_2 = 0\},$$

which is equivalent to

$$\max_{\beta, x} \{\beta \; : \; \beta = x_1, \, \beta = 1 - x_3, \, x \geq 0, \, x_2 = 0\}.$$

Clearly the maximum value of β is attained at $x_1 = 1$ and $x_3 = 0$. Thus we find the second break point and the optimal value at this break point:

$$\beta_2 = 1, \quad x^1 = (1, 0, 0), \quad f(\beta_2) = c^T x^2 = 1.$$

The second auxiliary problem becomes

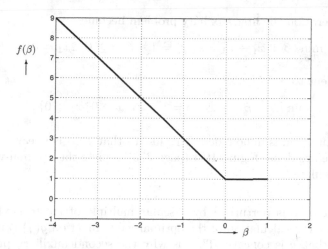

Figure 19.4 The optimal-value function $f(\beta)$.

$$\max_{y} \{y_1 - y_2 \ : \ -1 \le y_1 \le 1, \ y_2 \le 1, \ 1 - y_1 = 0\},$$

which is equivalent to

$$\max_{y} \{1 - y_2 \ : \ y_2 \le 1, \ y_1 = 1\}.$$

Clearly this problem is unbounded. Hence $f'_+(\beta_2) = \infty$ and we are done. For larger values of β the primal problem (P_β) becomes infeasible and the dual problem (D_β) unbounded.

We proceed by calculating $f(\beta)$ for negative values of β. Using Exercise 99 (page 380, the first auxiliary problem, in the first iteration, becomes simply

$$\min_{\beta,x} \{\beta \ : \ x_1 - x_2 = \beta, \ x_3 = 1 - \beta, \ x \ge 0, \ x_1 + x_2 = 0\}.$$

We can easily verify that this problem has the same solution as its counterpart, when we maximize β. This is due to the fact that $\beta = 0$ is a break point of f. We find, as before,

$$\beta_1 = 0, \quad x^1 = (0, 0, 1), \quad f(\beta_1) = c^T x^1 = 1.$$

We proceed with the second auxiliary problem:

$$\min_{y} \{y_1 - y_2 \ : \ -1 \le y_1 \le 1, \ y_2 \le 1, \ 1 - y_2 = 0\}.$$

Since $y_2 = 1$ we have $y_1 - y_2 = y_1 - 1$ and this is minimal if $y_1 = -1$. Thus we find an optimal solution (y^1, s^1) for the linearity interval just to the left of $\beta_1 = 0$ and the slope of f on this interval:

$$y^1 = (-1, 1), \quad s^1 = (2, 0, 0), \quad f'_-(\beta_1) = \Delta b^T y_1 = -2.$$

In the second iteration the first auxiliary problem becomes

$$\min_{\beta,x} \left\{ \beta \ : \ x_1 - x_2 = \beta, \ x_3 = 1 - \beta, \ x \geq 0, \ 2x_1 = 0 \right\},$$

which is equivalent to

$$\min_{\beta,x} \left\{ \beta \ : \ \beta = -x_2, \ \beta = 1 - x_3, \ x \geq 0, \ x_1 = 0 \right\}.$$

Obviously this problem is unbounded. This means that $f(\beta)$ is linear on the negative real line, and we are done. Figure 19.4 (page 383) depicts the optimal-value function $f(\beta)$ as just calculated. ◇

When the vector c is perturbed by a scalar multiple of Δc to $c(\gamma) = c + \gamma \Delta c$, the algorithm for the calculation of the optimal value function $g(\gamma)$ can be stated as follows. Recall that g is concave. That is why the second auxiliary problem in the algorithm is a minimization problem.[18]

The Optimal Value Function $g(\gamma)$, $\gamma \geq 0$

Input:
   An optimal solution x^* of (P);
   a perturbation vector Δc.
begin
   ready:=false;
   $k := 1; x^0 := x^*$;
   while not ready **do**
   begin
      Solve $\max_{\gamma,y,s} \left\{ \gamma \ : \ A^T y + s = c + \gamma \Delta c, \ s \geq 0, \ s^T x^{k-1} = 0 \right\}$;
      if this problem is unbounded: ready:=true
      else let (γ_k, y^k, s^k) be an optimal solution;
      begin
         Solve $\min_x \left\{ \Delta c^T x \ : \ Ax = b, \ x \geq 0, \ x^T s^k = 0 \right\}$;
         if this problem is unbounded: ready:=true
         else let x^k be an optimal solution;
         $k := k + 1$;
      end
   end
end

The above algorithm finds the successive break points of g on the nonnegative real line as well as the slopes of g on the successive linearity intervals. The proof uses

---
[18] **Exercise 100** When the maximization problem in the algorithm is changed into a minimization problem and the minimization into a maximization problem, the algorithm yields the break points and linearity intervals for negative values of γ. Prove this.

arguments similar to the arguments in the proof of Theorem IV.78 and is therefore omitted.

Theorem IV.80 *The algorithm terminates after a finite number of iterations. If K is the number of iterations upon termination then $\gamma_1, \gamma_2, \ldots, \gamma_K$ are the successive break points of g on the nonnegative real line. The optimal value at γ_k $(1 \leq k \leq K)$ is given by $b^T y^k$ and the slope of g on the interval (γ_k, γ_{k+1}) $(1 \leq k < K)$ by $\Delta c^T x^k$.*

$\square$

The next example illustrates the use of the above algorithm.

Example IV.81 In Example IV.72 we considered the primal problem

$$(P) \quad \min \{x_1 + 3x_2 + x_3 \ : \ x_1 + x_2 + x_3 = 4, \ x_1, x_2, x_3 \geq 0\}$$

and its dual problem

$$(D) \quad \max \{4y \ : \ y \leq 1, \ y \leq 3, \ y \leq 1\},$$

with the perturbation vector

$$\Delta c = (0, 1, -1)$$

and we calculated the linearity intervals from Lemma IV.67. This required the knowledge of an optimal primal solution for each interval. Theorem IV.80 enables us to find these intervals from the knowledge of an optimal solution x^* of (P) only.

Entering the first iteration of the above algorithm with $x^* = (4, 0, 0)$ we consider

$$\max_{\gamma, y} \{\gamma \ : \ y \leq 1, \ y \leq 3 + \gamma, \ y \leq 1 - \gamma, \ 4(1 - y) = 0\}.$$

We can easily see that $y = 1$ is optimal with $\gamma = 0$. Thus we find the first break point and the optimal value at this break point:

$$\gamma_1 = 0, \quad y^1 = 1, \quad s^1 = (0, 2, 0), \quad g(\gamma_1) = b^T y^1 = 4.$$

The second auxiliary problem is now given by:

$$\min_{x} \{x_2 - x_3 \ : \ x_1 + x_2 + x_3 = 4, \ x_1, x_2, x_3 \geq 0, \ 2x_2 = 0\}.$$

It follows that $x_2 = 0$ and $x_2 - x_3 = -x_3$ is minimal if $x_3 = 4$ and $x_1 = 0$. Thus we find an optimal solution x^1 for the linearity interval just to the right of γ_1 and the slope of g on this interval:

$$x^1 = (0, 0, 4), \quad g'_+(\gamma_1) = \Delta c^T x_1 = -4.$$

In the second iteration the first auxiliary problem is

$$\max_{\gamma, y} \{\gamma \ : \ y \leq 1, \ y \leq 3 + \gamma, \ y \leq 1 - \gamma, \ 4(1 - \gamma - y) = 0\}.$$

It follows that $y = 1 - \gamma$ and the problem becomes equivalent to

$$\max_{\gamma,y} \{\gamma \; : \; 1 - \gamma \leq 1, \, 1 - \gamma \leq 3 + \gamma, \, y = 1 - \gamma\}.$$

Clearly this problem is unbounded. Hence g is linear for values of γ larger than $\gamma_1 = 0$.

We proceed by calculating $g(\gamma)$ for negative values of γ. Using Exercise 100 (page 384), the first auxiliary problem, in the first iteration, becomes simply

$$\min_{\gamma,y} \{\gamma \; : \; y \leq 1, \, y \leq 3 + \gamma, \, y \leq 1 - \gamma, \, 4(1 - y) = 0\}.$$

Since $y = 1$ this is equivalent to

$$\min_{\gamma,y} \{\gamma \; : \; -2 \leq \gamma \leq 0, \, y = 1\},$$

so the first break point and the optimal value at this break point are given by

$$\gamma_1 = -2, \quad y^1 = 1, \quad s^1 = (0, 0, 2), \quad g(\gamma_1) = b^T y^1 = 4.$$

The second auxiliary problem is now given by:

$$\max_{x} \{x_2 - x_3 \; : \; x_1 + x_2 + x_3 = 4, \, x_1, x_2, x_3 \geq 0, \, 2x_3 = 0\},$$

which is equivalent to

$$\max_{x} \{x_2 \; : \; x_1 + x_2 = 4, \, x_1, x_2 \geq 0, \, x_3 = 0\}.$$

Since x_2 is maximal if $x_1 = 0$ and $x_2 = 4$ we find an optimal solution x^1 for the linearity interval just to the left of γ_1 and the slope of g on this interval:

$$x^1 = (0, 4, 0), \quad g'_-(\gamma_1) = \Delta c^T x^1 = 4.$$

In the second iteration the first auxiliary problem is

$$\min_{\gamma,y} \{\gamma \; : \; y \leq 1, \, y \leq 3 + \gamma, \, y \leq 1 - \gamma, \, 4(3 + \gamma - y) = 0\}.$$

It follows that $y = 3 + \gamma$ and the problem becomes equivalent to

$$\min_{\gamma,y} \{\gamma \; : \; 3 + \gamma \leq 1, \, 3 + \gamma \leq 1 - \gamma, \, y = 3 + \gamma\}.$$

Clearly this problem is unbounded. Hence g is linear for values of γ smaller than $\gamma_1 = -2$. This completes the calculation of the optimal-value function $g(\gamma)$ for the present example. We can easily check that the above results are in accordance with the graph of $g(\gamma)$ in Figure 19.3 (page 369).[19]   ◇

---

[19] **Exercise 101** In Example IV.81 the algorithm for the computation of the optimal-value function $g(\gamma)$ was initialized by the optimal solution $x^* = (4, 0, 0)$ of (P). Execute the algorithm once more now using the optimal solution $x^* = (2, 0, 2)$ of (P).

19.5   Sensitivity analysis

Sensitivity analysis is the special case of parametric analysis where only one coefficient of b, or c, is perturbed. This means that the perturbation vector is a unit vector. The derivative of the optimal-value function to a coefficient is called the *shadow price* and the corresponding linearity interval the *range* of the coefficient. When dealing with sensitivity analysis the aim is to find the shadow prices and ranges of all coefficients in b and c. Of course, the current value of a coefficient may or may not be a break point. In the latter case, when the current coefficient is not a break point, it belongs to an open linearity interval and the range of the coefficient is just this closed linearity interval and its shadow price the slope of the optimal-value function on this interval. If the coefficient is a break point, then we have two shadow prices, the *left-shadow price*, which is the left derivative of the optimal-value function at the current value, and the *right-shadow price*, the right derivative of the optimal-value function at the current value.[20]

19.5.1   Ranges and shadow prices

Let x^* be an optimal solution of (P) and (y^*, s^*) an optimal solution of (D). With e_i denoting the i-th unit vector $(1 \leq i \leq m)$, the range of the i-th coefficient b_i of b is simply the linearity interval of the optimal-value function $z_A(b + \beta e_i, c)$ that contains zero. From Theorem IV.73, the extreme points of this linearity interval follow by minimizing and maximizing β over the set

$$\left\{\beta \ : \ Ax = b + \beta e_i, \ x \geq 0, \ x^T s^* = 0\right\}.$$

With b_i considered as a variable, its range of b_i follows by minimizing and maximizing b_i over the set

$$\left\{b_i \ : \ Ax = b, \ x \geq 0, \ x^T s^* = 0\right\}. \tag{19.6}$$

The variables in this problem are x and b_i. For the shadow prices of b_i we use Theorem IV.62. The left- and right-shadow prices of b_i follow by minimizing and maximizing $e_i^T y = y_i$ over the set

$$\left\{y_i \ : \ A^T y + s = c, \ s \geq 0, \ s^T x^* = 0\right\}. \tag{19.7}$$

Similarly, the range of the j-th coefficient c_j of c is equal to the linearity interval of the optimal-value function $z_A(b, c + \gamma e_j)$ that contains zero. Changing c_j into a variable and using Theorem IV.75, we obtain the extreme points of this linearity interval by minimizing and maximizing c_j over the set

$$\left\{c_j \ : \ A^T y + s = c, \ s \geq 0, \ s^T x^* = 0\right\}. \tag{19.8}$$

---

[20] Sensitivity analysis is an important topic in the application oriented literature on LO. Some relevant references, in chronological order, are Gal [89], Gauvin [93], Evans and Baker [72, 73], Akgül [6], Knolmayer [173], Gal [90], Greenberg [128], Rubin and Wagner [247], Ward and Wendell [288], Adler and Monteiro [4], Mehrotra and Monteiro [207], Jansen, Roos and Terlaky [153], Jansen, de Jong, Roos and Terlaky [152] and Greenberg [129]. It is surprising that in the literature on sensitivity analysis it is far from common to distinguish between left- and right-shadow prices. One of the early exceptions was Gauvin [93]; this paper, however, is not mentioned in the historical survey on sensitivity analysis of Gal [90].

In this problem the variables are the vectors y and s and also c_j. For the shadow prices of c_j we use Theorem IV.68. The left- and right-shadow prices of c_j follow by minimizing and maximizing $e_j^T x = x_j$ over the set

$$\left\{ x_j \ : \ Ax = b, \ x \geq 0, \ x^T s^* = 0 \right\}. \tag{19.9}$$

Some remarks are in order. If b_i is not a break point, which becomes evident if the extreme values in (19.6) both differ from b_i, then we know that the left- and right-shadow prices of b_i are the same and these are given by y_i^*. In that case there is no need to solve (19.7). On the other hand, when b_i is a break point, it is clear from the discussion following the proof of Theorem IV.73 that there are three possibilities. When the range of b_i is determined by solving (19.6) the result may be one of the two linearity intervals surrounding b_i; in that case y_i^* is the shadow price of b_i on this interval. This happens if and only if the given optimal solution y^* is such that y_N^* is an extreme value in the set (19.7). The third possibility is that the extreme values in the set (19.6) are both equal to b_i. This certainly occurs if y^* is a strictly complementary solution of (D). In each of the three cases it becomes clear after (19.6) is solved, that b_i is a break point, and the left- and right-shadow prices at b_i can be found by determining the extreme values of (19.7). Clearly similar remarks apply to the ranges and shadow prices of the coefficients of the vector c.

19.5.2 Using strictly complementary solutions

The formulas for the ranges and shadow prices of the coefficients of b and c can be simplified when the given optimal solutions x^* of (P) and (y^*, s^*) of (D) are strictly complementary. Let (B, N) denote the optimal partition of (P) and (D). Then we have $x_B^* > 0$, $x_N^* = 0$ and $s_B^* = 0$, $s_N^* > 0$. As a consequence, we have $x^T s^* = 0$ in (19.6) and (19.9) if and only if $x_N = 0$. Similarly, $s^T x^* = 0$ holds in (19.7) and (19.8) if and only if $s_B = 0$.

Using this we can reformulate (19.6) as

$$\left\{ b_i \ : \ Ax = b, \ x_B \geq 0, \ x_N = 0 \right\}, \tag{19.10}$$

and (19.7) as

$$\left\{ y_i \ : \ A^T y + s = c, \ s_B = 0, \ s_N \geq 0 \right\}. \tag{19.11}$$

Similarly, (19.8) can be rewritten as

$$\left\{ c_j \ : \ A^T y + s = c, \ s_B = 0, \ s_N \geq 0 \right\}, \tag{19.12}$$

and (19.9) as

$$\left\{ x_j \ : \ Ax = b, \ x_B \geq 0, \ x_N = 0 \right\}. \tag{19.13}$$

We proceed with an example.[21]

Example IV.82 Consider the (primal) problem (P) defined by

$$
\begin{array}{rrrrrrrrr}
\min & x_1 & + & 4x_2 & + & x_3 & + & 2x_4 & + & 2x_5 \\
\text{s.t.} & -2x_1 & + & x_2 & + & x_3 & & & + & x_5 & -x_6 & & = 0 \\
& x_1 & + & x_2 & - & x_3 & + & x_4 & & & & -x_7 & = 1
\end{array}
$$

$$x_1, x_2, x_3, x_4, x_5, x_6, x_7 \geq 0.$$

The dual problem (D) is

$$
\begin{array}{rrrrll}
\max & y_2 & & & & \\
\text{s.t.} & -2y_1 & + & y_2 & \leq 1 & (1) \\
& y_1 & + & y_2 & \leq 4 & (2) \\
& y_1 & - & y_2 & \leq 1 & (3) \\
& & & y_2 & \leq 2 & (4) \\
& y_1 & & & \leq 2 & (5) \\
& -y_1 & & & \leq 0 & (6) \\
& & & -y_2 & \leq 0 & (7)
\end{array}
$$

Problem (D) can be solved graphically. Its feasible region is shown in Figure 19.5 (page 390).

Since we are maximizing y_2 in (D), the figure makes clear that the set of optimal solutions is given by

$$\mathcal{D}^* = \{(y_1, y_2) : 0.5 \leq y_1 \leq 2, y_2 = 2\},$$

and hence the optimal value is 2. Note that all slack values can be positive at an optimal solution except the slack value of the constraint $y_2 \leq 2$. This means that the set N in the optimal partition (B, N) equals $N = \{1, 2, 3, 5, 6, 7\}$. Hence, $B = \{4\}$. Therefore, at optimality only the variable x_4 can be positive. It follows that

$$\mathcal{P}^* = \{x \in \mathcal{P} : x_1 = x_2 = x_3 = x_5 = x_6 = x_7 = 0\} = \{(0, 0, 0, 1, 0, 0, 0)\},$$

and (P) has a unique solution: $x = (0, 0, 0, 1, 0, 0, 0)$.

_____

[21] **Exercise 102** The ranges and shadow prices can also be found by solving the corresponding dual problems. For example, the maximal value of b_i in (19.10) can be found by solving

$$\min \left\{ b^T y : A_B^T y \geq 0, y_i = -1 \right\}$$

and the minimal value by solving

$$\max \left\{ b^T y : A_B^T y \leq 0, y_i = -1 \right\}.$$

Formulate the dual problems for the other six cases.

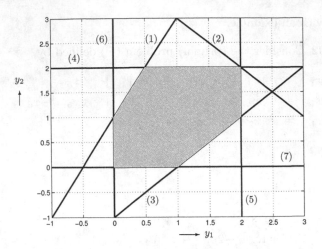

Figure 19.5 The feasible region of (D).

The next table shows the result of a complete sensitivity analysis. It shows the ranges and shadow prices for all coefficients of b and c, where these vectors have their usual meaning. For each coefficient that is a break point we give the shadow price as a closed interval; the extreme values of this interval are the left- and right-shadow prices of the coefficient. In this example this happens only for b_1. The range of a break point consists of the point itself; the table gives this point. On the other hand, for 'nonbreak points' the range is a proper interval and the shadow price is a number.

Coefficient	Range	Shadow prices
$b_1 = 0$	0	$[\frac{1}{2}, 2]$
$b_2 = 1$	$[0, \infty)$	2
$c_1 = 1$	$[-2, \infty)$	0
$c_2 = 4$	$[\frac{5}{2}, \infty)$	0
$c_3 = 1$	$[-\frac{3}{2}, \infty)$	0
$c_4 = 2$	$[0, 3]$	1
$c_5 = 2$	$[\frac{1}{2}, \infty)$	0
$c_6 = 0$	$[-2, \infty)$	0
$c_7 = 0$	$[-2, \infty)$	0

We perform the sensitivity analysis here for b_1 and c_4.

Range and shadow prices for b_1

Using (19.10) the range of b_1 follows by minimizing and maximizing b_1 over the system

$$
\begin{aligned}
0 &= b_1 \\
x_4 &= 1.
\end{aligned}
$$

The solution of this system is unique: $x_4 = 1$ and $b_1 = 0$, so the range of b_1 is the interval $[0, 0]$. This means that $b_1 = 0$ is a break point.

The left- and right-shadow prices of b_1 follow by minimizing and maximizing y_1 over $y \in \mathcal{D}^*$. The minimal value is 0.5 and the maximal value 2, so the left- and right-shadow prices 0.5 and 2.

Range and shadow price for c_4

The range of c_4 is found by using (19.12). This amounts to minimizing and maximizing c_4 over the system

$$
\begin{aligned}
-2y_1 + y_2 &\leq 1 \\
y_1 + y_2 &\leq 4 \\
y_1 - y_2 &\leq 1 \\
y_2 &= c_4 \\
y_1 &\leq 2 \\
y_1 &\geq 0 \\
y_2 &\geq 0.
\end{aligned}
$$

This optimization problem can easily be solved by using Figure 19.5. It amounts to the question of which values of y_2 are feasible when the fourth constraint is removed in Figure 19.5. We can easily verify that all values of y_2 in the closed interval $[0, 3]$ (and no other values) satisfy. Therefore, the range of c_4 is this interval. The shadow price of c_4 is given by $e_4^T x = x_4 = 1$. $\diamondsuit$

19.5.3 Classical approach to sensitivity analysis

Commercial optimization packages for the solution of LO problems usually offer the possibility of doing sensitivity analysis. The sensitivity analysis in many existing commercial optimization packages is based on the naive approach presented in first year textbooks. As a result, the outcome of the sensitivity analysis is often confusing. We explain this below.

The 'classical' approach to sensitivity analysis is based on the Simplex Method for solving LO problems.[22] The Simplex Method produces a so-called *basic solution* of

---

[22] With the word 'classical' we want to refer to the approach which dominates the literature, especially well known textbooks dealing with parametric and/or sensitivity analysis. This approach has led to the existing misuse of parametric optimization in commercial packages. This misuse is however a shortcoming of the packages and by no means a shortcoming in the whole existing theoretical literature. In this respect we want to refer the reader to Nožička, Guddat, Hollatz and Bank [228]. In this book the parametric issue is correctly handled in terms of the Simplex Method, polyhedrons, faces of polyhedra etc. Besides parameterizing either the objective vector or the right-hand side vector, much more general parametric issues are also discussed. The following citation is taken from this book: *Den qualitativen Untersuchungen in den meisten erschienenen Aufsätzen und Büchern liegt das Simplexverfahren zugrunde. Zwangsläufig unterliegen alle derartig gewonnenen Aussagen den Schwierigkeiren, die bei Beweisführungen mit Hilfe der Simplexmethode im Falle der Entartung auftreten. In einigen Arbeiten wurde ein rein algebraischer Weg verfolgt, der in gewisse Spezialfällen zu Resultaten führte, im allgemeinen aber bisher keine qualitative Analyse erlieferte.*

the problem. It suffices for our purpose to know that such a solution is determined by an *optimal basis*. Assuming that A is of size $m \times n$ and rank $(A) = m$, a *basis* is a nonsingular $m \times m$ submatrix $A_{B'}$ of A and the corresponding basic solution x is determined by

$$A_{B'} x_{B'} = b, \; x_{N'} = 0,$$

where N' consists of the indices not in B'. Defining a vector y by

$$A_{B'}^T y = c_{B'},$$

and denoting the slack vector of y by s, we have $s_{B'} = 0$. Since $x_{N'} = 0$, it follows that $xs = 0$, proving that x and s are complementary vectors. Hence, if $x_{B'}$ and $s_{N'}$ are nonnegative then x is optimal for (P) and (y, s) is optimal for (D). In that case $A_{B'}$ is called an optimal basis for (P) and (D). A main result in the Simplex based approach to LO is that such an optimal basis always exists — provided the assumption that rank $(A) = m$ is satisfied — and the Simplex Method generates such a basis. For a detailed description of the Simplex Method and its underlying theory we refer the reader to any (text-)book on LO.[23]

Any optimal basis leads to a natural division of the indices into m *basic indices* and $n - m$ *nonbasic indices*, thus yielding a partition (B', N') of the index set. We call this the *optimal basis partition* induced by the optimal basis B'. Obviously, an optimal basis partition need not be an optimal partition. In fact, this observation is crucial as we show below.

The classical approach to sensitivity analysis amounts to applying the 'formulas' (19.10) – (19.13) for the ranges and shadow prices, but with the optimal basis partition (B', N') instead of the optimal partition (B, N). It is clear that in general (B', N') is not necessarily the optimal partition because (P) and (D) may have more than one optimal basis. The outcome of the classical analysis will therefore depend on the optimal basis $A_{B'}$. Hence, correct implementations of the classical approach may give rise to different 'ranges' and 'shadow prices'.[24] The next example illustrates this phenomenon. In a subsequent section a further example is given, where we apply several commercial optimization packages to a small transportation problem.

Example IV.83 For problems (P) and (D) in Example IV.82 we have three optimal bases. These are given in the table below. The column at the right gives the 'ranges' for c_4 for each of these bases.

Basis	B'	'Range' for c_4
1	$\{1, 4\}$	$[1, 3]$
2	$\{2, 4\}$	$[2, 3]$
3	$\{4, 5\}$	$[1, 2]$

We get three different 'ranges', depending on the optimal basis. Let us do the calculations for the first optimal basis in the table. The 'range' of c_4 is found by

---

[23] See, e.g., Dantzig [59], Papadimitriou and Steiglitz [231], Chvátal [55], Schrijver [250], Fang and Puthenpura [74] and Sierksma [256].

[24] We put the words range and shadow price between quotes if they refer to ranges and shadow prices obtained from an optimal basis partition (which may differ from the unique optimal partition).

using (19.12) with (B, N) such that $B = B' = \{1, 4\}$. This amounts to minimizing and maximizing c_4 over the system

$$
\begin{array}{rcrcl}
-2y_1 & + & y_2 & = & 1 \\
y_1 & + & y_2 & \leq & 4 \\
y_1 & - & y_2 & \leq & 1 \\
& & y_2 & = & c_4 \\
y_1 & & & \leq & 2 \\
y_1 & & & \geq & 0 \\
& & y_2 & \geq & 0.
\end{array}
$$

Using Figure 19.5 we can easily solve this problem. The question now is which values of y_2 are feasible when the fourth constraint is removed in Figure 19.5 and the first constraint is active. We can easily verify that this leads to $1 \leq y_2 \leq 3$, thus yielding the closed interval $[1, 3]$ as the 'range' for c_4. The other two 'ranges' can be found in the same way by keeping the second and the fifth constraints active, respectively.

A commercial optimization package provides the user with one of the three ranges in the table, depending on the optimal basis found by the package. Observe that each of the three ranges is a subrange of the correct range, which is $[0, 3]$. Note that the current value 2 of c_4 lies in the open interval, whereas for two of the 'ranges' in the table, 2 is an extreme point. This might lead to the wrong conclusion that 2 is a break point of the optimal-value function. ◇

It can easily be understood that the 'range' obtained from an optimal basis partition is always a subinterval of the whole linearity interval. Of course, sometimes the subinterval may coincide with the whole interval. For the shadow prices a similar statement holds. At a 'nonbreak point' an optimal basis partition yields the correct shadow price. At a break point, however, an optimal basis partition yields one 'shadow price', which may be any number between the left- and the right-shadow price. The example in the next section demonstrates this behavior very clearly.

Before proceeding with the next section we must note that from a computational point of view, the approach using an optimal basis partition is much cheaper than using the optimal partition. In the latter case we need to solve some auxiliary LO problems — in the worst case four for each coefficient. When the optimal partition (B, N) is replaced by an optimal basis partition (B', N'), however, it becomes computationally very simple to determine the 'ranges' and 'shadow prices'.

For example, consider the 'range' problem for b_i. This amounts to minimizing and maximizing b_i over the set

$$\{b_i \ : \ Ax = b, \ x_{B'} \geq 0, \ x_{N'} = 0\}.$$

Since $A_{B'}$ is nonsingular, it follows that

$$x_{B'} = A_{B'}^{-1} b,$$

and hence the condition $x_{B'} \geq 0$ reduces to

$$A_{B'}^{-1} b \geq 0.$$

This is a system of m linear inequalities in the coefficient b_i, with i fixed, and hence its solution can be determined straightforwardly. Note that the system is feasible, because the current value of b_i is such that the system is satisfied. Hence, the solution is a closed interval containing the current value of b_i.

19.5.4 Comparison of the classical and the new approach

For the comparison we use a simple problem, arising when transporting commodities (of one type) from three distribution centers to three warehouses. The supply values at the three distribution centers are 2, 6 and 5 units respectively, and the demand value at each of the three warehouses is just 3. We assume that the costs for transportation of one unit of commodity from a distribution center to a warehouse is independent of the distribution center and the warehouse, and this cost is equal to one (unit of currency). The aim is to meet the demand at the warehouses at minimal cost. This problem is depicted in Figure 19.6 by means of a network. The left three nodes in this network represent the distribution centers and the right three nodes the three warehouses. The arcs represent the transportation routes from the distribution centers to the warehouses. The supply and demand values are indicated at the respective nodes. The transportation problem consists of assigning 'flow' values to the arcs in

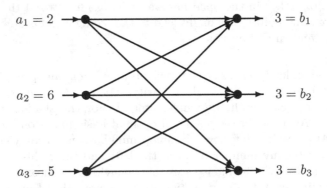

Figure 19.6   A transportation problem.

the network so that the demand is met and the supply values are respected; this must be done in such a way that the cost of the transportation to the demand nodes is minimized. Because of the choice of cost coefficients, the total cost is simply the sum of all arc flow values. Since the total demand is 9, this is also the optimal value for the total cost value. Note that there are many optimal flows; this is due to the fact that all arcs are equally expensive. So far, everything is trivial.

Sensitivity to demand and supply values

Now we want to determine the sensitivity of the optimal value to perturbations of the supply and demand values. Denoting the supply values by $a = (a_1, a_2, a_3)$ and the demand values by $b = (b_1, b_2, b_3)$, we can determine the ranges of these values by hand.

For example, when b_1 is changed, the total demand becomes $6 + b_1$ and this is the optimal value as long as such a demand can be met by the present supply. This leads to the condition

$$6 + b_1 \leq 2 + 6 + 5 = 13,$$

which yields $b_1 \leq 7$. For larger values of b_1 the problem becomes infeasible. When $b_1 = 0$, the arcs leading to the first demand node have zero flow value in any optimal solution. This means that 0 is a break point, and the range of b_1 is $[0, 7]$. Because of the symmetry in the network for the demand nodes, the range for b_2 and b_3 will be the same interval.

When a_1 is changed, the total supply becomes $11 + a_1$ and this will be sufficient as long as

$$11 + a_1 \geq 9,$$

which yields $a_1 \geq -2$. The directed arcs can only handle nonnegative supply values, and hence the range of a_1 is $[0, \infty)$. Similarly, the range for a_2 follows from

$$7 + a_2 \geq 9,$$

which yields the range $[2, \infty)$ for a_2, and the range for a_3 follows from

$$8 + a_3 \geq 9,$$

yielding the range $[1, \infty)$ for a_3.

To compare these ranges with the 'ranges' provided by the classical approach, we made a linear model of the above problem, solved it using five well-known commercial optimization packages, and performed a sensitivity analysis with these packages. We used the following linear standard model:

$$\min \sum_{i=1}^{3} \sum_{j=1}^{3} x_{ij}$$

$$
\begin{array}{rcllllllll}
\text{s.t.} & x_{11} & + & x_{12} & + & x_{13} & + & s_1 & = & 2 \\
& x_{21} & + & x_{22} & + & x_{23} & + & s_2 & = & 6 \\
& x_{31} & + & x_{32} & + & x_{33} & + & s_3 & = & 5 \\
& x_{11} & + & x_{21} & + & x_{31} & - & d_1 & = & 3 \\
& x_{12} & + & x_{22} & + & x_{32} & - & d_2 & = & 3 \\
& x_{13} & + & x_{23} & + & x_{33} & - & d_3 & = & 3
\end{array}
$$

$$x_{ij}, s_i, d_j \geq 0, \ i, j = 1, 2, 3.$$

The meaning of the variables is as follows:

x_{ij} : the amount of transport from supply node i to demand node j,

s_i : excess supply at supply node i,

d_j : shortage of demand at node j,

where i and j run from 1 to 3.

The result of the experiment is shown in the table below.[25] The columns correspond to the supply and the demand coefficients. Their current values are put between brackets. The rows in the table corresponding to the five packages[26] CPLEX, LINDO, PC-PROG, XMP and OSL show the 'ranges' produced by these packages. The last row contains the ranges calculated before by hand.[27]

LO package	'Ranges' of supply and demand values					
	$a_1(2)$	$a_2(6)$	$a_3(5)$	$b_1(3)$	$b_2(3)$	$b_3(3)$
CPLEX	$[0,3]$	$[4,7]$	$[1,\infty)$	$[2,7]$	$[2,5]$	$[2,5]$
LINDO	$[1,3]$	$[2,\infty)$	$[4,7]$	$[2,4]$	$[1,4]$	$[1,7]$
PC-PROG	$[0,\infty)$	$[4,\infty)$	$[3,6]$	$[2,5]$	$[0,5]$	$[2,5]$
XMP	$[0,3]$	$[6,7]$	$[1,\infty)$	$[2,3]$	$[2,3]$	$[2,7]$
OSL	$[0,3]$	$[4,7]$	$(-\infty,\infty)$	$[2,7]$	$[2,5]$	$[2,5]$
Correct range	$[0,\infty)$	$[2,\infty)$	$[1,\infty)$	$[0,7]$	$[0,7]$	$[0,7]$

The table clearly demonstrates the weaknesses of the classical approach. Sensitivity analysis is considered to be a tool for obtaining information about the bottlenecks and degrees of freedom in the problem. The information provided by the commercial optimization packages is confusing and hardly allows a solid interpretation. For example, in our example problem there is obvious symmetry between the demand nodes. None of the five packages gives evidence of this symmetry.

Remark IV.84 As stated before, the 'ranges' and 'shadow prices' provided by the classical approach arise by applying the formulas (19.10) – (19.13) for the ranges and shadow prices, but replacing the optimal partition (B, N) by the optimal basis partition (B', N'). Indeed, the 'ranges' in the table can be reconstructed in this way. We will not do this here, but to enable the interested reader to perform the relevant calculations we give the optimal basis partitions used by the packages. If the optimal basis partition is (B', N'), it suffices to know the variables in B' for each of the five packages. These 'basic variables' are given in the next table.

LO package	Basic variables					
CPLEX	x_{12}	x_{21}	x_{22}	x_{23}	x_{31}	s_3
LINDO	x_{11}	x_{23}	x_{31}	x_{32}	x_{33}	s_2
PC-PROG	x_{22}	x_{23}	x_{31}	x_{33}	s_1	s_2
XMP	x_{13}	x_{21}	x_{22}	x_{23}	x_{33}	s_3
OSL	x_{12}	x_{21}	x_{22}	x_{23}	x_{31}	s_3

---

[25] The dual problem has a unique solution in this example. These are the shadow prices for the demand and supply values. All packages return this unique solution, namely 0 for the supply values — due to the excess of supply — and 1 for the demand values.

[26] For more information on these packages we refer the reader to Sharda [253].

[27] The 'range' provided by the IBM package OSL (Optimization Subroutine Library) for a_3 is not a subrange of the correct range; this must be due to a bug in OSL. The correct 'range' for the optimal basis partition used by OSL is $[1, \infty)$.

Note that CPLEX and OSL use the same optimal basis. The output of their sensitivity analysis differs, however. As noted before, the explanation of this phenomenon is that the OSL implementation of the classical approach must contain a bug. •

The sensitivity analysis for the cost coefficients c_{ij} is considered next. The results are similar, as we shall see.

Sensitivity to cost coefficients

The current values of the cost coefficients c_{ij} are all 1. As a consequence each feasible flow on the network is optimal if the sum of the flow values x_{ij} equals 9. When one of the arcs becomes more expensive, then the flow on this arc can be rerouted over the other arcs and the optimal value remains 9. Hence the right-shadow price of each cost coefficient equals 0. On the other hand, if one of the arc becomes cheaper, then it becomes attractive to let this arc carry as much flow as possible. The maximal flow values for the arcs are 2 for the arcs emanating from the first supply node and 3 for the other arcs. Since for each arc there exists an optimal solution of the problem in which the flow value on that arc is zero, a decrease of 1 in the cost coefficient for the arcs emanating from the first supply node leads to a decrease of 2 in the total cost, and for the other arcs the decrease in the total cost is 3. Thus we have found the left- and right-shadow prices of the cost coefficients. Since the left- and right-shadow prices are all different, the current value of each of the cost coefficients is a break point. Obviously, the linearity interval to the left of this break point is $(-\infty, 1]$ and the linearity interval to the right of it is $[1, \infty)$.

In the next table the 'shadow prices' provided by the five commercial optimization packages are given. The last row in the table contains the correct values of the left- and right-shadow prices, as just calculated.

LO package	'Shadow prices' of cost coefficients								
	c_{11}	c_{12}	c_{13}	c_{21}	c_{22}	c_{23}	c_{31}	c_{32}	c_{33}
CPLEX	0	2	0	2	1	3	1	0	0
LINDO	2	0	0	0	0	2	1	3	1
PC-PROG	0	0	0	0	3	1	3	0	2
XMP	0	0	2	3	3	0	0	0	1
OSL	0	2	0	2	1	3	1	0	0
Correct values	[2,0]	[2,0]	[2,0]	[3,0]	[3,0]	[3,0]	[3,0]	[3,0]	[3,0]

Note that in all cases the 'shadow price' of a package lies in the interval between the left- and right-shadow prices.

The last table shows the 'ranges' of the packages and the correct left- and right-hand side ranges for the cost coefficients.[28] It is easy to understand the correct ranges. For example, if c_{11} increases then the corresponding arc becomes more expensive than the other arcs, and hence will not be used in an optimal solution. On the other hand, if

---

[28] In this table we use shorthand notation for the infinite intervals $[1, \infty)$ and $(-\infty, 1]$. The interval $[1, \infty)$ is denoted by $[1,)$ and the interval $(-\infty; 1]$ by $(, 1]$.

c_{11} decreases than it becomes attractive to use this arc as much as possible; due to the limited supply value (i.e., 2) in the first supply node a flow of value 2 will be sent along this arc whatever the value of c_{11} is. Considering c_{21}, we see the same behavior if c_{21} increases: the arc will not be used. But if $c_{21} \in [0, 1]$, then the arc will be preferred above the other arcs, and its flow value will be 3. If c_{21} would become negative, then it becomes attractive to send even a flow of value 6 along this arc, despite the fact that than the first demand node receives oversupply. So $c_{21} = 0$ is a break point.

Note that if a 'shadow price' of a package is equal to the left or right-shadow price then the 'range' provided by the package must be a subinterval of the correct range. Moreover, if the 'shadow price' of a package is not equal to the left or right-shadow price then the 'range' provided by the package must be the singleton $[1, 1]$. The results of the packages are consistent in this respect, as follows easily by inspection.

LO package	'Ranges' of the cost coefficients								
	c_{11}	c_{12}	c_{13}	c_{21}	c_{22}	c_{23}	c_{31}	c_{32}	c_{33}
CPLEX	[1,)	(,1]	[1,)	[1,1]	[1,1]	[0,1]	[1,1]	[1,)	[1,)
LINDO	(,1]	[1,)	[1,)	[1,)	[1,)	[1,1]	[1,1]	[0,1]	[1,1]
PC-PROG	[1,)	[1,)	[1,)	[1,)	[0,1]	[1,1]	[0,1]	[1,)	[1,1]
XMP[29]	–	–	(,1]	[0,1]	[0,1]	[1,1]	–	–	[1,1]
OSL[30]	[1,)	[1,1]	[1,)	[1,1]	[1,1]	[1,1]	[1,1]	[1,)	[1,)
Left range	(,1]	(,1]	(,1]	[0,1]	[0,1]	[0,1]	[0,1]	[0,1]	[0,1]
Right range	[1,)	[1,)	[1,)	[1,)	[1,)	[1,)	[1,)	[1,)	[1,)

19.6 Concluding remarks

In this chapter we developed the theory necessary for the analysis of one-dimensional parametric perturbations of the vectors b and c in the standard formulation of the primal problem (P) and its dual problem (D). Given a pair of optimal solutions for these problems, we presented algorithms in Section 19.4.5 for the computation of the optimal-value function under such a perturbation. In Section 19.5 we concentrated on the special case of sensitivity analysis. In Section 19.5.1 we showed that the ranges and shadow prices of the coefficients of b and c can be obtained by solving auxiliary LO problems. We also discussed how the ranges obtained in this way can be ambiguous, but that the ambiguity can be avoided by using strictly complementary solutions.

We proceeded in Section 19.5.3 by discussing the classical approach to sensitivity analysis, based on the use of an optimal basic solution and the corresponding optimal basis. We showed that this approach is much cheaper from a computational point of

---

[29] For some unclear reason XMP did not provide all ranges. The missing entries in its row are all equal to $[1, \infty)$.

[30] In Remark IV.84 it was established that OSL and CPLEX use the same optimal basis; nevertheless their 'ranges' for c_{12} and c_{23} are different. One may easily verify that these 'ranges' are $(-\infty, 1]$ and $[0, 1]$ respectively. Thus, the CPLEX 'ranges' are consistent with this optimal basis and the OSL 'ranges' are not.

view. On the other hand, much less information is usually obtained and the information is often confusing. In the previous section we provided a striking example by presenting the sensitivity information provided by five commercial optimization packages for a simple transportation problem.

The shortcomings of the classical approach are well known among experts in the field. At several places in the literature these experts raised their voices to warn of the possible implications of using the classical approach. By way of example we include a citation of Rubin and Wagner [247]:

> Managers who build their own microcomputer linear programming models are apt to misuse the resulting shadow prices and shadow costs. Fallacious interpretations of these values can lead to expensive mistakes, especially unwarranted capital investments.

As a result of the unreliability of the sensitivity information provided by computer packages, the reputation of sensitivity analysis as a tool for obtaining information about the bottlenecks and degrees of freedom has suffered a lot. Many potential users of such information do not use it, because they want to avoid the pitfalls that are inherent in the classical approach.

The theory developed in this chapter provides a solid base for reliable sensitivity modules in future generations of computer packages for LO.

20
Implementing Interior Point Methods

20.1 Introduction

Several polynomial interior-point algorithms were discussed in the previous chapters. Interior point algorithms not only provide the best theoretical complexity for LO problems but allow highly efficient implementations as well. Obviously not all polynomial algorithms are practically efficient. In particular, all full Newton step methods (see, e.g., Section 6.7) are inefficient in practice. However variants like the predictor-corrector method (see Section 7.7) and large-update methods (see Section 6.9) allow efficient implementations. The aim of this chapter is to give some hints on how some of these interior point algorithms can be converted into efficient implementations. To reach this goal several problems have to be dealt with. Some of these problems have been at least partially discussed earlier (e.g., the embedding problem in Chapter 2) but need further elaboration. Some other topics (e.g., methods of sparse numerical linear algebra, preprocessing) have not yet been touched.

By reviewing the various interior-point methods we observe that they are all based on similar assumptions and are built up from similar ingredients. We can extract the following essential elements of interior-point methods (IPMs).

Appropriate problem form. All algorithms assume that the LO problem satisfies certain assumptions. The problem must be in an appropriate form (e.g., the canonical form or the standard form). In the standard form the coefficient matrix A must have full row rank. Techniques to bring a given LO problem to the desired form, and at the same time to eliminate redundant constraints and variables, are called *preprocessing* and are discussed in Section 20.3.

Search direction. The search direction in interior-point methods is always a Newton direction. To calculate this direction we have to solve a system of linear equations. Except for the right-hand side and the scaling, this system is the same for all the methods. Computationally the solution of the system amounts to factorizing a square matrix and then solving the triangular systems by forward or backward substitution. The factorization is the most expensive part of an iteration. Without efficient sparse linear algebra routines, interior-point methods would not be practical. Various elements of sparse matrix techniques are discussed in Section 20.4. A straightforward idea for reducing the computational cost is to reuse the same factorization. This leads to the idea of second- and higher-order methods discussed in Section 20.4.3.

Interior point. The interior-point assumption is presupposed, i.e. that both the primal and the dual problem have a strictly positive (preferably centered) initial solution. Most LO problems do not have such a solution, but still have to be solved. A theoretically appealing and at the same time practical method is to embed the problem in a self-dual model, as discussed in Chapter 2. The embedding model is revisited and elaborated on in Section 20.5.

Reoptimization. In practice it often happens that several variants of the same LO problem need to be solved successively. One might expect that the solution of an earlier version would be a good starting point for a slightly modified problem. For this so-called warm start problem the embedding model also provides a good solution as discussed in Section 20.5.2.

Parameters: Step size, stopping criteria. The iterates in IPMs should stay in some neighborhood of the central path. Theoretically good step-sizes can result in hopelessly slow convergence in practice. A practical step-size selection rule is discussed. At some point, when the duality gap is small, the calculation is terminated. The theoretical criteria are typically far beyond today's machine precision. A practical criterion is presented in Section 20.6.

Optimal basis identification. It is not an essential element of interior-point methods, but sometimes it still might be important[1] to find an optimal basis. Then we need to provide the ability to 'cross over' from an interior solution to a basic solution. An elegant strongly polynomial algorithm is presented in Section 20.7.

20.2 Prototype algorithm

In most practical LO problems, in addition to the equality and inequality constraints, the variables have lower and upper bounds. Thus we deal with the primal problem in the following form:

$$\min_{x} \left\{ c^T x \ : \ Ax \geq b, \ x \leq b_u, \ x \geq 0 \right\}, \tag{20.1}$$

where $c, x, b_u \in \mathbb{R}^n, b \in \mathbb{R}^m$, and the matrix A is of size $m \times n$. Now its dual is

$$\max_{y, y_u} \left\{ b^T y - b_u^T y_u \ : \ A^T y - y_u \leq c, \ y \geq 0, \ y_u \geq 0 \right\}, \tag{20.2}$$

where $y \in \mathbb{R}^m$ and $y_u \in \mathbb{R}^n$. Let us denote the slack variables in the primal problem (20.1) by

$$z = Ax - b, \quad z_u = b_u - x$$

and in the dual problem (20.2) by

$$s = c + y_u - A^T y,$$

---

[1] Here we might think about linear integer optimization when cutting planes are to be generated to cut off the current nonintegral optimal solution.

respectively. Here we assume not only that the problem pair satisfies the interior-point assumption, but also that a strictly positive solution $(x, z_u, z, s, y_u, y) > 0$ is given, satisfying all the constraints in (20.1) and (20.2) respectively. How to solve these problems without the explicit knowledge of an interior point is discussed in Section 20.5.

The central path of the pair of problems given in (20.1) and (20.2) is defined as the set of solutions $(x(\mu), z_u(\mu), z(\mu)) > 0$ and $(s(\mu), y_u(\mu), y(\mu)) > 0$ for $\mu > 0$ of the system

$$
\begin{aligned}
Ax - z &= b, \\
x + z_u &= b_u, \\
A^T y - y_u + s &= c, \\
xs &= \mu e, \\
z_u y_u &= \mu e, \\
zy &= \mu e,
\end{aligned}
\tag{20.3}
$$

where e is the vector of all ones with appropriate dimension.

Observe that the first three of the above equations are linear and force primal and dual feasibility of the solution. The last three equations are nonlinear. They become the complementarity conditions when $\mu = 0$, which together with the feasibility constraints provides optimality of the solutions. The actual duality (or complementarity) gap g can easily be computed:

$$
g = x^T s + z_u^T y_u + z^T y,
$$

which equals $(2n + m)\mu$ on the central path .

One iteration of a primal-dual algorithm makes one step of Newton's method applied to system (20.3) with a given μ; then μ is reduced and the procedure is repeated as long as the duality gap is larger than a predetermined tolerance.

Given a solution $(x, z_u, z) > 0$ of (20.1) and $(s, y_u, y) > 0$ of (20.2) the Newton direction for (20.3) is obtained by solving a system of linear equations. This system can be written as follows, where the ordering of the variables is chosen so that the structure of the coefficient matrix becomes apparent.

$$
\begin{bmatrix}
0 & A & -I & 0 & 0 & 0 \\
A^T & 0 & 0 & -I & 0 & I \\
0 & I & 0 & 0 & I & 0 \\
0 & S & 0 & 0 & 0 & X \\
0 & 0 & 0 & Z_u & Y_u & 0 \\
Z & 0 & Y & 0 & 0 & 0
\end{bmatrix}
\begin{bmatrix}
\Delta y \\
\Delta x \\
\Delta z \\
\Delta y_u \\
\Delta z_u \\
\Delta s
\end{bmatrix}
=
\begin{bmatrix}
0 \\
0 \\
0 \\
\mu e - xs \\
\mu e - z_u y_u \\
\mu e - zy
\end{bmatrix}
.
\tag{20.4}
$$

In making a step, in order to preserve the positivity of (x, z_u, z) and (s, y_u, y), a step-size α usually smaller than one (a damped Newton step) is chosen.

Let us have a closer look at the Newton system. From the last four equations in (20.4) we can easily derive

$$
\begin{aligned}
\Delta z_u &= -\Delta x, \\
\Delta s &= x^{-1}(\mu e - xs - s\Delta x), \\
\Delta y_u &= z_u^{-1}(\mu e - z_u y_u - y_u \Delta z_u) = z_u^{-1}(\mu e - z_u y_u + y_u \Delta x), \tag{20.5} \\
\Delta z &= y^{-1}(\mu e - yz - z\Delta y).
\end{aligned}
$$

With these relations, (20.4) reduces to

$$\begin{bmatrix} D^2 & A \\ A^T & -\bar{D}^{-2} \end{bmatrix} \begin{bmatrix} \Delta y \\ \Delta x \end{bmatrix} = \begin{bmatrix} r \\ h \end{bmatrix}, \tag{20.6}$$

where

$$\begin{aligned} D^2 &= ZY^{-1} \\ \bar{D}^{-2} &= SX^{-1} + Y_u Z_u^{-1} \\ r &= y^{-1}(\mu e - yz) \\ h &= z_u^{-1}(\mu e - z_u y_u) - x^{-1}(\mu e - xs). \end{aligned}$$

The solution of the reduced Newton system (20.6) is the computationally most involved step of any interior point method. The system (20.6) in this form is a symmetric indefinite system and is referred to as the *augmented system*. If the second equation in the augmented system is multiplied by -1 a system with a positive definite (but unsymmetric) matrix is obtained.

The augmented system (20.6) is equivalent to

$$\Delta x = \bar{D}^2 (A^T \Delta y - h)$$

and

$$\left(A\bar{D}^2 A^T + D^2\right) \Delta y = r + A\bar{D}^2 h. \tag{20.7}$$

The last equation is referred to as the *normal equation*.[2] The way to solve the systems (20.6) and (20.7) efficiently is discussed in detail in Section 20.4.

After system (20.6) or (20.7) is solved, using formulas (20.5) we obtain the solution of (20.4). Now the maximal feasible step lengths α_P for the primal (x, z, z_u) and α_D for the dual (s, y, y_u) variables are calculated. Then these step-sizes are slightly reduced by a factor $\alpha_0 < 1$ to avoid reaching the boundary. The new iterate is computed as

$$\begin{aligned} x^{k+1} &:= x^k + \alpha_0 \alpha_P \Delta x, \\ z_u^{k+1} &:= z_u^k + \alpha_0 \alpha_P \Delta z_u, \\ z^{k+1} &:= z^k + \alpha_0 \alpha_P \Delta z, \\ s^{k+1} &:= s^k + \alpha_0 \alpha_D \Delta s, \\ y_u^{k+1} &:= y_u^k + \alpha_0 \alpha_D \Delta y_u, \\ y^{k+1} &:= y^k + \alpha_0 \alpha_D \Delta y. \end{aligned} \tag{20.9}$$

After the step, the parameter μ is updated and the process is repeated. A prototype algorithm can be summarized as follows.

---

[2] **Exercise 103** Show that if first Δy is calculated from the system (20.6) as a function of Δx the following formulas arise:

$$\Delta y = -D^{-2}(A\Delta x - r)$$

and

$$\left(A^T D^{-2} A + \bar{D}^{-2}\right) \Delta x = A^T D^{-1} r - h. \tag{20.8}$$

Observe that this symmetric formulation allows for further utilization of the structure of the normal equation. We are free to choose between (20.7) and (20.8) depending on which has a nicer sparsity structure.

Prototype Primal–Dual Algorithm

Input:
    An accuracy parameter $\varepsilon > 0$;
    (x^0, z_u^0, z^0) and (s^0, y_u^0, y^0); interior solutions for (20.1) and (20.2);
    parameter $\alpha^0 < 1$; $\mu^0 > 0$.

begin
    $x, z_u, z, s, y_u, y) = (x^0, z_u^0, z^0, s^0, y_u^0, y^0)$; $\mu := \mu^0$;
    while $(2n + m)\mu \geq \varepsilon$ **do**
    begin
        reduce μ
        solve (20.4) to obtain $(\Delta x, \Delta z_u, \Delta z, \Delta s, \Delta y_u, \Delta y)$;
        determine α_P and α_D;
        update (x, z_u, z, y_u, y, s) by (20.9)
    end
end

Before discussing all the ingredients in more detail we make an important observation. Solving a problem with individual upper bounds on the variables does not require significantly more computational effort than solving the same problem without such upper bounds. In both cases the augmented system (20.6) and the normal equation (20.7) have the same size. The extra costs per iteration arising from the upper bounds are just $\mathcal{O}(n)$, namely some extra ratio tests to determine the maximal possible steps sizes and some extra vector manipulations (see equations (20.5)).[3]

20.3 Preprocessing

An important issue for all implementations is to transform the problem into an *appropriate form*, e.g., to the canonical form with upper bounded variables (20.1), and to *reduce the problem size* in order to reach a minimal representation of the problem. This aim is quite plausible. A smaller problem needs less memory to store, usually fewer iterations of the algorithm, and if the transformation reduces the number of nonzero coefficients or improves the sparsity structure then fewer drithmetic operations per iteration are needed. A minimal representation should be free of redundancies, implied variables and inequalities. In general it is not realistic to strive to find the minimal representation of a given problem. But by analysing the structure of the problem it is often possible to reduce the problem size significantly. In fact, almost all large-scale LO problems contain redundancies in practice. The use of modeling languages and matrix generators easily allows the generation of huge models. Modelers choose to formulate models that are easy to understand and modify; this often leads to the introduction

---

[3] **Exercise 104** Check that the computational cost per iteration increases just by $\mathcal{O}(n)$ if individual upper bounds are imposed on the variables.

of superfluous variables and redundant constraints. To remove at least most of these redundancies is, however, a nontrivial task; this is the aim of preprocessing.

As we have already indicated, computationally the most expensive part of an interior-point iteration is calculating the search direction, to solve the normal equation (20.7) or the augmented system (20.6). With a compact formulation the speedup can be significant.[4]

20.3.1 Detecting redundancy and making the constraint matrix sparser

By analysing the sparsity pattern of the matrix A, one can frequently reduce the problem size. The aim of the sparsity analysis is to reduce the number of nonzero elements in the constraint matrix A; this is done by elementary matrix operations. In fact, as a consequence, the sparsity analysis mainly depends on just the nonzero structure of the matrix A and it is largely independent of the magnitude of the coefficients.

1. First we look for pairs of constraints with the same nonzero pattern. If we have two (in-)equality constraints which are identical — up to a scalar multiplier — then one of these constraints is removed from the problem. If one of them is an equality constraint and the other an inequality constraint then the inequality constraint is dropped. If they are opposite inequalities then they are replaced by one equality constraint.
2. Linearly dependent constraints are removed. (Dependency can easily be detected by using elimination.)
3. Duplicate columns are removed.
4. To improve the sparsity pattern of the constraint matrix A further we first put the constraints into equality form. Then by adding and subtracting constraints with appropriate multipliers, we can eliminate several nonzero entries.[5] During this process we have to make sure that the resulting sparser system is equivalent to the original one. Mathematically this means that we look for a nonsingular matrix $Q \in \mathbb{R}^{m \times m}$ such that the matrix QA is as sparse as possible. Such sparser constraints in the resulting equivalent formulation

$$QAx = Qb$$

might be much more suitable for a direct application of the interior-point solver.[6]

---

[4] Preprocessing is not a new idea, but has enjoyed much attention since the introduction of interior-point methods. This is due to the fact that the realized speedup is often larger than for the Simplex Method. For further reading we refer the reader to, e.g., Brearley, Mitra and Williams [49], Adler et al. [1], Lustig, Marsten and Shanno [191], Andersen and Andersen [9], Gondzio [113], Andersen [8], Bixby [42], Lustig, Marsten and Shanno [193] and Andersen et al. [10].

[5] As an illustration let us consider two constraints $a_k^T x = b_k$ and $a_j^T x = b_j$ where $\sigma(a_k) \subseteq \sigma(a_j)$. (Recall that $\sigma(x) = \{ i \mid x_i \neq 0 \}$.) Now if we define $\bar{a}_j = a_j + \lambda a_k$ and $\bar{b}_j = b_j + \lambda b_k$, where λ is chosen so that $\sigma(\bar{a}_j) \subset \sigma(a_j)$, then the constraint $a_j^T x = b_j$ can be replaced by $\bar{a}_j^T x = \bar{b}_j$ while the number of nonzero coefficients is reduced by at least one.

[6] Exact solution of this *Sparsity Problem* is an NP-complete problem (Chang and McCormick [54]) but efficient heuristics (Adler et al. [1], Chang and McCormick [54] and Gondzio [113]) usually produce satisfactory nonzero reductions in A. The algorithm of Gondzio [113], for example, looks for a row of A that has a sparsity pattern that is a subset of the sparsity pattern of other rows and uses it to eliminate nonzero elements from these rows.

20.3.2 Reducing the size of the problem

In general, finding all redundancies in an LO problem is a more difficult problem than solving the problem; hence, preprocessing procedures use a great variety of simple inspection techniques to detect obvious redundancies. These techniques are very cheap and fast, and are applied repeatedly until the problem cannot be reduced by these techniques any more. Here we discuss a small collection of commonly used reduction procedures.

1. Empty rows and columns are removed.
2. A fixed variable ($x_j = u_j$) can be substituted out of the problem.
3. A row with a single variable defines a simple bound; after an appropriate bound update the row can be removed.
4. We call variable x_j a free column singleton if it contains a single nonzero coefficient and there are neither lower nor upper bounds imposed on it. In this case the variable x_j can be substituted out of the problem. As a result both the variable x_j and the constraint in which it occurs are eliminated. The same holds for so-called *implied free variables*, i.e., for variables for which implied bounds (discussed later on) are at least as tight as the original bounds.
5. All the free variables can be eliminated by making them a free singleton column by eliminating all but one coefficient in their columns. Here we recall the techniques that were discussed in Theorem D.1 in which the LO problem was reduced to canonical form. In the elimination steps we have to pay special attention to the sparsity, by choosing elements in the elimination steps that reduce the number of nonzero coordinates in A or, at least, produce the smallest amount of new nonzero elements.
6. Trivial lower and upper bounds for each constraint i are determined. If

$$\underline{b_i} = \sum_{\{j:a_{ij}<0\}} a_{ij}b_{uj}, \quad \text{and} \quad \overline{b_i} = \sum_{\{j:a_{ij}>0\}} a_{ij}b_{uj}, \qquad (20.10)$$

then clearly

$$\underline{b_i} \leq \sum_j a_{ij}x_j \leq \overline{b_i}. \qquad (20.11)$$

Observe that due to the nonnegativity of x, for the bounds we have $\underline{b_i} \leq 0 \leq \overline{b_i}$. If the inequalities (20.11) are at least as tight as the original constraint, then the constraint i is *redundant*. If one of them contradicts the original i-th constraint, then the problem is infeasible. In some special cases (e.g.: 'less than or equal to' row with $\underline{b_i} = b_i$, 'greater than or equal to' row with $\overline{b_i} = b_i$, or equality row for which b_i is equal to one of the limits $\underline{b_i}$ or $\overline{b_i}$) the constraint in the optimization problem becomes a *forcing* one. This means that the only way to satisfy the constraint is to fix all variables that appear in it on their appropriate bounds. Then all of these variables can be substituted out of the problem.
7. From the constraint limits (20.10), implied variable bounds can be derived (remember, we have $0 \leq x \leq b_u$). Assume that for an inequality constraint the bounds (20.11) are derived. Then for each k such that $a_{ik} > 0$ we have

$$\underline{b_i} + a_{ik}x_k \leq \sum_j a_{ij}x_j \leq b_i$$

and for each k such that $a_{ik} < 0$ we have

$$\underline{b_i} + a_{ik}(x_k - u_k) \le \sum_j a_{ij} x_j \le b_i.$$

Now the new implied bounds from row i are easily derived as

$$
\begin{aligned}
x_k \le u_k' &= (b_i - \underline{b_i})/a_{ik} && \text{for all} \quad k: \; a_{ik} > 0, \\
x_k \ge l_k' &= u_k + (b_i - \underline{b_i})/a_{ik} && \text{for all} \quad k: \; a_{ik} < 0.
\end{aligned}
$$

If these bounds are tighter than the original ones, then the variable bounds are improved.

8. Apply the same techniques to the dual problem.

The application of all presolve techniques described so far often results in impressive reductions of the initial problem formulation. Once the solution for the reduced problem is found, we have to recover the complete primal and dual solutions for the original problem. This phase is called *postprocessing*.

20.4 Sparse linear algebra

As became clear in Section 20.2, the computationally most intensive part of an interior-point algorithm is to solve either the augmented system (20.6):

$$
\begin{bmatrix} D^2 & A \\ A^T & -\bar{D}^{-2} \end{bmatrix}
\begin{bmatrix} \Delta y \\ \Delta x \end{bmatrix} =
\begin{bmatrix} r \\ h \end{bmatrix}, \tag{20.12}
$$

or the normal equation (20.7):

$$
\left(A \bar{D}^2 A^T + D^2 \right) \Delta y = q, \tag{20.13}
$$

where $q = r + A\bar{D}^2 h$. At each iteration one of the systems (20.12) or (20.13) has to be solved. In the subsequent iterations only the diagonal scaling matrices D and $\bar{D}$ and the right-hand sides are changing. The nonzero structure of the augmented and normal matrices remains the same in all the iterations. For an efficient implementation it is absolutely necessary to design numerical routines that make use of this constant sparsity structure.

20.4.1 Solving the augmented system

To solve the augmented system (20.12) a well-established technique, the Bunch–Parlett factorization[7] may be used. Observe that the coefficient matrix of (20.12) is nonsingular, symmetric and indefinite. The Bunch–Parlett factorization of the symmetric indefinite matrix has the form

$$
P \begin{bmatrix} D^2 & A \\ A^T & -\bar{D}^{-2} \end{bmatrix} P^T = L \Lambda L^T, \tag{20.14}
$$

---

[7] For the original description of the algorithm we refer to Bunch and Parlett [53]. For further application to solving least squares problems we refer the reader to Arioli, Duff and de Rijk [27], Björk [44] and Duff [68].

for some permutation matrix P, where Λ is an indefinite block diagonal matrix with 1×1 and 2×2 blocks and L is a lower triangular matrix. The factorization is basically an elimination (Gaussian) algorithm, in which we have to specify at each stage which row and which column is used for the purpose of elimination.

In the Bunch–Parlett factorization, to produce a sparse and numerically stable L and Λ at each iteration the system is dynamically analyzed. Thus it may well happen that at each iteration structurally different factors are generated. This means that in the choice of the element that is used for the elimination, both the sparsity and stability of the triangular factor are considered. Within these stability and sparsity considerations we have a great deal of freedom in this selection; we are not restricted to the diagonal elements (one possible trivial choice) of the coefficient matrix. The efficiency depends strongly on the heuristics used in the selection strategy.

The relatively expensive so-called *analyze phase* is frequently skipped and the same structure is reused in subsequent iterations and updated only occasionally when the numerical properties make it necessary. A popular selection rule is detecting 'dense' columns and rows (with many nonzero coefficients) and eliminating first in the diagonal positions of D^2 and $\bar{D}^{-2}$ in the augmented matrix (20.12) corresponding to sparse rows and columns. The dense structure is pushed to the last stage of factorization as a dense window. In general it is unclear what threshold density should be used to separate dense and sparse structures. When the number of nonzeros in dense columns is significantly larger than the average number of entries in sparse columns then it is easy to determine a fixed threshold value. Whenever more complicated sparsity structures appear, more sophisticated heuristics are needed.[8]

20.4.2 Solving the normal equation

The other popular method for calculating the search direction is to solve the normal equation (20.13). The method of choice in this case is the sparse Cholesky factorization:

$$\bar{P}\left(A\bar{D}^2 A^T + D^2\right)\bar{P}^T = \bar{L}\bar{\Lambda}\bar{L}^T, \qquad (20.15)$$

for some permutation matrix $\bar{P}$, where $\bar{L}$ is a lower triangular matrix and $\bar{\Lambda}$ is a positive definite diagonal matrix. It is should be clear from the derivation of the normal equation that the normal equation approach can be considered as a special implementation of the augmented system approach. More concretely this means that we first eliminate either Δx or Δy by using all the diagonal entries of either $\bar{D}^{-2}$ or D^2. Thus the normal equation approach is less flexible but, on the other hand, the coefficient matrix to be factorized is symmetric positive definite, and both the matrix and its factors have a constant sparsity structure.

The Cholesky factorization of (20.15) exists for any positive D^2 and $\bar{D}^2$. The sparsity structure of $\bar{L}$ is independent of these diagonal matrices and hence is constant in all iterations if the same elimination steps are performed. Consequently it is sufficient to analyze the structure just once and determine a good ordering of the rows and

---

[8] To discuss these heuristics is beyond the scope of this chapter. The reader can find detailed discussion of the advantages and disadvantages of the normal equation approach in the next section and in the papers Andersen et al. [10], Duff et al. [69], Fourer and Mehrotra [78], Gondzio and Terlaky [116], Maros and Mészáros [195], Turner [275], Vanderbei [277] and Vanderbei and Carpenter [278].

columns in order to obtain sparse factors. To determine such an ordering involves considerable computational effort, but it is the basis of a successful implementation of the Cholesky factorization in interior-point methods. This is the *analyze phase*. More formally, we have to find a permutation matrix P such that the Cholesky factor of $P(A\bar{D}^2 A^T + D^2)P^T$ is the sparsest possible. Due to the difficulty of this problem, heuristics are used in practice to find such a good permutation.[9] Two efficient heuristics, namely the *minimum degree* and the *minimum local fill-in* orderings, are particularly useful in interior-point method implementations. These heuristics are described briefly below.

Minimum degree ordering

Since the matrix to be factorized is positive definite and symmetric the elimination can be restricted to the diagonal elements. This limitation preserves the symmetry and positive definiteness of the Schur complement. Let us assume that in the k-th step of the Gaussian elimination the i-th row of the Schur complement contains n_i nonzero entries. If this row is used for the elimination, then the elimination requires

$$f_i = \frac{1}{2}(n_i - 1)^2, \tag{20.16}$$

floating point operations (*flops*). The number f_i estimates the computational effort and gives an overestimate of the fill-in that can result from the elimination. The best choice of row i at step k is the one that minimizes f_i.[10]

Minimum local fill-in ordering

Let us observe that, in general, f_i in (20.16) considerably overestimates the number of fill-ins at a given iteration of the elimination process because it does not take into account the fact that in many positions of the predicted fill-in, nonzero entries already exist. It is possible that another candidate that seems to be worse in terms of (20.16) would produce less fill-in because in the elimination, mainly existing nonzero entries would be updated. The minimum local fill-in ordering takes locally the real fill-in into account. As a consequence, each step is more expensive but the resulting factors are sparser. This higher cost has to be paid once in the analyze phase.

Disadvantages of the normal equations approach

The normal equations approach shows uniformly good performance when applied to the solution of the majority of linear programs. Unfortunately, it suffers from a serious drawback. The presence of dense columns in A might be catastrophic if they are not treated with extra care. A dense column of A with k nonzero elements creates a $k \times k$ dense submatrix (dense window) of the normal matrix (20.13). Such dense columns do not seriously influence the efficiency of the augmented system approach.

---

[9] Yannakakis [302] proved that finding the optimal permutation is an NP-complete problem.

[10] The function f_i is Markowitz's merit function [194]. Interpreting this process in terms of the elimination graph (cf. George and Liu [94]), we can see that it is equivalent to the choice of the node in the graph that has the minimum degree (this gave the name to this heuristic).

In order to handle dense columns efficiently the first step is to identify them. This typically means to chose a threshold value. If the number of nonzeros in a column is larger than this threshold, the column is considered to be dense, the remaining columns as sparse. Denoting the matrix of the sparse columns in A by A_s and the matrix of the dense columns by A_d, the equation (20.12) can be written as follows.

$$\begin{bmatrix} D^2 & A_d & A_s \\ A_d^T & -\bar{D}_d^{-2} & 0 \\ A_s^T & 0 & -\bar{D}_s^{-2} \end{bmatrix} \begin{bmatrix} \Delta y \\ \Delta x_d \\ \Delta x_s \end{bmatrix} = \begin{bmatrix} r \\ h_d \\ h_s \end{bmatrix}. \tag{20.17}$$

After eliminating $\Delta x_s = -\bar{D}_s^{-2}(h_s - A_s^T d_y)$ we get the equation

$$\begin{bmatrix} D^2 + A_s\bar{D}_s^{-2}A_s^T & A_d \\ A_d^T & -\bar{D}_d^{-2} \end{bmatrix} \begin{bmatrix} \Delta y \\ \Delta x_d \end{bmatrix} = \begin{bmatrix} r + A_s\bar{D}_s^{-2}h_s \\ h_d \end{bmatrix}. \tag{20.18}$$

Here the left-upper block of the coefficient matrix is positive definite symmetric and sparse, thus it is easy to factorize efficiently. As the reader can easily see, this approach tries to combine the advantages of the normal equation approach and the augmented system approach.[11,12]

20.4.3   Second-order methods

An attempt to reduce the computational cost of interior-point methods is based on trying to reuse the same factorization of either the normal matrix or the augmented system. Both in theory and in practice, factorization is much more expensive than backsolve of triangular systems; so we can do additional backsolves in each iteration with different right-hand sides if these reduce the total number of interior-point iterations. This is the essential idea of *higher-order methods*. Our discussion here follows the present computational practice; so we consider only the second-order

---

[11] An appealing advantage of the symmetric formulation of the LO problem is that in (20.18) the matrix $D^2 + A_s\bar{D}_s^{-2}A_s^T$ is nonsingular. If one would use the standard $Ax = b$, $x \geq 0$ form, then we would have just $A_s\bar{D}_s^{-2}A_s^T$ which might be singular. To handle this unpleasant situation an extra trick is needed. For this we refer the reader to Andersen et al. [13] and also to Exercise 105.

[12] **Exercise 105** Verify that $(\Delta y, \Delta x_d)$ is the solution of

$$\begin{bmatrix} A_s\bar{D}_s^{-2}A_s^T & A_d \\ A_d^T & -\bar{D}_d^{-2} \end{bmatrix} \begin{bmatrix} \Delta y \\ \Delta x_d \end{bmatrix} = \begin{bmatrix} r + A_s\bar{D}_s^{-2}h_s \\ h_d \end{bmatrix}$$

if and only if $(\Delta y, \Delta x_d, u)$ solves

$$\begin{bmatrix} A_s\bar{D}_s^{-2}A_s^T + QQ^T & A_d & Q \\ A_d^T & -\bar{D}_d^{-2} & 0 \\ Q^T & 0 & I \end{bmatrix} \begin{bmatrix} \Delta y \\ \Delta x_d \\ u \end{bmatrix} = \begin{bmatrix} r + A_s\bar{D}_s^{-2}h_s \\ h_d \\ 0 \end{bmatrix}$$

with any matrix Q having appropriate dimension. Observe that by choosing Q properly (e.g. diagonal) the matrix $A_s\bar{D}_s^{-2}A_s^T + QQ^T$ is nonsingular.

predictor-corrector method that is implemented in several codes with great success.[13]

Predictor-corrector technique

This predictor-corrector method has two components. The first is an adaptive choice of the barrier parameter μ; the other is a second-order approximation of the central path .

The first step in the predictor-corrector algorithm is to compute the primal-dual affine-scaling (predictor) direction. This is the solution of the Newton system (20.4) with $\mu = 0$ and is indicated by Δ^a. It is easy to see that if a step of size α is taken in the affine-scaling direction, then the duality gap is reduced by α; i.e. if a large step can be made in this direction then significant progress is made in the optimization. If the feasible step-size in the affine-scaling direction is small, we expect that the current point is close to the boundary; thus centering is needed and μ should not be reduced too much.

In the predictor-corrector algorithm, first the predicted duality gap is calculated that results from a step along the primal-dual affine-scaling direction. To this end, when the affine-scaling direction is computed, the maximum primal (α_P^a) and dual (α_D^a) feasible step sizes are determined that preserve nonnegativity of (x, z_u, z) and (s, y_u, y). Then the predicted duality gap

$$
\begin{aligned}
g_a \quad = \quad & (x + \alpha_P^a \Delta^a x)^T (s + \alpha_D^a \Delta^a s) + (z_u + \alpha_P^a \Delta^a z_u)^T (y_u + \alpha_D^a \Delta^a y_u) \\
& + (z + \alpha_P^a \Delta^a z)^T (y + \alpha_D^a \Delta^a y)
\end{aligned}
$$

is computed and is used to determine a target point

$$
\mu = \left(\frac{g_a}{g}\right)^2 \frac{g_a}{n} \tag{20.19}
$$

on the central path . Here g_a/n relates to the central point with the same duality gap that the predictor affine step would produce, and the factor $(g_a/g)^2$ pushes the target further towards optimality in a way that depends on the achieved reduction of the predictor step. Now the second-order component of the predictor-corrector direction is computed. Ideally we would like to compute a step such that the next iterate is perfectly centered, i.e.,

$$
\begin{aligned}
(x + \Delta x)(s + \Delta s) \quad &= \quad \mu e, \\
(z_u + \Delta z_u)(y_u + \Delta y_u) \quad &= \quad \mu e, \\
(z + \Delta z)(y + \Delta y) \quad &= \quad \mu e,
\end{aligned}
$$

---

[13] The second-order predictor-corrector technique presented here is due to Mehrotra [205]; from a computational point of view the method is very successful. The higher than order 2 methods — discussed in Chapter 18 — are implementable too, but to date computational results with methods of order higher than 2 are quite disappointing. See Andersen et al. [10]. Mehrotra was motivated by the paper of Monteiro, Adler and Resende [220], who were the first to introduce the primal-dual affine-scaling direction and higher-order versions of the primal-dual affine-scaling direction; they elaborated on a computational paper of Adler, Karmarkar, Resende and Veiga [2] that uses the dual affine-scaling direction and higher-order versions of it.

or equivalently

$$\begin{aligned}
x\Delta s + s\Delta x &= -xs + \mu e - \Delta x\Delta s, \\
z_u\Delta y_u + y_u\Delta z_u &= -z_u y_u + \mu e - \Delta z_u\Delta y_u, \\
z\Delta y + y\Delta z &= -zy + \mu e - \Delta z\Delta y.
\end{aligned}$$

Usually, in the computation of the Newton direction the second-order terms

$$\Delta x\Delta s, \ \Delta z_u\Delta y_u, \ \Delta z\Delta y$$

are neglected (recall (20.4)). Instead of neglecting the second-order term, the affine directions

$$\Delta^a x, \Delta^a s, \ \Delta^a z_u\Delta^a y_u, \ \Delta^a z\Delta^a y$$

are used as the predictions of the second-order effect. One step of the algorithm can be summarized as follows.

- Solve (20.4) with $\mu = 0$, resulting in the affine step $(\Delta^a x, \Delta^a z_u, \Delta^a z)$ and $(\Delta^a s, \Delta^a y_u, \Delta^a y)$.
- Calculate the maximal feasible step lengths α_P^a and α_D^a.
- Calculate the predicted duality gap g_a and μ by (20.19).
- Solve the corrected Newton system

$$\begin{aligned}
A\Delta x - \Delta z &= 0 \\
\Delta x + \Delta z_u &= 0 \\
A^T\Delta y - \Delta y_u + \Delta s &= 0 \\
x\Delta s + s\Delta x &= -xs + \mu e - \Delta^a x\Delta^a s, \\
z_u\Delta y_u + y_u\Delta z_u &= -z_u y_u + \mu e - \Delta^a z_u\Delta^a y_u, \\
z\Delta y + y\Delta z &= -zy + \mu e - \Delta^a z\Delta^a y.
\end{aligned} \qquad (20.20)$$

- Calculate the maximal feasible step lengths α_P and α_D and make a damped step by using (20.9).[14]

Finally, observe that a single iteration of this second-order predictor-corrector primal-dual method needs two solves of the same large, sparse linear system (20.4) and (20.20) for two different right-hand sides. Thus the same factorization can be used twice.

20.5 Starting point

The self-dual embedding problem is an elegant theoretical construction for handling the starting point problem. At the same time it can also be the basis of an efficient implementation. In this section we show that solving the slightly larger embedding

---

[14] This presentation of the algorithm follows the paper of Mehrotra [205]. It differs from the 2-order method of Chapter 18.

problem does not increase the computational cost significantly.[15] Before presenting the embedding problem, we summarize some of its surprisingly nice properties.

1. The embedding problem is self-dual: the dual problem is identical to the primal one.
2. It is always feasible. Furthermore, the interior of the feasible set of the embedding problem is also nonempty; hence the optimal faces are bounded (from Theorem II.10). So interior-point methods always converge to an optimal solution.
3. Optimality of the original problem is detected by convergence, independently of the boundedness or unboundedness of the optimal faces of the original problem.
4. Infeasibility of the original problem is detected by convergence as well.[16] Primal, dual or primal and dual rays for the original problems are identified to prove dual, primal or dual and primal infeasibility (cf. Theorem I.26).
5. For the embedding problem a perfectly centered initial pair can always be constructed.
6. It allows an elegant handling of the warm start problem.
7. The embedding problem can be solved with any method that generates a strictly complementary solution; if the chosen method is polynomial, it solves the original problem with essentially the same complexity bound. Thus we can achieve the best possible complexity bounds for solving an arbitrary problem.

Self-dual embedding

We consider problems (20.1) and (20.2). To formulate the embedding problem we need to introduce some further vectors in a way similar to that of Chapter 2. We start with

$$x^0 > 0, \; z_u^0 > 0, \; z^0 > 0, \; s^0 > 0, \; y_u^0 > 0, \; y^0 > 0, \; \kappa^0 > 0, \; \vartheta^0 > 0, \; \rho^0 > 0, \; \nu^0 > 0,$$

where $x^0, z_u^0, s^0, y_u^0 \in \mathbb{R}^n$, $y^0, z^0 \in \mathbb{R}^m$ and $\kappa^0, \vartheta^0, \rho^0, \nu^0 \in \mathbb{R}$ are arbitrary. Then we define $\bar{b} \in \mathbb{R}^m$, $\bar{b}_u, \bar{c} \in \mathbb{R}^n$, the scaled error at the arbitrary initial interior solutions (recall the construction in Section 4.3), and parameters $\beta, \gamma \in \mathbb{R}$ as follows:

$$\bar{b}_u = \frac{1}{\vartheta^0}(b_u \kappa^0 - x^0 - z_u^0)$$

$$\bar{b} = \frac{1}{\vartheta^0}(b\kappa^0 - Ax^0 + z^0)$$

$$\bar{c} = \frac{1}{\vartheta^0}(c\kappa^0 + y_u^0 - A^T y^0 - s^0)$$

$$\beta = \frac{1}{\vartheta^0}(c^T x^0 - b^T y^0 + b_u^T y_u + \rho^0)$$

---

[15] Such embedding was first introduced by Ye, Todd and Mizuno [316] using the standard form problems (20.29) and (20.30). They discussed most of the advantages of this embedding and showed that Mizuno, Todd and Ye's [217] predictor-corrector algorithms solve the LO problem in $\mathcal{O}(\sqrt{n}L)$ iterations, yielding the first infeasible IPM with this complexity. Somewhat later Jansen, Roos and Terlaky [155] presented the self-dual problem for the symmetric form primal-dual pair in a concise introduction to the theory of LO based on IPMs.

[16] The popular so-called infeasible-start methods detect unboundedness or infeasibility of the original problem by divergence of the iterates.

$$\gamma \quad = \quad \beta\kappa^0 + \bar{b}^T y^0 - \bar{b}_u y_u^0 - \bar{c}^T x^0 + \nu^0$$

$$= \quad \frac{1}{\vartheta^0}[(x^0)^T s^0 + (y_u^0)^T z_u^0 + (y^0)^T z^0 + \kappa^0 \rho^0] + \nu^0 > 0.$$

It is worth noting that if x^0 is strictly feasible for (20.1), $\kappa^0 = 1$, $z^0 = Ax^0 - b$ and $z_u^0 = b_u - x^0$, then $\bar{b} = 0$ and $\bar{b}_u = 0$. Also if (y^0, y_u^0) is strictly feasible for (20.2), $\kappa^0 = 1$ and $s^0 = c - A^T y + y_u^0$, then $\bar{c} = 0$. In some sense the vectors $\bar{b}$, $\bar{b}_u$ and $\bar{c}$ measure the amount of scaled infeasibility of the given vectors x^0, z^0, z_u^0, s^0, y^0, y_u^0.

Now consider the following self-dual LO problem:

(SP)   min $\gamma\vartheta$

s.t.
$$
\begin{array}{cccccc}
 & -x & +b_u\kappa & -\bar{b}_u\vartheta & \geq 0 \\
 & Ax & -b\,\kappa & +\bar{b}\,\vartheta & \geq 0 \\
y_u & -A^T y & +c\,\kappa & -\bar{c}\,\vartheta & \geq 0 \\
-b_u^T y_u & +b^T y & -c^T x & +\beta\,\vartheta & \geq 0 \\
\bar{b}_u^T y_u & -\bar{b}^T y & +\bar{c}^T x & -\beta\,\kappa & \geq -\gamma \\
\end{array}
$$
$$y_u \geq 0, \quad y \geq 0, \quad x \geq 0, \quad \kappa \geq 0, \quad \vartheta \geq 0.$$

Let us denote the slack variables for the problem (SP) by z_u, z, s, ν and ρ respectively. By construction the positive solution $x = x^0$, $z = z^0$, $z_u = z_u^0$, $s = s^0$, $y = y^0$, $y_u = y_u^0$, $\kappa = \kappa^0$, $\vartheta = \vartheta^0$, $\nu = \nu^0$, $\rho = \rho^0$ is interior feasible for problem (SP). Also note that if, e.g., we choose $x = x^0 = e$, $z = z^0 = e$, $z_u = z_u^0 = e$, $s = s^0 = e$, $y = y^0 = e$, $y_u = y_u^0 = e$, $\kappa = \kappa^0 = 1$, $\vartheta = \vartheta^0 = 1$, $\nu = \nu^0 = 1$, $\rho = \rho^0 = 1$, then this solution with $\mu = 1$ is a perfectly centered initial solution for problem (SP). The following theorem follows easily in the same way as Theorem I.26.[17]

Theorem IV.85 *The embedding (SP) of the given problems (20.1) and (20.2) has the following properties:*

(i) *The self-dual problem (SP) is feasible and hence both primal and dual feasible. Thus it has an optimal solution.*

(ii) For any optimal solution of (SP), $\vartheta^* = 0$.

(iii) *(SP) always has a strictly complementary optimal solution $(y_u^*, y^*, x^*, \kappa^*, \vartheta^*)$.*

(iv) *If $\kappa^* > 0$, then x^*/κ^* and $(y^*, y_u^*)/\kappa^*$ are strictly complementary optimal solutions of (20.1) and (20.2) respectively.*

(v) *If $\kappa^* = 0$, then either (20.1) or (20.2) or both are infeasible.*

Solving the embedding model needs just slightly more computation per iteration than solving problem (20.1). This small extra effort is the cost of having several important advantages: having a centered initial starting point, detecting infeasibility by convergence, applicability of any IPM without degrading theoretical complexity. The rest of this section is devoted to showing that the computation of the Newton direction for the embedding problem (SP) reduces to almost the same sized augmented (20.6) or normal equation (20.7) systems as in the case of (20.1).

---

[17] **Exercise 106** Prove this theorem.

In Chapter 3 the self-dual problem

$$(SP) \qquad \min \left\{ \tilde{q}^T \tilde{x} \ : \ M\tilde{x} \geq -\tilde{q}, \ \tilde{x} \geq 0 \right\},$$

was solved, where M is of size $n \times n$ and skew-symmetric and $\tilde{q} \in \mathbb{R}_+^n$. Given an initial positive solution $(\tilde{x}, \tilde{s}) > 0$, where $\tilde{s} = M\tilde{x} + \tilde{q}$, a Newton step for problem (SP) with a value $\mu > 0$ was given as

$$\widetilde{\Delta s} = M\widetilde{\Delta x},$$

where $\widetilde{\Delta x}$ is the solution of the system

$$(M + \tilde{X}^{-1}\tilde{S})\widetilde{\Delta x} = \mu \tilde{x}^{-1} - \tilde{s}. \tag{20.21}$$

Now we have to analyze how the positive definite system (20.21) can be efficiently solved in the case of problem (SP). For this problem we have $\tilde{x} = (y_u, y, x, \kappa, \vartheta)$, $\tilde{s} = (z_u, z, s, \nu, \rho)$ and

$$M = \begin{bmatrix} 0 & 0 & -I & b_u & -\bar{b}_u \\ 0 & 0 & A & -b & \bar{b} \\ I & -A^T & 0 & c & -\bar{c} \\ -b_u^T & b^T & -c^T & 0 & \beta \\ \bar{b}_u^T & -\bar{b}^T & \bar{c}^T & -\beta & 0 \end{bmatrix} \quad \text{and} \quad \tilde{q} = \begin{bmatrix} 0 \\ 0 \\ 0 \\ 0 \\ \gamma \end{bmatrix}.$$

Hence the Newton equation (20.21) can be written as

$$\begin{bmatrix} Y_u^{-1} Z_u & 0 & -I & b_u & -\bar{b}_u \\ 0 & Y^{-1}Z & A & -b & \bar{b} \\ I & -A^T & X^{-1}S & c & -\bar{c} \\ -b_u^T & b^T & -c^T & \frac{\nu}{\kappa} & \beta \\ \bar{b}_u^T & -\bar{b}^T & \bar{c}^T & -\beta & \frac{\rho}{\vartheta} \end{bmatrix} \begin{bmatrix} \Delta y_u \\ \Delta y \\ \Delta x \\ \Delta \kappa \\ \Delta \vartheta \end{bmatrix}] = \begin{bmatrix} \mu y_u^{-1} - z_u \\ \mu y^{-1} - z \\ \mu x^{-1} - s \\ \mu \frac{1}{\kappa} - \nu \\ \mu \frac{1}{\vartheta} - \rho \end{bmatrix}. \tag{20.22}$$

From the first and the second equation it easily follows that

$$\Delta y_u = Y_u Z_u^{-1}(\Delta x - b_u \Delta\kappa + \bar{b}_u \Delta\vartheta + \mu y_u^{-1} - z_u)$$

and

$$\Delta y = YZ^{-1}(-A\Delta x + b\Delta\kappa - \bar{b}\Delta\vartheta + \mu y^{-1} - z).$$

We simplify the notation by introducing

$$W_u := Z_u^{-1} Y_u.$$

Then, by substituting the value of Δy_u in (20.22) we find[18]

$$
\begin{bmatrix}
Y^{-1}Z & A & -b & \bar{b} \\
-A^T & X^{-1}S + W_u & c - W_u b_u & -\bar{c} + W_u \bar{b}_u \\
b^T & -c^T - b_u^T W_u & \frac{\nu}{\kappa} + b_u^T W_u b_u & \beta - b_u^T W_u \bar{b}_u \\
-\bar{b}^T & \bar{c}^T + \bar{b}_u^T W_u & -\beta - \bar{b}_u^T W_u b_u & \frac{\rho}{\vartheta} + \bar{b}_u^T W_u \bar{b}_u
\end{bmatrix}
\begin{bmatrix}
\Delta y \\
\Delta x \\
\Delta \kappa \\
\Delta \vartheta
\end{bmatrix}
=
\begin{bmatrix}
r_1 \\
r_2 \\
r_3 \\
r_4
\end{bmatrix}, \quad (20.23)
$$

where for simplicity the right-hand side elements are denoted by $r_1, \ldots, r_4$. Now if we multiply the second block of equations (corresponding to the right-hand side r_2) in (20.23) by -1, a system analogous to the augmented system (20.6) of problem (20.1) is obtained. The difference is that here we have two additional constraints and variables. For the solution of this system, the factorization of the matrix may happen in the same way, but the last two rows and columns (these are typically dense) should be left to the last two steps of the factorization. A 2×2 dense window for $(\Delta \kappa, \Delta \vartheta)$ then remains.

If we further simplify (20.23) by substituting the value of Δy, the analogue to the normal equation system of the problem (SP) is produced. For simplicity the scalars here are denoted by $\eta_1, \ldots, \eta_8$ and r_5, r_6, r_7.[19],[20]

$$
\begin{bmatrix}
A^T Z^{-1} Y A + X^{-1} S + Z_u^{-1} Y_u & \eta_1 & \eta_2 \\
\eta_3 & \eta_4 & \eta_5 \\
\eta_6 & \eta_7 & \eta_8
\end{bmatrix}
\begin{bmatrix}
\Delta x \\
\Delta \kappa \\
\Delta \vartheta
\end{bmatrix}
=
\begin{bmatrix}
r_5 \\
r_6 \\
r_7
\end{bmatrix}. \quad (20.24)
$$

---

[18] **Exercise 107** Verify that

$$
\begin{bmatrix}
r_1 \\
r_2 \\
r_3 \\
r_4
\end{bmatrix}
=
\begin{bmatrix}
\mu y^{-1} - z \\
\mu x^{-1} - s - (\mu z_u^{-1} - y_u) \\
\mu \frac{1}{\kappa} - \nu + b_u^T (\mu z_u^{-1} - y_u) \\
\mu \frac{1}{\vartheta} - \rho - \bar{b}_u^T (\mu z_u^{-1} - y_u)
\end{bmatrix}.
$$

[19] **Exercise 108** Verify that

$$
\begin{aligned}
\eta_1 &= c - W_u b_u - A^T Z^{-1} Y b \\
\eta_2 &= -\bar{c} + W_u \bar{b}_u + A^T Z^{-1} Y \bar{b} \\
\eta_3 &= -c^T - b_u^T W_u - b^T Z^{-1} Y A \\
\eta_4 &= \nu \kappa^{-1} + b_u^T W_u b_u + b^T Z^{-1} Y b \\
\eta_5 &= \beta - b_u^T W_u \bar{b}_u - b^T Z^{-1} Y \bar{b} \\
\eta_6 &= \bar{c}^T + \bar{b}_u^T W_u + \bar{b}^T Z^{-1} Y A \\
\eta_7 &= -\beta - \bar{b}_u^T W_u b_u - \bar{b}^T Z^{-1} Y b \\
\eta_8 &= \rho \vartheta^{-1} + \bar{b}_u^T W_u \bar{b}_u + \bar{b}^T Z^{-1} Y \bar{b}
\end{aligned}
$$

and

$$
\begin{aligned}
r_5 &= \mu x^{-1} - s - (\mu z_u^{-1} - y_u) + A^T (\mu z^{-1} - y) \\
r_6 &= \mu \kappa^{-1} - \nu + b_u^T (\mu z_u^{-1} - y_u) - b^T (\mu z^{-1} - y) \\
r_7 &= \mu \vartheta^{-1} - \rho - \bar{b}_u^T (\mu z_u^{-1} - y_u) + \bar{b}^T (\mu z^{-1} - y).
\end{aligned}
$$

[20] **Exercise 109** Develop similar formulas for the normal equation if Δx is eliminated instead of Δy. Compare the results with (20.7) and (20.8).

20.5.1 Simplifying the Newton system of the embedding model

As mentioned with respect to the augmented system, we easily verify that the difference between the normal equations of problem (20.1) and the embedding problem (SP) is that here two additional constraints and variables are present. Note that the last two rows and columns in (20.23) and (20.24) are neither symmetric nor skew-symmetric. The reader might think that these two extra columns deteriorate the efficiency of the algorithm (it requires two additional back-solves for the computation of the Newton direction) and hence make the embedding approach less attractive in practice. However, the computational cost can easily be reduced by a simple observation. First, note that for any interior solution $(y_u, y, , x, \kappa, \vartheta)$ the duality gap (see also Exercise 10 on page 35) is equal to

$$2\gamma\vartheta.$$

Second, remember that in Lemma II.47 we have proved that in a primal-dual method the target duality gap is always reached after a full Newton step. Since the duality gap on the central path with the value μ equals to

$$2(m + 2n + 2)\mu$$

and thus, the target duality gap is determined by the target value $\mu^+ = (1 - \theta)\mu$, the step $\Delta\vartheta$ can directly be calculated.

$$\Delta\vartheta = \vartheta^+ - \vartheta = \frac{\mu^+ - \mu}{\gamma}(m + 2n + 2) = \frac{\theta\mu}{\gamma}(m + 2n + 2)$$

As a result we conclude that the value of $\Delta\vartheta$ in (20.24) is known, thus it can simply be substituted in the Newton system and the system (20.23) reduces to almost the original size. This simplification allows to implement IPMs based on the self-dual embedding model efficiently, the cost per iteration is only one extra back-solve.

20.5.2 Notes on warm start

Many practical problems need the solution of a sequence of similar linear programs where small perturbations are made to b and/or c (possibly also in A). As long as these perturbations are small, we naturally expect that the optimal solutions are not far from each other and restarting the optimization from the solution of the old problem (warm start) should be more efficient than solving the problem from scratch.[21]

The difficulty in the IPM warm start comes from the fact that the old optimal solution is very close to the boundary (this is a necessity since all optimal solutions in an LO problem are on the boundary of the feasible set) and well centered. This point, in the perturbed problem, still remains close to the boundary or becomes infeasible, but even if it remains feasible it is very poorly centered. Consequently, the IPM makes a long sequence of short steps because the iterates cannot get away from the boundary. Therefore for an efficient warm start we need a well-centered point close to

---

[21] Some early attempts to solve such problems are due to Freund [84] who uses shifted barriers, and Polyak [234] who applies modified barrier functions. For further elaboration of the literature see, e.g., Lustig, Marsten and Shanno [193], Gondzio and Terlaky [116] and Andersen et al. [10].

the old optimal one or an efficient centering method (to get far from the boundary) to overcome these difficulties. These two possibilities are discussed briefly below.

Independent of the approach chosen it would be wise to save a well-centered almost optimal solution (say, with 10^{-2} relative duality gap) that is still sufficiently far away from the boundary.

- **Centered solutions for warm start in (SP) embedding.** Among the spectacular properties of the (SP) embedding listed in the previous section, the ability to construct always perfectly centered initial interior points was mentioned. The old well-centered almost optimal solution x^*, z^*, z_u^*, s^*, y^*, y_u^*, κ^*, ϑ^*, ρ^*, ν^* can be used as the initial point for embedding the perturbed problem. As we have seen in Section 20.5, $\bar{b}, \bar{c}, \beta$ and γ can always be redefined so that the above solution stays well centered. The construction allows simultaneous perturbations of b, b_u, c and even the matrix A. Additionally, it extends to handling new constraints or variables added to the problem (e.g., in buildup or cutting plane schemes). In these cases, we can keep the solution for the old coordinates (let μ be the actual barrier parameter) and set the initial value of the new complementary variables equal to $\sqrt{\mu}$. This results in a perfectly centered initial solution.

- **Efficient centering.** If the old solution remains feasible, but is badly centered, we might proceed with this solution without making a new embedding. The common approach is to use a path-following method for the recentering process; it uses targets on the central path . Because of the weak performance of Newton's method far off the central path, this approach is too optimistic for a warm start. The target-following method discussed in Part III (Section 11.4) offers much more flexibility in choosing achievable targets, thus leading to efficient ways of centering. A target sequence that improves centrality allows larger steps and therefore speeds up the centering and, as a consequence, the optimization process.[22]

20.6 Parameters: step-size, stopping criteria

20.6.1 Target-update

The easiest way to ensure that all iterates remain close to the central path is to decrease μ by a very small amount at each iteration. This provides the best theoretical worst-case complexity, as we have seen in discussing *full Newton step methods*. These methods demonstrate hopelessly slow convergence in practice and their theoretical worst-case complexity is identical to their practical performance.

In *large-update methods* the barrier parameter is reduced much faster than the theory suggests. To preserve polynomial convergence of these methods in theory, several Newton steps are computed between two reductions of μ (update of the target) until the iterate is in a sufficiently small neighborhood of the central path . In practice this multistep strategy is ignored and at each reduction of μ, at each target-update, only one Newton step is made. A drawback of this strategy is that the iterates might get

---

[22] Computational results based on centering target sequences are presented in Gondzio [114] and Andersen et al. [10].

far away from the central path or from the target point, and the efficiency of the Newton method might deteriorate. A careful strategy for updating μ and for step-length selection reduces the danger of this negative scenario.

At an interior iterate the current duality gap is given by

$$g = x^T s + z_u^T y_u + z^T y,$$

which is equal to $(2n + m)\mu$ if the iterate is on the central path . The central point with the same duality gap as the current iterate belongs to the value

$$\mu = \frac{x^T s + z_u^T y_u + z^T y}{2n + m}.$$

The target μ value is chosen so that the target duality gap is significantly smaller, but does not put the target too far away. Thus we take

$$\mu_{\text{new}} = (1 - \theta) \frac{x^T s + z_u^T y_u + z^T y}{2n + m}, \tag{20.25}$$

where $\theta \in [0, 1]$. The value $\theta = 0$ corresponds to pure centering, while $\theta < 1$ aims to reduce the duality gap. A solid but still optimistic update is $\theta = \frac{3}{4}$.[23]

20.6.2 Step size

Although there is not much supporting theory, current implementations use very large and different step-sizes in the primal and dual spaces.[24] All implementations use a variant of the following strategy. First the maximum possible step-sizes are computed:

$$\alpha_P \quad := \max\left\{\alpha > 0 : (x, z, z_u) + \alpha(\Delta x, \Delta z, \Delta z_u) \geq 0\right\},$$

$$\text{and} \quad \alpha_D \quad := \max\left\{\alpha > 0 : (s, y, y_u) + \alpha(\Delta s, \Delta y, \Delta y_u) \geq 0\right\},$$

and these step-sizes are slightly reduced by a factor $\alpha_0 = 0.99995$ to ensure that the new point is strictly positive. Although this aggressive, i.e. very large, choice of α_0 is frequently reported to be the best, we must be careful and include a safeguard to handle the case when $\alpha_0 = 0.99995$ turns out to be too aggressive.

20.6.3 Stopping criteria

Interior point algorithms terminate when the duality gap is small enough and the current solution is feasible for the original problems (20.1) and (20.2), or when the

---

[23] In the published literature, iteration counts larger than 50 almost never occur and most frequently iteration numbers around 20 are reported. Taking this number as a target iteration count and assuming that (in contrast to the theoretical worst-case analysis) Newton's method provides iterates always close to the target point, we can calculate how large the target-update (how small $(1 - \theta)$) should be to reach the desired accuracy within the required number of iterations. Thus, for a problem with 10^4 variables and a centered initial solution with $\mu = 1$ and aiming for a solution with 8 digits of accuracy, we have to reduce the duality gap by a factor of 10^{12} in 20 iterations. By straightforward calculation we can easily verify that the value $\theta = \frac{3}{4}$ is an appropriate choice for this purpose.

[24] Kojima, Megiddo and Mizuno [174] proved global convergence of a primal-dual method that allows such large step-sizes in most iterations.

infeasibility is small enough. The practical tolerances are larger than the theoretical bounds that guarantee identification of an exact solution; this is a common drawback of all numerical algorithms for solving LO problems. To obtain a sensible solution the duality gap and the measure of infeasibility should be related to the problem data. Relative primal infeasibility is related to the length of the vectors b and b_u, dual infeasibility is related to the length of the vector c, and the duality gap is related to the actual objective value. A solution with p digits relative accuracy is guaranteed by the stopping criteria presented here:

$$\frac{||Ax - z - b||}{1 + ||b||} \leq 10^{-p} \quad \text{and} \quad \frac{||x + z_u - b_u||}{1 + ||b_u||} \leq 10^{-p}, \tag{20.26}$$

$$\frac{||c - A^T y + y_u - s||}{1 + ||c||} \leq 10^{-p}, \tag{20.27}$$

$$\frac{|c^T x - (b^T y - b_u^T y_u)|}{1 + |c^T x|} \leq 10^{-p}. \tag{20.28}$$

An 8-digit solution ($p = 8$) is typically required in the literature. Let us observe that conditions (20.26–20.28) still depend on the scaling of the problem and somehow use the assumption that the coefficients of the vectors b, b_u, c are about the same magnitude as those of the matrix A — preferably near 1.

An important note is needed here. The theoretical worst-case bound $\mathcal{O}(\sqrt{n}\log\frac{1}{\varepsilon})$ is still far from computational practice. It is still extremely pessimistic; in practice the number of iterations is something like $\mathcal{O}(\log n)$. It is rare that the current implementations of interior-point methods use more than 50 iterations to reach an 8-digit optimal solution.

20.7 Optimal basis identification

20.7.1 Preliminaries

An *optimal basis identification procedure* is an algorithm that generates an *optimal basis* and the related optimal basic solutions from an arbitrary primal-dual optimal solution pair. In this section we briefly recall the notion of an optimal basis. In order to ease the discussion we use the standard format:

$$\min\left\{c^T x \ : \ Ax = b, \ x \geq 0\right\}, \tag{20.29}$$

where $c, x, \in \mathbb{R}^n, b \in \mathbb{R}^m$, and the matrix A is of size $m \times n$. The dual problem is

$$\max\left\{b^T y \ : \ A^T y + s = c, \ s \geq 0\right\}, \tag{20.30}$$

where $y \in \mathbb{R}^m$ and $s \in \mathbb{R}^n$. We assume that A has rank m. A basis $A_\mathcal{B}$ is a nonsingular rank m submatrix of A, where the set of column indices of $A_\mathcal{B}$ is denoted by $\mathcal{B}$. A basic solution of the primal problem (20.29) is a vector x where all the coordinates in $\mathcal{N} = \{1, \ldots, n\} - \mathcal{B}$ are set to zero and the basis coordinates form the unique solution of the equation $A_\mathcal{B} x_\mathcal{B} = b$. The corresponding dual basic solution is defined as the unique solution of $A_\mathcal{B}^T y = c_\mathcal{B}$, along with $s_\mathcal{B} = 0$ and $s_\mathcal{N} = c_\mathcal{N} - A_\mathcal{N}^T y$. It is clear from this definition that a primal-dual pair (x, s) of basic solutions is always complementary, and hence, if both x and s are feasible, they are primal and dual optimal, respectively. A basic solution is called primal (dual) degenerate if at least one component of x_B (s_N) is zero.

There might be two reasons in practice to require an optimal basic solution for an LO problem.

1. If the given problem is a mixed integer LO problem then some or all of the variables must be integer. After solving the continuous relaxation we have to generate cuts to cut off the nonintegral optimal solution. To date, such cuts can be generated only if an optimal basic solution is available.[25] Up till now there has been only one attempt to design a cut generation procedure within the interior-point setting (see Mitchell [211]).

2. In practical applications of LO, a sequence of slightly perturbed problems often has to be solved. This is the case in combinatorial optimization when new cuts are added to the problem or if a branch and bound algorithm is applied. Also if, e.g., in production planning models the optimal solutions for different scenarios are calculated and compared, we need to solve a sequence of slightly perturbed problems. When such closely related problems are solved, we expect that the previous optimal solution can help to solve the new problem faster. Although some methods for potentially efficient warm start were discussed in Section 20.5.2, in some cases it might be advantageous in practice to use Simplex type solvers initiated with an old optimal basis.

In this section we describe how an optimal basis solution can be obtained from any optimal solution pair of the problem.

20.7.2 Basis tableau and orthogonality

We introduce briefly the notions of basis tableau and pivot transformation and we show how orthogonal vectors can be obtained from a basis tableau. Let A be the constraint matrix, with columns a_j for $1 \leq j \leq n$, and let A_B be a basis chosen from the columns of A. The basis tableau Q^B corresponding to B is defined by the equation

$$A_B Q^B = A. \tag{20.31}$$

Because this gives no rise to confusion we write below Q instead of Q^B. The rows of Q are naturally indexed by the indices in B and the columns by $1, 2, \ldots,$ n. If $i \in B$ and $j = 1, 2, \ldots,$ n the corresponding element of Q is denoted by q_{ij}. See Figure 20.1 (page 423). It is clear that q_{ij} is the coefficient of a_i in the unique basis representation of the vector a_j:

$$a_j = \sum_{i \in B} q_{ij} a_i.$$

For $j \in B$ this implies

$$q_{ij} = \begin{cases} 1 & \text{if } i = j, \\ 0 & \text{otherwise,} \end{cases}$$

_____

[25] The reader may consult the books of Schrijver [250] and Nemhauser and Wolsey [224] to learn about combinatorial optimization.

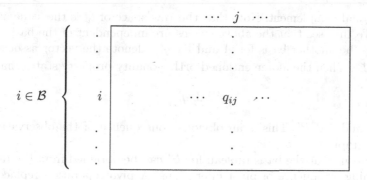

Figure 20.1 Basis tableau.

Thus, if $j \in \mathcal{B}$, the corresponding column in Q is a unit vector with its 1 in the row corresponding to j. Hence, by a suitable reordering of columns $Q_{\mathcal{B}}$ — the submatrix of Q consisting of the columns indexed by $\mathcal{B}$ — becomes an identity matrix. It is convenient for the reasoning if this identity matrix occupies the first m columns. Therefore, by permuting the columns of Q by a permutation matrix P we write

$$QP = \begin{bmatrix} I & Q_{\mathcal{N}} \end{bmatrix}, \tag{20.32}$$

where $Q_{\mathcal{N}}$ denotes the submatrix of Q arising when the columns of $Q_{\mathcal{B}}$ are deleted.

In the next section, where we present the optimal basis identification procedure, we will need a well-known *orthogonality* property of basis tableaus.[26] This property follows from the obvious matrix identity

$$\begin{bmatrix} I & Q_{\mathcal{N}} \end{bmatrix} \begin{bmatrix} Q_{\mathcal{N}} \\ -I \end{bmatrix} = 0.$$

Because of (20.32) this can be written as

$$QP \begin{bmatrix} Q_{\mathcal{N}} \\ -I \end{bmatrix} = 0. \tag{20.33}$$

Defining

$$R := P \begin{bmatrix} Q_{\mathcal{N}} \\ -I \end{bmatrix}, \tag{20.34}$$

we have $\operatorname{rank} Q = m$, $\operatorname{rank} R = n - m$ and $QR = 0$. We associate with each index a vector in $\mathbb{R}^n$ as follows. If $i \in \mathcal{B}$, $q^{(i)}$ will denote the corresponding row of Q and if $j \in \mathcal{N}$ then $q_{(j)}$ is the corresponding column of R.

Clearly, the vectors $q^{(i)}$, $i \in \mathcal{B}$, span the row space of $Q = Q^{\mathcal{B}}$ and the vectors $q_{(j)}$, $j \in \mathcal{N}$, span the column space of R. Since these spaces are orthogonal, they are each

---

[26] See, e.g., Rockafellar [238] or Klafszky and Terlaky [171].

others orthogonal complement. Note that the row space of Q is the same as the row space of A. We thus see that the above spaces are independent of the basis $\mathcal{B}$.

Now let $A_{\mathcal{B}'}$ be another basis for A and let $q'_{(j)}$ denote the vector associated with an index $j \notin \mathcal{B}'$. Then the aforementioned orthogonality property states that

$$q^{(i)} \perp q'_{(j)}$$

for all $i \in \mathcal{B}$ and $j \notin \mathcal{B}'$. This is an obvious consequence of the observation in the previous paragraph.

It is well known that the basis tableau for $\mathcal{B}'$ can be obtained from the tableau for $\mathcal{B}$ by performing a sequence of pivot operations. A pivot operation replaces a basis vector a_i, $i \in \mathcal{B}$ by a nonbasic vector a_j, $j \notin \mathcal{B}$.[27]

Example IV.86 For better understanding let us consider a simple numerical example. The following two basic tableaus can be transformed into each other by a single pivot.

	a_1	a_2	a_3	a_4	a_5
a_5	2	1	3	0	1
a_4	−1	−1	4	1	0

	a_1	a_2	a_3	a_4	a_5
a_2	2	1	3	0	1
a_4	1	0	7	1	1

It is easy to check that for the first tableau $q_{(3)} = (0, 0, -1, 4, 3)$ and for the second tableau $q^{(4)} = (1, 0, 7, 1, 1)$, and that these vectors are orthogonal.[28,29] □ ◇

20.7.3 The optimal basis identification procedure

Given any complementary solution, the algorithm presented below constructs an optimal basis in at most n iterations.[30] Since the iteration count and thus the number of necessary arithmetic operations depends only on the dimension of the problem and is independent of the actual problem data, the algorithm is called *strongly polynomial*.

The algorithm can be initialized with any optimal (and thus complementary) solution pair (x, s). This pair defines a partition of the index set as follows:

$$B = \{i \mid x_i > 0\}, \quad N = \{i : s_i > 0\}, \quad T = \{i : x_i = s_i = 0\}.$$

---

[27] **Exercise 110** Let $i \in \mathcal{B}$, where $A_{\mathcal{B}}$ is a basis. For any $j \notin \mathcal{B}$ show that $\mathcal{B}' = (\mathcal{B} \setminus \{i\}) \cup \{j\}$ also defines a basis, and the tableau for $\mathcal{B}'$ can be obtained from the tableau for $\mathcal{B}$ by one pivot operation.

[28] **Exercise 111** For each of the tableaus in Example IV.86, give the permutation matrix P and the matrix R according to (20.33) and (20.34).

[29] **Exercise 112** For each of the tableaus in Example IV.86, give a full bases of the row space of the tableau and of its orthogonal complement.

[30] The algorithm discussed here was proposed by Megiddo [201]. He has also proved that an optimal basis cannot be constructed only from a primal or dual optimal solution in strongly polynomial time unless there exists a strongly polynomial algorithm for solving the LO problem. The problem of constructing a vertex solution from an interior-point solution has also been considered by Mehrotra [203].

As we have seen in Section 3.3.6, interior-point methods produce a strictly complementary optimal solution and hence such a solution gives a partition with $T = \emptyset$. But
below we deal with the general case and we allow T to be nonempty.

The optimal basis identification procedure consists of three phases. In the first phase
a so-called *maximal basis* is constructed. A basis of A is called maximal with respect
to (x, s) if

- it has the maximum possible number of columns from A_B,
- it has the maximum possible number of columns from (A_B, A_T).

Then, in the second and third phases, independently of each other, primal and dual
elimination procedures are applied to produce primal and dual feasible basic solutions
respectively.

Note that a maximal basis is not unique and not necessarily primal and/or dual
feasible. A maximal basis can be found by the following simple pivot algorithm.
Because of the assumption rank $(A) = m$, all the artificial basis vectors $\{e_1, \cdots, e_m\}$

Initial basis

Input:
    Optimal solution pair (x, s) and the related partition (B, N, T);
    artificial basis $a_{n+1} = e_1, \cdots, a_{n+m} = e_m$;
    $\mathcal{B} = \{n + 1, \cdots, n + m\}$.

Output:
    A maximal basis $\mathcal{B} \subseteq \{1, \cdots, n\}$.

begin
    while $q_{ij} \neq 0, \; i > n, \; j \in A_B$ **do**
    begin
        pivot on position (i, j) (a_i leaves and a_j enters the basis);
        $\mathcal{B} := (\mathcal{B} \setminus \{i\}) \cup \{j\}$.
    end
    while $q_{ij} \neq 0, \; i > n, \; j \in A_T$ **do**
    begin
        pivot on position (i, j) (a_i leaves and a_j enters the basis);
        $\mathcal{B} := (\mathcal{B} \setminus \{i\}) \cup \{j\}$.
    end
    while $q_{ij} \neq 0, \; i > n, \; j \in A_N$ **do**
    begin
        pivot on position (i, j) (a_i leaves and a_j enters the basis);
        $\mathcal{B} := (\mathcal{B} \setminus \{i\}) \cup \{j\}$.
    end
end

---

are eliminated from the basis at termination. Since the A_B part is investigated first,
the number of basis vectors from A_B is maximal; similarly the number of basis vectors
from $[A_B, A_T]$ is also maximal. In a practical implementation, special attention must

be given to the selection of the pivot elements in the above algorithm. Typically there is lot of freedom in the pivot selection, since a large number of leaving and/or entering variables could be selected at each iteration.[31] The structure of the basis tableau resulting from the algorithm is visualized in Figure 20.2. Note that the tableau is never computed in practice; just the basis, in a factorized form. The tableau form is used just to ease the explanation and understanding.

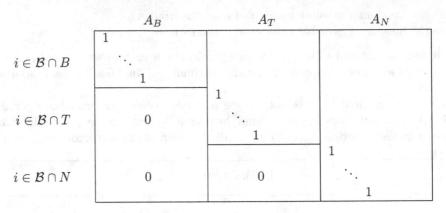

Figure 20.2   Tableau for a maximal basis.

We proceed by a primal and a dual phase, performed independently of each other. They make the basis primal and dual feasible, respectively.

Observe that in the elimination step of the first **while**-loop of the primal phase the columns of $A_{\widetilde{B}}$ are dependent. Hence there exists a nonzero solution of $A_{\widetilde{B}}\bar{x}_B = 0$.[32] In the elimination step the 'maximal' property of the basis is lost, but it is restored in the second **while**-loop. As we can see, the *Primal phase* works only on the $\left(A_{\widetilde{B}}, A_{\widetilde{T}}\right)$ part of the matrix A. In fact it reduces the $A_{\widetilde{B}}$ part to an independent set of column vectors. At termination the maximal basis is primal feasible and $\tilde{x}$ is the corresponding primal feasible basic solution, i.e., $\tilde{x}_\mathcal{B} = A_\mathcal{B}^{-1}b \geq 0$ and $\tilde{x}_\mathcal{N} = 0$. The number of eliminations in the first **while**-loop is at most $|B| - \text{rank}(B)$ and the number of pivots in the second **while**-loop is also at most $|B| - \text{rank}(B)$.

The *Dual phase* presented below works on the (A_T, A_N) part. It reduces A_N and extends A_T so that no vector from A_N remains in the basis.

Note that in the elimination step of the first **while**-loop the rank of $[A_B, A_{\widetilde{T}}]$ is less than m.[33] In the elimination step the 'maximal' property of the basis is

---

[31] We would like to choose always the pivot element that produces the least fill-in in the inverse basis. For this pivot selection problem many heuristics are possible. It can at least locally be optimized by, e.g., the heuristic Markowitz rule (recall Section 20.4). Implementation issues related to optimal basis identification procedures are discussed in Andersen and Ye [11], Andersen et al. [10] and Bixby and Saltzman [43].

[32] In fact, an appropriate $\bar{x}$ can be read from the tableau. Because of the orthogonality property any vector $q_{(j)}$ for $j \in \widetilde{B} - \mathcal{B}$ can be used. In a practical implementation the tableau is not available; only the (factorized) basis matrix Q_B is available. But then a vector $q_{(j)}$ can be obtained by computing a single nonbasic column of the tableau.

[33] For an appropriate $\bar{s}$ we can choose any vector $q^{(i)}$ for $i \in \widetilde{N} \cap \mathcal{B}$; so only one row of the tableau has to be computed at each execution of the first **while**-loop.

Primal phase

Input:
   Optimal solution pair $(\tilde{x}, s)$ and the related partition $(\widetilde{B}, N, \widetilde{T})$;
   maximal basis $\mathcal{B}$.
Output:
   A maximal basis $\mathcal{B} \subseteq \{1, \cdots, n\}$;
   optimal solution $(\tilde{x}, s)$, partition $(\widetilde{B}, N, \widetilde{T})$ with $\widetilde{B} \subset \mathcal{B}$.
begin
   while $\widetilde{B} \not\subseteq \mathcal{B}$ **do**
   begin
      begin
         let $\bar{x}$ be such that $A_{\widetilde{B}}\bar{x}_{\widetilde{B}} = 0$, $\bar{x}_{\widetilde{T} \cup N} = 0$, $\bar{x} \neq 0$;
         eliminate a(t least one) coordinate of $\tilde{x}$, let $\tilde{x} := \tilde{x} - \vartheta\bar{x} \geq 0$;
         $\widetilde{B} := \sigma(\tilde{x})$, $\widetilde{T} := \{1, \ldots, n\} \setminus \left(\widetilde{B} \cup N\right)$;
      end
      while $q_{ij} \neq 0$, $i \in \widetilde{T} \cap \mathcal{B}$, $j \in B$ **do**
      begin
         pivot on position (i, j) (a_i leaves, a_j enters the basis);
         $\mathcal{B} := (\mathcal{B} \setminus \{i\}) \cup \{j\}$.
      end
   end
end

---

lost but is restored in the second **while**-loop. At termination the maximal basis is dual feasible and $\tilde{s}$ is the corresponding dual feasible basic solution, i.e., $\tilde{s}_N = c_N - A_N^T(A_B^{-1})^T c_B$ and $\tilde{s}_B = 0$. The number of eliminations in the first **while**-loop is at most $m - \text{rank}(A_B, A_T)$ and the number of pivots in the second **while**-loop is also at most $m - \text{rank}(A_B, A_T)$.

To summarize, by first constructing a maximal basis and then performing the primal and dual phases, the above algorithm generates an optimal basis after at most n iterations. First we need at most m pivots to construct the maximal basis, then in the primal phase $|B| - \text{rank}(B)$ and in the dual phase $m - \text{rank}(A_B, A_T)$ pivots follow. Finally, to verify the n-step bound, observe that after the initial maximal basis is constructed, each variable might enter the basis at most once.

20.7.4   Implementation issues of basis identification

In the above basis identification algorithm it is assumed that a pair of exact primal/dual optimal solutions is known. This is never the case in practice. Interior point algorithms generate only a sequence converging to optimal solutions and because of the finite precision of computations the solutions are neither exactly feasible nor complementary. Somehow we have to make a decision about which variables are

---

Dual phase

Input:
    Optimal solutions $(x, \tilde{s})$, partition $(B, \widetilde{N}, \widetilde{T})$;
    maximal basis $\mathcal{B}$.

Output:
      A maximal basis $\mathcal{B} \subseteq \{1, \cdots, n\}$;
      optimal solution $(x, \tilde{s})$, partition $(B, \widetilde{N}, \widetilde{T})$ with $\widetilde{N} \cap \mathcal{B} = \emptyset$.

begin
  while $\widetilde{N} \cap \mathcal{B} \neq \emptyset$ **do**
  begin
    begin
      let $\bar{s}$ be such that $\bar{s} = A^T y$, $\bar{s}_{B \cup \widetilde{T}} = 0$, $\bar{s} \neq 0$;
      eliminate a(t least one) coordinate of s, let $\tilde{s} := \tilde{s} - \vartheta \bar{s} \geq 0$;
      $\widetilde{N} := \sigma(\tilde{s})$, $\widetilde{T} := \{1, \cdots, n\} \setminus \left(B \cup \widetilde{N}\right)$;
    end
    while $q_{ij} \neq 0$, $i \in \widetilde{N} \cap \mathcal{B}$, $j \in \widetilde{T}$ **do**
    begin
      pivot on position (i, j) (a_i leaves, a_j enters the basis);
      $\mathcal{B} := (\mathcal{B} \setminus \{i\}) \cup \{j\}$.
    end
  end
end

---

positive and which are zero at the optimum.

Let $(\bar{x}, \bar{y}, \bar{s})$ be feasible and $(\bar{x})^T \bar{s} \leq \varepsilon$. Let us make a guess for the optimal partition of the problem as

$$B = \{ i \,|\, \bar{x}_i \geq \bar{s}_i \} \quad \text{and} \quad N = \{ i \,|\, \bar{x}_i < \bar{s}_i \}.$$

Now we can define the following perturbed problem[34]

$$\text{minimize} \left\{ \bar{c}^T x \; : \; Ax = \bar{b}, \, x \geq 0 \right\}, \tag{20.35}$$

where

$$\bar{b} = A_B \bar{x}_B, \quad \bar{c}_B = A_B^T \bar{y} \quad \text{and} \quad \bar{c}_N = A_N^T \bar{y} + s_N.$$

Now the vectors (x, y, s), where $y = \bar{y}$ and

$$x_i = \begin{cases} \bar{x}_i & i \in B, \\ 0 & i \in N \end{cases} \quad \text{and} \quad s_i = \begin{cases} 0 & i \in B, \\ \bar{s}_i & i \in N \end{cases} \tag{20.36}$$

---

[34] This approach was proposed by Andersen and Ye [11].

are strictly complementary optimal solutions of (20.35).[35] If ε is small enough, then the partition (B, N) is the optimal partition of (20.29) (recall the results of Theorem I.47 and observe that the proof does not depend on the specific algorithm, just on the centrality condition and the stopping precision). Thus problems (20.29) and (20.35) share the same partition and the same set of optimal bases. As an optimal complementary solution for (20.35) is available, the above basis identification algorithm can be applied to this perturbed problem. The resulting optimal basis, within a small margin of error (depending on ε), is an optimal basis for (20.29).

20.8 Available software

After twenty years of intensive research, IPMs are now well understood both in theory and practice. As a result a number of sophisticated implementations exist of IPMs for LO. Below we give a list of some of these codes; some of them contain both a Simplex and an IPM solver. They are capable to solve linear problems on a PC in some minutes that were hardly solvable on a super computer fifteen years ago.

CPLEX (CPLEX/ BARRIER) (CPLEX Optimization, Inc.). For information contact http://www.cplex.com.

CPLEX is leading the market at this moment. It is a most complete and robust package. It contains a primal and a dual Simplex solver, an efficient interior-point implementation with cross-over,[36] a good mixed-integer code, a network and a quadratic programming solver. It is supported by most modelling languages and available for most platforms.

XPRESS-MP (DASH Optimization). For information contact the vendor's WEB page: http://www.dashoptimization.com.

An excellent package including Simplex and IPM solvers. It is almost as complete as CPLEX.

CLP (The LO solver on COIN-OR). For more information contact http://www.coin-or.org/cgi-bin/cvsweb.cgi/COIN/Clp/.

COIN-OR's LO package is written by the IBM Lo team. Like CPLEX, CLP contains both Simplex and IPM solvers. It is capable to solve linear and quadratic optimization problems.

LOQO. Available from http://www.princeton.edu/~rvdb/.

LOQO is developed by Vanderbei (Department of Operations Research and Financial Engineering, Princeton University, Princeton, NJ 08544, USA). It is a robust implementation of a primal-dual infeasible-start IPM for convex quadratic optimization. LOQO is a commercial package, like CPLEX and OSL, but it is available for academic purposes for a modest license fee.

---

[35] Producing a reliable guess for the optimal partition is a nontrivial task. The simple method presented by (20.36) seems to work reasonably well in practice. See El-Bakry, Tapia and Zhang [71, 70]. However, Andersen and Ye [11] report good results by using a more sophisticated indicator to predict the optimal partition (B, N) based on the primal-dual search direction.

[36] Close to optimality the solver rounds the IPM solution to a (not necessarily optimal) basic solution and switches to the Simplex solver, that generates an optimal basic solution.

HOPDM. Available from
http://www.maths.ed.ac.uk/~gondzio/software/hopdm.html.

HOPDM is developed by Gondzio (School of Mathematics, The University of Edinburgh, Edinburgh, Scotland). It implements a higher order primal-dual method. It is in the public domain — in a form of FORTRAN source files — for academic purposes.

BPMPD. Available from http://www.sztaki.hu/~meszaros/bpmpd/.

Mészáros' BPMPD, is an implementation of a primal-dual predictor-corrector IPM including both the normal and augmented system approach. The code is available as an executable file for academic purposes.

LIPSOL. Available from http://www.caam.rice.edu/~yzhang/.

Zhang's LIPSOL is written in MATLAB and FORTRAN. It is an implementation of the primal-dual predictor-corrector method. One of its features is the use of the MATLAB programming language, which makes its use relatively easy.

PCx. Available from http://www-fp.mcs.anl.gov/otc/Tools/PCx/.

This code was developed by Czyzyk, Mehrotra and Wrightat the Argonne National Lac, Chicago.. It is a stand alone C implementation of an infeasible primal-dual predictor corrector IPM. PCx is freely available, but is not public domain software

McIPM. Available from http://www.cas.mcmaster.ca/~oplab/index.html.

This code was developed at the Advanced Optimization Lab, McMaster University by Zhu, Peng and Terlaky. McIPM is written in MATLAB and C. It is a unique implementation of a Self-Regular primal-dual predictor-corrector IPM and it is based on the self-dual embedding model. The use of the MATLAB makes its use relatively easy. It is freely available under an open source license.

More information about codes for linear optimization, either for commercial or research purposes, are available at the World Wide Web site of LP FAQ (LP Frequently Asked Questions) at

• http://www-unix.mcs.anl.gov/otc/Guide/faq/linear-programming-faq.html

• ftp://rtfm.mit.edu/pub/usenet/sci.answers/linear-programming-faq

Appendix A

Some Results from Analysis

In Part II we need a result from convex analysis. We include its elementary proof in this appendix for the sake of completeness. A closely related result can be found in Bazaraa et al. [37] (Theorem 3.4.3 and Corollary 1, pp. 101–102). Recall that a subset C of $\mathbb{R}^k$ is called *relatively open* if C is open in the smallest affine subspace of $\mathbb{R}^k$ containing C.

Proposition A.1 *Let* $f : D \to \mathbb{R}$ *be a differentiable function, where* $D \subseteq \mathbb{R}^k$ *is an open set, and let* C *be a relatively open convex subset of* D *such that* f *is convex on* C. *Moreover, let* $\mathcal{L}$ *denote the subspace parallel to the smallest affine space containing* C. *Then,* $x^* \in C$ *minimizes* f *over* C *iff*

$$\nabla f(x^*) \perp \mathcal{L}. \tag{A.1}$$

Proof: Since f is convex on C, we have for any $x, x^* \in C$,

$$f(x) \geq f(x^*) + \nabla f(x^*)^T (x - x^*).$$

Since $x - x^* \in \mathcal{L}$, the sufficiency of condition (A.1) follows immediately. To prove the necessity of (A.1), consider $x_t = x^* + t(x - x^*)$, with $t \in \mathbb{R}$. The convexity of C implies that if $0 \leq t \leq 1$, then $x_t \in C$. Moreover, since C is open, we also have $x_t \in C$ when $t \geq -a$ for some positive a. Since f is differentiable, we have

$$\nabla f(x^*)^T (x - x^*) = \lim_{t \downarrow 0} \frac{f(x_t) - f(x^*)}{t} = \lim_{t \uparrow 0} \frac{f(x_t) - f(x^*)}{t}.$$

Now let $x^* \in C$ minimize f. Since $f(x_t) \geq f(x^*)$, letting $t \to 0$ we have that the first limit must be nonnegative, and the second nonpositive. Hence both limits are zero. So we have $\nabla f(x^*)^T (x - x^*) = 0$, $\forall x \in C$. Thus (A.1) follows. □

At several places in the book we mention the *implicit function theorem*. There exists many forms of this theorem. See, e.g., Franklin [82], Buck [52], Fiacco [76] or Rudin [248]. We cite here a version of Bertsekas [40] (Proposition A.25, pp. 554).[1]

Proposition A.2 (Implicit Function Theorem) *Let* $f : \mathbb{R}^{n+m} \to \mathbb{R}^m$ *be a function of* $w \in \mathbb{R}^n$ *and* $z \in \mathbb{R}^m$ *such that:*

---

[1] In fact, Proposition A.25 in Bertsekas [40] contains a typo. It says that $f : \mathbb{R}^{n+m} \to \mathbb{R}^n$ instead of $f : \mathbb{R}^{n+m} \to \mathbb{R}^m$.

(i) There exist $\bar{w} \in \mathbb{R}^n$ and $\bar{z} \in \mathbb{R}^m$ such that $f(\bar{w}, \bar{z}) = 0$.

(ii) f is continuous and has a continuous and nonsingular gradient matrix (or Jacobian) $\nabla_z f(w, z)$ in an open set containing $(\bar{w}, \bar{z})$.

Then there exists open sets $S_{\bar{w}} \subseteq \mathbb{R}^n$ and $S_{\bar{z}} \subseteq \mathbb{R}^m$ containing $\bar{w}$ and $\bar{z}$, respectively, and a continuous function $\phi : S_{\bar{w}} \to S_{\bar{z}}$ such that $\bar{z} = \phi(\bar{w})$ and $f(w, \phi(w)) = 0$ for all $w \in S_{\bar{w}}$. The function ϕ is unique in the sense that if $w \in S_{\bar{w}}$, $z \in S_{\bar{z}}$, and $f(w, z) = 0$, then $z = \phi(w)$. Furthermore, if for some $p > 0$, f is p times continuously differentiable the same is true for ϕ, and we have

$$\nabla \phi(w) = -\left(\nabla_z f(w, \phi(w))\right)^{-1} \nabla_w f(w, \phi(w)).$$

Appendix B

Pseudo-inverse of a Matrix

We are interested in the least norm solution of the linear system of equations

$$Ax = b,$$

where A is an $m \times n$ matrix of rank r, and $b \in \mathbb{R}^m$. We assume that a solution exists, i.e., b belongs to the column space of A.

First we consider the case where $r = n$. Then the columns of A are linearly independent and hence the solution is unique. It is obtained by premultiplication of the system by A^T: $A^T A x = A^T b$. Since $A^T A$ is nonsingular we find

$$x = (A^T A)^{-1} A^T b \quad (r = n).$$

We proceed with the case where $r = m < n$. Then $Ax = b$ has multiple solutions. The least norm solution is characterized by the fact that it is orthogonal to the null space of A. So in this case the solution belongs to the row space of A and hence can be written as $x = A^T \lambda$, $\lambda \in \mathbb{R}^m$. This implies that $AA^T \lambda = b$. This time AA^T is nonsingular, and we obtain that $\lambda = (AA^T)^{-1} b$, whence

$$x = A^T (AA^T)^{-1} b \quad (r = m).$$

Finally we consider the general case, without making any assumption on the rank of A. We start by decomposing A as follows:

$$A = A_1 A_2,$$

where A_1 is an $m \times r$ matrix of rank r, and A_2 is an $r \times n$ matrix of rank r. There are many ways to realize such a decomposition. One way is the well-known LU decomposition of A.[1]

Now $Ax = b$ can be rewritten as $A_1 A_2 x = b$. Since A_1 has full column rank we are in the first situation, and hence it follows that

$$A_2 x = (A_1^T A_1)^{-1} A_1^T b.$$

Thus our problem is reduced to finding a least norm solution of the last system. Since A_2 has full row rank we are now in the second situation, and hence it follows that

$$x = A_2^T (A_2 A_2^T)^{-1} (A_1^T A_1)^{-1} A_1^T b.$$

---

[1] See, e.g., the book of Strang [259].

Thus we have found the least norm solution of $Ax = b$. Defining the matrix A^+ according to

$$A^+ = A_2^T (A_2 A_2^T)^{-1} (A_1^T A_1)^{-1} A_1^T, \tag{B.1}$$

we conclude that the least norm solution of $Ax = b$ is given by $x = A^+ b$.

The matrix A^+ is called the *pseudo-inverse* of A. We can easily verify that A^+ satisfies the following relations:

$$
\begin{aligned}
AA^+A &= A, &\text{(B.2)}\\
A^+AA^+ &= A^+, &\text{(B.3)}\\
(AA^+)^T &= AA^+, &\text{(B.4)}\\
(A^+A)^T &= A^+A. &\text{(B.5)}
\end{aligned}
$$

Theorem B.1 *The equations (B.2) to (B.5) determine A^+ uniquely.*

Proof: We already have seen that a solution exists. Suppose that we have two solutions, X_1 and X_2 say. From (B.2) and (B.5) we derive that $X_1 AA^T = A^T$, and $X_2 AA^T = A^T$. So $(X_1 - X_2)AA^T = 0$. This implies $(X_1 - X_2)AA^T(X_1 - X_2)^T = 0$, and hence we must have $(X_1 - X_2)A = 0$. This means that the columns of $X_1 - X_2$ belong to the left null space of A. On the other hand (B.3) and (B.4) imply that $AX_1 X_1^T = X_1^T$, and $AX_2 X_2^T = X_2^T$. Hence $A(X_1 X_1^T - X_2 X_2^T) = X_1^T - X_2^T$. This means that the columns of $X_1 - X_2$ belong to the column space of A. Since the column space and the left null space of A are orthogonal this implies that $X_1 = X_2$. $\qquad\square$

There is another interesting way to describe the pseudo-inverse A^+ of A, which uses the so-called *singular value decomposition (SVD)* of A. Let r denote the rank of A, and let $\lambda_1, \lambda_2, \cdots, \lambda_r$ denote the nonzero (hence positive) eigenvalues of AA^T. Furthermore, let Q_1 and Q_2 denote orthogonal matrices such that the first r columns of Q_1 constitute a basis of the column space of A, and the first r columns of Q_2 constitute a basis of the row space of A. Then, if Σ denotes the $m \times n$ matrix whose only nonzero elements are $\Sigma_{11}, \Sigma_{22}, \cdots, \Sigma_{rr}$, with

$$\Sigma_{ii} = \sigma_i := \sqrt{\lambda_i}, \ 1 \le i \le r,$$

then we have

$$A = Q_1 \Sigma Q_2^T.$$

This is the *SVD* of A, and the numbers σ_i, $1 \le i \le r$ are called the *singular values* of A.

Using Theorem B.1 we can easily verify that Σ^+ is the $n \times m$ matrix whose only nonzero elements are the first r diagonal elements, and these are the inverses of the singular values of A. Then, using Theorem B.1 once more, we can easily check that A^+ is given by

$$A^+ = Q_2 \Sigma^+ Q_1^T.$$

Appendix C

Some Technical Lemmas

Lemma C.1 *Let A be an $m \times n$ matrix with columns A_j and b a vector of dimension m such that the set*

$$\mathcal{S} := \{x \: : \: Ax = b, \, x \geq 0\}$$

is bounded and contains a positive vector. Moreover, let all the entries in A and b be integral. Then for each i, with $1 \leq i \leq n$,

$$\max_x \{x_i \: : \: x \in \mathcal{S}\} \geq \frac{1}{\prod_{j=1}^n \|A_j\|}.$$

Proof: Observe that each column A_j of A must be nonzero, due to the boundedness of $\mathcal{S}$. Fixing the index i, let $x \in \mathcal{S}$ be such that x_i is maximal. Note that such an x exists since $\mathcal{S}$ is bounded. Moreover, since $\mathcal{S}$ contains a positive vector, we must have $x_i > 0$. Let J be the support of x:

$$J = \{j \: : \: x_j > 0\}.$$

We assume that x is such that the cardinality of its support is minimal. Then the columns of the submatrix A_J of A are linearly independent. This can be shown as follows. Let there exist a nonzero vector $\lambda \in \mathbb{R}^n$ such that

$$\sum_{j \in J} \lambda_j A_j = 0,$$

and $\lambda_k = 0$ for each $k \notin J$. Then $A\lambda = 0$. Hence, if ε is small enough, $x \pm \varepsilon\lambda$ has the same support as x and is positive on J. Moreover, $x \pm \varepsilon\lambda \in \mathcal{S}$. Since the i-th coordinate cannot exceed x_i it follows that $\lambda_i = 0$. Since $\mathcal{S}$ is bounded, at least one of the coordinates of λ must be negative, because otherwise $\mathcal{S}$ would contain the ray $x + \varepsilon\lambda, \varepsilon > 0$. By increasing the value of ε until one of its coordinates reaches zero we get a vector in $\mathcal{S}$ with less than $|J|$ nonzero coordinates and for which the i-th coordinate still has value x_i. This contradicts the assumption that x has minimal support among such vectors, thus proving that the columns of the submatrix A_J of A are linearly independent.

Now let A_{KJ} be any nonsingular submatrix of A_J. Here K denotes a suitable subset of the row indices $1, 2, \cdots, m$ of A. Then we have

$$A_{KJ}x_J = b_K,$$

since the coordinates x_j of x with $j \notin J$ are zero. We can solve x_i from this equation by using Cramer's rule. [1] This yields

$$x_i = \frac{\det A'_{KJ}}{\det A_{KJ}}, \tag{C.1}$$

where A'_{KJ} denotes the matrix arising from A_{KJ} by replacing the i-th column by b_K. We know that $x_i > 0$. This implies $|\det A'_{KJ}| > 0$. Since all the entries in the matrix A'_{KJ} are integral the absolute value of its determinant is at least 1. Thus we find

$$x_i \geq \frac{1}{|\det A_{KJ}|}.$$

Now using that $|\det A_{KJ}|$ is bounded above by the product of the norms of its columns, due to the well-known Hadamard inequality[2] for determinants, we find[3]

$$x_i \geq \frac{1}{\prod_{j \in J} \|A_{Kj}\|} \geq \frac{1}{\prod_{j \in J} \|A_j\|} \geq \frac{1}{\prod_{j=1}^{n} \|A_j\|}.$$

The second inequality is obvious and the last inequality follows since A has no zero columns and hence the norm of each column of A is at least 1. This proves the lemma.
□

We proceed with a proof of the two basic inequalities in (6.24) on page 134. The proof uses standard techniques for proving elementary inequalities.[4]

Lemma C.2 *Let $z \in \mathbb{R}^n$, and $\alpha \geq 0$. Then each of the two inequalities*

$$\psi\left(\alpha \|z\|\right) \leq \Psi\left(\alpha z\right) \leq \psi\left(-\alpha \|z\|\right)$$

holds whenever the involved expressions are well defined. The left (right) inequality holds with equality if and only if one of the coordinates of z equals $\|z\|$ ($-\|z\|$, respectively) and the remaining coordinates are zero.

Proof: Fixing z we introduce

$$g(\alpha) := \psi\left(\alpha \|z\|\right)$$

and

$$G(\alpha) := \Psi\left(\alpha z\right) = \sum_{i=1}^{n} \psi\left(\alpha z_i\right),$$

where α is such that $\alpha z > -e$ and $\alpha \|z\| > -1$. Both functions are twice differentiable with respect to α. Using that $\psi'(t) = 1 - 1/t$ we obtain

$$g'(\alpha) = \frac{\alpha \|z\|^2}{1 + \alpha \|z\|}, \quad G'(\alpha) = \sum_{i=1}^{n} \frac{\alpha z_i^2}{1 + \alpha z_i}$$

---

[1] The idea of using Cramer's rule in this way was applied first by Khachiyan [167].

[2] cf. Section 1.7.3.

[3] The idea of using Hadamard's inequality for deriving bounds on the coordinates of x_i from (C.1) was applied earlier by Klafszky and Terlaky [170] in a similar context.

[4] The proof is due to Jiming Peng [232].

and

$$g''(\alpha) = \frac{\|z\|^2}{(1 + \alpha \|z\|)^2}, \quad G''(\alpha) = \sum_{i=1}^{n} \frac{z_i^2}{(1 + \alpha z_i)^2}.$$

Now consider the case where $\alpha \geq 0$. Then using $z_i \leq \|z\|$ we may write

$$G''(\alpha) = \sum_{i=1}^{n} \frac{z_i^2}{(1 + \alpha z_i)^2} \geq \sum_{i=1}^{n} \frac{z_i^2}{(1 + \alpha \|z\|)^2} = \frac{\|z\|^2}{(1 + \alpha \|z\|)^2} = g''(\alpha).$$

So $G(\alpha) - g(\alpha)$ is convex for $\alpha \geq 0$. Since

$$g(0) = G(0) = 0, \quad g'(0) = G'(0) = 0 \tag{C.2}$$

it follows that $G(\alpha) \geq g(\alpha)$ if $\alpha \geq 0$. This proves the left hand side inequality in the lemma.

The right inequality follows in the same way. Let $\alpha \geq 0$ be such that $e + \alpha z > 0$ and $1 - \alpha \|z\| > 0$. Using $1 + \alpha z_i \geq 1 - \alpha \|z\| > 0$ we may write

$$G''(\alpha) = \sum_{i=1}^{n} \frac{z_i^2}{(1 + \alpha z_i)^2} \leq \sum_{i=1}^{n} \frac{z_i^2}{(1 - \alpha \|z\|)^2} = \frac{\|z\|^2}{(1 - \alpha \|z\|)^2} = g''(-\alpha).$$

This implies that $G(\alpha) - g(-\alpha)$ is concave for $\alpha \geq 0$. Using (C.2) once more we obtain $G(\alpha) \leq g(-\alpha)$ if $\alpha \geq 0$, which is the right hand side inequality in the lemma.

Note that in both cases equality occurs only if $z_i^2 = \|z\|^2$ for some i. Since the remaining coordinates are zero in that case, the lemma follows. $\square$

We proceed with another technical lemma that is used in the proof of Lemma IV.15 in Chapter 17 (page 325).

Lemma C.3 *Let p be a positive number and let $f : \mathbb{R}_+ \to \mathbb{R}_+$ be defined by*

$$f(x) := |1 - x| + \left|1 - \frac{p}{x}\right|.$$

If $p \geq 1$ then f attains its minimal value at $x = \sqrt{p}$, and if $0 < p \leq 1$ then f attains its minimal value at $x = 1$ and at $x = p$.

Proof: First consider the case $p \geq 1$. If $x \leq 1$ then we have

$$f(x) = 1 - x + \frac{p}{x} - 1 = \frac{p}{x} - x.$$

Hence, if $x \leq 1$ the derivative of f is given by

$$f'(x) = -\frac{p}{x^2} - 1 < 0.$$

Thus, f is monotonically decreasing if $x \leq 1$. If $x \geq p$ then we have

$$f(x) = x - 1 + 1 - \frac{p}{x} = x - \frac{p}{x}$$

and the derivative of f is given by

$$f'(x) = 1 + \frac{p}{x^2} > 0,$$

proving that f is monotonically increasing if $x \geq p$. For $1 \leq x \leq p$ we have

$$f(x) = x - 1 + \frac{p}{x} - 1 = x + \frac{p}{x} - 2.$$

Now the derivative of f is given by

$$f'(x) = 1 - \frac{p}{x^2}$$

and the second derivative by

$$f''(x) = \frac{2p}{x^3} > 0.$$

Hence f is convex if $x \in [1, p]$. Putting $f'(x) = 0$ we get $x = \sqrt{p}$, proving the first part of the lemma.

The case $p \leq 1$ is treated as follows. If $x \leq p$ then

$$f(x) = 1 - x + \frac{p}{x} - 1 = \frac{p}{x} - x,$$

and, as before, f is monotonically decreasing. If $x \geq 1$ then

$$f(x) = x - 1 + 1 - \frac{p}{x} = x - \frac{p}{x}$$

and f is monotonically increasing. Now let $p \leq x \leq 1$. Then

$$f(x) = 1 - x + 1 - \frac{p}{x} = 2 - x - \frac{p}{x}.$$

Hence f is concave if $x \in [p, 1]$, and f has local minima at $x = p$ and $x = 1$. Since $f(1) = f(p) = 1 - p$ the second part of the lemma follows. $\qquad\square$

The rest of this appendix is devoted to some properties of the componentwise product uv of two orthogonal vectors u and v in $\mathbb{R}^n$. The first two lemmas give some upper bounds for the 2-norm and the infinity norm of uv.

Lemma C.4 (First uv-lemma) *If u and v are orthogonal in $\mathbb{R}^n$, then*

$$\|uv\|_\infty \leq \frac{1}{4} \|u + v\|^2, \quad \|uv\| \leq \frac{\sqrt{2}}{4} \|u + v\|^2.$$

Proof: We may write

$$uv = \frac{1}{4} \left((u + v)^2 - (u - v)^2 \right). \tag{C.3}$$

From this we derive the componentwise inequality

$$-\frac{1}{4}(u - v)^2 \leq uv \leq \frac{1}{4}(u + v)^2.$$

This implies

$$-\frac{1}{4}\|u-v\|^2 \, e \le uv \le \frac{1}{4}\|u+v\|^2 \, e.$$

Since u and v are orthogonal, the vectors $u-v$ and $u+v$ have the same norm, and hence the first inequality in the lemma follows. For the second inequality we derive from (C.3) that

$$\|uv\|^2 = e^T (uv)^2 = \frac{1}{16} e^T \left((u+v)^2 - (u-v)^2\right)^2 \le \frac{1}{16} e^T \left((u+v)^4 + (u-v)^4\right).$$

Since $e^T z^4 \le \|z\|^4$ for any $z \in \mathbb{R}^n$, we obtain

$$\|uv\|^2 \le \frac{1}{16}\left(\|u+v\|^4 + \|u-v\|^4\right). \tag{C.4}$$

Using again that $\|u-v\| = \|u+v\|$, we confirm the second inequality. $\square$

The next lemma provides a second upper bound for $\|uv\|$.

Lemma C.5 (Second uv-lemma) [5] *If u and v are orthogonal in $\mathbb{R}^n$, then $\|uv\| \le \frac{1}{\sqrt{2}}\|u\|\,\|v\|$.*

Proof: Recall from (C.4) that

$$\|uv\|^2 \le \frac{1}{16}\left(\|u+v\|^4 + \|u-v\|^4\right).$$

Now first assume that u and v are unit vectors, i.e., $\|u\| = \|v\| = 1$. Then the orthogonality of u and v implies that $\|u+v\|^4 = \|u-v\|^4 = 4$, whence $\|uv\|^2 \le 1/2$. In the general case, if u or v is not a unit vector, then if one of the two vectors is the zero vector, the lemma is obvious. Else we may write

$$\|uv\| = \|u\|\,\|v\| \left\|\frac{u}{\|u\|}\frac{v}{\|v\|}\right\|.$$

Now applying the above result for the case of unit vectors to $u/\|u\|$ and $v/\|v\|$ we obtain the lemma. $\square$

The bound for $\|uv\|$ in Lemma C.5 is stronger than the corresponding bound in Lemma C.4. This easily follows by using $ab \le \frac{1}{2}\left(a^2 + b^2\right)$ with $a = \|u\|$ and $b = \|v\|$. It may be noted that the last inequality provides also an alternative proof for the bound for $\|uv\|_\infty$ in Lemma C.5.

For the proof of the third uv-lemma we need the next lemma.

Lemma C.6 [6] *Let γ be a vector in $\mathbb{R}^p$ such that $\gamma > -e$ and $e^T\gamma = \sigma$. Then if either $\gamma \ge 0$ or $\gamma \le 0$,*

$$\sum_{i=1}^p \frac{-\gamma_i}{1+\gamma_i} \le \frac{-\sigma}{1+\sigma};$$

equality holds if and only if at most one of the coordinates of γ is nonzero.

---

[5] For the case in which u and v are unit vectors, this lemma has been found by several authors. See, e.g., Mizuno [214], Jansen et al. [154], Gonzaga [125]. The extension to the general case in Lemma C.5 is due to Gonzaga (private communication). We will refer to this lemma as the second uv-lemma.

[6] This lemma and the next lemma are due to Ling [182, 183].

Proof: The lemma is trivial if $\gamma = 0$, so we may assume that γ is nonzero. For the proof of the lemma we use the function $f : (-1, \infty)^p \to \mathbb{R}$ defined by

$$f(\gamma) := \sum_{i=1}^{p} \frac{-\gamma_i}{1 + \gamma_i}.$$

We can easily verify that f is convex (its Hessian is positive definite). Observe that $\sum_{i=1}^{p} \gamma_i / \sigma = 1$ and, since either $\gamma \geq 0$ or $\gamma \leq 0$, $\gamma_i / \sigma \geq 0$. Therefore we may write

$$f(\gamma) = f\left(\sum_{i=1}^{p} \frac{\gamma_i}{\sigma} \sigma e_i\right) \leq \sum_{i=1}^{p} \frac{\gamma_i}{\sigma} f(\sigma e_i) = \sum_{i=1}^{p} \frac{\gamma_i}{\sigma}\left(\frac{-\sigma}{1+\sigma}\right) = \frac{-\sigma}{1+\sigma},$$

where e_i denotes the i-th unit vector in $\mathbb{R}^p$. This proves the inequality in the lemma. Note that the inequality holds with equality if $\gamma = \sigma e_i$, for some i, and that in all other cases the inequality is strict since the Hessian of f is positive definite. Thus the lemma has been proved. $\qquad\square$

Using the above lemmas we prove the next lemma.

Lemma C.7 (Third uv-lemma) *Let u and v be orthogonal in $\mathbb{R}^n$, and suppose $\|u + v\| = 2r$ with $r < 1$. Then*

$$e^T\left(\frac{e}{e + uv} - e\right) \leq \frac{2r^4}{1 - r^4}.$$

Proof: The first uv-lemma implies that $\|uv\|_\infty \leq r^2 < 1$. Hence, putting $\beta := uv$ we have $e^T\beta = 0$ and $-e < \beta < e$. Now let

$$I_+ := \{i : \beta_i > 0\},$$
$$I_- := \{i : \beta_i < 0\}.$$

Then

$$\sum_{i \in I_+} \beta_i = -\sum_{i \in I_-} \beta_i.$$

Let σ denote this common value. Using Lemma C.6 twice, with respectively $\gamma_i = b_i$ for $i \in I_+$ and $\gamma_i = b_i$ for $i \in I_-$, we obtain

$$
\begin{aligned}
e^T\left(\frac{e}{e + uv} - e\right) &= e^T\left(\frac{e}{e + \beta} - e\right) = \sum_{i=1}^{n} \frac{-\beta_i}{1 + \beta_i} \\
&= \sum_{i \in I_+} \frac{-\beta_i}{1 + \beta_i} + \sum_{i \in I_-} \frac{-\beta_i}{1 + \beta_i} \\
&\leq \frac{-\sigma}{1 + \sigma} + \frac{\sigma}{1 - \sigma} = \frac{2\sigma^2}{1 - \sigma^2}.
\end{aligned}
$$

The last expression is monotonically increasing in σ. Hence we may replace it by an upper bound, which can be obtained as follows:

$$\sigma = \frac{1}{2}\sum_{i=1}^{n} |\beta_i| = \frac{1}{2}\sum_{i=1}^{n} |u_i v_i| \leq \frac{1}{4}\sum_{i=1}^{n} (u_i^2 + v_i^2) = \frac{1}{4}\|u + v\|^2 = r^2.$$

Substitution of this bound for σ yields the lemma. □

Lemma C.8 (Fourth uv-lemma) *Let u and v be orthogonal in $\mathbb{R}^n$ and suppose $\|u + v\| \leq \sqrt{2}$ and $\delta = \|u + v + uv\| \leq 1/\sqrt{2}$. Then*

$$\|u\| \leq \sqrt{1 - \sqrt{1 - 2\delta^2}}.$$

Proof: It is convenient for the proof to introduce the vector

$$z = u + v,$$

and to denote $r := \|z\|$. Since u and v are orthogonal there exists a $\varphi, 0 \leq \varphi \leq \pi/2$, such that

$$\|u\| = r \cos \varphi, \qquad \|v\| = r \sin \varphi. \tag{C.5}$$

Note that if the angle φ equals $\pi/4$ then $r = \|z\| \leq \sqrt{2}$ implies that $\|u\| = \|v\| < 1$. But for the general case we only know that $0 \leq \varphi \leq \pi/2$ and hence at first sight we should expect that the norms of $\|u\|$ and $\|v\|$ may well exceed 1. However, it will turn out below that the second condition in the lemma, namely $\delta = \|u + v + uv\| \leq 1/\sqrt{2}$, restricts the values of φ to a small neighborhood of $\pi/4$, depending on δ, thus yielding the tighter upper bound for $\|u\|$ in the lemma. Of course, the symmetry with respect to u and v implies the same upper bound for $\|v\|$.

We have

$$\delta = \|u + v + uv\| \geq \|u + v\| - \|uv\| = \|z\| - \|uv\|. \tag{C.6}$$

Applying the second uv-lemma (Lemma C.5) we find

$$\|uv\| \leq \frac{1}{\sqrt{2}} \|u\| \, \|v\| = \frac{1}{\sqrt{2}} r^2 \cos \varphi \sin \varphi = \frac{r^2 \sin 2\varphi}{2\sqrt{2}}.$$

Substituting this in (C.6) we obtain

$$\delta \geq r - \frac{r^2 \sin 2\varphi}{2\sqrt{2}}. \tag{C.7}$$

The lemma is trivial if either $\varphi = 0$ or $\varphi = \pi/2$, because then either $u = 0$ or $u = z$. In the latter case, $v = 0$, whence $\|u\| = \delta$. Since (cf. Figure 6.12, page 138)

$$\delta \leq \sqrt{1 - \sqrt{1 - 2\delta^2}},$$

the claim follows. Therefore, from now on it is assumed that

$$0 < \varphi < \frac{\pi}{2}.$$

Thus, $\sin 2\varphi > 0$ and (C.7) is equivalent to

$$(\sin 2\varphi) r^2 - 2r\sqrt{2} + 2\varphi\sqrt{2} \geq 0.$$

The left-hand side expression is quadratic in r and vanishes if

$$r = \frac{\sqrt{2}}{\sin 2\varphi}\left(1 \pm \sqrt{1 - \delta\sqrt{2}\sin 2\varphi}\right).$$

The plus sign gives a value larger than $\sqrt{2}$. Thus we obtain

$$r \leq \frac{\sqrt{2}}{\sin 2\varphi}\left(1 - \sqrt{1 - \delta\sqrt{2}\sin 2\varphi}\right) = \frac{2\delta}{1 + \sqrt{1 - \delta\sqrt{2}\sin 2\varphi}}.$$

Consequently, using $0 < \varphi < \pi/2$,

$$\|u\| = r\cos\varphi \leq \frac{2\delta\cos\varphi}{1 + \sqrt{1 - \delta\sqrt{2}\sin 2\varphi}}.$$

We proceed by considering the function

$$f(\varphi) := \frac{2\delta\cos\varphi}{1 + \sqrt{1 - \delta\sqrt{2}\sin 2\varphi}}, \quad 0 \leq \varphi \leq \pi/2,$$

with $\delta\sqrt{2} \leq 1$. Clearly this function is nonnegative and differentiable on the interval $[0, \pi/2]$. Moreover, $f(0) = \delta$ and $f(\pi/2) = 0$. On the open interval $(0, \pi/2)$ the derivative of f with respect to φ is zero if and only if

$$-\sin\varphi\left(1 + \sqrt{1 - \delta\sqrt{2}\sin 2\varphi}\right) + \frac{\delta\sqrt{2}\cos\varphi\cos 2\varphi}{\sqrt{1 - \delta\sqrt{2}\sin 2\varphi}} = 0.$$

This reduces to

$$-\sin\varphi\sqrt{1 - \delta\sqrt{2}\sin 2\varphi} - \sin\varphi\left(1 - \delta\sqrt{2}\sin 2\varphi\right) + \delta\sqrt{2}\cos\varphi\cos 2\varphi = 0,$$

which can be rewritten as

$$\delta\sqrt{2}\cos\varphi - \sin\varphi = \sin\varphi\sqrt{1 - \delta\sqrt{2}\sin 2\varphi}.$$

Taking squares we obtain

$$2\delta^2\cos^2\varphi + \sin^2\varphi - \delta\sqrt{2}\sin 2\varphi = \sin^2\varphi - \delta\sqrt{2}\sin^2\varphi\sin 2\varphi,$$

which simplifies to

$$2\delta^2\cos^2\varphi = \delta\sqrt{2}\sin 2\varphi\left(1 - \sin^2\varphi\right) = \delta\sqrt{2}\sin 2\varphi\cos^2\varphi.$$

Dividing by $\delta\sqrt{2}\cos^2\varphi$ we find the surprisingly simple expression

$$\sin 2\varphi = \delta\sqrt{2}.$$

We assume that δ is positive, because if $\delta = 0$ the lemma is trivial. Then $\sin 2\varphi = \delta\sqrt{2}$ admits two values for φ on the interval $[0, \pi/2]$, one at each side of $\pi/4$. Since we are maximizing f we have to take the value to the left of $\pi/4$. For this value, $\cos 2\varphi$ is positive. Therefore we may write

$$f(\varphi) = \frac{2\delta \cos \varphi}{1 + \sqrt{1 - \sin^2 2\varphi}} = \frac{2\delta \cos \varphi}{1 + \cos 2\varphi} = \frac{2\delta \cos \varphi}{2 \cos^2 \varphi} = \frac{\delta}{\cos \varphi}.$$

Now $\cos \varphi$ can be solved from the equation $2 \cos \varphi \sin \varphi = \delta\sqrt{2}$. Taking the larger of the two roots we obtain

$$\cos \varphi = \frac{1}{\sqrt{2}} \sqrt{1 + \sqrt{1 - 2\delta^2}}.$$

For this value of φ we have

$$f(\varphi) = \frac{\delta\sqrt{2}}{\sqrt{1 + \sqrt{1 - 2\delta^2}}} = \frac{\delta\sqrt{2}}{\sqrt{2\delta^2}} \sqrt{1 - \sqrt{1 - 2\delta^2}} = \sqrt{1 - \sqrt{1 - 2\delta^2}}.$$

Clearly this value is larger than the values at the boundary points $\varphi = 0$ and $\varphi = \pi/2$. Hence it gives the maximum value of $r \cos \varphi$ on the whole interval $[0, \pi/2]$. Thus the lemma follows. $\qquad\square$

Appendix D

Transformation to canonical form

D.1 Introduction

It is almost obvious that every LO problem can be rewritten in the canonical form given by (P). To see this, some simple observations are sufficient. First, any maximization problem can be turned into a minimization problem by multiplying the objective function by -1. Second, any equality constraint $a^T x = b$ can be replaced by the two inequality constraints $a^T x \leq b$, $a^T x \geq b$, and any inequality constraint $a^T x \leq b$ is equivalent to $-a^T x \geq -b$. Third, any free variable x, with no sign requirements, can be written as $x = x^+ - x^-$, with x^+ and x^- nonnegative. By applying these transformations to any given LO problem, we get an equivalent problem that has the canonical form of (P). The new problem is equivalent to the given problem in the sense that the new problem is feasible if and only if the given problem is feasible, and unbounded if and only if the given problem is unbounded, and, moreover, if the given problem has (one or more) optimal solutions then these can be found from the optimal solution(s) of the new problem.

The approach just sketched is quite popular in textbooks,[1] despite the fact that in practice, when dealing with solution methods, it has a number of obvious shortcomings. First, it increases the number of constraints and/or variables in the problem description. Each equality constraint is removed at the cost of an extra constraint, and each free variable is removed at the cost of an extra variable. Especially when the given problem is a large-scale problem it may be desirable to keep the dimensions of the problem as small as possible. Apart from this shortcoming the approach is even more inappropriate when dealing with an interior-point solution method. It will become clear later on that it is then essential to have a feasible region with a nonempty interior so that the level sets for the duality gap are bounded. However, when an equality constraint is replaced by two inequality constraints, these two inequalities cannot have positive slack values for any feasible point. This means that the interior of the feasible region is empty after the transformation. Moreover, the nonnegative variables introduced by eliminating a free variable are unbounded: when the same constant is added to the two new variables their difference remains the same. Hence, if in the original problem the level sets of the duality gap were bounded, we would lose this property in the new formulation of the problem.

For deriving theoretical results, the above properties of the described transformations may give no problems at all. In fact, an example of an application of this type is

---

[1] See, e.g., Schrijver [250], page 91, and Padberg [230], page 23.

given in Section 2.10. However, when it is our aim to solve a given LO problem, the approach cannot be recommended, especially if the solution method is an interior-point method.

The purpose of this section is to show that there exists an alternative approach that has an opposite effect on the problem size: it reduces the size of the problem. Moreover, if the original problem has a nonempty interior feasible region then so has the transformed problem, and if the level sets in the original problem are bounded then they are bounded after the transformation as well. In this approach, outlined below, *each equality constraint and each free variable in the original problem reduces the number of variables or the number of constraints by one*. Stated more precisely, we have the following result.

Theorem D.1 *Let* (P) *be an LO problem with* m *constraints and* n *variables. Moreover let* (P) *have* m_0 *equality constraints and* n_0 *free variables. Then there exists an equivalent canonical problem for which the sum of the number of constraints and the number of variables is not more than* $n + m - n_0 - m_0$.

Proof: In an arbitrary LO problem we distinguish between the following types of variable: nonnegative variables, free variables and nonpositive variables.[2] Similarly, three types of constraints can occur: equality constraints, inequality constraints of the less-than-or-equal-to ($\leq$) type and inequality constraints of the greater-than-or-equal-to ($\geq$) type. It is clear that nonpositive variables can be replaced by nonnegative variables at no cost by taking the opposites as new variables. Also, inequality constraints of the less-than-or-equal-to type can be turned into inequality constraints of the greater-than-or-equal-to type through multiplication by -1. In this way we can transform the problem to the following form at no cost:

$$(P) \quad \min\left\{ \begin{bmatrix} c^0 \\ c^1 \end{bmatrix}^T \begin{bmatrix} x^0 \\ x^1 \end{bmatrix} : \begin{array}{cc} A_0 x^0 + A_1 x^1 & = & b^0 \\ B_0 x^0 + B_1 x^1 & \geq & b^1 \end{array}, \quad x^1 \geq 0 \right\},$$

where, for $i = 0, 1$, A_i and B_i are matrices and b^i, c^i and x^i are vectors. The vector x^0 contains the n_0 free variables, and there are m_0 equality constraints. The variables in x^1 are nonnegative, and their number is $n - n_0$, whereas the number of inequality constraints is $m - m_0$. The sizes of the matrices and the vectors in (P) are such that all expressions in the problem are well defined and need no further specification.

D.2   Elimination of free variables

In this section we discuss the elimination of free variables, thus showing how to obtain a problem in which all variables are nonnegative. We may assume that the matrix

---

[2] A variable x_i in (P) is called a nonnegative variable if (P) contains an explicit constraint $x_i \geq 0$ and a nonpositive variable if there is a constraint $x_i \leq 0$ in (P); all remaining variables are called free variables. For the moment this classification of the variables is sufficient for our goal. But it may be useful to discuss the role of bounds on the variables. In this proof we consider any constraint of the form $x_i \geq \ell$ or $x_i \leq u$, with ℓ and u nonzero, as an inequality constraint. If the problem requires a variable x_i to satisfy $\ell \leq x_i \leq u$ then we can save one constraint by a simple shift of x_i: defining $x_i' := x_i - \ell$, the new variable is nonnegative and is bounded above by $x_i' \leq u - \ell$.

$[A_0 \ A_1]$ has full row rank. Otherwise the set of equality constraints is redundant or inconsistent. If the system is not inconsistent, we can eliminate some of these constraints until the above condition on the rank is satisfied, i.e., rank $(A_0 \ A_1) = m_0$. Introducing a surplus vector x^2, we can write the inequality constraints as

$$B_0 x^0 + B_1 x^1 - x^2 = b^1, \quad x^2 \geq 0.$$

The constraints in the problem are then represented by the equality system

$$\begin{bmatrix} A_0 & A_1 & 0 \\ B_0 & B_1 & -I_{m-m_0} \end{bmatrix} \begin{bmatrix} x^0 \\ x^1 \\ x^2 \end{bmatrix} = \begin{bmatrix} b^0 \\ b^1 \end{bmatrix}, \quad x^1 \geq 0, x^2 \geq 0,$$

where I_{m-m_0} denotes the identity matrix of size $(m - m_0) \times (m - m_0)$. We now have m equality constraints and $n + m - m_0$ variables. Grouping together the nonnegative variables, we may write the last system as

$$[F \ \ G] \begin{bmatrix} x^0 \\ z \end{bmatrix} = \begin{bmatrix} b^0 \\ b^1 \end{bmatrix}, \quad z = \begin{bmatrix} x^1 \\ x^2 \end{bmatrix} \geq 0,$$

where x^0 contains the free variables, as before, and the variables in z are nonnegative. Note that, as a consequence of the above rank condition, the matrix $[F \ G]$ has full row rank. The size of F is $m \times n_0$ and the size of G is $m \times (n - n_0 + m - m_0)$.

Let us denote the rank of F by r. The we obviously have $r \leq n_0$. Then, using Gaussian elimination, we can express r free variables in the remaining variables. We simply have to pivot on free variables as long as possible. So, as long as free variables occur in the problem formulation we choose a free variable and a constraint in which it occurs. Then, using this (equality) constraint, we express the free variable in the other variables and by substitution eliminate it from the other constraints and from the objective function. Since F has rank r, we can do this r times, and after reordering variables and equations if necessary, the constraints get the form

$$\begin{bmatrix} I_r & H & D_r \\ 0 & 0 & D \end{bmatrix} \begin{bmatrix} \bar{x}^0 \\ \tilde{x}^0 \\ z \end{bmatrix} = \begin{bmatrix} d^r \\ d \end{bmatrix}, \quad x^1 = \begin{bmatrix} \bar{x}^0 \\ \tilde{x}^0 \end{bmatrix}, \quad z \geq 0, \quad \text{(D.1)}$$

where I_r is the $r \times r$ identity matrix, which is multiplied with $\bar{x}^0$, the vector of the eliminated free variables, and H is an $r \times (n_0 - r)$ matrix, which is multiplied with $\tilde{x}^0$, the vector of free variables that are not eliminated; the columns of D_r and D correspond to the nonnegative variables in z. Moreover, since the variables $\bar{x}^0$ have been eliminated from the objective function, there exist vectors c_H and c_D such that the objective function has the form

$$c_H^T \tilde{x}^0 + c_D^T z. \quad \text{(D.2)}$$

We are left with m equalities. The first r equalities express the free variables in $\bar{x}^0$ in the remaining variables, while the remaining $m - r$ equalities contain no free variables. Observe that the first r equalities do not impose a condition on the feasibility of the vector z; they simply tell us how the values of the free variables in $\bar{x}^0$ can be calculated from the remaining variables.

We conclude that the problem is feasible if and only if the system

$$Dz = d, \quad z \geq 0 \tag{D.3}$$

is feasible. Assuming this, for an any z satisfying (D.3) we can choose the vector $\tilde{x}^0$ arbitrarily and then compute $\bar{x}^0$ such that the resulting vector satisfies (D.1). So fixing z, and hence also fixing its contribution $c_D^T z$ to the objective function (D.2), we can make the objective value arbitrary small if the vector c_H is nonzero. Since the variables in $\bar{x}^0$ do not occur in the objective function, it follows from this that the problem is unbounded if c_H is nonzero.

So, if the problem is not unbounded then $c_H = 0$. In that case it remains to solve the problem

$$(P') \quad \min \left\{ c_D^T z \ : \ Dz = d, z \geq 0 \right\},$$

where D is an $(m-r) \times (n - n_0 + m - m_0)$ matrix and this matrix has rank $m - r$. Note that (P') is in standard format.

D.3 Removal of equality constraints

We now show how problem (P') can be reduced to canonical form. This goes by using the same pivoting procedure as above. Choose a variable and an equality constraint in which it occurs. Use the constraint to express the chosen variable in the other variables and then eliminate this variable from the other constraints and the objective function. Since A has rank $m - r$ we can repeat this process $m - r$ times and then we are left with expressions for the $m - r$ eliminated variables in the remaining (nonnegative) variables. The number of the remaining variables is

$$n - n_0 + m - m_0 - (m - r) = n - n_0 + r - m_0.$$

Now the nonnegativity conditions on the $m - r$ eliminated variables result in $m - r$ inequality constraints for the remaining $n - n_0 + r - m_0$ variables. So we are left with $m - r$ inequality constraints that contain $n - n_0 + r - m_0$ variables. The sum of these numbers being $n + m - n_0 - m_0$, the theorem has been proved. $\qquad \square$

Before giving an example of the above reduction we make some observations.

Remark D.2 When dealing with an LO problem, it is most often desirable to have an economical representation of the problem. Theorem D.1 implies that whenever the model contains equality constraints or free variables, then the size of the constraint matrix can be reduced by transforming the problem to a canonical form. As a consequence, when we consider the dimension of the constraint matrix as a measure of the size of the model, then any minimal representation of the problem has a canonical form. Of course, here it is assumed that in any such representation, nonpositive variables are replaced by nonnegative variables and $\leq$ inequalities by $\geq$ inequalities; these transformations do not change the dimension of the constraint matrix. In this connection it may be useful to point out that the representation obtained by the transformation in the proof of Theorem D.1 may be far from a minimal representation. Any claim of this type is poorly founded. For example, if the given problem is infeasible then a representation with one constraint and one variable exists. But to find out whether the problem is infeasible one really has to solve it.

Remark D.3 It may happen that after the above transformations we are left with a canonical problem

$$(P) \qquad \min \left\{ c^T x \ : \ Ax \geq b, \ x \geq 0 \right\},$$

for which the matrix A has a zero row. In that case we can reduce the problem further. If the i-th row of A is zero and $b_i \leq 0$ then the i-th row of A and the i-th entry of b can be removed. If $b_i > 0$ then we may decide that the problem is infeasible.

Remark D.4 Also if A has a zero column further reduction is possible. If the j-th column of A is zero and $c_j > 0$ then we have $x_j = 0$ in any optimal solution and this column and the corresponding entry of c can be deleted. If $c_j < 0$ then the problem is unbounded. Finally, if $c_j = 0$ then x_j may be given any (nonnegative) value. For the further analysis of the problem we may delete the j-th column of A and the entry c_j in c.

Example D.5 By way of example we consider the problem

$$(EP) \qquad \max \left\{ y_1 + y_2 \ : \ -1 \leq y_1 \leq 1, \ y_2 \leq 1 \right\}. \qquad (D.4)$$

This problem has two variables and three constraints, so the constraint matrix has size 3×2. Since the two variables are free (cf. Footnote 2), Theorem D.1 guarantees the existence of a canonical description of the problem for which the sum of the numbers of rows and columns in the constraint matrix is at most 3 $(= 5 - 2)$. Following the scheme of the proof of Theorem D.1 we construct such a canonical formulation. First, by introducing nonnegative slack variables for the three inequality constraints, we change all constraints into equality constraints:

$$\begin{aligned} -y_1 \quad &+ \ s_1 \qquad\qquad\qquad = \ 1 \\ y_1 \qquad &\quad\;\; + \ s_2 \qquad\quad = \ 1 \\ y_2 \qquad\qquad &\qquad\quad + \ s_3 \ = \ 1. \end{aligned}$$

The free variables y_1 and y_2 can be eliminated by using

$$\begin{aligned} y_1 \ &= \ s_1 - 1 \\ y_2 \ &= \ 1 - s_3, \end{aligned}$$

and since $y_1 + y_2 = s_1 - s_3$ we obtain the equivalent problem

$$\max \left\{ s_1 - s_3 \ : \ s_1 + s_2 = 2, \ s_1, s_2, s_3 \geq 0 \right\}.$$

By elimination of s_2 this reduces to

$$\max \left\{ s_1 - s_3 \ : \ s_1 \leq 2, \ s_1, s_3 \geq 0 \right\}. \qquad (D.5)$$

The problem is now reduced to the dual canonical form, as given by (2.2), with the following constraint matrix A, right-hand side vector c and objective vector b:

$$A = \begin{bmatrix} 1 \\ 0 \end{bmatrix}, \quad c = \begin{bmatrix} 2 \end{bmatrix}, \quad b = \begin{bmatrix} 1 \\ -1 \end{bmatrix}.$$

Note that the constraint matrix in this problem has size 2×1, and the sum of the dimensions is 3, as expected. ◇

In the above example the optimal solution $y = (1, 1)$ is unique. We consider below two modifications of the sample problem (EP) by changing the objective function. In the first modification we use the objective function y_1; then the optimal set consists of all $y = (1, y_2)$ with $y_2 \leq 1$. The optimal solution is no longer unique. The second modification has objective function $y_1 - y_2$; then the problem is unbounded, as can easily be seen.

Example D.6 In this example we consider the problem

$$\max \{ y_1 : -1 \leq y_1 \leq 1, y_2 \leq 1 \}. \tag{D.6}$$

As in the previous example we can introduce nonnegative slack variables s_1, s_2 and s_3 and then eliminate the variables y_1, y_2 and s_2, arriving at the canonical problem

$$\max \{ s_1 : s_1 \leq 2, s_1, s_3 \geq 0 \}. \tag{D.7}$$

Here we have replaced the objective $y_1 = s_1 - 1$ simply by s_1, thereby omitting the constant -1, which is irrelevant for the optimization. The dependence of the eliminated variables on the variables in this problem is the same as in the previous example:

$$
\begin{aligned}
y_1 &= s_1 - 1 \\
y_2 &= 1 - s_3 \\
s_2 &= 2 - s_1.
\end{aligned}
$$

The constraint matrix A and the right-hand side vector c in the dual canonical formulation are the same as before; only the objective vector b has changed:

$$A = \begin{bmatrix} 1 \\ 0 \end{bmatrix}, \quad c = \begin{bmatrix} 2 \end{bmatrix}, \quad b = \begin{bmatrix} 1 \\ 0 \end{bmatrix}. \qquad ◇$$

Example D.7 Finally we consider the unbounded problem

$$\max \{ y_1 - y_2 : -1 \leq y_1 \leq 1, y_2 \leq 1 \}. \tag{D.8}$$

In this case the optimal set is empty. To avoid repetition we immediately state the canonical model:

$$\max \{ s_1 + s_3 : s_1 \leq 2, s_3 \geq 0 \}. \tag{D.9}$$

The dependence of the eliminated variables on the variables in this problem is the same as in the previous example. The matrix A and vectors c and b are now

$$A = \begin{bmatrix} 1 \\ 0 \end{bmatrix}, \quad c = \begin{bmatrix} 2 \end{bmatrix}, \quad b = \begin{bmatrix} 1 \\ 1 \end{bmatrix}. \qquad ◇$$

Appendix E

The Dikin step algorithm

E.1   Introduction

In this appendix we reconsider the self-dual problem

$$(SP) \qquad \min \left\{ q^T z \; : \; Mz \geq -q, \, z \geq 0 \right\}. \qquad (E.1)$$

as given by (2.16) and we present a simple algorithm for solving (SP) different from the full-Newton step algorithm of Section 3. Recall that we may assume without loss of generality that $x = e$ is feasible and $s(e) = Me + q = e$, so e is the point on the central path of (SP) corresponding to the value 1 of the barrier parameter. Moreover, at this point the objective value equals n, the order of the skew-symmetric matrix M.

The algorithm can be described roughly as follows. Starting at $x^0 = e$ the algorithm approximately follows the central path until the objective value reaches some (small) target value ε. This is achieved by moving from x^0 along a direction — more or less tangent to the central path — to the next iterate x^1, in such a way that x^1 is close to the central path again, but with a smaller objective value. Then we repeat the same procedure until the objective has become small enough.

In the next section we define the *search direction* used in the algorithm.[1] Then, in Section E.3 the algorithm is defined and in subsequent sections the algorithm is analyzed. This results in an iteration bound, in Section E.5.

E.2   Search direction

Let x be a positive solution of (SP) such that its surplus vector $s = s(x)$ is positive, and let Δx denote a displacement in the x-space. For the moment we neglect the nonnegativity conditions in (SP). Then, the new iterate x^+ is given by

$$x^+ := x + \Delta x,$$

and the new surplus vector s^+ follows from

$$s^+ = s(x^+) = M(x + \Delta x) + q = s + M \Delta x.$$

---

[1] After the appearance of Karmarkar's paper in 1984, Barnes [34] and Vanderbei, Meketon and Freedman [279] proposed a simplified version of Karmarkar's algorithm. Later, their algorithm appeared to be just a rediscovery of the *primal affine-scaling method* proposed by Dikin [63] in 1967. See also Barnes [35]. The search direction used in this chapter can be considered as a primal-dual variant of the affine-scaling direction of Dikin (cf. the footnote on page 339) and is therefore named the *Dikin direction*. It was first proposed by Jansen, Roos and Terlaky [156].

The displacement Δs in the s-space is simply given by

$$\Delta s = s^+ - s = M\Delta x,$$

and, hence, the two displacements are related by

$$M\Delta x - \Delta s = 0. \tag{E.2}$$

This implies, by the orthogonality property (2.22), that Δx and Δs are orthogonal:

$$(\Delta x)^T \Delta s = (\Delta x)^T M\Delta x = 0. \tag{E.3}$$

The inequality constraints in (SP) require that

$$x + \Delta x \geq 0, \quad s + \Delta s \geq 0.$$

In fact, we want to stay in the interior of the feasible region, so we need to find displacements Δx and Δs such that

$$x + \Delta x > 0, \quad s + \Delta s > 0.$$

Following an idea of Dikin [63, 65], we replace the nonnegativity conditions by requiring that the next iterates $(x + \Delta x, s + \Delta s)$ belong to a suitable ellipsoid. We define this ellipsoid by requiring that

$$\left\| \frac{\Delta x}{x} + \frac{\Delta s}{s} \right\| \leq 1, \tag{E.4}$$

and call this ellipsoid in $\mathbb{R}^{2n}$ the *Dikin ellipsoid*.

Remark E.1 It may be noted that when there are no additional conditions on the displacements Δx and Δs, then the Dikin ellipsoid is highly degenerate in the sense that it contains a linear space. For then the equation $s\Delta x + x\Delta s = 0$ determines an n-dimensional linear space that is contained in it. However, when intersecting the Dikin ellipsoid with the linear space (E.2), we get a bounded set. This can be seen as follows. The pair $(\Delta x, \Delta s)$ belongs to the Dikin ellipsoid if and only if (E.4) holds. Now (E.4) can be rewritten as

$$\left\| \frac{s\Delta x + x\Delta s}{xs} \right\| \leq 1.$$

By substitution of $\Delta s = M\Delta x$ this becomes

$$\left\| \frac{s\Delta x + xM\Delta x}{xs} \right\| \leq 1,$$

which is equivalent to

$$\left\| (XS)^{-1} (S + XM) \Delta x \right\| \leq 1.$$

The matrix $(XS)^{-1} (S + XM)$ is nonsingular, and hence Δx is bounded. See also Exercise 9 (page 29) and Exercise 113 (page 453). •

Our aim is to minimize the objective value $q^T x = x^T s$. The new objective value is

$$(x + \Delta x)^T (s + \Delta s) = x^T s + x^T \Delta s + s^T \Delta x.$$

Here we have used that Δx and Δs are orthogonal, from (E.3). Now minimizing the new objective value over the Dikin ellipsoid amounts to solving the following optimization problem:

$$\min \left\{ s^T \Delta x + x^T \Delta s \; : \; M\Delta x - \Delta s = 0, \; \left\| \frac{\Delta x}{x} + \frac{\Delta s}{s} \right\| \leq 1 \right\}. \qquad (E.5)$$

We proceed by showing that this problem uniquely determines the search direction vectors. For this purpose we rewrite (E.5) as follows.

$$\min \left\{ (xs)^T \left(\frac{\Delta x}{x} + \frac{\Delta s}{s} \right) \; : \; M\Delta x - \Delta s = 0, \; \left\| \frac{\Delta x}{x} + \frac{\Delta s}{s} \right\| \leq 1 \right\}. \qquad (E.6)$$

The vector

$$\xi := \frac{\Delta x}{x} + \frac{\Delta s}{s}$$

must belong to the unit ball. When we neglect the affine constraint $\Delta s = M\Delta x$ in (E.6) we get the relaxation

$$\min \left\{ (xs)^T \xi \; : \; \|\xi\| \leq 1 \right\}.$$

This problem has a trivial (and unique) solution, namely

$$\xi = -\frac{xs}{\|xs\|}.$$

Thus, if we can find Δx and Δs such that

$$\frac{\Delta x}{x} + \frac{\Delta s}{s} = -\frac{xs}{\|xs\|} \qquad (E.7)$$

$$\Delta s = M\Delta x \qquad (E.8)$$

then Δx and Δs will solve (E.5). Multiplying both sides of (E.7) with xs yields

$$s\Delta x + x\Delta s = -\frac{x^2 s^2}{\|xs\|}. \qquad (E.9)$$

Now substituting (E.8) we get[2,3]

$$(S + XM)\, \Delta x = -\frac{x^2 s^2}{\|xs\|}.$$

Thus we have found the solution of (E.5), namely

$$\Delta x = -(S + XM)^{-1} \frac{x^2 s^2}{\|xs\|} \qquad (E.10)$$

$$\Delta s = M\Delta x. \qquad (E.11)$$

_____

[2] As usual, $X = \mathrm{diag}\,(x)$ and $S = \mathrm{diag}\,(s)$.

[3] **Exercise 113** If we define $d := \sqrt{x/s}$ then show that the Dikin step Δx can be rewritten as

$$\Delta x = -D\,(I + DMD)^{-1} \frac{x^{\frac{3}{2}} s^{\frac{3}{2}}}{\|xs\|}.$$

We call Δx the *Dikin direction* or *Dikin step* at x for the self-dual model (SP). In the next section we present an algorithm that is based on the use of this direction, and in subsequent sections we prove that this algorithm solves (SP) in polynomial time.

E.3   Algorithm using the Dikin direction

The reader should be aware that we have so far not discussed whether the Dikin step yields a feasible point. Before stating our algorithm we need to deal with this. For the moment it suffices to point out that in the algorithm we use a *step-size* parameter α. Starting at x we move in the direction along the Dikin step Δx to $x + \alpha \Delta x$. The value of α is specified later on. The algorithm can now be described as follows.

Dikin Step Algorithm for the Self-dual Model

Input:
    An accuracy parameter $\varepsilon > 0$;
    a step-size parameter α, $0 < \alpha \leq 1$;
    $x^0 > 0$ such that $s(x^0) > 0$.
begin
    $x := x^0$; $s := s(x)$;
    while $x^T s \geq \varepsilon$ **do**
    begin
        $x := x + \alpha \Delta x$ (with Δx from (E.10));
        $s := s(x)$;
    end
end

Below we analyze this algorithm and provide a default value for the step-size parameter α for which the Dikin step is always feasible. This makes the algorithm well defined. In the analysis of the algorithm we need a measure for the 'distance' of an iterate x to the central path . To this end, for each positive feasible vector x with $s(x) > 0$, we use the number $\delta_c(x)$ as introduced in (3.20):

$$\delta_c(x) := \frac{\max(xs(x))}{\min(xs(x))}. \tag{E.12}$$

Below, in Theorem E.5 we show that the algorithm needs no more than

$$\left\lceil \tau n \log \frac{q^T x^0}{\varepsilon} \right\rceil$$

iterations to produce a solution x with $x^T s(x) \leq \varepsilon$, where τ depends on x^0 according to

$$\tau = \max\left(2, \delta_c(x^0)\right).$$

Recall that it may be assumed without loss of generality that x^0 lies on the central path , in which case $\delta_c(x^0) = 1$ and $\tau = 2$.

E.4 Feasibility, proximity and step-size

We proceed by a condition on the step-size that guarantees the feasibility of the new iterates. Let us say that the *step-size α is feasible* if the new iterate and its surplus vector are positive. Then we may state the following result.

Lemma E.2 *Let $\alpha \geq 0$, $x^\alpha = x + \alpha\Delta x$ and $s^\alpha = s + \alpha\Delta s$. If $\bar{\alpha}$ is such that $x^\alpha s^\alpha > 0$ for all α satisfying $0 \leq \alpha \leq \bar{\alpha}$, then the step-size $\bar{\alpha}$ is feasible.*

Proof: If $\bar{\alpha}$ satisfies the hypothesis of the lemma then the coordinates of x^α and s^α cannot vanish for any $\alpha \in [0, \bar{\alpha}]$. Hence, since $x^0 s^0 > 0$, by continuity, x^α and s^α must be positive for any such α. □

We use the superscript $^+$ to refer to entities after the Dikin step of size α at x:

$$x^+ \quad := \quad x + \alpha\Delta x,$$
$$s^+ \quad := \quad s + \alpha\Delta s.$$

Consequently,

$$x^+ s(x^+) = (x + \alpha\Delta x)(s + \alpha\Delta s) = xs + \alpha\,(x\Delta x + s\Delta s) + \alpha^2\Delta x\Delta s.$$

Since, by (E.9,

$$s\Delta x + x\Delta s = -\frac{x^2 s^2}{\|xs\|},$$

we obtain

$$x^+ s(x^+) = xs - \alpha\frac{x^2 s^2}{\|xs\|} + \alpha^2\Delta x\Delta s. \qquad (E.13)$$

Observe that Lemma E.2 implies that the step-size $\bar{\alpha}$ is feasible if

$$xs - \alpha\frac{x^2 s^2}{\|xs\|} + \alpha^2\Delta x\Delta s > 0$$

for all α satisfying $0 \leq \alpha \leq \bar{\alpha}$. Recall that the objective value is given by $q^T x = x^T s(x)$. In the next lemma we investigate the reduction of the objective value during a Dikin step with size α.

Lemma E.3 *If the step-size α is feasible then*

$$\left(x^+\right)^T s^+ \leq \left(1 - \frac{\alpha}{\sqrt{n}}\right) x^T s.$$

Proof: Using (E.13) and the fact that Δx and Δs are orthogonal, the objective value $\left(x^+\right)^T s^+$ after the step can be expressed as follows.

$$\left(x^+\right)^T s^+ = x^T s - \alpha e^T\frac{x^2 s^2}{\|xs\|} = x^T s - \alpha\,\|xs\|.$$

The Cauchy–Schwarz inequality implies

$$x^T s = e^T(xs) \leq \|e\|\,\|xs\| = \sqrt{n}\,\|xs\|.$$

Substitution gives

$$\left(x^+\right)^T s^+ \leq \left(1 - \frac{\alpha}{\sqrt{n}}\right) x^T s.$$

Hence the lemma follows. $\qquad\square$

Now let $\tau > 1$ be some constant. We assume that we are given a feasible x such that $\delta_c(x) \leq \tau$, and we establish a bound for the step-size α such that this property is maintained after the Dikin step. Note that $\delta_c(x) \leq \tau$ implies the existence of positive numbers τ_1 and τ_2 such that

$$\tau_1 e \leq xs \leq \tau_2 e, \quad \text{with } \tau_2 = \tau\tau_1.$$

The numbers τ_1 and τ_2 are used in the next lemma.

Lemma E.4 *Let $\tau > 1$. Suppose that $x > 0$ is feasible so that $s := s(x) > 0$ and $\delta_c(x) \leq \tau$. Then, any step-size α satisfying*

$$\alpha \leq \frac{\|xs\|}{2\tau_2} \quad \text{and} \quad \alpha < \frac{4\tau_1}{\|xs\|}$$

is feasible, and after a step of this size we have $\delta_c(x^+) \leq \tau$.

Proof: Recall from (E.13) that

$$x^+ s(x^+) = xs - \alpha \frac{x^2 s^2}{\|xs\|} + \alpha^2 \Delta x \Delta s. \tag{E.14}$$

Using the first bound on α in the lemma, we can easily verify that the map

$$t \mapsto t - \alpha \frac{t^2}{\|xs\|}$$

is an increasing function for $t \in [0, \tau_2]$. Application of this map to each component of the vector xs gives

$$\left(\tau_1 - \alpha \frac{\tau_1^2}{\|xs\|}\right) e \leq xs - \alpha \frac{x^2 s^2}{\|xs\|} \leq \left(\tau_2 - \alpha \frac{\tau_2^2}{\|xs\|}\right) e.$$

Substitution in (E.14) gives

$$\left(\tau_1 - \alpha \frac{\tau_1^2}{\|xs\|}\right) e + \alpha^2 \Delta x \Delta s \leq x^+ s(x^+) \leq \left(\tau_2 - \alpha \frac{\tau_2^2}{\|xs\|}\right) e + \alpha^2 \Delta x \Delta s, \tag{E.15}$$

thus yielding lower and upper bounds for the entries of $x^+ s(x^+)$. It follows that if the Dikin step with size α is feasible, then we certainly have $\delta_c(x^+) \leq \tau$ if

$$\tau\left(\left(\tau_1 - \alpha \frac{\tau_1^2}{\|xs\|}\right) e + \alpha^2 \Delta x \Delta s\right) > \left(\tau_2 - \alpha \frac{\tau_2^2}{\|xs\|}\right) e + \alpha^2 \Delta x \Delta s. \tag{E.16}$$

On the other hand, if (E.16) holds, then this implies feasibility of the step-size α. This follows by substituting (E.15) into (E.16), which gives

$$\tau\left(\left(\tau_1 - \alpha\frac{\tau_1^2}{\|xs\|}\right)e + \alpha^2\Delta x\Delta s\right) > \left(\tau_1 - \alpha\frac{\tau_1^2}{\|xs\|}\right)e + \alpha^2\Delta x\Delta s.$$

Since $\tau > 1$ this implies

$$\left(\tau_1 - \alpha\frac{\tau_1^2}{\|xs\|}\right)e + \alpha^2\Delta x\Delta s > 0.$$

By (E.15), this makes clear that the coordinates of x^+s^+ do not vanish for any step-size satisfying the bound in the lemma. By Lemma E.2 this implies that any such step-size α is feasible. It remains to show that α satisfies (E.16).

The inequality (E.16) can be simplified by using $\tau_2 = \tau\tau_1$, and then dividing by α. Thus we find that (E.16) is equivalent to

$$\left(\frac{\tau_2^2 - \tau\tau_1^2}{\|xs\|}\right)e + \alpha(\tau - 1)\Delta x\Delta s > 0.$$

This can be further simplified by using

$$\tau_2^2 - \tau\tau_1^2 = (\tau - 1)\,\tau_1\tau_2.$$

Thus the condition that guarantees $\delta_c(x^+) \leq \tau$ reduces to

$$\frac{\tau_1\tau_2}{\|xs\|}e + \alpha\Delta x\Delta s > 0. \tag{E.17}$$

Note that the orthogonality of Δx and Δs implies that not all coordinates of the vector $\Delta x\Delta s$ can be positive. The most negative element of $\Delta x\Delta s$ gives the strongest bound on α in the above inequality. We can find a lower bound for this element by using Lemma C.4 in Appendix C. The first statement in this lemma (with $u = d^{-1}\Delta x$ and $v = d\Delta s$, where $d = \sqrt{x/s}$) gives

$$\|\Delta x\Delta s\|_\infty = \left\|\left(d^{-1}\Delta x\right)\left(d\Delta s\right)\right\|_\infty \leq \frac{1}{4}\left\|d^{-1}\Delta x + d\Delta s\right\|^2 = \frac{1}{4}\left\|\frac{s\Delta x + x\Delta s}{\sqrt{xs}}\right\|^2.$$

Using (E.9) once more we get

$$\|\Delta x\Delta s\|_\infty \leq \frac{1}{4}\left\|\sqrt{xs}\,\frac{xs}{\|xs\|}\right\|^2 \leq \frac{1}{4}\|xs\|_\infty\left\|\frac{xs}{\|xs\|}\right\|^2 = \frac{1}{4}\|xs\|_\infty \leq \frac{\tau_2}{4}.$$

Thus it follows by substitution that (E.17) is certainly satisfied if

$$\frac{\tau_1\tau_2}{\|xs\|} - \frac{\alpha\tau_2}{4} > 0.$$

This is equivalent to

$$\alpha < \frac{4\tau_1}{\|xs\|},$$

which is the second bound on α in the lemma. This completes the proof. $\qquad\square$

E.5 Convergence analysis

The previous section contains the ingredients necessary for deriving an upper bound for the number of iterations needed by the algorithm.

Theorem E.5 *Let $\tau := \max\left(2, \delta_c(x^0)\right)$ and $\alpha = 1/(\tau\sqrt{n})$. Then, if $n \geq 2$, the Dikin Step Algorithm for the self-dual Model requires at most*

$$\left\lceil \tau n \log \frac{q^T x^0}{\varepsilon} \right\rceil$$

iterations. The output is a feasible solution x such that $\delta_c(x) \leq \tau$ and $q^T x \leq \varepsilon$.

Proof: Initially we are given a feasible $x = x^0 > 0$ such that $\delta_c(x) \leq \tau$. The choice of the step-size α guarantees that after each iteration these properties are maintained. This can be deduced from Lemma E.4, as we now show. It suffices to show that the specified value of α meets the bound in Lemma E.4. Since $n \geq 2$ we have

$$\alpha = \frac{1}{\tau\sqrt{n}} = \frac{\tau_1}{\tau_2\sqrt{n}} \leq \frac{\tau_1\sqrt{n}}{2\tau_2} = \frac{\|\tau_1 e\|}{2\tau_2} \leq \frac{\|xs\|}{2\tau_2},$$

where we have also used that $0 \leq \tau_1 e \leq xs$. Furthermore, using $\|xs\| \leq \tau_2\sqrt{n}$, we may write

$$\frac{4\tau_1}{\|xs\|} \geq \frac{4\tau_1}{\tau_2\sqrt{n}} = \frac{4}{\tau\sqrt{n}} > \alpha.$$

Thus, Lemma E.4 implies that after each iteration the iterate x satisfies $\delta_c(x) \leq \tau$. Initially the objective value equals $q^T x^0$. Each iteration reduces the objective by a factor $1 - 1/(n\tau)$, from Lemma E.3. Hence, after k iterations the objective value is smaller than ε if

$$\left(1 - \frac{1}{n\tau}\right)^k q^T x^0 \leq \varepsilon.$$

Taking logarithms, this becomes

$$k \log\left(1 - \frac{1}{n\tau}\right) + \log(q^T x^0) \leq \log\varepsilon.$$

Since

$$-\log\left(1 - \frac{1}{n\tau}\right) \geq \frac{1}{n\tau},$$

this is certainly satisfied if

$$\frac{k}{n\tau} \geq \log(q^T x^0) - \log\varepsilon = \log\frac{q^T x^0}{\varepsilon}.$$

This implies the theorem. $\qquad\qquad\qquad\qquad\qquad\qquad\qquad\qquad\qquad\qquad\qquad$ □

Example E.6 In this example we demonstrate the behavior of the Dikin Step Algorithm by applying it to the problem (SP) in Example I.7, as given in (2.19) (page 23). The same problem was solved earlier by the Full-Newton Step Algorithm in Example I.38.

We initialize the algorithm with $z = e$. Then Theorem E.5, with $\tau = 2$ and $n = 5$, yields that the algorithm requires at most

$$\left\lceil 10 \log \frac{5}{\varepsilon} \right\rceil$$

iterations. For $\varepsilon = 10^{-2}$ we have $\log(5/\varepsilon) = \log 500 = 6.2146$, and we get 63 as an upper bound for the number of iterations. When running the algorithm with this ε the actual number of iterations is 58. The output of the algorithm is

$$z = (1.5985, 0.0025, 0.7998, 0.8005, 0.0020)$$

and

$$s(z) = (0.0012, 0.8005, 0.0025, 0.0025, 1.0000).$$

The left plot in Figure E.1 shows how the coordinates of the vector z develop in the course of the algorithm. The right plot does the same for the coordinates of the surplus vector $s = s(z)$. Observe that z and $s(z)$ converge to the same solution as found in

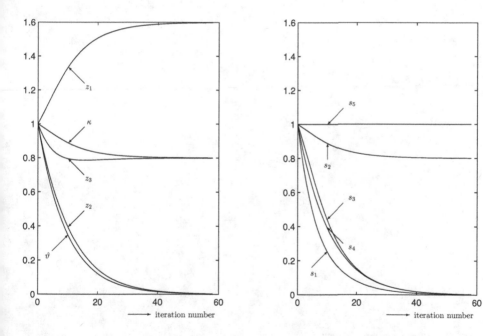

Figure E.1 Output of the Dikin Step Algorithm for the problem in Example I.7.

Example I.38 by the Full-Newton Step Algorithm, but the number of iterations is higher. ◇

Bibliography

[1] I. Adler, N.K. Karmarkar, M.G.C. Resende, and G. Veiga. Data structures and programming techniques for the implementation of Karmarkar's algorithm. *ORSA J. on Computing*, 1:84–106, 1989.

[2] I. Adler, N.K. Karmarkar, M.G.C. Resende, and G. Veiga. An implementation of Karmarkar's algorithm for linear programming. *Mathematical Programming*, 44:297–335, 1989. Errata in *Mathematical Programming*, 50:415, 1991.

[3] I. Adler and R.D.C. Monteiro. Limiting behavior of the affine scaling continuous trajectories for linear programming problems. *Mathematical Programming*, 50:29–51, 1991.

[4] I. Adler and R.D.C. Monteiro. A geometric view of parametric linear programming. *Algorithmica*, 8:161–176, 1992.

[5] A.V. Aho, J.E. Hopcroft, and J.D. Ullman. *The Design and Analysis of Computer Algorithms*. Addison-Wesley, Reading, Mass., 1974.

[6] M. Akgül. A note on shadow prices in linear programming. *J. Operational Research Society*, 35:425–431, 1984.

[7] A. Altman and K.C. Kiwiel. A note on some analytic center cutting plane methods for convex feasibility and minimization problems. *Computational Optimization and Applications*, 5, 1996.

[8] E.D. Andersen. Finding all linearly dependent rows in large-scale linear programming. *Optimization Methods and Software*, 6:219–227, 1995.

[9] E.D. Andersen and K.D. Andersen. Presolving in linear programming. *Mathematical Programming*, 71:221–245, 1995.

[10] E.D. Andersen, J. Gondzio, Cs. Mészáros, and X. Xu. Implementation of interior point methods for large scale linear programming. In T. Terlaky, editor, *Interior Point Methods of Mathematical Programming*, pp. 189–252. Kluwer Academic Publishers, Dordrecht, The Netherlands, 1996.

[11] E.D. Andersen and Y. Ye. Combining interior-point and pivoting algorithms for linear programming. *Management Science*, 42(12):1719–1731, 1996.

[12] K.D. Andersen. A modified Schur complement method for handling dense columns in interior-point methods for linear programming. *ACM Transactions Mathematical Software*, 22(3):348–356, 1996.

[13] E.D. Andersen, C. Roos, T. Terlaky, T. Trafalisand J.P. Warners. The use of low-rank updates in interior-point methods. In: Ed. Y. Yuan, *Numerical Linear Algebra and Optimization*, pp. 1–12. Science Press, Beijing, China, 2004.

[14] K.M. Anstreicher. A monotonic projective algorithm for fractional linear programming. *Algorithmica*, 1(4):483–498, 1986.

[15] K.M. Anstreicher. A strengthened acceptance criterion for approximate projections in Karmarkar's algorithm. *Operations Research Letters*, 5:211–214, 1986.

[16] K.M. Anstreicher. A combined Phase I – Phase II projective algorithm for linear programming. *Mathematical Programming*, 43:209–223, 1989.

[17] K.M. Anstreicher. Progress in interior point algorithms since 1984. *SIAM News*, 22:12–14, March 1989.

[18] K.M. Anstreicher. The worst-case step in Karmarkar's algorithm. *Mathematics of Operations Research*, 14:294–302, 1989.

[19] K.M. Anstreicher. Dual ellipsoids and degeneracy in the projective algorithm for linear programming. *Contemporary Mathematics*, 114:141–149, 1990.

[20] K.M. Anstreicher. A standard form variant and safeguarded linesearch for the modified Karmarkar algorithm. *Mathematical Programming*, 47:337–351, 1990.

[21] K.M. Anstreicher. A combined phase I – phase II scaled potential algorithm for linear programming. *Mathematical Programming*, 52:429–439, 1991.

[22] K.M. Anstreicher. On the performance of Karmarkar's algorithm over a sequence of iterations. *SIAM J. on Optimization*, 1(1):22–29, 1991.

[23] K.M. Anstreicher. On interior algorithms for linear programming with no regularity assumptions. *Operations Research Letters*, 11:209–212, 1992.

[24] K.M. Anstreicher. Potential reduction algorithms. In T. Terlaky, editor, *Interior Point Methods of Mathematical Programming*, pp. 125–158. Kluwer Academic Publishers, Dordrecht, The Netherlands, 1996.

[25] K.M. Anstreicher and R.A. Bosch. Long steps in a $O(n^3 L)$ algorithm for linear programming. *Mathematical Programming*, 54:251–265, 1992.

[26] K.M. Anstreicher and R.A. Bosch. A new infinity-norm path following algorithm for linear programming. *SIAM J. on Optimization*, 5:236–246, 1995.

[27] M. Arioli, I.S. Duff, and P.P.M. de Rijk. On the augmented system approach to sparse least-squares problems. *Numer. Math.*, 55:667–684, 1989.

[28] M.D. Asić, V.V. Kovačević-Vujčić, and M.D. Radosavljević-Nikolić. A note on limiting behavior of the projective and the affine rescaling algorithms. *Contemporary Mathematics*, 114:151–157, 1990.

[29] D.S. Atkinson and P.M. Vaidya. A scaling technique for finding the weighted analytic center of a polytope. *Mathematical Programming*, 57:163–192, 1992.

[30] D.S. Atkinson and P.M. Vaidya. A cutting plane algorithm that uses analytic centers. *Mathematical Programming*, 69(69), 1995.

[31] O. Bahn, J.-L. Goffin, O. du Merle, and J.-Ph. Vial. A cutting plane method from analytic centers for stochastic programming. *Mathematical Programming*, 69(1):45–73, 1995.

[32] Y.Q. Bai, M. Elghami, and C. Roos. A comparative study of kernel functions for primal-dual interior-point algorithms in linear optimization. *SIAM J. on Optimization*, 15(1):101–128, 2004.

[33] M.L. Balinski and A.W. Tucker. Duality theory of linear programs: a constructive approach with applications. *SIAM Review*, 11:499–581, 1969.

[34] E.R. Barnes. A variation on Karmarkar's algorithm for solving linear programming problems. *Mathematical Programming*, 36:174–182, 1986.

[35] E.R. Barnes. Some results concerning convergence of the affine scaling algorithm. *Contemporary Mathematics*, 114:131–139, 1990.

[36] E.R. Barnes, S. Chopra, and D.J. Jensen. The affine scaling method with centering. Technical Report, Dept. of Mathematical Sciences, IBM T. J. Watson Research Center, P. O. Box 218, Yorktown Heights, NY 10598, USA, 1988.

[37] M.S. Bazaraa, H.D. Sherali, and C.M. Shetty. *Nonlinear Programming: Theory and Algorithms*. John Wiley & Sons, New York (second edition), 1993.

[38] R. Bellman. *Introduction to Matrix Analysis*. Volume 12 of *SIAM Classics in Applied Mathematics*. SIAM, Philadelphia, 1995.

[39] A. Ben-Israel and T.N.E. Greville. *Generalized Inverses: Theory and Applications*. John Wiley & Sons, New York, USA, 1974.

[40] D.P. Bertsekas. *Nonlinear Programming*. Athena Scientific, Belmont, Massachusetts, 1995.

[41] G. Birkhoff and S. MacLane. *A Survey of Modern Algebra*. Macmillan, New York, 1977.

[42] R.E. Bixby. Progress in linear programming. *ORSA J. on Computing*, 6:15–22, 1994.

[43] R.E. Bixby and M.J. Saltzman. Recovering an optimal LP basis from an interior point solution. *Operations Research Letters*, 15(4):169–178, 1994.

[44] Å. Björk. *Numerical Methods for Least Squares Problems*. SIAM, Philadelphia, 1996.

[45] J.F. Bonnans and F.A. Potra. Infeasible path-following algorithms for linear complementarity problems. *Mathematics of Operations Research*, 22(2), 378–407, 1997.

[46] R.A. Bosch. On Mizuno's rank one updating algorithm for linear programming. *SIAM J. on Optimization*, 3:861–867, 1993.

[47] R.A. Bosch and K.M. Anstreicher. On partial updating in a potential reduction linear programming algorithm of Kojima, Mizuno and Yoshise. *Algorithmica*, 9(1):184–197, 1993.

[48] S.E. Boyd and L. Vandenberghe. Semidefinite programming. *SIAM Review*, 38(1):49–96, 1996.

[49] A.L. Brearley, G. Mitra, and H.P. Williams. Analysis of mathmetical programming problems prior to applying the simplex algorithm. *Mathematical Programming*, 15:54–83, 1975.

[50] M.G. Breitfeld and D.F. Shanno. Computational experiene with modified log-barrier methods for nonlinear programming. *Annals of Operations Research*, 62:439–464, 1996.

[51] J. Brinkhuis and G. Draisma. Schiet Op^{TM}. Special Report, Econometric Institute, Erasmus University, Rotterdam, 1996.

[52] R.C. Buck. *Advanced Calculus*. International Series in Pure and Apllied Mathematics. Mac-Graw Hill Book Company, New York (third edition), 1978.

[53] J.R. Bunch and B.N. Parlett. Direct methods for solving symmetric indefinit systems of linear equations. *SIAM J. on Numerical Analysis*, 8:639–655, 1971.

[54] S.F. Chang and S.T. McCormick. A hierachical algorithm for making sparse matrices sparse. *Mathematicaal Programming*, 56:1–30, 1992.

[55] V. Chvátal. *Linear programming*. W.H. Freeman and Company, New York, USA, 1983.

[56] S.A. Cook. The complexity of theorem-proving procedures. In *Proceedings of Third Annual ACM Symposium on Theory of Computing*, pp. 151–158. ACM, New York, 1971.

[57] J.P. Crouzeix and C. Roos. On the inverse target map of a linear programming problem. Unpublished Manuscript, University of Clermont, France, 1994.

[58] I. Csiszár. *I*-divergence geometry of probability distributions and minimization problems. *Annals of Probability*, 3:146–158, 1975.

[59] G.B. Dantzig. *Linear Programming and Extensions*. Princeton Univ. Press, Princeton, New Jersey, 1963.

[60] G.B. Dantzig. Linear programming. In J.K. Lenstra, A.H.G. Rinnooy Kan, and A. Schrijver, editors, *History of mathmetical programming. A collection of personal reminiscences*. CWI, North–Holland, The Netherlands, 1991.

[61] A. Deza, E. Nematollahi, R. Peyghami and T. Terlaky. The central path visits all the vertices of the Klee-Minty cube. AdvOl-Report #2004/11. McMaster Univ., Hamilton, Ontario, Canada.

[62] A. Deza, E. Nematollahi and T. Terlaky. How good are interior point methods? Klee-Minty cubes tighten iteration-complexity bounds. AdvOl-Report #2004/20. McMaster Univ., Hamilton, Ontario, Canada.

[63] I.I. Dikin. Iterative solution of problems of linear and quadratic programming. *Doklady Akademii Nauk SSSR*, 174:747–748, 1967. Translated in *Soviet Mathematics Doklady*, 8:674–675, 1967.

[64] I.I. Dikin. On the convergence of an iterative process. *Upravlyaemye Sistemi*, 12:54–60, 1974. (In Russian).

[65] I.I. Dikin. Letter to the editor. *Mathematical Programming*, 41:393–394, 1988.

[66] I.I. Dikin and C. Roos. Convergence of the dual varaibles for the primal affine scaling method with unit steps in the homogeneous case. *J. of Optimization Theory and Applications*, 95:305–321, 1997.

[67] J. Ding and T.Y. Li. An algorithm based on weighted logarithmic barrier functions for linear complementarity problems. *Arabian J. for Science and Engineering*, 15(4):679–685, 1990.

[68] I.S. Duff. The solution of large-scale least-squares problems on supercomputers. *Annals of Operations Research*, 22:241–252, 1990.

[69] I.S. Duff, N.I.M. Gould, J.K. Reid, J.A. Scott, and K. Turner. The factorization of sparse symmetric indefinite matrices. *IMA J. on Numerical Analysis*, 11:181–204, 1991.

[70] A.S. El-Bakry, R.A. Tapia, and Y. Zhang. A study of indicators for identifying zero variables in interior-point methods. *SIAM Review*, 36(1):45–72, 1994.

[71] A.S. El-Bakry, R.A. Tapia, and Y. Zhang. On the convergence rate of Newton interior-point methods in the absence of strict complementarity. *Computational optimization and Applications*, 6:157-167, 1996.

[72] J.R. Evans and N.R. Baker. Degeneracy and the (mis)interpretation of sensitivity analysis in linear programming. *Decision Sciences*, 13:348–354, 1982.

[73] J.R. Evans and N.R. Baker. Reply to 'On ranging cost coefficients in dual degenerate linear programming problems'. *Decision Sciences*, 14:442–443, 1983.

[74] S.-Ch. Fang and S. Puthenpura. *Linear optimization and extensions: theory and algorithms*. Prentice Hall, Englewood Cliffs, New Jersey 07632, 1993.

[75] J. Farkas. Theorie der einfachen Ungleichungen. *J. Reine und Angewandte Mathematik*, 124:1–27, 1902.

[76] A.V. Fiacco. *Introduction to Sensitivity and Stability Analysis in Nonlinear Programming*, volume 165 of *Mathematics in Science and Engineering*. Academic Press, New York, 1983.

[77] A.V. Fiacco and G.P. McCormick. *Nonlinear Programming: Sequential Unconstrained Minimization Techniques*. John Wiley & Sons, New York, 1968. Reprint: Volume 4 of *SIAM Classics in Applied Mathematics*, SIAM Publications, Philadelphia, PA 19104–2688, USA, 1990.

[78] R. Fourer and S. Mehrotra. Solving symmetric indefinite systems in an interior-point method for linear programming. *Mathematical Programming*, 62:15–40, 1993.

[79] C. Fraley. Linear updates for a single-phase projective method. *Operations Research Letters*, 9:169–174, 1990.

[80] C. Fraley and J.-Ph. Vial. Numerical study of projective methods for linear programming. In S. Dolecki, editor, *Optimization: Proceedings of the 5th French-German Conference in Castel-Novel, Varetz, France, October 1988*, volume 1405 of *Lecture Notes in Mathematics*, pp. 25–38. Springer Verlag, Berlin, West-Germany, 1989.

[81] C. Fraley and J.-Ph. Vial. Alternative approaches to feasibility in projective methods for linear programming. *ORSA J. on Computing*, 4:285–299, 1992.

[82] P. Franklin. *A Treatise on Advanced Calculus*. John Wiley & Sons, New York (fifth edition), 1955.

[83] R.M. Freund. An analogous of Karmarkar's algorithm for inequality constrained linear programs, with a 'new' class of projective transformations for centering a polytope. *Operations Research Letters*, 7:9–13, 1988.

[84] R.M. Freund. Theoretical efficiency of a shifted barrier function algorithm for linear programming. *Linear Algebra and Its Applications*, 152:19–41, 1991.

[85] R.M. Freund. Projective transformation for interior-point algorithms, and a superlinearly convergent algorithm for the w-center problem. *Mathematical Programming*, 58:385–414, 1993.

[86] K.R. Frisch. Principles of linear programming—the double gradient form of the logarithmic potential method. Memorandum, Institute of Economics, University of Oslo, Oslo, Norway, October 1954.

[87] K.R. Frisch. La resolution des problemes de programme lineaire par la methode du potential logarithmique. *Cahiers du Seminaire D'Econometrie*, 4:7–20, 1956.

[88] K.R. Frisch. The logarithmic potential method for solving linear programming problems. Memorandum, University Institute of Economics, Oslo, 1955.

[89] T. Gal. *Postoptimal analyses, parametric programming and related topics*. Mac-Graw Hill Inc., New York/Berlin, 1979.

[90] T. Gal. Shadow prices and sensitivity analysis in linear programming under degeneracy, state-of-the-art-survey. *OR Spektrum*, 8:59–71, 1986.

[91] D. Gale. *The Theory of Linear Economic Models*. McGraw–Hill, New York, USA, 1960.

[92] M.R. Garey and D.S. Johnson. *Computers and Intractability: a Guide to the Theory of NP-completeness*. Freeman, San Francisco, 1979.

[93] J. Gauvin. Quelques precisions sur les prix marginaux en programmation lineaire. *INFOR*, 18:68–73, 1980. (In French).

[94] A. George and J.W.-H. Liu. *Computing Solution of Large Sparse Positive Definite Systems*. Prentice-Hall, Englewood Cliffs, NJ, 1981.

[95] G. de Ghellinck and J.-Ph. Vial. A polynomial Newton method for linear programming. *Algorithmica*, 1(4):425–453, 1986.

[96] G. de Ghellinck and J.-Ph. Vial. An extension of Karmarkar's algorithm for solving a system of linear homogeneous equations on the simplex. *Mathematical Programming*, 39:79–92, 1987.

[97] P.E. Gill, W. Murray, M.A. Saunders, J.A. Tomlin, and M.H. Wright. On projected Newton barrier methods for linear programming and an equivalence to Karmarkar's projective method. *Mathematical Programming*, 36:183–209, 1986.

[98] J.-L. Goffin, J. Gondzio, R. Sarkissian, and J.-Ph. Vial. Solving nonlinear multicommodity flow problems by the analytic center cutting plane method. *Mathematical Programming*, 76:131–154, 1997.

[99] J.-L. Goffin, A. Haurie, and J.-Ph. Vial. Decomposition and nondifferentiable optimization with the projective algorithm. *Management Science*, 38(2):284–302, 1992.

[100] J.-L. Goffin, Z.-Q. Luo, and Y. Ye. Complexity analysis of an interior cutting plane for convex feasibility problems. *SIAM J. on Optimization*, 6(3), 1996.

[101] J.-L. Goffin and F. Sharifi-Mokhtarian-Mokhtarian. Primal-dual-infeasible Newton approach for the analytic center deep-cutting plane method. *J. Optim. Theory Appl.*, 101(1):35–58, 1999.

[102] J.-L. Goffin and J.-Ph. Vial. On the computation of weighted analytic centers and dual ellipsoids with the projective algorithm. *Mathematical Programming*, 60:81–92, 1993.

[103] J.-L. Goffin and J.-Ph. Vial. Short steps with Karmarkar's projective algorithm for linear programming. *SIAM J. on Optimization*, 4:193–207, 1994.

[104] J.-L. Goffin and J.-Ph. Vial. Shallow, deep and very deep cuts in the analytic center cutting plane method. *Math. Program.*, 84(1, Ser. A):89–103, 1999.

[105] D. Goldfarb and S. Mehrotra. Relaxed variants of Karmarkar's algorithm for linear programs with unknown optimal objective value. *Mathematical Programming*, 40:183–195, 1988.

[106] D. Goldfarb and S. Mehrotra. A relaxed version of Karmarkar's method. *Mathematical Programming*, 40:289–315, 1988.

[107] D. Goldfarb and S. Mehrotra. A self-correcting version of Karmarkar's algorithm. *SIAM J. on Numerical Analysis*, 26:1006–1015, 1989.

[108] D. Goldfarb and D.X. Shaw. On the complexity of a class of projective interior point methods. *Mathematics of Operations Research*, 20:116–134, 1995.

[109] D. Goldfarb and M.J. Todd. Linear Programming. In G.L. Nemhauser, A.H.G. Rinnooy Kan, and M.J. Todd, editors, *Optimization*, volume 1 of *Handbooks in Operations Research and Management Science*, pp. 141–170. North Holland, Amsterdam, The Netherlands, 1989.

[110] D. Goldfarb and D. Xiao. A primal projective interior point method for linear programming. *Mathematical Programming*, 51:17–43, 1991.

[111] A.J. Goldman and A.W. Tucker. Theory of linear programming. In H.W. Kuhn and A.W. Tucker, editors, *Linear Inequalities and Related Systems*, Annals of Mathematical Studies, No. 38, pp. 53–97. Princeton University Press, Princeton, New Jersey, 1956.

[112] G.H. Golub and C.F. Van Loan. *Matrix Computations*. Johns Hopkins University Press, Baltimore (second edition), 1989.

[113] J. Gondzio. Presolve analysis of linear programs prior to applying the interior point method. *INFORMS J. on Computing*, 9:73–91,1997.

[114] J. Gondzio. Multiple centrality corrections in a primal-dual method for linear programming. *Computational Optimization and Applications*, 6:137–156, 1996.

[115] J. Gondzio, O. du Merle, R. Sarkissian, and J.-Ph. Vial. ACCPM - a library for convex optimization based on an analytic center cutting plane method. *European J. of Operational Research*, 94:206–211, 1996.

[116] J. Gondzio and T. Terlaky. A computational view of interior point methods for linear programming. In J.E. Beasley, editor, *Advances in Linear and Integer Programming*, pp. 103–185. Oxford University Press, Oxford, Great Britain, 1996.

[117] C.C. Gonzaga. A simple representation of Karmarkar's algorithm. Technical Report, Dept. of Systems Engineering and Computer Science, COPPE Federal University of Rio de Janeiro, 21941 Rio de Janeiro, RJ, Brazil, May 1988.

[118] C.C. Gonzaga. An algorithm for solving linear programming problems in $O(n^3 L)$ operations. In N. Megiddo, editor, *Progress in Mathematical Programming: Interior Point and Related Methods*, pp. 1–28. Springer Verlag, New York, 1989.

[119] C.C. Gonzaga. Conical projection algorithms for linear programming. *Mathematical Programming*, 43:151–173, 1989.

[120] C.C. Gonzaga. Convergence of the large step primal affine-scaling algorithm for primal nondegenerate linear programs. Technical Report ES–230/90, Dept. of Systems Engineering and Computer Science, COPPE Federal University of Rio de Janeiro, 21941 Rio de Janeiro, RJ, Brazil, September 1990.

[121] C.C. Gonzaga. Large step path-following methods for linear programming, Part I: Barrier function method. *SIAM J. on Optimization*, 1:268–279, 1991.

[122] C.C. Gonzaga. Large step path-following methods for linear programming, Part II: Potential reduction method. *SIAM J. on Optimization*, 1:280–292, 1991.

[123] C.C. Gonzaga. Search directions for interior linear programming methods. *Algorithmica*, 6:153–181, 1991.

[124] C.C. Gonzaga. Path-following methods for linear programming. *SIAM Review*, 34(2): 167–227, 1992.

[125] C.C. Gonzaga. The largest step path following algorithm for monotone linear complementarity problems. *Mathematical Programming*, 76(2):309–332, 1997.

[126] C.C. Gonzaga and R.A. Tapia. On the convergence of the Mizuno–Todd–Ye algorithm to the analytic center of the solution set. *SIAM J. on Optimization*, 7: 47–65, 1997.

[127] C.C. Gonzaga and R.A. Tapia. On the quadratic convergence of the simplified Mizuno–Todd–Ye algorithm for linear programming. *SIAM J. on Optimization*, 7:66–85, 1997.

[128] H.J. Greenberg. An analysis of degeneracy. *Naval Research Logistics Quarterly*, 33:635–655, 1986.

[129] H.J. Greenberg. The use of the optimal partition in a linear programming solution for postoptimal analysis. *Operations Research Letters*, 15:179–186, 1994.

[130] O. Güler, 1994. Private communication.

[131] O. Güler. Limiting behavior of the weighted central paths in linear programming. *Mathematical Programming*, 65(2):347–363, 1994.

[132] O. Güler, D. den Hertog, C. Roos, T. Terlaky, and T. Tsuchiya. Degeneracy in interior point methods for linear programming: A survey. *Annals of Operations Research*, 46:107–138, 1993.

[133] O. Güler, C. Roos, T. Terlaky, and J.-Ph. Vial. Interior point approach to the theory of linear programming. Cahiers de Recherche 1992.3, Faculte des Sciences Economique et Sociales, Universite de Geneve, Geneve, Switzerland, 1992.

[134] O. Güler, C. Roos, T. Terlaky, and J.-Ph. Vial. A survey of the implications of the behavior of the central path for the duality theory of linear programming. *Management Science*, 41:1922–1934, 1995.

[135] O. Güler and Y. Ye. Convergence behavior of interior-point algorithms. *Mathematical Programming*, 60(2):215–228, 1993.

[136] W. W. Hager. Updating the inverse of a matrix. *SIAM Review*, 31(2):221–239, June 1989.

[137] M. Halická. Analytical properties of the central path at the boundary point in linear programming. *Mathematical Programming*, 84:335-355, 1999.

[138] L.A. Hall and R.J. Vanderbei. Two-third is sharp for affine scaling. *Operations Research Letters*, 13:197–201, 1993.

[139] G.H. Hardy, J.E. Littlewood, and G. Pólya. *Inequalities*. Cambridge University Press, Cambridge, Cambridge, 1934.

[140] D. den Hertog. *Interior Point Approach to Linear, Quadratic and Convex Programming*, volume 277 of *Mathematics and its Applications*. Kluwer Academic Publishers, Dordrecht, The Netherlands, 1994.

[141] D. den Hertog, J.A. Kaliski, C. Roos, and T. Terlaky. A logarithmic barrier cutting plane method for convex programming. *Annals of Operations Reasearch*, 58:69–98, 1995.

[142] D. den Hertog and C. Roos. A survey of search directions in interior point methods for linear programming. *Mathematical Programming*, 52:481–509, 1991.

[143] D. den Hertog, C. Roos, and T. Terlaky. A potential reduction variant of Renegar's short-step path-following method for linear programming. *Linear Algebra and Its Applications*, 152:43–68, 1991.

[144] D. den Hertog, C. Roos, and T. Terlaky. On the monotonicity of the dual objective along barrier paths. *COAL Bulletin*, 20:2–8, 1992.

[145] D. den Hertog, C. Roos, and T. Terlaky. Adding and deleting constraints in the logarithmic barrier method for LP. In D.-Z. Du and J. Sun, editors, *Advances in Optimization and Approximation*, pp. 166–185. Kluwer Academic Publishers, Dordrecht, The Netherlands, 1994.

[146] D. den Hertog, C. Roos, and J.-Ph. Vial. A complexity reduction for the long-step path-following algorithm for linear programming. *SIAM J. on Optimization*, 2:71–87, 1992.

[147] R.A. Horn and C.R. Johnson. *Matrix Analysis*. Cambridge University Press, Cambridge, UK, 1985.

[148] P. Huard. Resolution of mathmetical programming with nonlinear constraints by the method of centers. In J. Abadie, editor, *Nonlinear Programming*, pp. 207–219. North Holland, Amsterdam, The Netherlands, 1967.

[149] P. Huard. A method of centers by upper-bounding functions with applications. In J.B. Rosen, O.L. Mangasarian, and K. Ritter, editors, *Nonlinear Programming: Proceedings of a Symposium held at the University of Wisconsin*, Madison, Wisconsin, USA, May 1970, pp. 1–30. Academic Press, New York, USA, 1970.

[150] P. Hung and Y. Ye. An asymptotically $\mathcal{O}(\sqrt{n}L)$-iteration path-following linear programming algorithm that uses long steps. *SIAM J. on Optimization*, 6:570–586, 1996.

[151] B. Jansen. *Interior Point Techniques in Optimization. Complexity, Sensitivity and Algorithms*, volume 6 of *Applied Optimization*. Kluwer Academic Publishers, Dordrecht, The Netherlands, 1997.

[152] B. Jansen, J.J. de Jong, C. Roos, and T. Terlaky. Sensitivity analysis in linear programming: just be careful! *European Journal of Operations Research*, 101:15–28, 1997.

[153] B. Jansen, C. Roos, and T. Terlaky. An interior point approach to postoptimal and parametric analysis in linear programming. Technical Report 92–21, Faculty of Technical Mathematics and Computer Science, TU Delft, NL–2628 CD Delft, The Netherlands, April 1992.

[154] B. Jansen, C. Roos, and T. Terlaky. A family of polynomial affine scaling algorithms for positive semi-definite linear complementarity problems. *SIAM J. on Optimization*, 7(1):126–140, 1997.

[155] B. Jansen, C. Roos, and T. Terlaky. The theory of linear programming : Skew symmetric self-dual problems and the central path. *Optimization*, 29:225–233, 1994.

[156] B. Jansen, C. Roos, and T. Terlaky. A polynomial Dikin-type primal-dual algorithm for linear programming. *Mathematics of Operations Research*, 21:341–353, 1996.

[157] B. Jansen, C. Roos, T. Terlaky, and J.-Ph. Vial. Primal-dual algorithms for linear programming based on the logarithmic barrier method. *J. of Optimization Theory and Applications*, 83:1–26, 1994.

[158] B. Jansen, C. Roos, T. Terlaky, and J.-Ph. Vial. Long-step primal-dual target-following algorithms for linear programming. *Mathematical Methods of Operations Research*, 44:11–30, 1996.

[159] B. Jansen, C. Roos, T. Terlaky, and J.-Ph. Vial. Primal-dual target-following algorithms for linear programming. *Annals of Operations Research*, 62:197–231, 1996.

[160] B. Jansen, C. Roos, T. Terlaky, and Y. Ye. Improved complexity using higher-order correctors for primal-dual Dikin affine scaling. *Mathematical Programming*, 76:117–130, 1997.

[161] F. Jarre. Interior-point methods for classes of convex programs. In T. Terlaky, editor, *Interior Point Methods of Mathematical Programming*, pp. 255–296. Kluwer Academic Publishers, Dordrecht, The Netherlands, 1996.

[162] F. Jarre, M. Kocvara, and J. Zowe. Optimal truss design by interior-point methods. *SIAM J. on Optimization*, 8(4):1084–1107, 1998.

[163] F. Jarre and M.A. Saunders. An adaptive primal-dual method for linear programming. *COAL Newsletter*, 19:7–16, August 1991.

[164] J. Kaliski, D. Haglin, C. Roos, and T. Terlaky. Logarithmic barrier decomposition methods for semi-infinite programming. *International Transactions in Operations Research* 4:285–303, 1997.

[165] N.K. Karmarkar. A new polynomial-time algorithm for linear programming. *Combinatorica*, 4:373–395, 1984.

[166] R.M. Karp. Reducibility among combinatorial problems. In R.E. Miller and J.W. Thatcher, editors, *Complexity of computer computations*, pp. 85–103. Plenum Press, New York, 1972.

[167] L.G. Khachiyan. A polynomial algorithm in linear programming. *Doklady Akademiia Nauk SSSR*, 244:1093–1096, 1979. Translated into English in *Soviet Mathematics Doklady* 20, 191–194.

[168] K.C. Kiwiel. Complexity of some cutting plane methods that use analytic centers. *Mathematical Programming*, 74(1), 1996.

[169] E. Klafszky, J. Mayer, and T. Terlaky. Linearly constrained estimation by mathmetical programming. *European J. of Operational Research*, 34:254–267, 1989.

[170] E. Klafszky and T. Terlaky. On the ellipsoid method. *Szigma*, 20(2–3):196–208, 1988. In Hungarian.

[171] E. Klafszky and T. Terlaky. The role of pivoting in proving some fundamental theorems of linear algebra. *Linear Algebra and its Applications*, 151:97–118, 1991.

[172] E. de Klerk, C. Roos, and T. Terlaky. A nonconvex weighted potential function for polynomial target following methods. *Annals of Operations Reasearch*, 81:3–14, 1998.

[173] G. Knolmayer. The effects of degeneracy on cost-coefficient ranges and an algorithm to resolve interpretation problems. *Decision Sciences*, 15:14–21, 1984.

[174] M. Kojima, N. Megiddo, and S. Mizuno. A primal-dual infeasible-interior-point algorithm for linear programming. *Mathematical Programming*, 61:263–280, 1993.

[175] M. Kojima, N. Megiddo, T. Noma, and A. Yoshise. *A unified approach to interior point algorithms for linear complementarity problems*, volume 538 of *Lecture Notes in Computer Science*. Springer Verlag, Berlin, Germany, 1991.

[176] M. Kojima, S. Mizuno, and T. Noma. Limiting behavior of trajectories by a continuation method for monotone complementarity problems. *Mathematics of Operations Research*, 15(4):662–675, 1990.

[177] M. Kojima, S. Mizuno, and A. Yoshise. A polynomial-time algorithm for a class of linear complementarity problems. *Mathematical Programming*, 44:1–26, 1989.

[178] M. Kojima, S. Mizuno, and A. Yoshise. A primal-dual interior point algorithm for linear programming. In N. Megiddo, editor, *Progress in Mathematical Programming: Interior Point and Related Methods*, pp. 29–47. Springer Verlag, New York, 1989.

[179] E. Kranich. Interior point methods for mathmetical programming: A bibliography. Discussion Paper 171, Institute of Economy and Operations Research, FernUniversität Hagen, P.O. Box 940, D–5800 Hagen 1, West–Germany, May 1991. Available through *NETLIB*, see Kranich [180].

[180] E. Kranich. Interior-point methods bibliography. *SIAG/OPT Views-and-News, A Forum for the SIAM Activity Group on Optimization*, 1:11, 1992.

[181] P. Lancester and M. Tismenetsky. *The Theory of Matrices with Applications*. Academic Press, Orlando (second edition), 1985.

[182] P.D. Ling. A new proof of convergence for the new primal-dual affine scaling interior-point algorithm of Jansen, Roos and Terlaky. Working paper, University of East-Anglia, Norwich, England, 1993.

[183] P.D. Ling. A predictor-corrector algorithm. Working Paper, University of East Anglia, Norwich, England, 1993.

[184] C.L. Liu. *Introduction to Combinatorial Mathematics*. Mac-Graw Hill Book Company, New York, 1968.

[185] F.A. Lootsma. *Numerical Methods for Nonlinear Optimization*. Academic Press, London, UK, 1972.

[186] Z.-Q. Luo. Analysis of a cutting plane method that uses weighted analytic center an multiple cuts. *SIAM J. on Optimization*, 7(3):697–716, 1997.

[187] Z.-Q. Luo, C. Roos, and T. Terlaky. Complexity analysis of a logarithmic barrier decomposition method for semi-infinite linear programming. *Applied Numerical Mathematics*, 29:379–394, 1999.

[188] Z.-Q. Luo and Y. Ye. A genuine quadratically convergent polynomial interior point algorithm for linear programming. In D.-Z. Du and J. Sun, editors, *Advances in Optimization and Approximation*, pp. 235–246. Kluwer Academic Publishers, Dordrecht, The Netherlands, 1994.

[189] I.J. Lustig. An analysis of an available set of linear programming test problems. *Computers and Operations Research*, 16:173–184, 1989.

[190] I.J. Lustig. Phase 1 search directions for a primal-dual interior point method for linear programming. *Contemporary Mathematics*, 114:121–130, 1990.

[191] I.J. Lustig, R.E. Marsten, and D.F. Shanno. Computational experience with a primal-dual interior point method for linear programming. *Linear Algebra and Its Applications*, 152:191–222, 1991.

[192] I.J. Lustig, R.E. Marsten, and D.F. Shanno. On implementing Mehrotra's predictor-corrector interior point method for linear programming. *SIAM J. on Optimization*, 2:435–449, 1992.

[193] I.J. Lustig, R.E. Marsten, and D.F. Shanno. Interior point methods for linear programming: Computational state of the art. *ORSA J. on Computing*, 6(1):1–14, 1994.

[194] H.M. Markowitz. The elimination form of the inverse and its application to linear programming. *Management Science*, 3:255–269, 1957.

[195] I. Maros and Cs. Mészáros. The role of the augmented system in interior point methods. *European J. of Operational Research*, 107(3):720–736, 1998.

[196] R.E. Marsten, D.F. Shanno and E.M. Simantiraki. Interior point methods for linear and nonlinear programming. In I.A. Duff and A. Watson, editors, *The state of the art in numerical analysis (York, 1996)*, volume 63 of *Inst. Math. Appl. Conf. Ser. New Ser.*, pages 339–362. Oxford Univ. Press, New York, 1997.

[197] L. McLinden. The analogue of Moreau's proximation theorem, with applications to the nonlinear complementarity problem. *Pacific J. of Mathematics*, 88:101–161, 1980.

[198] L. McLinden. The complementarity problem for maximal monotone multifunctions. In R.W. Cottle, F. Giannessi, and J.L. Lions, editors, *Variational Inequalities and Complementarity Problems*, pp. 251–270. John Wiley and Sons, New York, 1980.

[199] K.A. McShane, C.L. Monma, and D.F. Shanno. An implementation of a primal-dual interior point method for linear programming. *ORSA J. on Computing*, 1:70–83, 1989.

[200] N. Megiddo. Pathways to the optimal set in linear programming. In N. Megiddo, editor, *Progress in Mathematical Programming: Interior Point and Related Methods*, pp. 131–158. Springer Verlag, New York, 1989. Identical version in *Proceedings of the 6th Mathematical Programming Symposium of Japan, Nagoya, Japan*, pp. 1–35, 1986.

[201] N. Megiddo. On finding primal- and dual-optimal bases. *ORSA J. on Computing*, 3:63–65, 1991.

[202] S. Mehrotra. Higher order methods and their performance. Technical Report 90–16R1, Dept. of Industrial Engineering and Management Science, Northwestern University, Evanston, IL 60208, USA, 1990. Revised July 1991.

[203] S. Mehrotra. Finding a vertex solution using an interior point method. *Linear Algebra and Its Applications*, 152:233–253, 1991.

[204] S. Mehrotra. Deferred rank-one updates in $O(n^3 L)$ interior point algorithm. *J. of the Operations Research Society of Japan*, 35:345–352, 1992.

[205] S. Mehrotra. On the implementation of a (primal-dual) interior point method. *SIAM J. on Optimization*, 2(4):575–601, 1992.

[206] S. Mehrotra. Quadratic convergence in a primal-dual method. *Mathematics of Operations Research*, 18:741–751, 1993.

[207] S. Mehrotra and R.D.C. Monteiro. Parametric and range analysis for interior point methods. *Mathematical Programming*, 74:65–82, 1996.

[208] S. Mehrotra and Y. Ye. On finding the optimal facet of linear programs. *Mathematical Programming*, 62:497–515, 1993.

[209] O. du Merle. *Mise en œuvre et développements de la méthode de plans coupants basés sur les centres analytiques*. PhD thesis, Faculté des Sciences Economiques et Sociales, Université de Genève, 1995. In French.

[210] H.D. Mills. Marginal values of matrix games and linear programs. In H.W. Kuhn and A.W. Tucker, editors, *Linear Inequalities and Related Systems*, Annals of Mathematical Studies, No. 38, pp. 183–193. Princeton University Press, Princeton, New Jersey, 1956.

[211] J.E. Mitchell. Fixing variables and generating classical cutting planes when using an interior point branch and cut method to solve integer programming problems. *European J. of Operational Research*, 97:139–148, 1997.

[212] S. Mizuno. An $O(n^3 L)$ algorithm using a sequence for linear complementarity problems. *J. of the Operations Research Society of Japan*, 33:66–75, 1990.

[213] S. Mizuno. A rank-one updating interior algorithm for linear programming. *Arabian J. for Science and Engineering*, 15(4):671–677, 1990.

[214] S. Mizuno. A new polynomial time method for a linear complementarity problem. *Mathematical Programming*, 56:31–43, 1992.

[215] S. Mizuno. A primal-dual interior point method for linear programming. *The Proceeding of the Institute of Statistical Mathematics*, 40(1):27–44, 1992. In Japanese.

[216] S. Mizuno and M.J. Todd. An $O(n^3 L)$ adaptive path following algorithm for a linear complementarity problem. *Mathematical Programming*, 52:587–595, 1991.

[217] S. Mizuno, M.J. Todd, and Y. Ye. On adaptive-step primal-dual interior-point algorithms for linear programming. *Mathematics of Operations Research*, 18:964–981, 1993.

[218] R.D.C. Monteiro and I. Adler. Interior-path following primal-dual algorithms: Part I: Linear programming. *Mathematical Programming*, 44:27–41, 1989.

[219] R.D.C. Monteiro and I. Adler. Interior path-following primal-dual algorithms: Part II: Convex quadratic programming. *Mathematical Programming*, 44:43–66, 1989.

[220] R.D.C. Monteiro, I. Adler, and M.G.C. Resende. A polynomial-time primal-dual affine scaling algorithm for linear and convex quadratic programming and its power series extension. *Mathematics of Operations Research*, 15:191–214, 1990.

[221] R.D.C. Monteiro and S. Mehrotra. A general parametric analysis approach and its implications to sensitivity analysis in interior point methods. *Mathematical Programming*, 72:65–82, 1996.

[222] R.D.C. Monteiro and T. Tsuchiya. Limiting behavior of the derivatives of certain trajectories associated with a monotone horizontal linear complementarity problem. *Mathematics of Operations Research* 21(4):793–814, 1996.

[223] M. Muramatsu and T. Tsuchiya. Convergence analysis of the projective scaling algorithm based on a long-step homogeneous affine scaling algorithm. *Mathematical Programming*, 72:291–305, 1996.

[224] G.L. Nemhauser and L.A. Wolsey. *Integer and Combinatorial Optimization*. J. Wiley & Sons, New York, 1988.

[225] Y. Nesterov. Cutting plane algorithms from analytic centers: efficiency estimates. *Mathematical Programming*, 69(1), 1995.

[226] Y. Nesterov and A.S. Nemirovskii. *Interior Point Polynomial Methods in Convex Programming: Theory and Algorithms*. SIAM Publications. SIAM, Philadelphia, USA, 1993.

[227] J. von Neumann. On a maximization problem. Manuscript, Institute for Advanced Studies, Princeton University, Princeton, NJ 08544, USA, 1947.

[228] F. Nožička, J. Guddat, H. Hollatz, and B. Bank. *Theorie der linearen parametrischen Optimierung*. Akademie-Verlag, Berlin, 1974.

[229] M.R. Osborne. *Finite Algorithms in Optimization and Data Analysis*. John Wiley & Sons, New York, USA, 1985.

[230] M. Padberg. *Linear Optimization and Extensions*, volume 12 of *Algorithmis and Combinatorics*. Springer Verlag, Berlin, West–Germany, 1995.

[231] C.H. Papadimitriou and K. Steiglitz. *Combinatorial optimization. Algorithms and complexity*. Prentice–Hall, Inc., Englewood Cliffs, New Jersey, 1982.

[232] J. Peng. Private communication.

[233] J. Peng, C. Roos and T. Terlaky. *Self-Regularity: A New Paradigm for Primal-Dual Interior Point Methods*. Princeton University Press, 2002.

[234] R. Polyak. Modified barrier functions (theory and methods). *Mathematical Programming*, 54:177–222, 1992.

[235] F.A. Potra. A quadratically convergent predictor-corrector method for solving linear programs from infeasible starting points. *Mathematical Programming*, 67(3):383–406, 1994.

[236] M.V. Ramana and P.M. Pardalos. Semidefinite programming. In T. Terlaky, editor, *Interior Point Methods of Mathematical Programming*, pp. 369–398. Kluwer Academic Publishers, Dordrecht, The Netherlands, 1996.

[237] J. Renegar. A polynomial-time algorithm, based on Newton's method, for linear programming. *Mathematical Programming*, 40:59–93, 1988.

[238] R.T. Rockafellar. The elementary vectors of a subspace of $\mathbb{R}^N$. In R.C. Bose and T.A. Dowling, editors, *Combinatorial Mathematics and Its Applications: Proceedings North Caroline Conference, Chapel Hill, 1967*, pp. 104–127. The University of North Caroline Press, Chapel Hill, North Caroline, 1969.

[239] C. Roos. New trajectory-following polynomial-time algorithm for linear programming problems. *J. of Optimization Theory and Applications*, 63:433–458, 1989.

[240] C. Roos. An $O(n^3 L)$ approximate center method for linear programming. In S. Dolecki, editor, *Optimization: Proceedings of the 5th French–German Conference in Castel-Novel, Varetz, France, October 1988*, volume 1405 of *Lecture Notes in Mathematics*, pp. 147–158. Springer Verlag, Berlin, West–Germany, 1989.

[241] C. Roos and D. den Hertog. A polynomial method of approximate weighted centers for linear programming. Technical Report 89–13, Faculty of Mathematics and Computer Science, TU Delft, NL–2628 BL Delft, The Netherlands, 1989.

[242] C. Roos and T. Terlaky. Advances in linear optimization. In M. Dell'Amico, F. Maffioli, and S. Martello, editors, *Annotated Bibliographies in Combinatorial Optimization*, pp. 95–114. John Wiley & Sons, New York, USA, 1997.

[243] C. Roos and J.-Ph. Vial. Analytic centers in linear programming. Technical Report 88–74, Faculty of Mathematics and Computer Science, TU Delft, NL–2628 BL Delft, The Netherlands, 1988.

[244] C. Roos and J.-Ph. Vial. Long steps with the logarithmic penalty barrier function in linear programming. In J. Gabszevwicz, J.F. Richard, and L. Wolsey, editors, *Economic Decision–Making: Games, Economics and Optimization, dedicated to J.H. Drèze*, pp. 433–441. Elsevier Science Publisher B.V., Amsterdam, The Netherlands, 1989.

[245] C. Roos and J.-Ph. Vial. A polynomial method of approximate centers for linear programming. *Mathematical Programming*, 54:295–305, 1992.

[246] C. Roos and J.-Ph. Vial. Achievable potential reductions in the method of Kojima et al. in the case of linear programming. *Revue RAIRO–Operations Research*, 28:123–133, 1994.

[247] D.S. Rubin and H.M. Wagner. Shadow prices: tips and traps for managers and instructors. *Interfaces*, 20:150–157, 1990.

[248] W. Rudin. *Principles of Mathematical Analysis*. Mac-Graw Hill Book Company, New York, 1978.

[249] R. Saigal. *Linear Programming, A modern integrated analysis*. International series in operations research & management. Kluwer Academic Publishers, Dordrecht, The Netherlands, 1995.

[250] A. Schrijver. *Theory of Linear and Integer Programming*. John Wiley & Sons, New York, 1986.

[251] D.F. Shanno. Computing Karmarkar projections quickly. *Mathematical Programming*, 41:61–71, 1988.

[252] D.F. Shanno, M.G. Breitfeld, and E.M. Simantiraki. Implementing barrier methods for nonlinear programming. In T. Terlaky, editor, *Interior Point Methods of Mathematical Programming*, pp. 369–398. Kluwer Academic Publishers, Dordrecht, The Netherlands, 1996.

[253] R. Sharda. Linear programming software for personal computers: 1995 survey. *OR/MS Today*, pp. 49–57, October 1995.

[254] D.X. Shaw and D. Goldfarb. A path-following projective interior point method for linear programming. *SIAM J. on Optimization*, 4:65–85, 1994.

[255] N.Z. Shor. Quadratic optimization problems. *Soviet J. of Computer and System Sciences*, 25:1–11, 1987.

[256] G. Sierksma. *Linear and integer programming*, volume 245 of *Monographs and Textbooks in Pure and Applied Mathematics*. Marcel Dekker Inc., New York, second edition, 2002. Theory and practice, With 1 IBM-PC floppy disk ("INTPM, a version of Karmarkar's Interior Point Method") by J. Gjaltema and G. A. Tijssen (3.5 inch; HD).

[257] Gy. Sonnevend. An "analytic center" for polyhedrons and new classes of global algorithms for linear (smooth, convex) programming. In A. Prékopa, J. Szelezsán, and B. Strazicky, editors, *System Modelling and Optimization: Proceedings of the 12th IFIP-Conference held in Budapest, Hungary, September 1985*, volume 84 of *Lecture Notes in Control and Information Sciences*, pp. 866–876. Springer Verlag, Berlin, West–Germany, 1986.

[258] Gy. Sonnevend, J. Stoer, and G. Zhao. On the complexity of following the central path by linear extrapolation in linear programming. *Methods of Operations Research*, 62:19–31, 1990.

[259] G. Strang. *Linear Algebra and its Applications*. Harcourt Brace Jovanovich, Orlando, Florida, USA, 1988.

[260] J.F. Sturm and S. Zhang. An $O(\sqrt{n}L)$ iteration bound primal-dual cone affine scaling algorithm. *Mathematical Programming*, 72:177–194, 1996.

[261] K. Tanabe. *Centered Newton methods and Differential Geometry of Optimization*. Cooperative Research Report 89. The Institute of Statistical Mathematics, Tokyo, Japan, 1996. (Contains 38 papers related to the subject).

[262] M.J. Todd. Recent developments and new directions in linear programming. In M. Iri and K. Tanabe, editors, *Mathematical Programming: Recent Developments and Applications*, pp. 109–157. Kluwer Academic Press, Dordrecht, The Netherlands, 1989.

[263] M.J. Todd. The effects of degeneracy, null and unbounded variables on variants of Karmarkar's linear programming algorithm. In T.F. Coleman and Y. Li, editors, *Large-Scale Numerical Optimization*. Volume 46 of *SIAM Proceedings in Applied Mathematics*, pp. 81–91. SIAM, Philadelphia, PA, USA, 1990.

[264] M.J. Todd. A lower bound on the number of iterations of primal-dual interior-point methods for linear programming. In G.A. Watson and D.F. Griffiths, editors, *Numerical Analysis 1993*, volume 303 of *Pitman Research Notes in Mathematics*, pp. 237–259. Longman Press, Harlow, 1994. See also [267].

[265] M.J. Todd. Potential-reduction methods in mathmetical programming. *Mathematical Programming*, 76 (1), 3–45, 1997.

[266] M.J. Todd and B.P. Burrell. An extension of Karmarkar's algorithm for linear programming using dual variables. *Algorithmica*, 1(4):409–424, 1986.

[267] M.J. Todd and Y. Ye. A lower bound on the number of iterations of long-step and polynomial interior-point linear programming algorithms. *Annals of Operations Research*, 62:233–252, 1996.

[268] T. Tsuchiya. Global convergence of the affine scaling methods for degenerate linear programming problems. *Mathematical Programming*, 52:377–404, 1991.

[269] T. Tsuchiya. Degenerate linear programming problems and the affine scaling method. *Systems, Control and Information*, 34(4):216–222, April 1990. (In Japanese).

[270] T. Tsuchiya. Global convergence property of the affine scaling methods for primal degenerate linear programming problems. *Mathematics of Operations Research*, 17(3):527–557, 1992.

[271] T. Tsuchiya. Quadratic convergence of Iri–Imai algorithm for degenerate linear programming problems. *J. of Optimization Theory and Applications*, 87(3):703–726, 1995.

[272] T. Tsuchiya. Affine scaling algorithm. In T. Terlaky, editor, *Interior Point Methods of Mathematical Programming*, pp. 35–82. Kluwer Academic Publishers, Dordrecht, The Netherlands, 1996.

[273] T. Tsuchiya and M. Muramatsu. Global convergence of the long-step affine scaling algorithm for degenerate linear programming problems. *SIAM J. on Optimization*, 5(3):525–551, 1995.

[274] A.W. Tucker. Dual systems of homogeneous linear relations. In H.W. Kuhn and A.W. Tucker, editors, *Linear Inequalities and Related Systems, Annals of Mathematical Studies, No. 38*, pp. 3–18. Princeton University Press, Princeton, New Jersey, 1956.

[275] K. Turner. Computing projections for the Karmarkar algorithm. *Linear Algebra and Its Applications*, 152:141–154, 1991.

[276] P.M. Vaidya. An algorithm for linear programming which requires $\mathcal{O}((m + n)n^2 + (m + n)^{1.5}nL)$ arithmetic operations. *Mathematical Programming*, 47:175–201, 1990.

[277] R.J. Vanderbei. Symmetric quasi-definite matrices. *SIAM J. on Optimization*, 5(1): 100–113, 1995.

[278] R.J. Vanderbei and T.J. Carpenter. Symmetric indefinite systems for interior point methods. *Mathematical Programming*, 58:1–32, 1993.

[279] R.J. Vanderbei, M.S. Meketon, and B.A. Freedman. A modification of Karmarkar's linear programming algorithm. *Algorithmica*, 1(4):395–407, 1986.

[280] S.A. Vavasis and Y. Ye. Condition numbers for polyhedra with real number data. *Operations Research Letters*, 17:209–214, 1995.

[281] S.A. Vavasis and Y. Ye. A primal-dual interior point method whose running time depends only on the constraint matrix. *Mathematical Programming*, 74:79–120, 1996.

[282] J.-Ph. Vial. A fully polynomial time projective method. *Operations Research Letters*, 7(1), 1988.

[283] J.-Ph. Vial. A unified approach to projective algorithms for linear programming. In S. Dolecki, editor, *Optimization: Proceedings of the 5th French–German Conference in Castel–Novel, Varetz, France, October 1988*, volume 1405 of *Lecture Notes in Mathematics*, pp. 191–220. Springer Verlag, Berlin, West–Germany, 1989.

[284] J.-Ph. Vial. A projective algorithm for linear programming with no regularity condition. *Operations Research Letters*, 12(1), 1992.

[285] J.-Ph. Vial. A generic path-following algorithm with a sliding constraint and its application to linear programming and the computation of analytic centers. Technical Report 1996.8, LOGILAB/Management Studies, University of Geneva, Switzerland, 1996.

[286] J.-Ph. Vial. A path-following version of the Todd-Burrell procedure for linear programming. *Mathematical Methods of Operations Research*, 46(2):153–167, 1997.

[287] G.R. Walsh. *An Introduction to Linear Programming*. John Wiley & Sons, New York, USA, 1985.

[288] J.E. Ward and R.E. Wendell. Approaches to sensitivity analysis in linear programming. *Annals of Operations Research*, 27:3–38, 1990.

[289] D.S. Watkins. *Fundamentals of Matrix Computations*. John Wiley & Sons, New York, 1991.

[290] M. Wechs. The analyticity of interior-point-paths at strictly complementary solutions of linear programs. *Optimization Methods and Software*, 9:209–243, 1998.

[291] A.C. Williams. Boundedness relations for linear constraints sets. *Linear Algebra and Its Applications*, 3:129–141, 1970.

[292] A.C. Williams. Complementarity theorems for linear programming. *SIAM Review*, 12:135–137, 1970.

[293] H.P. Williams. *Model Building in Mathematical Programming*. John Wiley & Sons, New York, USA (third edition), 1990.

[294] C. Witzgall, P.T. Boggs, and P.D. Domich. On the convergence behavior of trajectories for linear programming. *Contemporary Mathematics*, 114:161–187, 1990.

[295] S.J. Wright. An infeasible-interior-point algorithm for linear complementarity problems. *Mathematical Programming*, 67(1):29–52, 1994.

[296] S.J. Wright and D. Ralph. A superlinear infeasible-interior-point algorithm for monotone nonlinear complementarity problems. *Mathematics of Operations Research*, 21(4):815–838, 1996.

[297] S.J. Wright. A path-following infeasible-interior-point algorithm for linear complementarity problems. *Optimization Methods and Software*, 2:79–106, 1993.

[298] S.J. Wright. *Primal-Dual Interior-Point Methods*. SIAM, Philadelphia, 1996.

[299] F. Wu, S. Wu, and Y. Ye. On quadratic convergence of the $O(\sqrt{n}L)$-iteration homogeneous and self-dual linear programming algorithm. *Annals of Operations Research*, 87: 393–406, 1999.

[300] S.R. Xu, H.B. Yao, and Y.Q. Chen. An improved Karmarkar algorithm for linear programming and its numerical tests. *Mathematica Applicata*, 5(1):14–21, 1992. (In Chinese, English summary).

[301] H. Yamashita. A polynomially and quadratically convergent method for linear programming. Working Paper, Mathematical Systems Institute, Inc., Tokyo, Japan, 1986.

[302] M. Yannakakis. Computing the minimum fill-in is NP-complete. *SIAM J. on Algebraic Discrete Methods*, pp. 77–79, 1981.

[303] Y. Ye. *Interior algorithms for linear, quadratic, and linearly constrained convex programming*. PhD thesis, Dept. of Engineering Economic Systems, Stanford University, Stanford, CA 94305, USA, 1987.

[304] Y. Ye. Karmarkar's algorithm and the ellipsoid method. *Operations Research Letters*, 6:177–182, 1987.

[305] Y. Ye. A class of projective transformations for linear programming. *SIAM J. on Computing*, 19:457–466, 1990.

[306] Y. Ye. An $O(n^3 L)$ potential reduction algorithm for linear programming. *Mathematical Programming*, 50:239–258, 1991.

[307] Y. Ye. Extensions of the potential reduction algorithm for linear programming. *J. of Optimization Theory and Applications*, 72(3):487–498, 1992.

[308] Y. Ye. On the finite convergence of interior-point algorithms for linear programming. *Mathematical Programming*, 57:325–335, 1992.

[309] Y. Ye. On the q-order of convergence of interior-point algorithms for linear programming. In Wu Fang, editor, *Proceedings Symposium on Applied Mathematics*. Chinese Academy of Sciences, Institute of Applied Mathematics, 1992.

[310] Y. Ye. A potential reduction algorithm allowing column generation. *SIAM J. on Optimization*, 2:7–20, 1992.

[311] Y. Ye. Toward probabilistic analysis of interior-point algorithms for linear programming. *Mathematics of Operations Research*, 19:38–52, 1994.

[312] Y. Ye. Complexity analysis of the analytic center cutting plane method that uses multiple cuts. *Mathematical Programming*, 76(1):211–221, 1997.

[313] Y. Ye, O. Güler, R.A. Tapia, and Y. Zhang. A quadratically convergent $O(\sqrt{n}L)$-iteration algorithm for linear programming. *Mathematical Programming*, 59:151–162, 1993.

[314] Y. Ye and P.M. Pardalos. A class of linear complementarity problems solvable in polynomial time. *Linear Algebra and Its Applications*, 152:3–17, 1991.

[315] Y. Ye and M.J. Todd. Containing and shrinking ellipsoids in the path-following algorithm. *Mathematical Programming*, 47:1–10, 1990.

[316] Y. Ye, M.J. Todd, and S. Mizuno. An $O(\sqrt{n}L)$-iteration homogeneous and self-dual linear programming algorithm. *Mathematics of Operations Research*, 19:53–67, 1994.

[317] Y. Ye, O. Güler, R.A. Tapia, and Y. Zhang. A quadratically convergent $O(\sqrt{n}L)$ iteration algorithm for linear programming. *Mathematical Programming*, 59:151–162, 1993.

[318] L. Zhang and Y. Zhang. On polynomiality of the Mehrotra-type predictor-corrector interior-point algorithms. *Mathematical Programming*, 68:303–318, 1995.

[319] Y. Zhang and R.A. Tapia. Superlinear and quadratic convergence of primal-dual interior-point methods for linear programming revisited. *J. of Optimization Theory and Applications*, 73(2):229–242, 1992.

[320] G. Zhao. Interior point algorithms for linear complementarity problems based on large neighborhoods of the central path. *SIAM J. on Optimization*, 8(2), 397–413, 1998.

[321] G. Zhao and J. Zhu. Analytical properties of the central trajectory in interior point methods. In D-Z. Du and J. Sun, editors, *Advances in Optimization and Approximation*, pp. 362–375. Kluwer Academic Publishers, Dordrecht, The Netherlands, 1994.

Author Index

Subject Index

Symbol Index